Transportation Engineering

R Srinivasa Kumar
Department of Civil Engineering
University College of Engineering
Osmania University, Hyderabad

Universities Press

TRANSPORTATION ENGINEERING

UNIVERSITIES PRESS (INDIA) PRIVATE LIMITED

Registered office
3-6-747/1/A & 3-6-754/1, Himayatnagar, Hyderabad 500 029, Telangana, India
info@universitiespress. com; www.universitiespress.com

Distributed by
Orient Blackswan Private limited

Registered office
3-6-752 Himayatnagar, Hyderabad 500 029, Telangana, India

Other offices
Bengaluru, Bhopal, Chennai, Guwahati, Hyderabad, Jaipur, Kolkata
Lucknow, Mumbai, New Delhi, Noida, Patna, Visakhapatnam

501653

ISBN: 978-93-89211-16-0

Typeset in Minion Pro 10/12 by SRS Technologies, Puducherry

Printed at
B.B. Press, Noida

Published by
Universities Press (India) Private Limited 3-6-747/1/A & 3-6-754/1, Himayatnagar,
Hyderabad 500 029, Telangana, India

To all the researchers who contributed to the development of various methods, designs and models that are being used for Transportation Engineering and to solve traffic-related problems

Table of Contents

Preface

Transportation engineering is a branch of civil engineering that focuses on the planning, design, operation and maintenance of infrastructure facilities for any mode of transportation in order to provide safe, efficient, economical and environmentally friendly movement of people and goods. Hence, it is imperative to understand the interaction of all the related elements to identify methods to ensure free flow of vehicular and pedestrian traffic on pavements. To this end, the book incorporates contemporary knowledge in the field of roadway transportation and explains the concepts in clear and simple terms.

The objective of this book is to present the design guidelines of highways, materials, pavements and discuss the standard ways to control and regulate traffic flow in a safe and efficient order. The book conforms to the 2018 model curriculum for Transportation Engineering (core subject) prescribed by AICTE for undergraduate students of civil engineering and can also be useful to students preparing for competitive examinations such as GATE, IES and Public Service Commission selection exams. Further, it familiarises the reader with the latest designs by citing IRC standards, wherever appropriate. The text emphasises on exposing the student to real-life scenarios of material testing, mechanistic analysis of flexible and rigid pavements, and traffic design principles by providing several solved examples that demonstrate how theoretical knowledge can be put into practical use.

Transportation Engineering helps the reader gain an insight on the functional system characteristics of highway development and planning. It dwells on a wide range of topics that are pertinent for undergraduate students and the presentation is spread across eleven chapters. Chapter 1 traces the evolution of road development and planning in India. Chapter 2 discusses the design aspects of highway layouts, emphasising the importance of good geometric design for safe and uninterrupted movement of pedestrians and vehicles. Chapter 3 describes the guidelines for designing road intersections while Chapter 4 provides an overview of the use of traffic engineering techniques to promote safety and efficiency in vehicular and pedestrian traffic flow. Chapter 5 elaborates on the control and regulation of traffic flow with optimal utilisation of road space including road painting, signs and signals. Chapter 6 articulates on the various materials used for road construction and their engineering properties that determine their suitability. Chapter 7 explains the various tests conducted on bitumen used in pavement as binder and paving mixes and examines the properties of bitumen that make it suitable for the construction of wearing courses. Chapter 8 deals with the tests done on Portland cement and cement concrete mix for use as pavement material. Chapter 9 expounds on the factors affecting pavement design and performance. The last two chapters, Chapters 10 and 11, elucidate the concepts related to stress analyses for the design of flexible and rigid pavements, respectively.

The book is replete with figures and photographs that enable the reader to visualise the concepts and have the feel of actual field study. The review/exercise questions and multiple choice questions (MCQs) at the end of each chapter are designed to help students self-evaluate their level of understanding of the concepts. An exhaustive list of references given at the end of the book will be useful to those who may want to glean additional information on their topic of interest.

I welcome suggestions and ideas from students and teachers to improve this book further.

R Srinivasa Kumar

Acknowledgements

During the preparation of this textbook, I drew liberally from many publications (as cited in the list of References) and contacted several persons and organisations as acknowledged below.

In addition, special thanks are due to the following agencies and their representatives, who generously supported and shared information essential for the development of this book.

- The Indian Roads Congress (IRC), New Delhi, India
- The American Association of State Highway and Transportation Officials (AASHTO)
- Institute of Transportation Engineers (ITE)
- The Federal Highway Administration (FHWA); Lydia C Conrad, HOP – Office of Operations, Federal Highway Administration, Washington, DC
- The Australian Road Research Board (ARRB), www.austroads.com.au; Elaena Gardner, Communications Manager, Austroads
- The US National Cooperative Highway Research Program, The National Academy of Sciences (NAS); Maria Crawford, Editor, National Academy of Sciences, USA.
- United States Access Board, www.access-board.gov
- Unified Traffic and Transportation Infrastructure (Planning & Engineering) Centre, New Delhi.
- Japan Traffic Management Technology Association
- Masahiro Ono, President, and Yoshimi Tanaka, Deputy Manager, Research & Development Department, Japan Traffic Management Technology Association, Tokyo, Japan.
- The Oregon Department of Transportation, Dave Joe Polly, Senior Standards Engineer, and www.oregon.gov/odot/hwy/engservices
- Sheila Lyons, Pedestrian and Bicycle Program Manager, ODOT.
- Nancy Brickman, Oregon State University, http://cce.oregonstate.edu/traffic-safety-workshops
- Washington State DOT, www.wsdot.wa.gov; Chris Schroedel, Policy Research and Publications Engineer, WSDOT Development Division, Washington, DC.
- Transportation Research Board of National Scientific Council, Washington, DC.
- Georgia Department of Transportation
- Jamar Technologies Inc. and Mike Overholt, www.jamartech.com
- Harald Kolodziej, General Manager of TCS People Counting, www.tcsforsurveys.com.au
- Mikros Systems Pty Ltd and Rob Sik, www.mikros.co.za
- TWD of Essae Digitronics Pvt. Ltd and Samuel Praveen Wood, Manager, Transient Weighing Division(TWD); www.essae.in
- Kistler Group; Christoph Gilgen, Product Manager – SBF Road & Traffic Kistler Instrumente AG, Switzerland , www.kistler.com

- Traffic and Parking Control Co. Inc.; Andrew Bergholz, VP of Sales, www.tapconet.com
- Xtralis Pty Ltd; Ian Ehrenberg Sr. VP and Gen. Mgr. www.xtralis.com
- SmarTek Systems™ Inc.; Greg Pieper, www.smarteksys.com
- www.standardsforhighways.co.uk
- California Department of Transportation
- National Roads Authority, www.dttas.ie
- www.codot.gov; Tina Littleton, Customer Service Representative, Colorado DOT, USA.
- American Traffic Safety Services Association, www.atssa.com
- Office of Traffic, Security and Technology; Sue Zarling, OTST, Traffic Electrical Systems Engineer, Mn/DOT.
- East Nippon Expressway Company Limited; Kumada, Manager of Nexco-East International Department, Tokyo, Japan, www.e-nexco.co.jp
- Central Nippon Expressway Company Limited; Mitsuho O'Toole, International Business Department, Central Nippon Expressway Company Limited, Nagoya, Japan, http://global.c-nexco.co.jp
- Sachin Bhatia of www.metroinfrasys.com
- International Road Dynamics Inc.; Kerry Mitchell, Technical Design Co-ordinator, www.irdinc.com
- VICS Centre, Tokyo, Japan, www.vics.or.jp
- Mettler Toledo; Brian Howell, Product Manager – MTWT, www.mtwim.com
- Oliver Wang, Regional Marketing Manager, Moxa Americas Inc., USA., www.moxa.com
- Prof. Dusan Teodorovic and Prof. Antonio A. Trani, Transportation Infrastructure and Systems Engineering, Virginia Tech.
- Head, Department of Civil Engineering, Osmania University, Hyderabad, India.

I am indebted to several people who helped me accomplish the task of bringing out this book. Dr D Venkateshwar Rao, Professor of Civil Engineering (retired), has been the source of my inspiration and has guided me at every stage of work on this text.

Special thanks due to Mr Madhu Reddy of Universities Press Pvt Ltd, and his team members, Mr Thomas Mathew Rajesh, Mr Kallol Das, Mr Ramesh and Mr Janardhan Reddy.

Thanks are due to Prof. N Suresh Kumar, Department of Civil Engineering, UCE, Osmania University, Hyderabad, for his encouragement and support.

I also thank my students Mr Rohith Mane and Mr Shashi Kumar (currently pursuing PhD) for assisting me during the preparation of a few parts of the manuscript.

Finally, I am grateful to my family members, including, Mrs Rangoju Saritha, Ms Rangoju Gayatrisai, Mr Rangoju Vivek Dhir and Mrs NVL Narasamma.

I have referred to many published articles on different topics. It is difficult to quote all of them. However, I thank them profusely.

Although efforts were made to check and improve the text while writing, editing and proofing there still may be a few omissions and errors. I would be very grateful for any positive and constructive suggestions from the readers for further improvements to the text.

R Srinivasa Kumar

CHAPTER 1 Highway Development Planning, Programs and Surveys

1.0 GENERAL

A good transportation system facilitates the mobility of persons and goods to different places. If rural and urban areas were connected by a systematic network of roads, it would give an additional boost to developmental activities in the civil, military and recreational fronts. Trade of agricultural and industrial products and interchange of human resources takes place rapidly with increased road transportation. Thus, with a good network of road systems, the nation prospers. Keeping this relationship between good roads and prosperity in mind, Indians and Indian rulers through the years have considered road development works an important part of administration. In this chapter we will discuss the evolution of Indian roads through the ages.

1.1 ROAD DEVELOPMENT AND PLANNING IN INDIA DURING THE BRITISH RULE

After the decline of the Mughals, the road network in India was severely damaged and deteriorated. During the British rule, only those roads which were considered of strategic military use were built and maintained for political advantage. The modern Grand Trunk Road (the Grand Trunk Road already existed during the Mauryan era) was built from Delhi to Kolkata by William Bentinck. Later, Lord Dalhousie took steps to improve the service condition of roads. He formulated an organisation called the Public Works Department (PWD) in 1885, which is similar to the present PWD in India. After the introduction of railways in India, major road development works ceased except work on feeder roads from port, harbours, railway stations and other important places.

1.2 THE JAYAKAR COMMITTEE

In 1927, based on the recommendation of the Indian Roads and Transport Development Association (IRTDA), a committee chaired by Dr M R Jayakar was instituted by the central government. The committee studied the existing conditions of roads, the probable increase in volume of mixed traffic and the anticipated needs of road development. They prepared a report (1928) which delineated the methodology for the future development of roads in India. The following recommendations were made (RRDC, 1928).

- Road development in India should be considered a national interest. Major initiatives should be taken by the central government in this regard.
- An extra tax should be levied on petrol charges, and the money collected should be used for road development.
- A technical advisory organisation has to be formed to collect technical knowledge related to design, maintenance and other aspects related to road development and for its dissemination.
- An institution is to be set up to conduct research and offer consultancy services related to the development of road networks.

The central government accepted and implemented the above recommendations.

- In 1929, an extra tax of 2 annas/gallon of motor fuel was collected and termed the Central Road Fund (CRF). The CRF account was maintained by the Accountant General of the Central Reserve, New Delhi. Twenty per cent of the CRF collected was set apart as Central Reserve. It was used for administrative purposes, maintenance and dissemination of technical data, research and related purposes of road development. The remaining 80 per cent of the CRF was allotted to different states for the development of their respective roads based on the petrol/fuel consumption of the state.
- This fund has been legitimated under the Central Road Fund Act, which was enacted in December 2000.

The breakup details of the collected cess (Rs. 2/litre) are as follows.

(a) Rs 0.5/litre is allocated exclusively for the development and maintenance of national highways.

(b) The balance amount of Rs 1.5/litre is earmarked as follows:

(i) Fifty per cent of the cess collected on high-speed diesel (HSD) is set aside for the development of rural roads, under Ministry of Rural Development (MoRD).

(ii) The balance amount (50% on HSD) plus the entire cess collected on petrol is distributed as given below:

- A proportion of 57.5% sum is used for development and maintenance of NH, under MoRTH
- A proportion of 12.5% sum is used for construction of road under-bridges (RUB) or over-bridges (ROB) and safety works at unmanned railway crossing, under Ministry of Railways (MoR) and
- The remaining 30% of the sum is used for development and maintenance of State Roads. Out of this amount, 10% is reserved by the Central Government for allocation to States for implementation of Inter-state Connectivity and Economic Importance Roads, which will be approved by the Central Government under Ministry of Road Transport and Highways (MoRTH)

The Act's 2005 amendment instructs the use of Rs 2 cess collected from the sale of petrol and diesel. In addition, education cess as 2% and higher education cess as 1% (i.e., total 3%) is also collected on excise duty and customs duty on petroleum products.

Note: Apart from the above, some of the State Governments also run separate road funds.

The Central Road Fund (Amendment) Bill, 2017

As the NHDP and other prestigious projects are in different stages of completion, the government is assigning the use of CRF to other priority areas of transportation. Towards this objective, the CRF Act 2000 was amended in 2017 and 2.5% of the CRF sanctioned for the development of the inland waterways and declared as 'National Waterways (NW)' under the National Waterways Act, 2016. Currently, there are 111 NW under the 2016 Act. The Act decreased the allocation of cess towards the development and maintenance of NH from 41.5% to 39%.

The funds under the CRF scheme will be used for the development of State Roads and allocated to the states/UTs based on 30% weightage to fuel consumption and 70% weightage to the geographical area. The revenue collected through the cess is first credited to Consolidated Fund of India and later, after adjusting for the cost of tax collection, the amount is credited as CRF and thereafter distributed among the MoRD, MoR and MoRTH as prescribed in the CRF Act, 2000.

The CRF collected during the year 2018–19 was about Rs 83,374 crores. This fund may increase with increased consumption of fuel.

Other Important Organizational/Institutional Developments

- In 1930, the Central Road Organisation (CRO) was established.
- In 1934, a semi-official technical body named the Indian Roads Congress (IRC) was set up to act as a platform for providing all technical knowledge and experience related to road construction, materials, design specifications, maintenance, economics and overall development of roads within the country.
- In 1935, the Transport Advisory Committee was formed.
- In 1939, the Motor Vehicle Act was passed in the parliament to set standards for traffic rules, vehicle standards, driver characteristics and regulations. This act was revised in 1988.
- Long-term road development programmes were also planned for 20-year periods.
- In 1950, the Central Road Research Institute (CRRI) was set up at New Delhi by the Council of Scientific and Industrial Research (CSIR) to carry out research and consultancy works related to roads planning, construction, design, evaluation, traffic, maintenance, economics, financing, instrumentation and use of alternative materials for roads.
- In 1956, the National Highway Act was approved by the central government. As per the act, the central government reserves all rights to develop, maintain and designate any road as a national highway (NH) or declare an NH as any other type of road. In 1995, the act was amended to permit the central government to collaborate with any state government or private organisation for the development and maintenance of NHs. This amendment allows private participation in building any required road infrastructure.
- The Indian Roads Congress (IRC) was established in December 1934 by the Government of India to collect, organise and disseminate all information related to the development, design, construction and maintenance of roads and bridges in the country.
- To promote road research and related development activities in the country, the Highway Research Board (HRB) was set up in October 1973 with the assistance of IRC.
- The NHAI, which was constituted by the Parliament under the Act of the National Highways Authority of India 1998, has been operational since February 1995. It is entrusted with the task of implementing all projects that are related to the development, maintenance and management of all national highways and expressways in the country.
- The National Institute of Training for Highway Engineers (NITHE) is also a registered society like the IRC. It works under the administrative control of the Department of Road Transport and Highways (DORTH) and was set up as a collaborative body in 1983 to cater to the needs of the road and transport sector.
- In May 1960, the Border Roads Organization (BRO) was established to develop roads for strategic purposes which are used by the military in the northern, northeastern and other border areas.

1.3 INITIAL MAJOR ROAD DEVELOPMENT PLANS IN INDIA

The major plans that evolved since 1943 up to 2001 for the development of road networks in the country are given below.

1. Nagpur Road Plan: The First 20-year Road Development Plan (1943–1963).
2. Bombay Road Plan: The Second 20-year Road Development Plan (1961–1981).
3. Lucknow Road Plan: The Third 20-year Road Development Plan (1981–2001).

During the formulation of the Lucknow Road Plan, due consideration was given to road infrastructure requirements for cities, towns and villages. The economic growth pattern of the country, economical usage of fuels, improvement of living standards and safety aspects related to traffic operations were also taken into account. The policies and goals set up in these plans are briefed below:

1.3.1 The Nagpur Road Plan: The First 20-year Road Development Plan (1943–1963)

After the Second World War, the roads in India had deteriorated significantly since the responsibility of maintaining them were confined to provincial or local governments. It was not possible to satisfy the increased traffic needs due to poor condition of the roads (NRP, 1943). Therefore, in 1943, the central government and IRC convened a meeting of the chief engineers of all the states and provinces at Nagpur. The aim of the meeting was to chalk out a comprehensive plan for the development of a sophisticated road network. Following were the recommendations made by the conference for the target period, 1943 to 1963.

- A target road density (ratio of the length of road network to land area) of about 16 km/100 sq. km is to be achieved by the end of the plan period.
- A long-term plan is desired for road development work to facilitate greater communication and mobility all over the country.
- Required widths and lengths of NH, SH, ODR and VR are to be formulated based on the coverage of agricultural areas, non-agricultural areas and number of habitations within the area, and the existing length of railway tracks. An allowance of 15 per cent must be made for future developments in the next 20-year period.

The first 20-year road development plan (1943–1963) was systematically implemented during the first two five-year plans from 1951 to 1961 and attained the target road density of about 16 km/100 sq. km two years before the end of the plan period. Therefore, the second 20-year road development programme was planned to start from 1961.

1.3.2 The Bombay Road Plan: The Second 20-year Road Development Plan (1961–1981)

In 1959, a meeting of the chief engineers of all the states was convened at Bombay under the aegis of the IRC, to formulate the second 20-year road development plan; the plan was approved by the central government. The experience gained while implementing the Nagpur Road Plan was reviewed and a strategy for future road development was conceived based on the increased volume of traffic, agriculture, industries, population and anticipated traffic necessities for the next 20 years. The following recommendations were made in this plan (RDPI).

- A target road density of 32 km/100 sq. km is to be achieved by the end of the plan period.
- Due emphasis should be placed on the rural road communication system to provide effective connectivity of rural villages to nearby towns. This will discourage migration to urban areas.
- Separate equations were prescribed for length calculation of different road classes, namely, NH, (NH + SH), (NH + SH + MDR), (NH + SH + MDR + ODR) and (NH + SH + MDR + ODR + VR) based on the type of developed area, the number of towns with their population count and allowance for future developments.

By 1981, the density of road length achieved was of the order of 34.8 km/100 sq. km, which was more than the targeted value of 32 km/100 sq. km. The third 20-year plan was delayed due to change in policies from the 5th to 6th five-year plans (1974–1985). During this period, the Highway Research Board (HRB) and the National Transport Policy Committee (1978) were established.

1.3.3 The Lucknow Road Plan: The Third 20-year Road Development Plan (1981–2001)

To finalise the recommendations for the third 20-year road development plan covering the period 1981 to 2001, the IRC convened a meeting in 1985 at Lucknow. It constituted the chief engineers of all the states along with road experts from different organisations. During the formulation of the plan, due consideration was given to road infrastructure requirements for cities, towns and villages. The economic growth pattern of the country, economical usage of fuels, improvement of living standards and safety aspects related to traffic operations were also taken into account. The policies and goals set up in the plan are mentioned below (RDPI 1984, RDPV 2001).

- A target length of road density 82 km/100 sq. km by the end of the plan period was set up.
- All villages with a population of 500 and above should be connected with all-weather roads by the end of the plan period.
- More length of NH should be constructed to form an NH square-grid of size 100 km.
- On main traffic routes, expressways should be constructed to facilitate faster movement of vehicles.
- All towns and villages having a population above 1500 should be connected by MDR of length 3,00,000 km.
- All villages having population ranging from 1000 to 1500 should be connected by ODR and VR of a target length of 2189000 km.

1.3.4 Classification of Roads

The roads in India can be classified into different categories based on the seasonal use, purpose, plan, location and materials used for road construction. Figures 1.1 (a–c) show how roads can be classified according to plans and seasonal use.

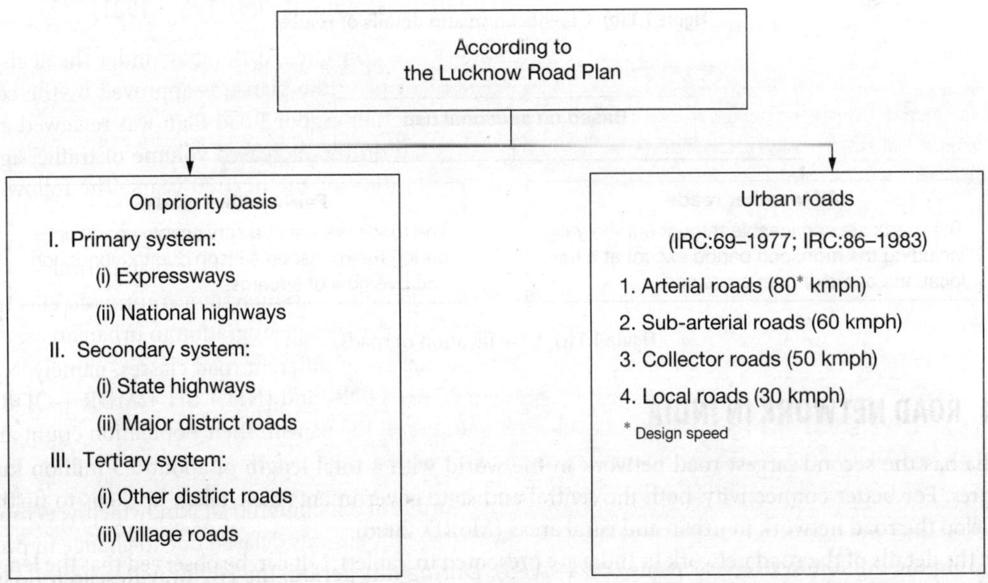

Figure 1.1 (a) Classification of roads based on the Lucknow road plan.

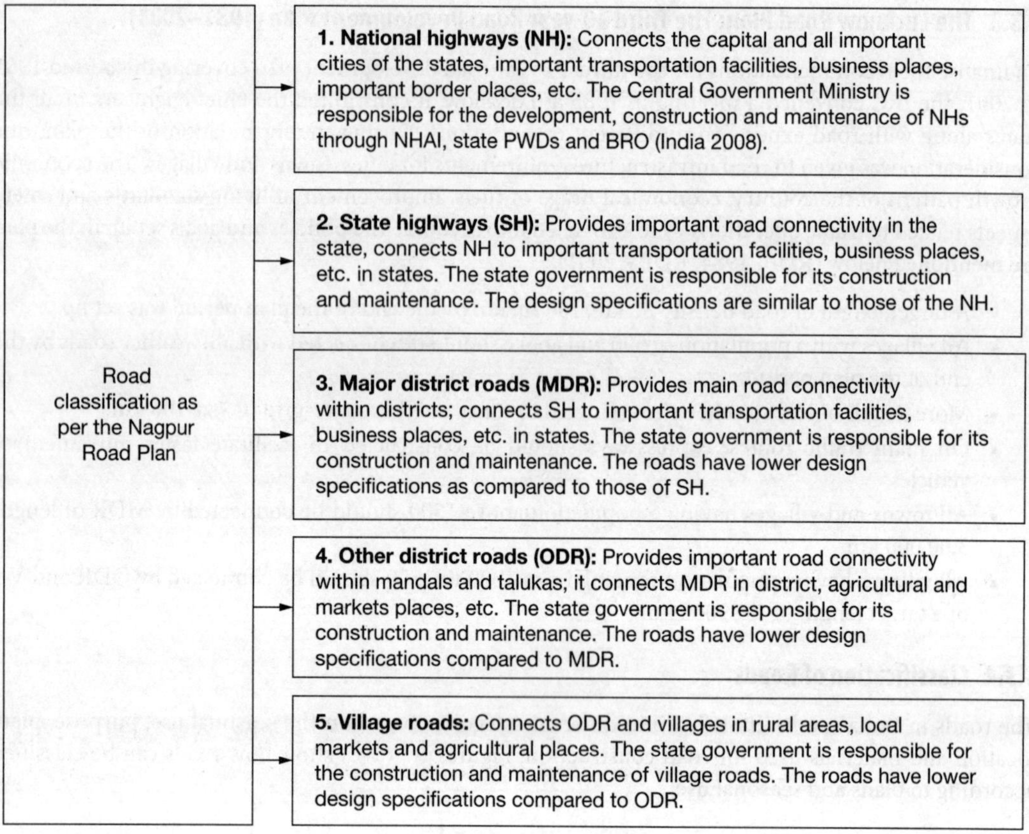

Road classification as per the Nagpur Road Plan
1. National highways (NH): Connects the capital and all important cities of the states, important transportation facilities, business places, important border places, etc. The Central Government Ministry is responsible for the development, construction and maintenance of NHs through NHAI, state PWDs and BRO (India 2008).
2. State highways (SH): Provides important road connectivity in the state, connects NH to important transportation facilities, business places, etc. in states. The state government is responsible for its construction and maintenance. The design specifications are similar to those of the NH.
3. Major district roads (MDR): Provides main road connectivity within districts; connects SH to important transportation facilities, business places, etc. in states. The state government is responsible for its construction and maintenance. The roads have lower design specifications as compared to those of SH.
4. Other district roads (ODR): Provides important road connectivity within mandals and talukas; it connects MDR in district, agricultural and markets places, etc. The state government is responsible for its construction and maintenance. The roads have lower design specifications compared to MDR.
5. Village roads: Connects ODR and villages in rural areas, local markets and agricultural places. The state government is responsible for the construction and maintenance of village roads. The roads have lower design specifications compared to ODR.

Figure 1.1 (b) Classification and details of roads.

Based on seasonal use
All-weather roads
The roads are serviceable throughout the year including the monsoon period except at a few locations of major river crossings.
Fair-weather roads
The roads are not in a serviceable condition during the monsoon season due to inundation and overflow of streams.

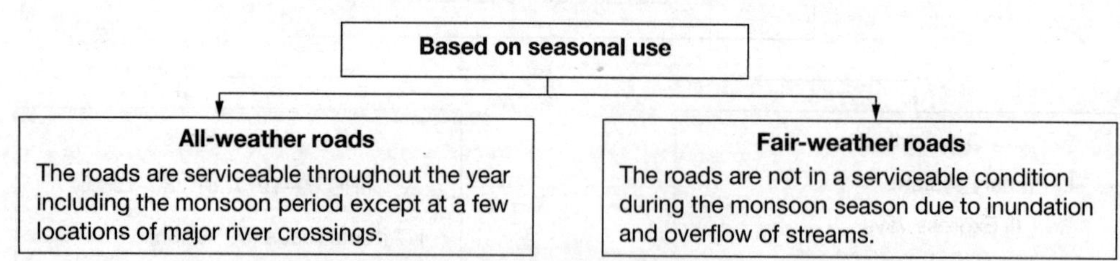

Figure 1.1 (c) Classification of roads.

1.4 ROAD NETWORK IN INDIA

India has the second largest road network in the world with a total length of about 5.5 million kilometres. For better connectivity, both the central and state governments have taken up steps to further develop the road network in urban and rural areas (MoRD, 2006).

The details of the road network in India are presented in Table 1.1. It can be observed that the length of the national highways consists of about 2% of the total length of the road network, but carries 40% of the total traffic in the country which is alarming and requires immediate attention. It calls for further development of the road network on a priority basis.

Table 1.1 Details of the Indian road network.

S. No.	Description of road		Length * (km)	Percentage of	
				Total road length	Total road † traffic carried
1	Expressways		1885		
	NH	4/6/8 lane: 23 % of length	132499 ‡	≈ 2%	≈ 40%
		Double and intermediate lane: 50% of length			
		Single lane: 27% of length			
2	SH	4 lane: 1% of length	1,76,166	≈ 3%	≈ 40%
		2 lane : 19% of length			
		Single lane: 80% of length			
3	MDR, VR and ODR		≈ 5.6 million Km	95%	
4	**Total**		**≈ 5.9 million Km**		

* The above numbers may change from time to time. The latest and more accurate detailed Indian road network with maps can be obtained from http://www.nhai.gov.in and http://morth.nic.in

† Indian roads carry about 80% of passenger traffic and 65% of the freight traffic (Manorama 2008). Average vehicle growth rate reported as 10.1% per annum during the period, 1999–2000 to 2003–2004 (Annual Report 2007–08, DORTH).

‡ At the time of independence (1947), the length of national highways was 21,440 km (Manorama 2008).

Note: Road infrastructure is used to transport 85% of total passenger traffic and over 60% of total goods.

1.5 RECENT ROAD DEVELOPMENTAL SCHEMES/ACTIVITIES IN INDIA

1.5.1 The National Highway Development Project (NHDP)

The central government of India has taken initiatives under various schemes to develop the road network. One of the objectives of the 9th plan period (1997–2002) was the coordinated and balanced development of road network in the country. The National Highway Development Project (NHDP) was evolved to realise this objective. Under the schemes of NHDP-I and -II, 14,357 km of national highways was to be converted to high-quality four/six lanes at an estimated cost of Rs. 64,639 crore at 2005 prices and planned to be completed by the end of the 10th Plan (December 2007).

As per the central government decision taken on April 2007, the NHDP consists of 7 phases, with an estimated cost of Rs.2,35,690 crore and was scheduled to be completed by 2015 (Annual Report 2007–2008, DoRTH). It was decided to implement all the NHDP programmes from Phase III onwards and a part of Phase II by the PPP (public–private partnership) route, first through BOT (toll), failing which, by BOT (annuity). If all else fails, the projects was to be completed through EPC (engineering, procurement and construction) contracts with the approval of the ministry. The NHAI is responsible for implementation of the projects. The NHDP is the largest and most prestigious road development project undertaken so far in India.

The NHDP scheme consists of the following major road networks (www.nhai.org):

NHDP Phase-I

(i) *Golden Quadrangle (GQ)*: Comprising a total National Highway length of 5,846 km, connecting four metropolitan cities of India, viz. Delhi-Kolkata-Chennai-Mumbai-Delhi.

(ii) *Port Connectivity and Other Projects*: To enhance accessibility and for improved connectivity to the existing ten major ports in the country, viz., Haldia, Cochin, Jawaharlal Nehru Port (Mumbai), Kandla, Mormugao, Mangalore, Visakapatnam, Chennai and Ennore. 380 km of 4-lane National Highways and 831 km of other important National Highways are being considered. Responsibility of implementation of this scheme has been entrusted to NHAI.

NHDP Phase-II

The North–South and East–West (NS–EW) Corridors: The north–south regions are connected by a national highway running from Srinagar to Kanyakumari including the Kochi–Salem spur. The East–West Corridor comprises a national highway running from Silchar to Porbandar. For a detailed highway map of India showing the golden quadrilateral, the East-West and North-South corridors, the readers may refer to https://nhai.gov.in/nhdp-english-map.htm

NHAI's main source of finance for the implementation of the NHDP Phase-I and -II is the cess collected on petrol and diesel. The other phases of the NHDP and the summary of the proposed highway length are presented in Table 1.2 (source:www.nhai.gov.in as of Jan. 2019):

Table 1.2 Summary of highway length proposed by NHDP as of Jan. 2019.

Scheme		Total length-proposed (Km.)	Already 4-/6-laned (Km.)
NHDP	GQ	5,846	5,846
	NS–EW, Phases I and II	7,142	6,568
	Port connectivity	435	383
	NHDP Phase III	12109	7,621
	NHDP Phase IV	20,000	4,058
	NHDP Phase V	6,500	2,564
	NHDP Phase VI	1,000	–
	NHDP Phase VII	700	22
Others (Ph.-I, Ph.-II and Misc.)		2,048	1,743
SARDP -NE		110	110

Apart from these schemes, a number of other projects for the development of new roads as well as for the up gradations of existing road networks have been undertaken by the central and the different state governments. Funds for the schemes have been borrowed from domestic markets through bonds that provide capital gain tax exemption and also by loans from various financial international institutions such as World Bank, Asian Development Bank (ADB), Overseas Economic Cooperation Fund, Japan (OECF), National Bank for Agriculture and Rural Development (NABARD) and Japan Bank for International Cooperation (JBIC) (Spectrum 2008).

In order to increase the development of the road network, the Indian government has announced numerous incentives for the active participation of the private sector—these incentives are meant not only to induce the industry to invest in road construction but also help in the maintenance operations. Several road projects were taken up for implementation on build, operate and transfer (BOT) basis and a few are considered on an annuity basis. Apart from NHDP, a few more new projects are in pipeline to be awarded soon (www.nhai.org).

1.6 SPECIAL ACCELERATED ROAD DEVELOPMENT (SARD) PROGRAMME FOR THE NORTHEASTERN REGION

The Ministry of Road Transport and Highways (MoRTH) initiated the SARD-NE programme for improving road connectivity to the remote places from strategically important towns of the northeastern region. The plan was to use budgetary support and private–public participation (PPP) on BOT (annuity) mode (MORTH 2006; India 2008). Under the SARD-NE programme, it is proposed to widen two/four-laning of about 3,228 km length of national highways; to two-lane or improve about 2,500 km of state highways and 1,888 km length of roads that are strategically important. With this development, road connectivity will be ensured to 34 out of 85 district headquarters which are not yet connected to the national highways in the eight N-E States viz., Arunachal Pradesh, Assam, Manipur, Meghalaya, Mizoram, Nagaland, Sikkim and Tripura.

The Department of Development of North-eastern Region (DoNER) was set up in 2001 to coordinate the developmental activities in the North-eastern Region (NER). It was converted into a full-fledged Ministry in 2004.

National Highways and Infrastructure Development Corporation Limited (NHIDCL) was set up in the year 2014 as a part of MoRTH by the Government of India to expedite construction of National Highway projects with specific focus on NER.

The central government is implementing the Kaladan Multimodal Transit Transport Project to connect Kolkata and Sittwe Port in Myanmar by constructing around 200 km long road (http://mdoner.gov.in).

1.7 RURAL ROADS

Roads are a key component of rural development. They provide connectivity to various business places, improve accessibility to various facilities like education, health, security and foster the economic well-being of the region (www.pmgsy.nic.in).

It is expected that in India by the year 2020, the urban population will rise from 28 per cent to 40 per cent (India Vision 2020, 2004). To discourage this migration of population from rural to urban areas for employment, it becomes necessary to develop adequate rural transport infrastructure. Comprehensive planning is needed to balance and distribute the development of road network, and to narrow down the disproportion between urban and rural areas.

During the process of planning, the government of India has formulated various schemes, viz., Minimum Needs Programme, National Rural Employment Programme, Command Area Agency and Employment Assurance; the objective was to construct earthen roads to connect clusters of villages in rural areas and nearby all-weather roads (AWR) that are negotiable throughout the year. At present, these schemes are clubbed together and allocated under the scheme of Sampurna Grameen Rozgar Yojana (SGRY).

In 2000, the government of India constituted a National Rural Roads Development Committee (NRRDC) to identify the total road length required for rural connectivity, specifications of AWR, sources of funds, and to construct and maintain rural roads. The recommendations of the NRRDC were accepted by the Ministry of Rural Development (MORD) and the ministry launched a scheme called the Pradhana Manthri Gram Sadak Yojana (PMGSY) in December 2000, to provide rural connectivity in a phase-wise manner and on a priority basis. The MORD is looking after 12 divisions—one of them is PMGSY. Following are the objectives of the PMGSY programme (India Vision 2020, 2004).

- To provide all-weather road (AWR) connectivity to all unconnected villages with a population of more than 1000 by the year 2003 and villages with a population between 1000 and 500 by the year 2007.
- To connect by all-weather roads (AWR) all unconnected habitations with a population of 250 persons and above in desert areas, tribal areas and hill states of the northeast, Sikkim, Himachal Pradesh, Uttaranchal, Jammu and Kashmir.
- It is envisaged that all-weather roads (AWR) are to be built to connect all villages with a population below 500 by the year 2010 (RDPV 2001).

PMGSY is sponsored (100%) by the central government by using 50% of the cess on high-speed diesel (HSD). The aim was to attain the above first two objectives in all the states and six union territories by end of the tenth plan (2007), with an estimated amount of Rs.58,200 crore (in the year 2005).

PMGSY works for upgradation/new connectivity are assessed by preparing (i) core network and (ii) district rural roads plans (DRRP), also known as district master plans for rural roads. The core network provides information about AWR-connectivity between villages/habitations and essential nearby services. Details of habitations, the existing core network, and the existing and proposed roads are sketched on to a map for each block. Such maps of all blocks are integrated at the district level to form the DRRP, which is necessary for getting approval from the district panchayat. Only such approved works are considered under PMGSY. Detailed guidelines can be obtained from MoRD.

National Rural Roads Development Agency (NRRDA)

The National Rural Roads Development Agency (NRRDA) works under the guidance of MoRD to provide managerial and technical support for the implementation of PMGSY. Its major activities consists of preparation of DRRPs, identification of the core network (CN), project scrutiny, programme monitoring, quality management and financial management. The NRRDA has taken active initiatives to develop a separate set of specifications and a standard data book for rate analysis of the cost of construction of rural roads in India. PMGSY works are implemented and monitored through the World Wide Web, on a site termed as On-line Management, Monitoring and Accounting System (OMMAS). Steps were taken for training the concerned engineers and technicians of all levels (www.omms.nic.in). Few R&D activities are also in progress for developing traffic growth models and performance-based designs.

PMGSY Phase II (Bharath Nirman)

Another scheme that has been recently started is termed as Bharat Nirman. It is a time-bound business plan under the framework of the PMGSY intended to accelerate and further develop/upgrade rural roads. Objectives of this scheme are listed below:

Consolidation of existing RR network in each district in respective states and up-gradation of "Select-through Routes" providing access to "Growth Poles" and "Rural Hubs" and "Other Important Places of Interest". This is anticipated to provide access to socio-economic facilities like health, education, marketing, tourist interest places.

All States and UTs are eligible under PMGSY-II and can avail 100% of funding for new connectivity and 90% of funding for upgradation of targeted works cleared by MoRD. Funding up to 95% for new connectivity and 85% for upgradation works is awarded. The cost-sharing would be up to 75% by Centre and 25% by State/UT for normal areas and up to 90% by Centre and 10% by State/UT for special areas. The allocation of fund for PMGSY-II would be within 15–20% of the annual allocated PMGSY budget of MoRD (BPC, 1029).

These projects would not only be funded by the CRF but also by multinational financial institutions, viz., World Bank, Asian Development Bank, NABARD, etc. For more information about the different schemes related to rural connectivity, the reader is advised to visit GoI Web sites (pmgsy; pmindia; planningcommission; cag).

About 29,000 km of new habitations were envisaged to be connected between the years 2015 to 2017. The annual allocation during year 2015–16 was enhanced by around Rs. 5,000 crore and with this, the total Central allocation to States would be Rs. 15,100 crore.

The Central Government has brought forward the target date by three years to 2019 from 2022 to achieve complete rural road connectivity under PMGSY. This accelerated implementation plan will be achieved by providing enhanced funding and through a structured funding pattern in the Bharath Nirman scheme.

1.8 BHARATHMALA PARIYOJANA

Bharathmala Pariyojana is a new planning/development program of the highways sector, which focuses on optimising and enhancing of freight and passenger movement across the country by bridging critical connectivity gaps through Phase-I development of different categories of highway links as (www.india.gov.in):

- *Economic Corridor* – 9000 km of highway corridors will be constructed by the central government.
- *Feeder Route or Inter Corridor* – Feeder routes will be constructed to the extent of 6000 km.
- *National Corridor Efficiency Improvement* – 5000 km of roads will be constructed for better connection between the existing roads.
- *Border Road and International Connectivity* – 2000 km of roads will be built to connect the cities and remote areas in the border regions.
- *Port Connectivity and Coastal Road* – 2000 km of roads to be constructed along the shorelines and important ports.
- *Green Field Expressway* – For better management of traffic and freight in the country.
- *Balance NHDP Works* – Construction and maintenance of about 10,000km of new roads.

Bharathmala Pariyojana is a centrally sponsored and funded highways project. A total highway length of 24,800 km is being considered as Phase-I. In addition, 10,000 km of balance road length under NHDP is also considered at a projected cost of Rs.5.35 lakh crore. Bharatmala Phase-I is planned to be implemented over the five-year period from 2017–18 to 2021–22.

1.9 HIGHWAY ALIGNMENT SURVEYS

Various surveys are required for the alignment of a new highway to decide on the precise location of the highway's centre line. This is in order to minimise cost of construction, maintenance, transportation, vehicle operating cost and to have the least impact on the surrounding environment. Horizontal and vertical controls of the proposed highway play a vital role in aligning the centre line to suit terrain conditions. Before a new highway project is started, the feasibility of the proposal should be assessed with the following factors in mind—the spatial conditions prevailing in the locality, economic considerations, environmental impact, availability of materials and various other engineering aspects. New highways as well as stage-wise up gradation of existing highways should be planned by adopting a systematic approach (Fig. 1.2). The following important elements of planning should be kept in mind during preliminary planning of a project.

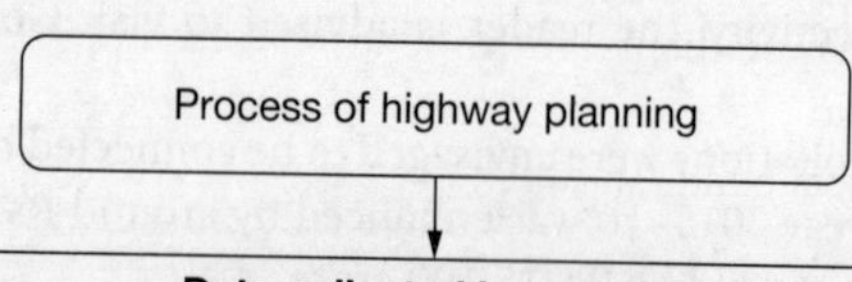

Data collected by surveys

General: Details of existing road network, agricultural areas, non-agricultural areas, number of towns, cities and villages to be served with census data, income per capita, length of existing railway lines, current and future developments envisaged.

Financial: Available and anticipated revenue from vehicle registration, taxes, government fees, other sources of income by private participation, loans from banks, cost of living, etc.

Engineering: Various surveys to collect data related to traffic volume, growth rate, desired road links, existing transport facilities, anticipated traffic demands, soil characteristics, topography, drainage, river crossings, materials availability and climatic conditions.

Graphical presentation of above data

- Topographic sheets showing all existing and spatial features of the terrain in the area under consideration and market facilities.
- Spatial distribution of habitation, industries, offices, agricultural areas, existing transport network and the proposed or desired road network.

Study of the above data and preparation of alternative routes for exigency.

Master plan

1. Analyses, comparison and prioritisation of alternative routes based on productivity, density of population with number of villages, intensity and demand of traffic, road development targets and funds allotment made available in annual and five-year plans.
2. Calculation of desired length of different classes (NH, SH, MDR, ODR and VR) of roads based on latest information available from road development plans, maximum utility by the locality or utility of road/km.
3. Final selection of a route and prioritisation or stage-wise construction of different links of the selected route based on funds availability, population, villages, industrial and agricultural productivity to be served by the proposed link or a part of the selected route.

Figure 1.2 Outline of the highway planning process.

- Standard design specifications should be adhered to as much as possible.
- There should be no alteration of the finalised alignment at later stages as it may be difficult and expensive to alter completed work (Losh, 1932).

1.9.1 Factors considered for alignment of a new highway

Places and towns that are to be connected should first be listed.

- The shortest route length is to be proposed.
- Gentle ground slopes or easy gradients are to be chosen to facilitate safe movement of animal-drawn and motor vehicular traffic.
- Both horizontal and vertical bends should have gentle curves with large radii without hampering design speeds.
- Adequate sight distance has to be maintained to avoid collisions.
- Sub-grade soil and foundations should be favourable for culverts and bridges.
- Good drainage conditions should be provided along the alignment to drain off surface run-off and seepage flow.
- Steps are to be taken to ensure that there is minimum earth work, balanced excavation and embankment fill along the alignment.
- There should be availability of adequate right-of-way.
- Appropriate locations have to be maintained at crossings of rivers, roads and railway lines.
- Construction materials should be available at reasonable and economical distances.
- The highway location should be free from landslides, avalanches and other natural hazards.
- The cost of construction, maintenance and operations has to be kept minimal for roads, especially in mountainous and desert areas.
- In the case of desert roads, the direction of sand dunes or ridges, terrain of the country and adequate space of land along alignments should be considered.
- Anticipated annual interest, maintenance and operations costs have to be taken into account (Kerr 1932)
- Provision has to be made for second-stage planning for construction of bypass roads due to paucity of funds or need-based highway construction (Giffin, 1932).
- Economical feasibility with ground improvement techniques has to be studied (Rao 2008).

The concept of minimum or shortest distance between any two places is always difficult to decide but suitable diversions are to be conceived for the following cases (Bishop, 1932).

- Marshy and swampy areas
- Location of river meanders and multiple crossings
- Wider and high-altitude river banks
- Archeological and historical monuments
- Religious places such as temples, mosques, churches, graveyards, etc.
- Well-developed areas such as agriculture, industrial, schools, hospitals, business places and residential areas
- Hills and valleys, where tunnelling work is needed
- Forest and other natural valuable resources
- Mining areas and other prohibited areas, viz., military and nuclear hazardous places
- Watershed and liquefaction zones (Rao, 2008).

1.9.2 Control, Mandatory and Obligatory Locations

The following locations along a new highway alignment are to be strictly connected to avoid inconvenience (Losh, 1932).

- *Terminal and intermediate points*: Main terminal points should be connected using suitable alignments taking into account intermediate places such as towns, villages and places of productivity. For example, a typical area showing two terminal places and a few intermediate points (Fig. 1.3 (a)) may be connected by alternative alignments as shown in Figs. 1.3 (b) and (c).
- *Crossings*: At crossing points of canals, streams, rivers and railway lines, the alignment of the road should be carefully decided to avoid difficulties in design, construction, maintenance and operational costs of structures. Multiple crossings are avoided as far as possible. Also, skew crossings are generally avoided and perpendicular crossings with straight approaches are preferred. A typical river crossing is shown in Fig. 1.3 (d).
- *Developed areas*: Built-up areas such as agricultural, industrial, residential areas, valuable natural resources, graveyards and religious places should be avoided or the alignment should be averted from these areas. In case of ponds (Fig. 1.3 (e)), the alignment should be at a safe embankment height above the maximum water level. Near hillocks, the alignment should be at a safe distance from the toe to avoid points of slips and rock falls. Figure 1.3 (f) presents a typical road diversion from the shortest route at religious places to avoid disturbances and prohibitive compensation costs. Also, public sentiment must be given due consideration (Carpenter, 1932) and a minimum safe distance should be left from religious places for keeping zones of silence.
- *Easy gradient*: Vertical alignment for undulating, rolling grounds, hills and valleys should be carefully studied to avoid hairpin bends with steep gradients. Length of the highway may be increased to get suitable gradient as shown in Fig. 1.3 (g). Generally, a gentle gradient of 1 in 40 for flat areas and 1 in 20 for hill roads is provided as a standard gradient. But for shorter distances, these gradients may be slightly relaxed and modified.
- *Location to prevent snow drifting*: Regions subject to heavy snowfall should be avoided (Bishop, 1932).
- As far as possible tunnel operations should be avoided (Rao, 2008).

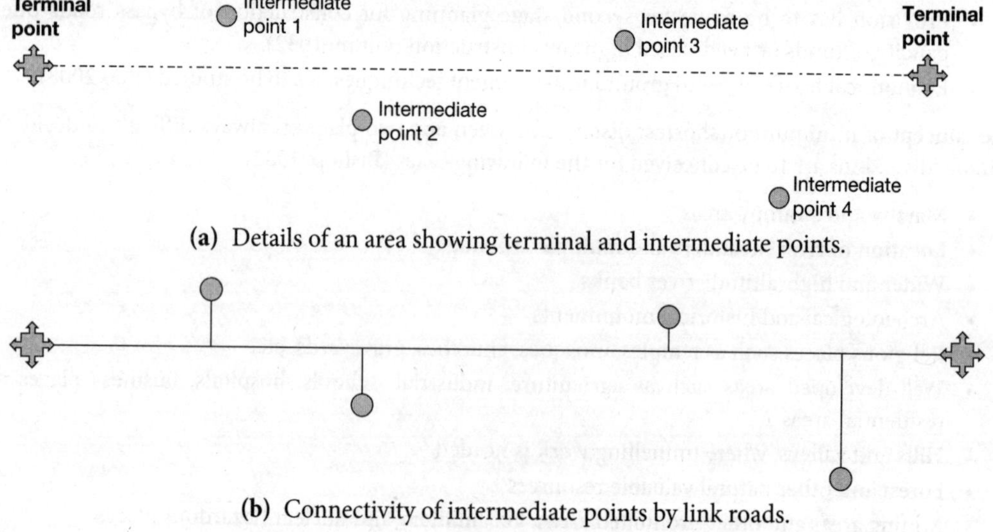

(a) Details of an area showing terminal and intermediate points.

(b) Connectivity of intermediate points by link roads.

Figure 1.3 Deviations of alignment from shortest route.

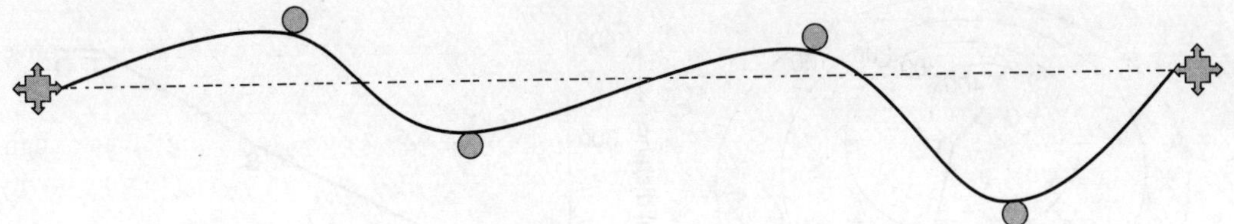

(c) Alternatively proposed connectivity of intermediate points by suitable curves.

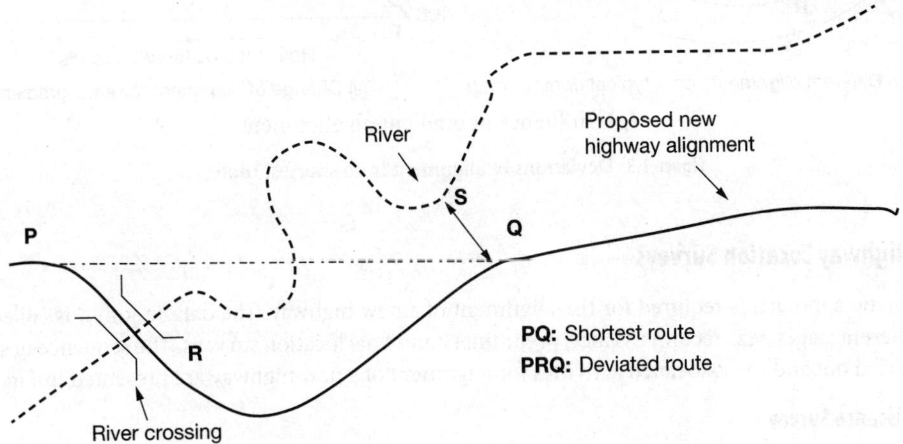

(d) Deviated alignment for suitable river crossing.

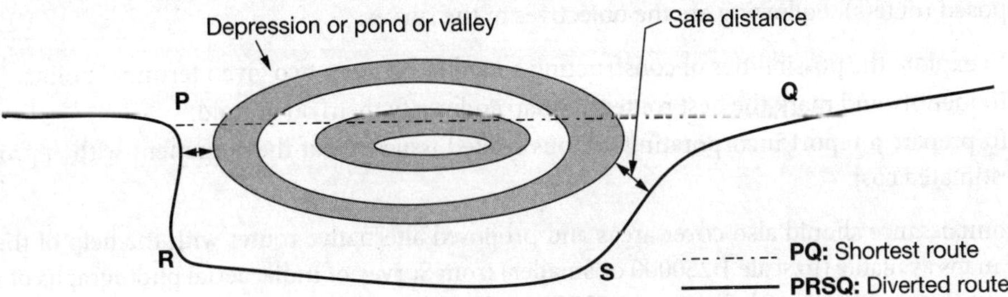

(e) Diverted alignment due to intermediate natural elements.

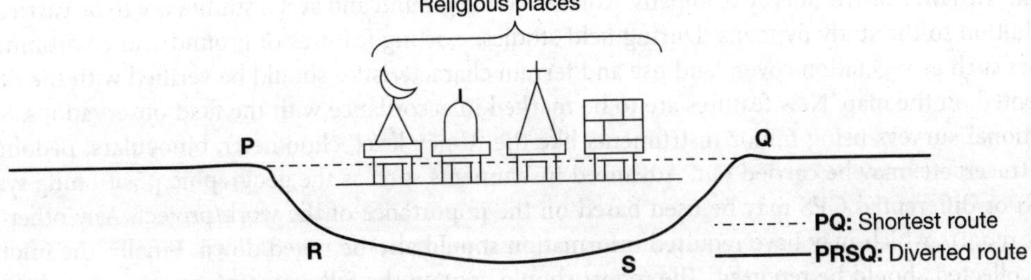

(f) Alignment avoiding elements of religious important places.

Figure 1.3 Deviations of alignment from shortest route.

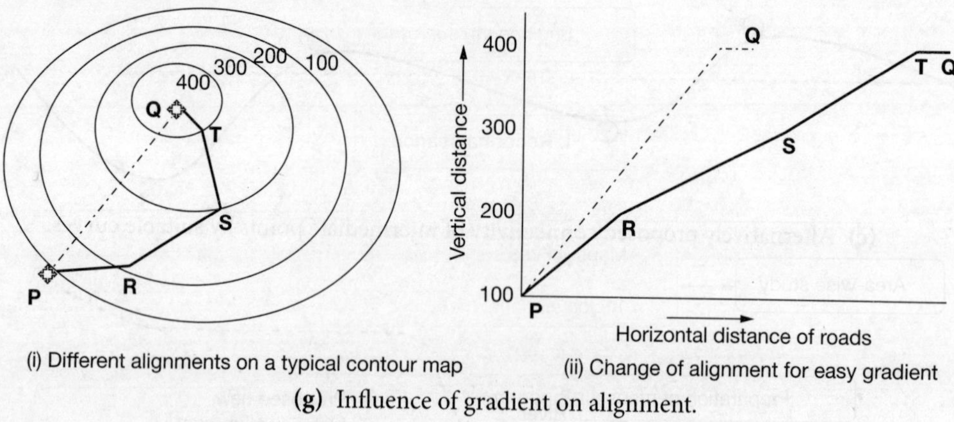

(i) Different alignments on a typical contour map (ii) Change of alignment for easy gradient

(g) Influence of gradient on alignment.

Figure 1.3 Deviations of alignment from shortest route.

1.9.3 Highway Location Surveys

A systematic approach is required for the alignment of a new highway. The data required is collected in three different stages, viz., reconnaissance, preliminary and final location surveys. The sequence of surveys to be carried out and the associated activities for alignment of a new highway are presented in Fig. 1.4.

Reconnaissance Survey

Information related to the existing geographical features of ground and all man-made structures in the study area is to be collected during reconnaissance survey. Such data is used for further detailed study of the proposed route(s). Following are the objectives of the survey.

- To explore the possibilities of constructing a road in between two given terminal points.
- To identify and mark the best routes on map and verify them on ground.
- To prepare a report incorporating various related issues about the alignment with approximate estimated cost.

The reconnaissance should also cover areas and proposed alternative routes with the help of the topographic maps available (to scale 1:250000 or smaller) from Survey of India, aerial photographs of similar scale or resolution (to any scale between 1:5000 to 1:20000), satellite remote sensing imagery (on scales of 1:12500, 1:25000 and 1:50000) from the National Remote Sensing Agency (NRSA), Hyderabad, and small-format aerial photographs (in original or enlarged scale of 1:1000 to 1:10000) from NRSA or Indian Air Force or Air Survey Company, Kolkata. Some ground and aerial studies are to be carried out in addition to the study by maps. During field studies, existing features of ground and environmental factors such as vegetation cover, land use and terrain characteristics should be verified with the details presented on the map. New features are to be marked in accordance with the field observations. Some traditional surveys using minor instruments like the Abney level, clinometer, binoculars, pedometer, ghat tracer, etc. may be carried out. Advanced instruments such as the geographic positioning system (GPS) or differential GPS may be used based on the importance of the work/project. Any other previous reports which may have required information should also be noted down. Finally, the information collected should be reported. The report should contain the following information to aid further investigations in finalising route(s).

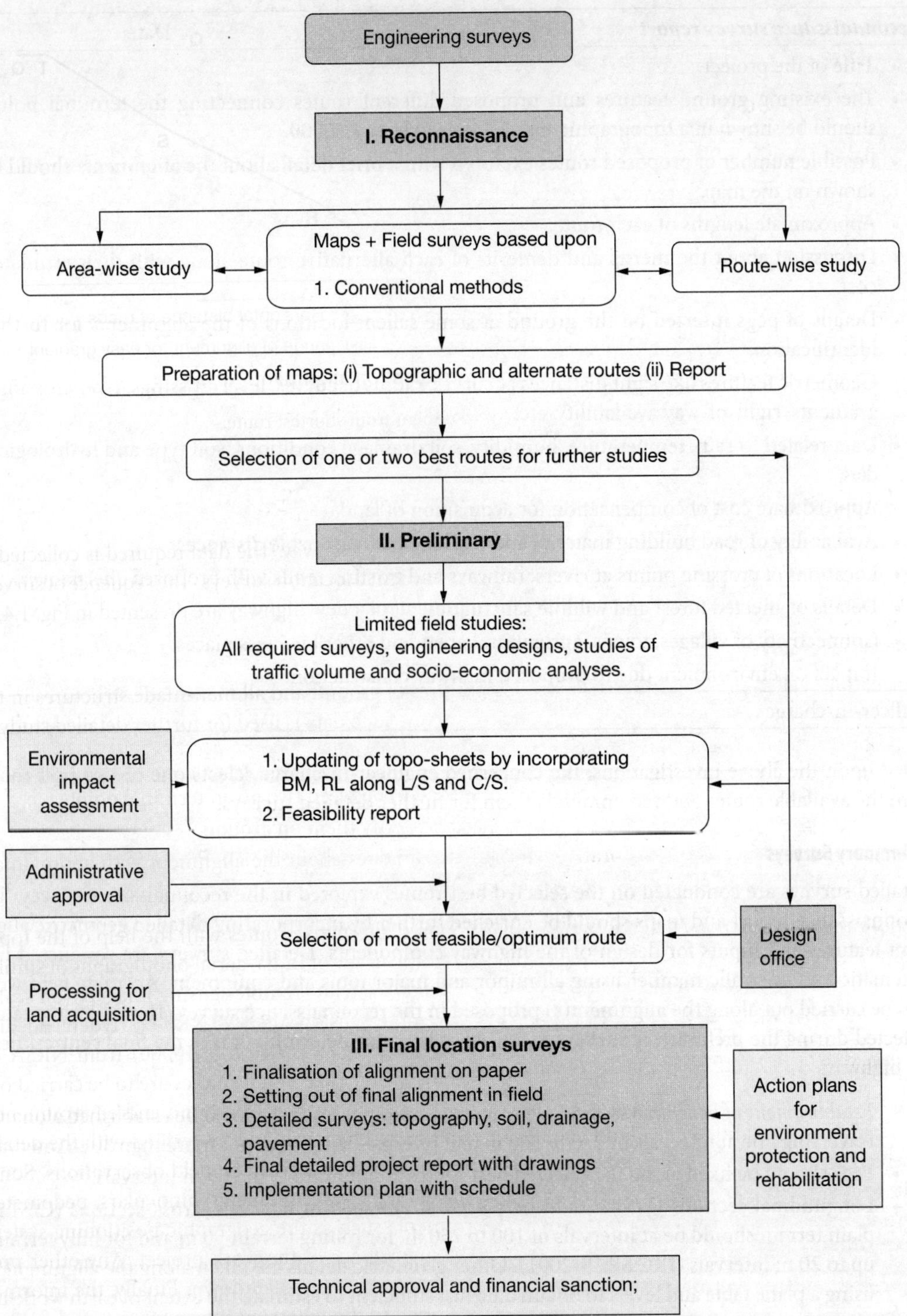

Figure 1.4 Details of highway location surveys.

Reconnaissance survey report	Date:
• Title of the project • The existing ground features and proposed different routes connecting the terminal points should be shown in a topographic map of scale at least 1:50000. • Possible number of proposed routes explored with a brief detail about the alignments should be shown on the map. • Approximate lengths of each route. • Discussion about the merits and demerits of each alternative route along with their estimated cost. • Details of pegs inserted on the ground at some salient locations of the alignments for further identification. • Geometric features like sight distances at curves, radius of curves, level crossings, river crossings, gradients, right-of-way availability, etc. • Data related to rain, temperature, humidity, soil drainage conditions, soil type and hydrological data. • Approximate cost of compensation for acquisition of land. • Availability of road building material and workmen at reasonable distances. • Locations of crossing points at rivers, railways and existing roads with proposed alternatives. • Details of affected forest and wildlife sanctuaries. • Connectivity of villages, towns, agricultural areas and other business places. • Impact on environment due to proposed highway construction.	
Officer-in-charge	

Based upon the above investigations, the concerned engineer-in-charge selects one or two best routes from the available routes and recommends them for further detailed surveys.

Preliminary Surveys

Detailed surveys are conducted on the selected best routes explored in the reconnaissance survey. The reconnaissance report and maps should be enriched further by incorporating detailed geometric alignment features and inputs for design of the highway components. Detailed surveys are conducted in a systematic and scientific manner using all minor and major tools and equipment. Accurate field work is to be carried out along the alignment(s) proposed in the reconnaissance survey. The following data is collected during the preliminary surveys and used as basis for determination of the final centre-line of the highway.

- *Establishment of benchmarks (BM)*: Permanent or temporary BM should be established along the traversing line at intervals of 250 to 500 m and intermediate BM at not more than 30 m intervals.
- Pegs should be fixed along the selected best routes at not more than 100 m intervals.
- Longitudinal sections (L/S) should be placed at every 50 m intervals, cross-sections (C/S) for plain terrain should be at intervals of 100 to 250 m, for rolling terrain 50 m and for hilly terrains, up to 20 m intervals (IRC:SP:19-2001). Other geometric details are taken along the alignment by using a plane table and levels to obtain data just sufficient to estimate the extent of earth work and approximate cost of the road per km.
- *Hydrological data*: Rainfall, flood levels, number of cross-drainage works with their geometrical details.

- *Soil data*: Classification and distribution of soil types along the alignment; Surface and sub-surface condition of the area; Estimation of earth work quantities; Ground water table levels; Forest and vegetation covered.
- All existing spatial features along the alignment such as buildings, ponds, agriculture area, communication lines, river crossings, fencings and rock strata.
- *Traffic surveys*: Origin and destination survey data, traffic volume, type of traffic, axle load, traffic projection, etc.
- In order to mitigate the harmful effects of road-building projects on the environment and the surrounding areas of the proposed projects, it was made mandatory to have an Environmental Impact Assessment (EIA) for 32 categories of development activities. The EIA study includes
  1. Environmental analysis dealing with protective measures and the positive and negative impacts of the project with reference to cost-effective preventive measures
  2. Environmental design for reducing the adverse impact of the proposed road such as erosion of soil due to floods; submergence of surrounding area by accumulated water; loss of vegetation, reserve forest, wild life, etc. Implementation of cost-effective measures to maintain environmental quality in terms of reduction of air pollution, water pollution and soil pollution due to spillages from automobiles, and to reduce noise pollution
  3. Environmental management action plan for execution of the schedule prepared under this category
  4. Resettlement and rehabilitation action plan to quantify loss sustained by tribes and habitation surrounding the project, further formulation of an action plan or policy of rehabilitation, employment and compensation package to the people affected. More details about EIA are available elsewhere (IRC: 104-1989; http://envfor.nic.in; UNEP 2002).

The above surveys may be conducted by using different procedures.

By Conventional Method: Theodolite, chain, steel tap, clinometer, plane table, staves, compass and in case of hilly area, ghat tracers are used for detailed surveying of the alignments with respect to longitudinal and cross-sectional view of the proposed routes.

By Modern Method: Aerial and satellite maps are used to supplement the available topographic maps. Remote sensing, geographic positioning system (GPS), geographic information system (GIS) and total station are also used for three-dimensional modelling and precise quantification of various ground features. These soft copies of maps are suitably converted into AutoCad format or similar files and the proposed route's alignment with respect to the longitudinal section and cross-section is plotted.

The following details are reported in the preliminary survey:

Report of the preliminary survey or feasibility report	Date:
• Title of the project • Location of terminal and intermediate points. • Drawings showing all existing topographic, spatial features, L/S and C/S along the alignments. The drawings are to be prepared for a contour interval of 1 to 3 m. • Required number and locations of cross-drainage works along L/S. • Soil and hydrological data. • Traffic details with anticipated grow rate. • Proposed pavement type with approximate thickness of the pavement crust. • Location and approximate transport cost of the road building materials.	

• Available and required road width along the alignments. • Details of cost estimate of each alternate route with economic justification. Economic analysis can be made with reference to IRC: SP:30 of India or HDM-III of World Bank or RTIM III of TRL UK. • A discussion of the merits and demerits of each route. • Special factors affecting the final location of the alignment. • Hydrological data • All drawings.
Officer-in-charge

A comparative statement of the routes surveyed has to be prepared and the most appropriate and economical route selected. The following major impact factors may be considered for evaluation of the alternatives (Kulkarni et al., 1993).

- Land use
- Relocation impact
- Agricultural production
- Biological/natural resources
- Cultural resources (with potential historic importance)
- Air quality
- Noise

The details given in the above feasibility report are used as a basis by competent authority to accord administrative approval for further detailed surveys.

After obtaining approval, the final selection of the alignment is marked on the map with details of earth work and costs involved. A team of engineers are then deployed for field verification and comparison with the alignment marked on the map. During the field inspection, an assessment is made to identify the possible problems and difficulties during construction, maintenance and operation of the highway. Based upon the examination in the field with reference to the map, required alterations are marked on the map for further survey. The process for land acquisition should be immediately started at this stage since it will take time to get approval from the private owners and government officials concerned.

Final Location Surveys

As discussed in the earlier sections, comprehensive investigations are made in the selection of the final alignment during the preliminary surveys. The centre-line of the final alignment should be marked in the field by fixing stakes at intervals of 20 m in hilly terrain and 50 m in plain and rolling terrains (IRC: SP:19-2001). Further studies are made along the centre-line to precisely work out and document the following data.

- Executive summary and socio-economic profile of the locality
- Availability of the required right-of-way along the centre-line.
- List of the associated problems for land acquisition and compensation costs

In addition data is also required for:

Geometric design of the highway such as right-of-way, design speed, minimum sight distances, overtaking sight distances, radius of curves, super-elevation, side drains and cross-drainage works.

Pavement design: traffic data, sub-grade soil support, foundation strength, soil type, surface and sub-surface drainage details.

Preparation of drawings: cross-drainage works, evaluation of earth-work quantifications, preparation of working drawings for construction, scheduling of the project work.

Cost calculations/evaluation: item-wise cost estimate, overall cost, average cost per km of the highway, economic and financial analyses and construction programming. The calculated cost should also include the expenses incurred for:

1. Action plans for mitigating environmental impacts and related issues such as rehabilitation and resettlement of effected people and habitat.
2. Construction materials, equipment, labour and related issues.
3. Miscellaneous and petty work
4. Conclusions and recommendations for successful implementation and maintenance of the project.

Field work: All required instruments used for marking and driving pegs on the final alignment in field, as well as instruments for marking boundaries of the highway along the alignment should be available. During driving of pegs, offsets or salient points or important permanent structures or points of BM (at every 250 m interval) are taken as reference. Offset distance from these reference points should be 10 m in case of straight and 20 m for curved paths along the centre line. If any alterations or modifications are to be made during pegging the centre line, they should be immediately marked on the map. Permanent reference poles and wooden pegs are fixed firmly on the ground at an interval of 100 m and 30 m respectively along the centre line by using any precise angle-measuring instrument (such as total station) with a steel tape. Cross-sectional details should be preferably taken at intervals of 50–100 mm, 50–75 m, 50 m and 20 m in plain terrain, rolling terrain, built-up areas and hilly terrains respectively (IRC: SP:19-2001). In addition to this, cross-sectional details should also be taken at abrupt changes in alignment, starting and ending point of curves and at other geometrically significant locations.

Precise levelling survey has to be performed along the reference poles and wooden pegs with reference to the permanent and temporary BM established during the preliminary survey. Cross-sections should be taken at right angles to the centre line. The desired width of right-of-way should be taken on both sides of the wooden pegs at an interval of not more than 20 m. If there are any considerable changes in the direction of the centre line or ground profile, additional sets of cross-sections should be taken.

Soil explorations should be made to collect data related to bearing capacity, soil classification, level of water table, rock strata and drainage conditions. Soil samples are collected by using an auger or any other tools and tested in the laboratory for determination of gradation (as per IS: 2720, Part IV), liquid and plastic limits (as per IS: 2720, Part V), density and optimum moisture content (as per IS: 2720 Part VII and VIII), contents of deleterious materials (as per IS: 2720 Part XXVII), and to determine soil engineering properties (as per IS: 2720 Parts XV and XI). Based on the project requirement, information should be collected for low embankments (as per IRC: 10 and IRC: SP: 13), high embankments (as per IRC: 75), landslide-prone areas (as per IRC: SP: 15), flexible pavement design (as per IRC: 37), and rigid pavement design (as per IRC: 58 and IRC: SP: 17). Non-destructive tests such as the electrical resistivity method or nuclear tests are to be conducted to assess the suitability of the site for construction of the highway or foundation for construction of cross-drainage works.

In case of upgradation or strengthening of existing highways, additional information such as road inventory, road riding quality condition, toll collection (for BOT projects) and future economic growth in the area affected by the project are required to be documented.

Location of crossings through reserved forest areas are to be specifically marked on the map of the final alignment. Similarly, the locations of hazardous areas and water-logged areas are also to be marked.

Any other data required for the estimation of the proposed project should be collected and suitably reported in the detailed project report (DPR) with detailed drawings.

Review Questions

1. Briefly explain how roads were developed during the British rule in India.
2. State the significance of the Jayakar Committee for development of the road network in India. What were its recommendations? Name the institutions set up and the road-related developments that took place.
3. What is the Central Road Fund? Mention the amendments and state how the fund is being distributed for different schemes of roads development activities.
4. Enumerate, along with the year of establishment, the different institutions/organisations related to road development in India.
5. What are the major road development plans that evolved since 1943? Briefly explain the salient features of each road network development plan in India.
6. Discuss in detail, the salient features of the Nagpur Road Plan or the First 20-Year Road Development Plan (1943–1963).
7. Discuss in detail, the salient features of the Bombay Road Plan or the Second 20-Year Road Development Plan (1961–1981).
8. Discuss in detail, the salient features of the Lucknow Road Plan or the Third 20-Year Road Development Plan (1981–2001).
9. What are the different types of classification of roads in India?
10. Write a short note on the scenario of the present road network and road-carried traffic.
11. Discuss the National Highway Development Project (NHDP) in India. State its objectives.
12. In detail, explain the NHDP Phases I and II. Also state the ten major ports of India.
13. Write a short note on the SARD-NE programme of road development in India.
14. State the significance of rural roads development in India.
15. What is the PMGSY programme of rural road development and state its objectives?
16. Write an overview on PMGSY Phases I and II
17. What is the need for systematic planning in highway alignment surveys? State the important elements of planning that should be kept in mind during preliminary planning of a new road project.
18. State the factors affecting alignment of a new highway.
19. At what locations is the concept of minimum or shortest distance between two terminal stations not to be considered and why?
20. What are the control, mandatory and obligatory locations to be considered for alignment of a new highway project? Draw figures wherever applicable.
21. Draw a flowchart describing the process of activities involved in highway location surveys?
22. What is the significance of reconnaissance survey in locating a new highway's centre line? Discuss the different types of reconnaissance surveys.
23. Discuss the different maps (with scale) used for reconnaissance surveys.
24. In detail, explain the surveys required at the preliminary stage.
25. State the significance and detail of surveys required under the Environmental Impact Assessment (EIA) for a new highway/road project.
26. State the major impact factors considered for evaluating alternative routes of road proposals.
27. Discuss in detail the surveys required for the final location of a new highway.

Multiple Choice Questions

1. The first 20 year road plan in India is known as
   (a) Delhi Road Plan
   (b) Nagpur Road Plan
   (c) Bombay Road Plan
   (d) Lucknow Road Plan
2. What was the target road density of The Bombay Road Plan?
   (a) 32 km/100 sq. km
   (b) 16 km/100 sq. km
   (c) 82 km/100 sq. km
   (d) 42 km/100 sq. km
3. In which of the following road plan was the target set for all towns and villages with population exceeding 1500 to be connected by MDR of length 3,00,000 km?
   (a) PMGSY Road Plan
   (b) Nagpur Road Plan
   (c) Bombay Road Plan
   (d) Lucknow Road Plan.
4. In which year did Dr M R Jayakar Committee prepare a report that delineated the methodology for the future development of roads in India
   (a) 1928 (b) 1930
   (c) 1927 (d) 1932
5. In which year was the semi-official technical body named Indian Roads Congress (IRC) set up?
   (a) 1934 (b) 1935
   (c) 1930 (d) 1939
6. In which year was the Central Road Research Institute (CRRI) set up at New Delhi?
   (a) 1930 (b) 1939
   (c) 1950 (d) 1956
7. In which year was The Indian Roads Congress (IRC) established?
   (a) 1934 (b) 1935
   (c) 1973 (d) 1939
8. In which year was the Border Roads Organization (BRO) established to develop roads for strategic purposes to be used by the military in the northern, northeastern and other border areas?
   (a) 1927 (b) 1933
   (c) 1960 (d) 1973
9. The National Highway Development Project (NHDP) plan evolved in India during the ________ five-year plan period.
   (a) 8th (b) 9th
   (c) 10th (d) 11th
10. In which year was National Highways and Infrastructure Development Corporation Limited (NHIDCL) set up as a part of MoRTH by the Government of India to expedite construction of National Highway projects with specific focus on North-eastern Region (NER)?
    (a) 2000 (b) 2001
    (c) 2004 (d) 2014

Answers to Multiple Choice Questions

1. (b) **2.** (a) **3.** (d) **4.** (a) **5.** (a) **6.** (c) **7.** (a) **8.** (c) **9.** (b) **10.** (d)

CHAPTER 2

Highway Geometric Design

2.0 INTRODUCTION

Developing a highway geometric design involves determining the physical dimensions of the various geometric elements necessary for a highway, based on scientific and practical requirements. A geometric design is deemed to be excellent if it allows the safe, smooth and economic movement of pedestrians and vehicles – both motorised and non-motorised. If the layout of different geometric elements and their dimensions are defective, it poses serious problems which may be impossible to rectify at a later stage due to developments along the roadside (Losh, 1932). Therefore, the need for good geometric design becomes more pressing. Appropriate specifications and the recommendations of various standard institutions should be suitably implemented in the design.

Design details of the various geometric elements of a highway are discussed below:

2.1 VEHICLE DIMENSIONS

Vehicle dimensions such as length, width, height and gross weight play an important role in the design of various geometric elements of highways (Table 2.1).

Table 2.1 Vehicle parameters useful for designing geometric elements of highways.

Vehicle parameter	Description
Width	Width and number of lanes or lane capacity Width of shoulder and design of signals Parking area, cross drainage works Lateral placement of vehicles
Height	Location of electric and other cables Flyovers and underpasses
Length	Bridges and underpasses Highway junctions, hill roads, curves and hairpin bends Overtaking zones, highway capacity
Engine capacity	Design speed, skid resistance and braking distance Gradient, grade compensation and length of vertical curve Super-elevation on curves Safety measures at salient locations

Table 2.2 lists the standard vehicle dimensions according to the IRC. The standard weights of axles and their configurations are discussed in Chapter 3.

Table 2.2 Vehicle parameters useful for designing geometric elements of highways (IRC: 86-1983 and IRC: 3-1983).

Vehicle parameter	Maximum allowable value (m)
Width	2.44
Height	For single deck: 3.81 For double deck: 4.72
Length†	With two axles: 10.67 More than two axles: 12.19 Tractor with semi-trailer: 15.24 Tractor with trailer: 18.29

† Maximum allowable length of vehicles is 31 m in the USA and 25 m in Canada (Wang and Cartmell, 1998).

2.2 TERRAIN CLASSIFICATION

The geometric design of highways is based on the topography of the terrain (general nature of the ground slope) on which the highway is to be laid and the corresponding design speeds specified. The cost of construction and maintenance of highways depends mainly on the type of the terrain. For example, the cost of road construction in a mountainous area is higher than that on flat or rolling ground. The cross-slope of a terrain may be classified into four types and illustrated in Fig. 2.1.

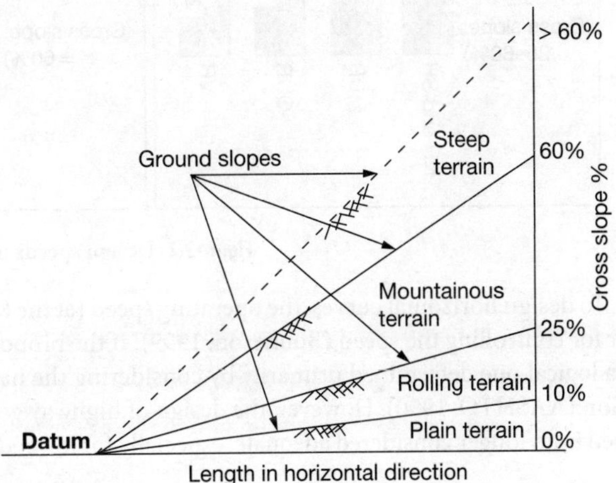

Figure 2.1 Terrain classification based on cross slope (IRC: 73-1980).

2.3 DESIGN SPEED

Design speed may be defined as the maximum average safe operating speed of vehicles that is either specified or maintained by most drivers travelling through all the features of the roadway. It is one of the prime parameters that dictate the overall geometry of highways in plain as well as hilly areas.

The dimensions and layout of most geometric elements are derived from design speed considerations.

The design speeds prescribed by IRC (Fig. 2.2) depend on the type of cross slope of the country, the terrain classification and the importance of the highway. According to IRC (IRC: 73-1980), the ruling design speed is generally considered a design criterion and taken as a primary input parameter for the design of various geometric elements of highways. The minimum design speed is considered if the ruling design speed cannot be adopted for practical reasons and other considerations. Appropriate design speeds may be selected based on site conditions, feasibility, economics of construction and traffic operations. However, care should be taken to maintain uniform vehicular speed throughout the length of the road, even when terrain conditions differ.

If the design must satisfy British and German design practices, estimates of the 85th percentile of operating speeds should be adopted (Kanellaidis, 1996). This means that on any given segment of roadway, the design speed should not be less than the estimated 85th percentile operating speed (Nicholson, 1998).

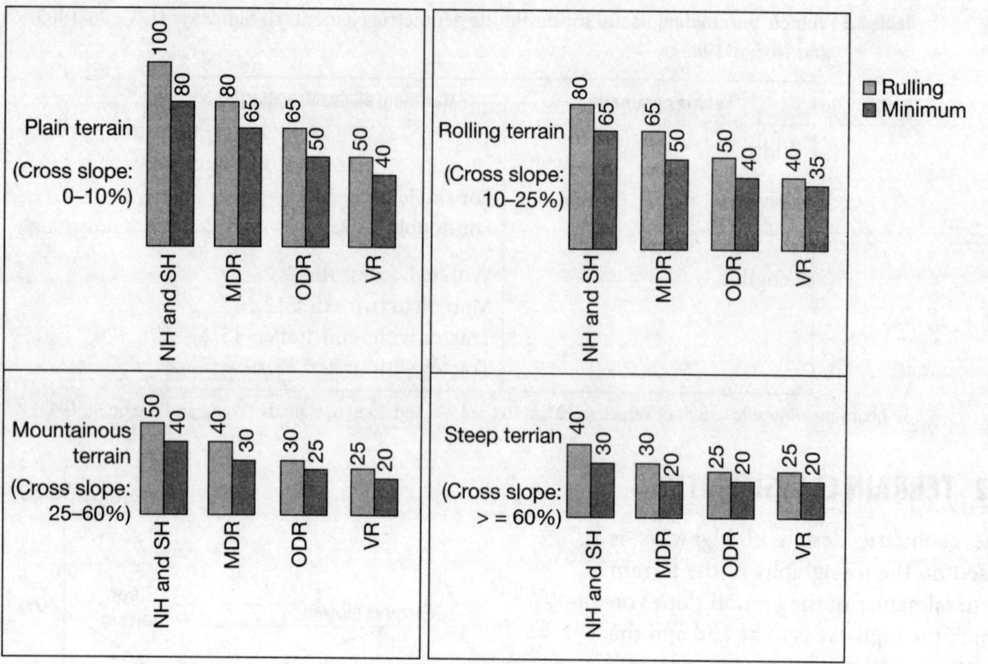

Figure 2.2 Design speeds in India.

To design horizontal curves, the operating speed (at the 85th percentile) is generally used as a parameter for controlling the speed (Bonneson, 1999). If the proposed design speed is to be assumed, it should be a logical one determined primarily by considering the natural terrain, land use and highway classification (AASHTO, 1990). However, the design of highway geometric elements based on assumed design speed is no longer considered adequate, especially for lower design speed of highways (Hirsh et al., 1986).

2.4 CROSS-SECTIONAL ELEMENTS

Cross-sectional elements include all the visible physical features of a highway (Fig. 2.3). Some of the elements important in highway geometric design are briefly discussed in the following sections. For more details on the cross-sections of urban roads, please refer to IRC: 69-1977.

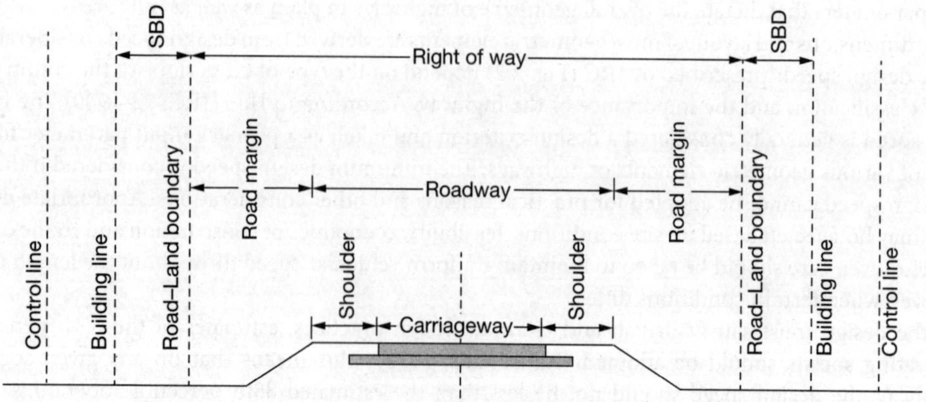

Figure 2.3 Salient cross-sectional details of a highway (IRC: 73-1980).

2.4.1 Boundary Lines

Boundary lines can be classified as road boundaries, building lines and control lines. Each boundary line serves a particular purpose (Fig. 2.2). No construction should be made within the limits of the building lines, to ensure uniformity of road width. Within the control lines, the government has limited control over the construction of structures. Based on the terrain and type of road to be laid, IRC: SP: 15-1974 recommends standard overall widths between these respective lines. These widths of land should be specified for new highway alignments to help in land acquisition. They should also be monitored to prevent illegal developments along the highways. The land width between the road boundary and the building line is called the **set-back distance** (SBD). The standard values of SBD recommended for national and state highways are 3–6 m in plain and rolling terrain and 3–5 m for the remaining classes in all terrains (IRC: 73-1980).

2.4.2 Right-of-Way or Road Land Width

Right-of-way is the width of land acquired to accommodate the width of the carriageway, the shoulders and road margins. It is generally acquired for at least five decades of the service life of a highway. Necessary provision has to be made for future road-widening or anticipated developments along the alignment. This is because, land acquisition, especially in built-up or urban areas, is very difficult and leads to legal problems and payment of a heavy compensation (Losh, 1932).

The required right-of-way for different terrain and road classes as recommended by the IRC is furnished in Table 2.3.

- **Side slopes**: Earthwork should be provided with side slopes during construction of embankment or cutting in order to keep it stable and free from slides. The side slopes are provided based on the type of soil and the height of the construction. As per IRC specifications, side slopes for embankments should be inclined in the ratio 2:1. For excavation zones, the slope ranges from 1:1 for ordinary soils to 1/16:1 for rocks. As the constituent material in cutting is hard strata or rock, a steep slope may be provided. In the case of high embankments, to restrict the land width, reinforced earth techniques are adopted to eliminate occupation of land for side slopes.
- **Kerbs:** It is a raised structure or a step-like component. A series of cut stones, pre-cast cement blocks and brick masonry walls are provided along the edges of a carriageway. Generally, kerbs are provided along footpaths in urban roads to separate or restrict the movement of vehicles and provide firm support to the footpath. Kerbs with a depth of about 100 200 mm are constructed at carriageway edges.
- **Road margin**: It is the width (or strip) of land provided along the sides of shoulders to accommodate side slopes, berms, side drains, spoil banks and arboriculture (if any).
- **Side drains**: Side drains are provided within the width of the road margin, on either side of the alignment, to drain off surface water to a safe distance. Suitable longitudinal gradients are

Table 2.3 Desirable right-of-way (m) for different road classes in plain and rolling terrain

Road class	Open areas(m)	Built-up areas (m)
NH and SH	30–60 (normal: 45)	30–60 (normal: 30)
MDR	25–30 (normal: 25)	15–25 (normal: 20)
ODR	15–25 (normal: 15)	15–25 (normal: 15)
VR	12–18 (normal: 42)	10–15 (normal: 10)

provided according to the terrain. Side drains are designed to carry away storm water and also to control the groundwater table (GWT) at a safe depth below the sub-grade by preventing capillary rise.

2.4.3 Carriageway Width and Roadway

The carriageway (CW) is the main width of the road responsible for providing a strong, stable and smooth riding surface. It is designed based on traffic volume, design life, lane capacity, vehicle dimensions and the clearance to be maintained between adjacent vehicles and the road edge. The IRC specifications regarding the width of carriageways and roadways, based on the number of lanes, are given in Tables 2.4 and 2.5, respectively.

Table 2.4 Width of carriageway.

Lane	Width of carriageway (m)
Single lane	3.75
Double lanes (without raised kerbs)	7.0
Double lanes (with raised kerbs)	7.5
Double lanes of less important roads	5.5
Multi-lane	3.5 per lane
Single lane (less important or village roads)	3.0

Table 2.5 Width of roadway or formation.

Road class		Roadway width (m) for terrain*	
		Plain and rolling	Mountainous and steep
NH and SH	Single lane	12.0	6.25
	Double lane	12.0	8.8
MDR	Single lane	9.0	4.75
	Double lane	9.0	
ODR	Single lane	7.5	4.75
	Double lane	9.0	
VR (Single lane)		7.5	4.0

* These values may vary subject to site conditions as recommended by IRC: 73-1980.

2.4.4 Shoulder

Shoulder is the width of paved surface laid along both sides of the carriageway to facilitate overtaking, temporary parking, walking and for emergency halting. Shoulders provide a lateral edge support to the carriageway along its length, a level surface during overtaking operations and facilitate draining of the surface water to the side drains. The width of the shoulder should be equal to one-half the difference between the widths of the roadway and the carriageway.

Depending on requirements, shoulders are paved using earthen surfaces, a mechanically compacted gravel–soil mix or stabilised soil using bituminous materials or cement or lime. In India, the outward minimum slope of shoulders for earthen surfaces is ½% steeper than the pavement camber (which should be a minimum of 3%). For paved shoulders, a suitable value is selected from Table 2.7. On super-elevated pavement sections, the same value of cross fall is provided as that of the pavement (IRC: 73-1980).

2.4.5 Camber

The cross-slope or slope in the transverse direction provided to the carriageway surface on either side of its centre-line is called a camber or cross-fall (Fig. 2.4). With reference to the level of carriageway edges, the middle portion of the carriageway surface is raised. This cross-fall or transverse slope of the carriageway surface facilitates the quick drain-off of rain water on to the shoulders and prevents stagnation on the carriageway.

Sufficient camber should be provided for safe surface drainage. The value for the camber should be carefully specified based on rainfall intensity, type of materials used for construction, required smoothness and imperviousness of the carriageway. Generally, more camber is provided for unpaved or earthen surfaces than for paved surfaces. This is because, paved surfaces are smooth and do not allow rain water to percolate beneath the layers. At the same time, care should be taken not to provide more camber than is required for unpaved surfaces. It may create erosion or rutting or potholes.

Table 2.6 shows the maximum allowable camber for different surfaces of pavements recommended by the IRC.

2.5 SIGHT DISTANCE

Sight distance is the maximum distance visible to the driver along the centre-line of a highway. The desirable length of sight distance depends on many factors such as characteristics of the driver and vehicles, design speed, super-elevation, and horizontal and vertical curvature. Driver characteristics include

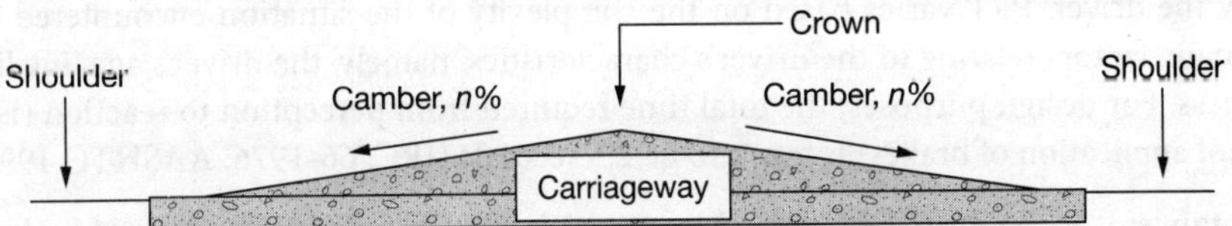

Figure 2.4 Details of transverse slope or camber.

Table 2.6 Values of camber for different surfaces of pavements.

Type of pavement surface	Symbolic profile shape	Range of camber	
		For light rainfall intensity	For heavy rainfall intensity
Impervious surfaces of high-type bituminous surfacing and cement concrete		1.7% (1 in 60)	2.0% (1 in 50)
Thin bituminous surfacing		2.0% (1 in 50)	2.5% (1 in 40)
Water-bound macadam (WBM) and gravel		2.5% (1 in 40)	3.0% (1 in 33)
Earthern road		3.0% (1 in 33)	4.0% (1 in 25)

physical and perception capabilities to react to eventualities during driving. Vehicle characteristics consist of efficiency of the braking system, engine capacity and dimensions (Table 2.2). Sight distance is classified as follows:

- Safe stopping sight distance (SSD) or non-passing sight distance
- Overtaking sight distance (OSD) or passing sight distance
- Intersection sight distance (ISD)

2.5.1 Safe Stopping Sight Distance (SSD)

The minimum distance travelled by a vehicle after the driver perceives an obstruction (distance travelled during perception time), acts on his observation (lag distance), applies the brakes and the vehicle stops (braking distance) is called the safe stopping sight distance. The distance and definition varies depending on the terrain on which the highway is laid. For example, in the case of a summit curve, according to IRC, SSD may be defined as the safe minimum visible straight distance of a driver along the centre-line of a highway when the height of the driver's eye and an object in the opposite direction above the road surface is 1.2 m and 0.15 m, respectively (Fig. 2.5) (IRC: 66-1976). A more detailed discussion on curves can be found in Section 2.11.2.

Perception–Reaction Time (PRT): It is the time required for a driver, while driving the vehicle, to observe, notice, assess and mentally analyse a given situation – the driver will then decide whether to react, apply the brake or steer away. In other words, during driving, it is the minimum time required for a driver to react instantly and exercise control. PRT may be explained by the PIEV theory.

PIEV theory: According to this theory, a driver perceives a situation on the road, analyses it using his intelligence, emotionally feels it and reacts with volition.

$$\text{Stopping sight distance (SSD)} = \text{Lag distance (LD)} + \text{Braking distance (BD)} \quad (2.1a)$$

Lag distance: It is the distance travelled by a vehicle from perception of a situation to the application of brakes by the driver. PRT varies based on the complexity of the situation encountered (AASHTO, 1990) and many factors relating to the driver's characteristics, namely, the driver's age, intelligence and physical fitness. For design purposes, the total time required from perception to reaction (i.e., the time comprising of application of brakes) is taken to be 2.5 seconds (IRC: 66-1976; AASHTO 1994).

Braking distance: It is the distance travelled by a vehicle just after the application of brakes, either to bring the vehicle to a complete halt or to steer it away. A vehicle moves even after the application of brakes since the kinetic energy of the vehicle cannot be instantly reduced to zero. Following are the factors affecting braking distance:

- Velocity of the travelling vehicle and its inertial force
- Brake efficiency of the vehicle and the type of braking system
- Surface friction or road surface characteristics
- Longitudinal slope or gradient of a road

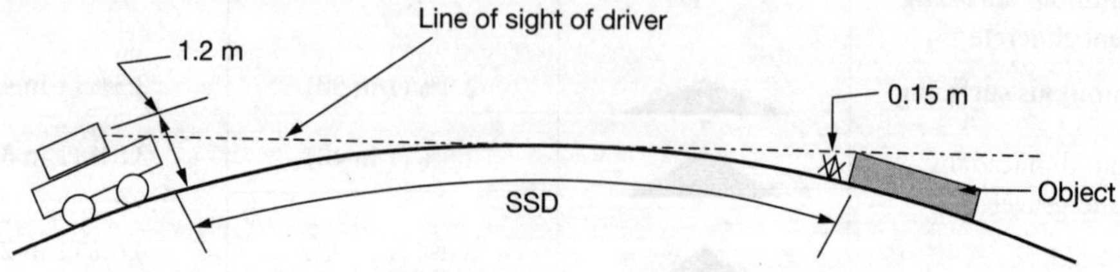

Figure 2.5 Safe stopping sight distance (SSD) on summit curve.

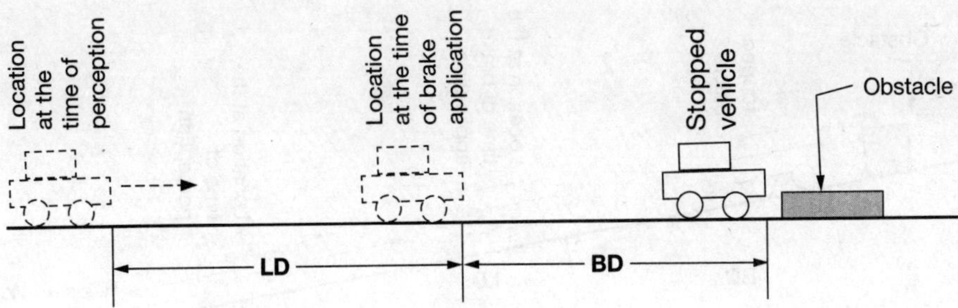

Figure 2.6 Safe stopping sight distance on a plain road.

According to IRC, the value of longitudinal friction factor (f_s) varies from 0.35 at 100 kmph to 0.40 at 20 kmph.

We will now calculate SSD on different terrains.

Case 1: On a level road surface (Fig. 2.6)

$$\text{Lag distance (LD)} = v \times t$$

Braking distance (BD): It may be derived by equating the kinetic energy of the moving vehicle to the work done by the frictional force (F) of the road surface to stop the vehicle.

$$\frac{1}{2}mv^2 = f_s WL$$

$$\frac{W}{2g}v^2 = f_s WL \qquad \left(\because\ W = mg;\ m = \frac{W}{g}\right)$$

$$L = \frac{v^2}{2g\,f_s}$$

$$\text{SSD} = vt + L$$

$$\text{SSD} = vt + \frac{v^2}{2g\,f_s} \qquad (2.1b)$$

Where L = banking distance (m); v = speed of the vehicle (m/s); f_s = coefficient of surface friction (0.35 to 0.40 for a speed range of 100 to 20 kmph); and g = acceleration due to gravity (9.81 m/s^2)

Case 2: On a sloped road surface (Fig. 2.7)

$$\text{Lag distance (LD)} = vt$$

Braking distance (BD), with reference to Fig. 3.10

$$\frac{1}{2}mv^2 = f_s WL + WL \tan\theta$$

$$\frac{W}{2g}v^2 = f_s WL + WL \tan\theta$$

$$\frac{v^2}{2g} = L(f_s + \tan\theta)$$

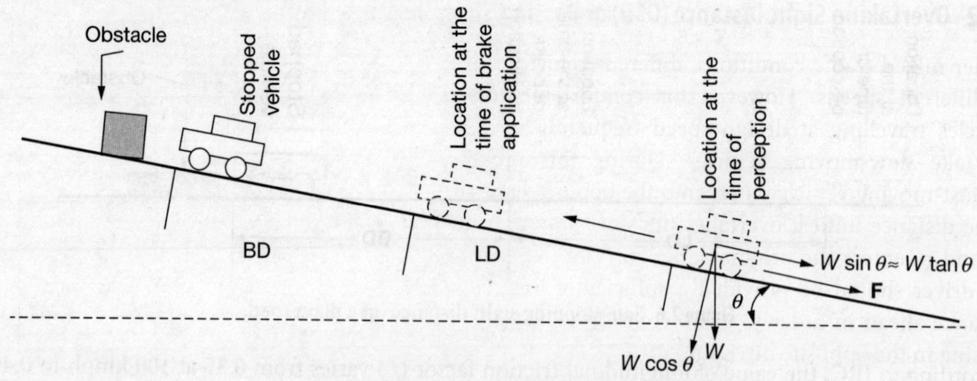

Figure 2.7 Stopping sight distance on a sloped road.

$$L = \frac{v^2}{2g(f_s + \tan\theta)}$$

$$L = \frac{v^2}{2g(f_s + 0.01n)}$$

[since tan $\theta = +0.01n$; if the ascending gradient is taken as $+n\%$ or 1 in n]

$$L = \frac{v^2}{2g(f_s - 0.01n)}$$

[since tan $\theta = -0.01n$; if the ascending gradient is taken as $-n\%$ or 1 in n]

$$\text{SSD} = (LD + BD) \ = \ vt + L$$

$$\therefore \quad \text{SSD} = vt + \frac{v^2}{2g(f_s \pm 0.01\,n)} \tag{2.1c}$$

EXAMPLE 2.1

A vehicle is travelling at an average speed of 100 km/h under the following conditions: (i) level surface, (ii) upward gradient of 1.98% and (iii) downward gradient of 2%. Assume perception and brake reaction time = 2.5 seconds and coefficient of longitudinal friction between vehicle tyres and road surface = 0.35. Determine the safe stopping sight distances.

Answer

Vehicle velocity = 100 km/h = 0.278 ×100 = 27.8 m/s

(i) On a level surface: By Eq. (2.1b), $\text{SSD} = 27.8 \times 2.5 + \dfrac{27.8^2}{2 \times 9.81 \times 0.35}$

$$= 69.5 + 112.5 = 182 \text{ m}$$

(ii) Upward gradient = 1.98%: SSD by Eq. (2.1c)

$$= 27.8 \times 2.5 + \frac{27.8^2}{2 \times 9.81(0.35 + 0.01 \times 1.98)} = 69.5 + 106.5 = 176 \text{ m}$$

(iii) Downward gradient of 2%: $\text{SSD} = 27.8 \times 2.5 + \dfrac{27.8^2}{2 \times 9.81(0.35 - 0.01 \times 2)}$

$$= 69.5 + 119.5 = 189 \text{m}$$

2.5.2 Overtaking Sight Distance (OSD) or Passing Sight Distance (PSD)

Under mixed traffic conditions, different vehicles travel at different speeds. However, for comfortable travel, vehicles travelling at design speed frequently need to overtake slow-moving vehicles. During this process, the fast-moving vehicle enters into the opposite lane for some distance until it overtakes the slow-moving vehicle and re-enters the original lane. For safe overtaking, the driver should be provided a sufficiently long sight distance ahead in order to avoid collision with vehicles coming in the opposite direction.

Overtaking sight distance may be defined as the minimum distance on an undivided highway required by a vehicle/driver to overtake a slow-moving vehicle that is travelling in the same direction safely against approaching traffic in the opposite direction.

The following standard conditions are assumed (AASHTO, 1994). The average speed differential between the impeding and overtaking vehicles passing at design speed is assumed to be 16 km/h. That means, if an overtaking vehicle is travelling at 100 km/h, the impeding vehicle is assumed to be travelling at (100 – 16) = 84 km/h. The maximum allowable difference of 16 km/h between the design and operating speeds is taken as a widely used assumption in highway design in order to achieve operational consistency and improve driving comfort and safety (Kanellaidis, 1991).

Let three vehicles travel on a two-lane undivided straight road, as shown in Fig. 2.8. The fast-moving vehicle F is travelling at design speed and the slow-moving vehicle S is travelling at a uniform speed slower than F. The third vehicle O is travelling at design speed in the opposite direction on the opposite lane.

L_1 = Distance travelled during a PRT of t_1 seconds and during the initial acceleration to the point of encroachment on the right lane. The perception time is the duration of time required for the driver of vehicle F to make the decision to overtake the slow-moving vehicle S.

L_2 = Total distance travelled by vehicle F from the time it overtakes S, enters onto the opposite (right) lane, accelerates while travelling along the side of vehicle S for some distance and re-enters onto the left lane after finishing overtaking.

L_3 = Distance between vehicle F and vehicle O at the end of its manoeuvre. It usually ranges from 30 to 90 m (AASHTO, 1994).

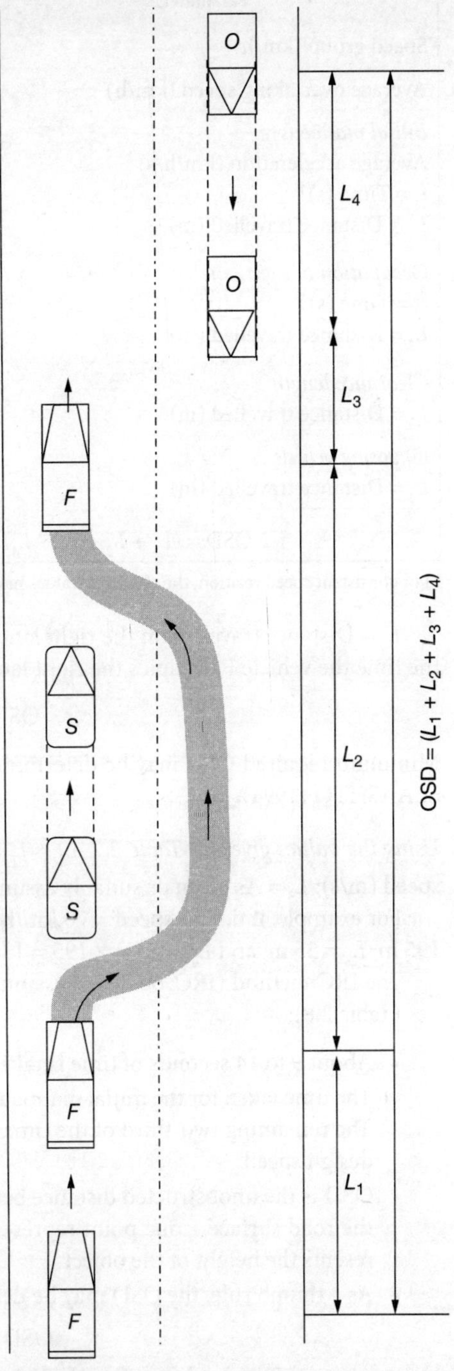

Figure 2.8 Details of overtaking on a two-lane undivided carriageway.

Table 2.7 Elements of safe OSD on undivided two-lane two-way carriageways.

Parameter	Value			
Speed group (km/h)	50–65	66–80	81–95	96–110
Average overtaking speed (km/h)	56.2	70.0	84.5	99.8
Initial manoeuvre				
Average acceleration (km/h/s) †	2.25	2.30	2.37	2.41
t_1 = Time (s)†	3.6	4.0	4.3	4.5
L_1 = Distance travelled (m)	45	65	90	110
Occupation of right lane				
t_2 = Time (s)†	9.3	10.0	10.7	11.3
L_2 = Distance travelled (m)	145	195	250	315
Clearance length				
L_3 = Distance travelled (m)†	30	55	75	90
Opposing vehicle				
L_4 = Distance travelled (m)	95	130	165	210
OSD = $(L_1 + L_2 + L_3 + L_4)$ =	315	445	580	725

† For consistent speed relation, the observed values have been adjusted slightly.

L_4 = Distance travelled on the right lane by the opposing vehicle O at design speed for two-thirds of the time the vehicle F occupies the right lane, or equal to $2L_2/3$.

$$\text{OSD} = L_1 + L_2 + L_3 + L_4 \tag{2.2}$$

Minimum required OSD may be determined using standard values (Table 2.7) or empirical equations of AASHTO (1994).

Using the values given in Table 2.7 $L_1 = t_1$ (s) × Speed of impeding vehicle (m/s); $L_2 = t_2$ (s) × Design speed (m/s); L_3 = As given or suitably assumed; and L_4 = 2/3rd of L_2.

For example, if design speed = 70 km/h, $L_1 = 4.0 \times (70 - 16) \times 0.278 = 60$ m; $L_2 = 10 \times 70 \times 0.278 = 195$ m; $L_3 = 55$ m; and $L_4 = (2/3) \times 195 = 130$, OSD = 60 + 195 + 55 + 130 = 440 m.

The IRC method (IRC: 66-1976) assumes the following conditions and recommends safe OSD values (Table 2.8):

- About 9 to 14 seconds of time is taken for completion of an overtaking manoeuvre.
- The time taken for the initial manoeuvre is about one-third the total time required for overtaking. The remaining two-third of the time is accounted for by the opposing vehicle that is travelling at design speed.
- OSD is the unobstructed distance between two points on the carriageway which are 1.2 m above the road surface – one point represents the position of the driver's eye and the other point represents the height of the object.
- As a thumb rule, the OSD may be determined as

$$\text{OSD} = (5/3) \times (T\,V_d) + L_3$$

where V_d (m/s) = design speed; T (s) = the total time required to complete the overtaking manoeuvre (assumed as 9 to 14 s) and L_3 (m) = distance between the overtaking and impeding vehicles at the end of manoeuvre (assume suitably).

Table 2.8 Overtaking sight distance.

Speed (km/h)	Safe OSD (m) as per Indian Roads Congress (from IRC: 66-1976)	Minimum OSD ‡ (m) as per the *Manual of Geometric Design Standards for Canadian Roads*, Transportation Association of Canada (TAC) (from Sparks et al., 1993)
40	165	
50	235	340
60	300	420
65	340	
70		480
80	470	560
90		620
100	640	680
110		740
120		800
130		860

‡ TAC recommends that the operating speed be 10 km/h less than the highway design speed

(Reproduced with permission from *Policy on Geometric Design of Highways and Streets*, 1994, published by the American Association of State Highway and Transportation Officials, Washington, DC. Documents may be purchased from the AASHTO bookstore at 1-800-231-3475 or online at http://bookstore.transportation.org)

The minimum OSD required for various design speeds according to the *Manual of Geometric Design Standards for Canadian Roads*, Transportation Association of Canada (TAC) (Sparks et al., 1993) is listed in Table 2.8.

The following factors/salient features are considered to be helpful in **determining the minimum OSD** (Wang and Cartmell, 1998):

- The longer the passing and/or impeding (slow-moving) vehicle, the longer the OSD. It has been shown that an increase in the length of the impeding vehicle has the same affect on OSD as that of an increase in the length of the passing vehicle.
- The greater the acceleration of the overtaking vehicle, the smaller the OSD. This means that if the overtaking vehicle has an engine of greater capacity, it requires a shorter OSD.
- If the speed of the impeding vehicle is greater, the required OSD is much larger and extra care should be taken.
- As the speed of the opposing vehicle increases, there is a linear increase in OSD.
- It was observed by another study (Sparks et al., 1993) that the values of OSD are significantly (≈ 40%) greater when overtaking a long truck (23–25 m) than when overtaking a car (4.5 m) at an operating speed of 100 km/h. It was also noted that whether a truck is 23 m or 25 m, it does not have a significant impact on the OSD requirements.
- The differential speed between the overtaken and the overtaking vehicles has the greatest influence in determining the required OSD (Sparks et al., 1993).

2.5.3 Intermediate Sight Distance (ISD)

At approaches to intersections between the main highway and intermediate or single or unimportant roads, intermediate sight distance (ISD) is provided. ISD is also used when provision of OSD is not possible due to economic reasons or space limitations at the site. According to Indian practices, ISD is taken as two times the value of the safe stopping sight distance (SSD). It is measured between two points on the carriageway which are 1.2 m above the road surface – one point represents the position of the driver's eye and the other point represents the height of the object (IRC: 66-1976). For practical reasons, ISDs are not provided on summits and horizontal curves. If there is no space for SSD, restrictive pavement markings (IRC: 35-1970) or 'No Overtaking' signboards should be installed at appropriate places along the road.

2.5.4 Intersection Sight Distance (ISSD)

Estimation of intermediate sight distance becomes crucial at intersections where no traffic control devices are installed. At such locations, minimum sight distance (*xz*) should be provided with a suitable margin or limit beyond which buildings should not be constructed, to avoid obstruction of vision along the *xy* line (Fig. 2.9). The area covered by the *xyz* triangle is called the sight triangle. As stated before, there should not be any obstructions to vision in this area. The sight distance (*xz*) is a function of the time gap (t_g) needed for a stopped vehicle on a minor road to turn or pass through a major road (Harwood et al., 2000). At point B, a total stop board is provided for vehicles approaching the junction from the minor road. The time-gap acceptance method is specified by AASHTO (2001) for determination of ISD based on the assumption that (i) the time gap is independent of the approach speed on the major road, (ii) the intersection angle is 90° and (iii) both major and minor roads are straight and have no curvature in the horizontal and vertical planes. The required sight distance (*xz*) is calculated by the following relationship.

$$\text{Sight distance } (xz) = v \times t_g \tag{2.3}$$

where v = the design speed on the major road (m/s) and t_g = the time gap (s) required for a vehicle on the minor road to proceed. For different types of vehicles, the value of t_g may be obtained from *A Policy on the Geometric Design of Highways and Streets* (AASHTO, 2001).

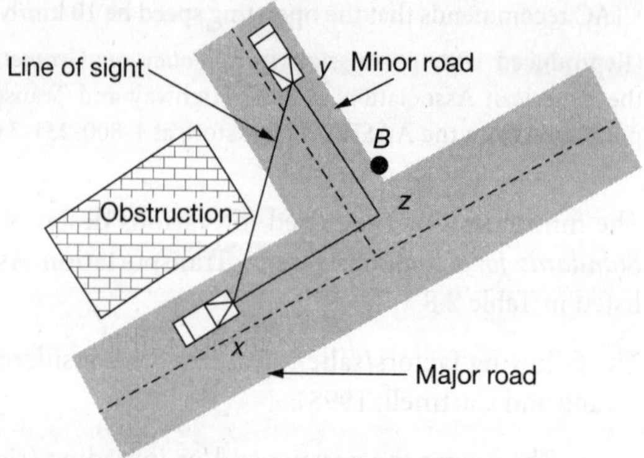

Figure 2.9 Details of intermediate sight distance.

Adequate sight distance should be provided at intersections to enable drivers to view vehicles approaching and decide whether to stop, reduce speed or cross the main road smoothly without collision (in case of un-controlled intersections). According to Indian practices (IRC: 66-1976), the minimum visibility distance (*yz*) along the minor road is taken as 15 m and the average time gap to travel the distance (*xz*) along the major road is assumed to be 8 seconds. For example, for a vehicle standing on a minor road at 15 m before approaching a major road, the minimum visibility distance required along the major road is equal to $v \times 8$, where v is the design speed in m/s on the major road.

2.5.5 Sight Distance at Medians

In case of median crossing, vehicles generally stop adjacent to the median and cross or take a U-turn when sufficient distance is available from the opposite on-coming vehicle (B) (Fig. 2.10). For crossing a median, the minimum required sight distance (SD) may be determined from the relationship, SD = $v \times T$, where v = design speed of the vehicle on the opposite lane (m/s); $T = t_p + t_c$; t_p = perception time (s) and t_c = time (s) required to cross or perform a U-turn of a stopped vehicle adjacent to the median.

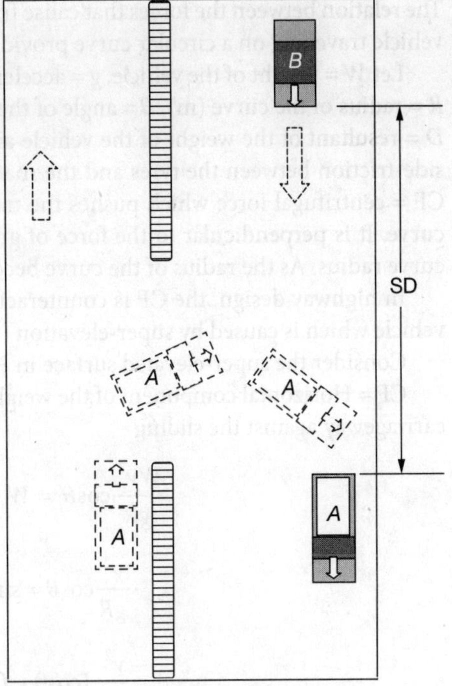

Figure 2.10 Sight distance at a median turning.

2.6 SUPER-ELEVATION

If a vehicle is travelling along a horizontal curve, there is a chance that it may fall, either by overturning or by skidding, due to centrifugal force. The intensity of the centrifugal force depends on the radius of the curve, and the weight and velocity of the travelling vehicle. To counteract the centrifugal force acting on the vehicle while travelling on the horizontal curve, the outer edge of the carriageway surface is elevated with respect to the inner side. This elevated outer edge of the carriageway or the provision of a slope in the transverse direction of the carriageway along a horizontal curve is called **super-elevation** (e) or cant or banking. The different forces acting on a vehicle travelling on a curved path with respect to its centre of gravity (cg) are presented in Fig. 2.11.

The super-elevation may be expressed as given below.

(i) Rate of super-elevation (in decimal form) = $e = \tan\theta = \frac{\text{MN}}{\text{OM}}$.

(ii) 1 in n = 1 unit rise for n units of horizontal distance; for example, 1 in 700 = 1 cm rise for 700 cm of horizontal distance.

(iii) n per cent = n unit rise for 100 units of horizontal distance; for example, 7% super-elevation for a 7-m-wide carriageway means that the outer edge of the carriageway is raised by [(7/100) × 700 cm)] = 49 cm above its inner edge.

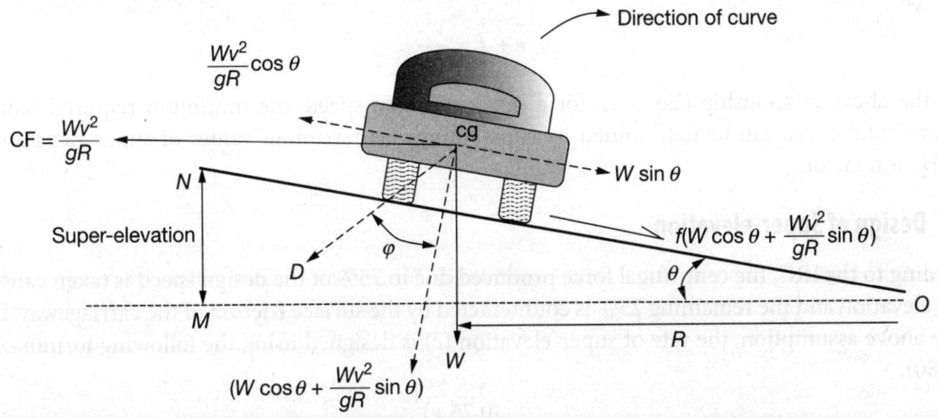

Figure 2.11 Forces acting on a vehicle travelling on a circular curve provided with super-elevation (not to scale).

The relation between the forces that cause friction across the carriageway and the centrifugal force of the vehicle travelling on a circular curve provided with super-elevation is given in Fig. 2.11.

Let W = weight of the vehicle; g = acceleration due to gravity (m/s^2); v = velocity of the vehicle (m/s); R = radius of the curve (m); θ = angle of the carriageway surface or transverse slope of the carriageway; D = resultant of the weight of the vehicle and the centrifugal force; and $f = \tan(\varphi)$ = coefficient of the side-friction between the tyres and the road surface or side friction factor (a non-dimensional value). CF = centrifugal force which pushes the travelling vehicle outwards from the centre of the horizontal curve. It is perpendicular to the force of gravity. The centrifugal force is inversely proportional to the curve radius. As the radius of the curve becomes smaller, the CF becomes greater and vice versa.

In highway design, the CF is counteracted by side friction and the component of the weight of the vehicle which is caused by super-elevation (Awadallah, 2005).

Consider the super-elevated surface in Fig. 2.11.

CF = Horizontal component of the weight of the vehicle + Force caused due to surface friction of the carriageway against the sliding

$$\frac{Wv^2}{gR}\cos\theta = W\sin\theta + f\left(W\cos\theta + \frac{Wv^2}{gR}\sin\theta\right)$$

$$\frac{v^2}{gR}\cos\theta = \sin\theta + f\cos\theta + f\frac{v^2}{gR}\sin\theta$$

$$\frac{v^2}{gR}(\cos\theta - f\sin\theta) = \sin\theta + f\cos\theta$$

Dividing both sides by $\cos\theta$

$$\frac{v^2}{gR}(1 - f\tan\theta) = \tan\theta + f$$

$$\frac{v^2}{gR}(1 - f\times e) = e + f \qquad \text{(since } \tan\theta = e\text{)}$$

$$\frac{v^2}{gR}(1 - 0) = e + f \text{ (since } f\times e \approx 0 \text{ when the maximum allowable values of } f \text{ and } e \text{ are taken)}$$

$$e + f = \frac{v^2}{gR} \tag{2.4}$$

From the above relationship (Eq. 2.4), for a selected design speed, the minimum required radius of the horizontal curve can be determined by substituting the maximum values of super-elevation and side-friction factor.

2.6.1 Design of Super-elevation

According to the IRC, the centrifugal force produced due to 75% of the design speed is taken care of by super-elevation and the remaining 25% is counteracted by the surface friction of the carriageway. Based on the above assumption, the rate of super-elevation (e) is designed using the following formula (IRC: 73-1980).

$$e = \frac{(0.75\,v)^2}{gR} \tag{2.5}$$

According to Indian (IRC: 73-1980) and US guidelines (AASHTO, 1984), design speeds are used for planning the super-elevation, whereas in Germany, operating speeds are used (Kanellaidis, 1991). Operating speeds are usually expressed as the 85th percentile of passenger cars in free-flow conditions (Lamm and Choueiri 1987).

2.6.2 Design Rate of Super-elevation

In India, the permissible e_{max} values recommended by IRC (IRC: 73-1980 and IRC: 86-1983) are given below.

- In plain and rolling terrain, including snowbound areas : 7% (1 in 16)
- In hilly areas not bounded by snow : 10% (1 in 10)
- In urban areas with frequent intersections : 4% (1 in 25)

When the super-elevation determined from Eq. (2.5) is less than the road camber, the normal camber should be provided along the curved section; the super-elevation need not be provided (IRC: 73-1980).

EXAMPLE 2.2

Determine the curve radii beyond which no super-elevation is provided when the design speed is 70 km/h and the camber is 2%.

Answer

$$v = 0.278 \times 70 = 19.46 \text{ m/s}$$

Using Eq. (2.5), $e = \dfrac{(0.75\, v)^2}{gR}$. Substitute the e value as equal to camber value.

$$R \geq \frac{(0.75\, v)^2}{g \times (\text{camber})} = \frac{(0.75 \times 19.46)^2}{9.81 \times \left(\dfrac{2}{100}\right)} = 1085.7\text{m}$$

∴ For the given conditions, no super-elevation is required when the radius is greater than 1086 m

2.6.3 Design Side-friction Factor

The optimal design side-friction factor (f_{max}) is one that assures safe driving on a horizontal curve having minimum permissible radius, maximum super-elevation and design speed (AASHTO, 2001). f_{max} is generally used to define the minimum radius available for the design of the horizontal curve.

In India, the value of f_{max} recommended for all design speeds is 0.15 (IRC:73-1980).

2.6.4 Analysis of Curve Radii

Ruling and minimum curve radii can be determined using the ruling and minimum design speeds, respectively. Depending on the type of terrain and design speed (Table 2.3), suitable equations may be used (either (2.6) or (2.7), given below). The IRC has recommended radii beyond which no super-elevation is required and also R_{min} for different terrain conditions (IRC:73-1980).

We have,
$$R = \frac{v^2}{g\,(e + f)}$$

The above equation may be expressed as follows:

$$R_{min} = \frac{v^2}{g\,(e_{max} + f_{max})} \tag{2.6a}$$

For plain, rolling terrian and hilly areas, including snowbound locations,

$$R_{min} = \frac{(v \text{ m/sec})^2}{9.81 \text{ m/sec}^2 (0.07 + 0.15)}$$

$$R_{min} = 0.46335 \, (v_{min})^2 \text{ m} \quad \text{and} \tag{2.6b}$$

$$R_{ruling} = 0.46335 \, (v_{ruling})^2 \text{ m} \tag{2.6c}$$

Similarly, for hilly areas not bounded by snow:

$$R_{min} = \frac{(v \text{ m/sec})^2}{9.81 \text{ m/sec}^2 (0.10 + 0.15)}$$

$$R_{min} = 0.40771 \, (v_{min})^2 \text{ m} \quad \text{and} \tag{2.7a}$$

$$R_{ruling} = 0.40771 \, (v_{ruling})^2 \text{ m} \tag{2.7b}$$

If the curve radius is greater than R_{min}, the values of super-elevation and side-friction factor should be suitably reduced, according to AASHTO. The limiting speed (V_L) may then be determined for the given values of the curve radius (R), design super-elevation (e) and maximum side-friction factor (f_{max}) using the following relationship (Nicholson, 1998).

$$V_L = \sqrt{gR(e + f_{max})} \tag{2.8}$$

EXAMPLE 2.3

For a proposed national highway (NH), determine the minimum required value of the radius when the design speed = 100 km/h, f_{max} = 0.15 and e_{max} = 0.07.

Answer

Let $v = 100 \times 0.278 = 27.8$ m/s. The minimum radius required for the curve corresponding to V is determined by Eq. (2.6b).

$$R_{min} = 0.46335 \, (v)^2 = 0.46335 \, (27.8)^2 = 358 \text{ m}$$

EXAMPLE 2.4

Determine the design speed margin of safety (DSMS) when the design speed = 90 km/h and the horizontal circular curve radius = 520 m. Assume suitable required values.

Answer

Assume f_{max} = 0.15 and e_{max} = 0.07. Let $v = 90 \times 0.278 = 25.02$ m/s.

From Eq. (2.5)

$$e = \frac{(0.75 \times 25.02)^2}{(9.81 \times 520)} = 0.069 < e_{max}; \quad \text{provide } e = 6.9\%$$

From Eq. (2.4), the actual required value of f is determined as

$$f = \frac{v^2}{gR} - e = \frac{(25.02)^2}{9.81 \times 520} - 0.069 = 0.0537$$

Friction utilisation ratio = $100 \times f/f_{max} = 100 \times 0.0537/0.15 = 35.8\%$

It is determined that the required side-friction is less than the assumed value of maximum side-friction. Therefore, the design is safe.

The corresponding limiting speed may be determined by Eq. (2.8).

$$V_L = \sqrt{gR(e + f_{max})} = \sqrt{9.81 \times 520(0.069 + 0.15)} = 33.42 \text{ m/s} = 120.32 \text{ km/h}$$

Therefore, the design speed margin of safety (DSMS) = V_L – Design speed (v)

$$= 120.23 - 90 = 30.32 \text{ km/h}$$

EXAMPLE 2.5

Determine the available and demanded side-friction factors for the given conditions: design speed = 90 km/h; operating speed = 80 km/h; and design value of super-elevation = 0.069.

Answer

$$v_d = 90 \times 0.278 = 25.02 \text{ m/s}; \ v_o = 22.24 \text{ m/s}.$$

From Eq. (2.4)

$$\text{Available (or assumed) side friction} = \frac{(25.02)^2}{(9.81 \times 520)} - 0.069 - 0.0537$$

$$\text{Demanded side friction} = \frac{(22.24)^2}{(9.81 \times 520)} - 0.069 = 0.028$$

These values and their difference may be used to evaluate the performance of horizontal curves against the recommended standard driving safety criteria (Lamm et al., 1991).

EXAMPLE 2.6

A national highway (NH) passes through plain, rolling and hilly areas. According to IRC guidelines, the design super-elevation for the given conditions is:

(i) For plain terrain: Ruling design speed = 80 kmph; horizontal curve radius = 215 m.
(ii) For rolling terrain: Ruling design speed = 80 kmph; horizontal curve radius = 300 m.
(iii) For hilly areas: Ruling design speed = 100 kmph; horizontal curve radius = 137 m.

Answer

Assume $e_{max} = 0.07$ and $f_{max} = 0.15$.

(i) For plain terrain

$$v = 80 \times 0.278 \text{ kmph} = 22.24 \text{ m/s}$$

From Eq. (2.5)

$$e = \frac{(0.75 \times 22.24)^2}{(9.81 \times 215)} = 0.1319 > e_{max}$$

Therefore, provide $e = e_{max}$ and check the value of f_{max}.

From Eq. (2.4)

$$\text{Available } f = \frac{v^2}{gR} - e_{max} = \frac{22.24^2}{(9.81 \times 215)} - 0.07 = 0.164 > f_{max}$$

Therefore, determine maximum allowable speed when $e = e_{max}$ and $f = f_{max}$ values are taken

$$V_L = \sqrt{9.81 \times 215(0.07 + 0.15)} = 21.54 \text{ m/s} = 77.48 \text{ kmph}$$

(ii) For rolling terrain

$$v = 80 \times 0.278 \text{ kmph} = 22.24 \text{ m/s}$$

From Eq. (2.5)

$$e = \frac{(0.75 \times 22.24)^2}{(9.81 \times 300)} = 0.094 > e_{max}$$

Therefore, take $e = e_{max} = 0.07$ and check the value of f_{max}.

From Eq. (2.4)

$$\text{Available } f = \frac{v^2}{gR} - e = \frac{22.24^2}{(9.81 \times 300)} - 0.07 = 0.098 < f_{max}$$

Provide $e = 0.07$ for the given speed.

(iii) For hilly areas

$$v = 100 \times 0.278 \text{ kmph} = 27.8 \text{ m/s}$$

From Eq. (2.5)

$$e = \frac{(0.75 \times 27.8)^2}{(9.81 \times 137)} = 0.080 < e_{max} \text{ where } e_{max} = 0.1.$$

Therefore, provide super-elevation rate = $0.08 \times 100 = 8\%$.

EXAMPLE 2.7

For achieving the best surface drainage conditions, economy in construction and environmental compatibility with surroundings, it was proposed to provide negative super-elevation on a horizontal curve of a multi-lane highway. Determine the minimum radius of the horizontal curve given the following conditions: operating speed of vehicles = 60 km/h, $e_{max} = -2.4\ \%$ and $f_{max} = 0.08$.

Answer

$$v = 60 \times 0.278 = 16.68 \text{ m/s};\ e_{max} = -2.4\ \% = -0.024 \text{ and } f_{max} = 0.08.$$

From Eq. (2.6a)

$$R_{min} = \frac{16.68^2}{9.81 \times (-0.024 + 0.08)} = 506.4 \text{ m}$$

2.6.5 Attaining Super-elevation

The choice of method depends on the situation encountered. A reference line or axis of roadway along the inner edge or outer edge or centre-line is considered the datum line or axis of rotation. About the axis of rotation, the respective points of the roadway surface are revolved in two stages.

In the first stage, the outer half of the cambered roadway surface is gradually rotated to the horizontal (i.e., normal cross slope = 0) and the inner half of the roadway surface is kept at the normal camber. This means that the normal camber on the outer half of the roadway is removed (Hinkle, 1930) and gradually made horizontal over a length of road. This is termed as a **tangent run out** (L_{TR}). In the second stage, the horizontally placed outer half of the roadway surface is gradually rotated about the axis of rotation until it matches the slope of the inner half of the roadway surface (at Q in Fig. 2.12). Further, the whole inclined roadway surface is revolved about the axis of rotation until full super-elevation is attained at the point when the circular curve begins. The curve between the end of the tangent run-out and the beginning of the circular curve is called the **spiral curve** (or transition curve) and its length is termed the **super-elevation run-off** or length of run-off (L_R). This length is determined based on the

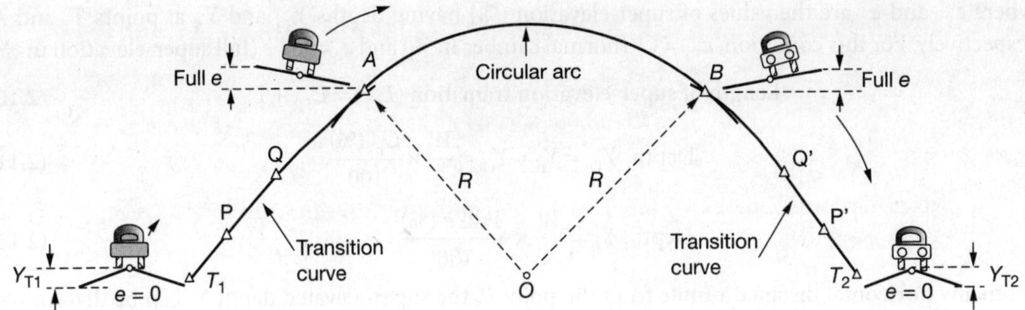

Figure 2.12 Plan view of a highway with a combined curve (roadway rotated about the centre-line).

super-elevation rate and design speed. Details of how super-elevation is attained by rotating the roadway surface about the centre-line of a highway with a combined curve are explained below.

From Fig. 2.13, it is observed that the super-elevation provided is gradually increased to its full value in stages – (i) point T_1 to point P and (ii) point P to junction point A of transition and the circular curves. The full value of super-elevation is provided on the circular curve portion from point A to junction point B of the circular and transition curves. Similarly, on the down slope, the full value of super-elevation (at B) is deceased to zero from points B to Q', P' and T_2.

Let us consider the combined curve (Fig. 2.13), where a circular curve is augmented with similar spiral curves at either ends. A normal camber of C (in per cent) is provided up to the point T_1. The width of the carriageway is termed W (in metres). The maximum allowable longitudinal slope is G (%).

The relationships required for super-elevation design with reference to Fig. 2.13(b) may be obtained as From the fundamental principles of rectangular hyperbola (Taiganidis and Kanellaidis, 1991)

$$\frac{L_{\mathrm{TR}}}{L_{\mathrm{R}}} = \frac{e_{\mathrm{T1}}}{\text{full } e} = \frac{Y_{\mathrm{T1}}}{Y_{\mathrm{A}}} \tag{2.9}$$

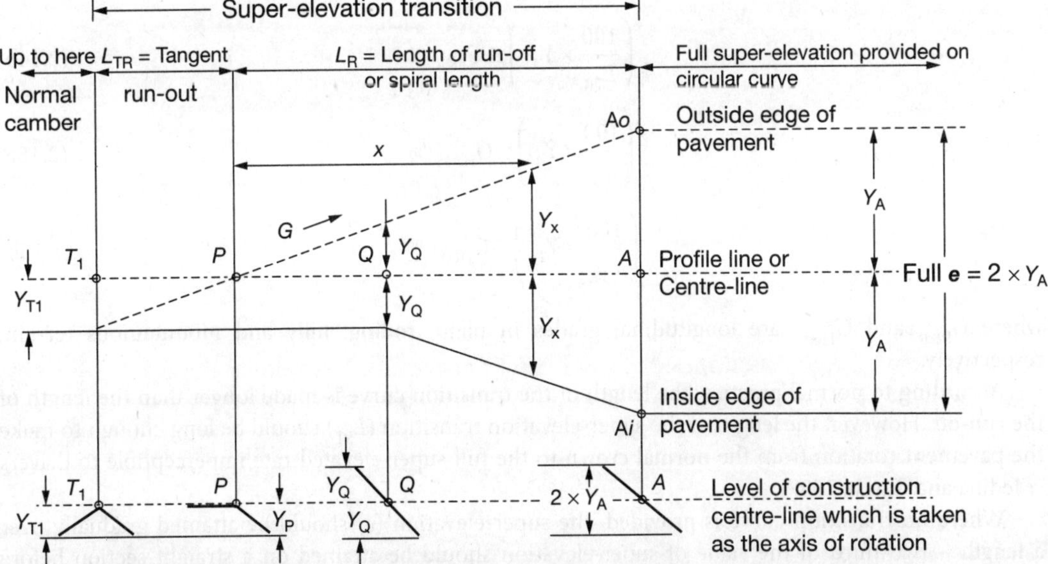

Figure 2.13 Profile of the roadway along the highway as given in Fig. 2.12 (not to scale).

where e_{T1} and e_A are the values of super-elevation (%) having depths Y_{T1} and Y_A at points T_1 and A, respectively. For this condition, $e_{T1} = C_n$ (normal camber in %) and e_A= full e (full super-elevation in %).

$$\text{Length of super-elevation transition } (L_{ST}) = L_{TR} + L_R \tag{2.10}$$

$$\text{Depth, } Y_{T1} = Y_P = Y_Q = \frac{W}{2} \times \frac{C_n\,(\%)}{100} \tag{2.11}$$

$$\text{Depth, } Y_A = \frac{W}{2} \times \frac{\text{Full } e\,(\%)}{100} \tag{2.12}$$

From any horizontal distance x units from the point P, the super-elevated depth Y_x can be determined based on the principle of similar triangles by considering the triangle PAA_0.

$$\frac{Y_x}{Y_A} = \frac{x}{L_R} \tag{2.13}$$

The depth values between points T_1 and P may be determined in a similar fashion.

2.6.6 Rate of Change of Super-elevation

While attaining super-elevation, there should be no abrupt change in the longitudinal gradient of the curve. Abrupt changes create inconvenience to passengers. It may also cause slow-moving vehicles to slide and the inner side tyres of the vehicles to undergo greater wear than the tyres on the outer side. Sufficient curve length, especially in the case of transition curves, should be maintained to keep the desired gradient along the curves. According to IRC:73, the rate of change of super-elevation should not be more than 1 in 150 (≈ 0.67 %) for plain and rolling terrain ($G_{P\&R}$) and 1 in 60 (≈ 1.67 %) for hilly or mountainous terrain ($G_{H\&M}$). The above criteria may be checked by Eqs. (2.14 a and b) and (2.15 a and b). If these conditions are not satisfied, the lengths of run-off or/and tangent run-out should be suitably increased.

$$\left(\frac{100}{L_R} \times Y_{A'}\right) \le G_{P\&R}\ \% \tag{2.14a}$$

$$\left(\frac{100}{L_{TR}} \times Y_{T1}\right) \le G_{P\&R}\ \% \tag{2.14b}$$

$$\left(\frac{100}{L_R} \times Y_A\right) \le G_{H\&M}\ \% \tag{2.15a}$$

$$\left(\frac{100}{L_{TR}} \times Y_{T1}\right) \le G_{H\&M}\ \% \tag{2.15b}$$

where $G_{P\&R}$ and $G_{H\&M}$ are longitudinal grades in plain, rolling, hilly and mountainous terrain, respectively.

According to normal practice, the length of the transition curve is made longer than the length of the run-off. However, the length of the super-elevation transition (L_{ST}) should be long enough to make the pavement rotation from the normal crown to the full super-elevated rate imperceptible to drivers (Medina and Tarko, 2007).

Where no transition curve is provided, the super-elevation (e) should be attained gradually over a length – two-third of the value of super-elevation should be attained on a straight section before the beginning of the circular curve and the remaining one-third value of super-elevation should be attained on the curve (AASHTO and IRC: 52-2001). For better speed operating conditions near circular curves, 70–90% of the run off length should be placed over the tangent length (AASHTO, 2004). To view super-elevation designs for multi-lane divided highways, AASHTO recommendations may be referred.

EXAMPLE 2.8

Determine the minimum length of the tangent run out, when camber = 2%, rate of design super-elevation = 7% and length of spiral = 90 m.

Answer

From Eq. (2.9), let us assume

$$\text{Length of spiral} = \text{Length of super-elevation run off } (L_R)$$

$$\frac{L_{TR}}{90\,\text{m}} = \frac{2\,\%}{7\,\%}; \qquad \therefore\ \text{LTR} = 25.7\ \text{m}$$

EXAMPLE 2.9

Determine the values of super-elevation at salient locations along a combined curve with reference to Fig. 2.14 (i) and check for maximum allowable longitudinal gradient according to the IRC guidelines. The pavement consists of an undivided dual lane in rolling terrain and has a width of 7.0 m. The other conditions are the same as given in the above problem.

Answer

From the above problem, we get $C = 2\%$; full $e = 7\%$; $L_R = 90$ m and $L_{TR} = 25.7$ m. Assume that full super-elevation should be provided throughout the length of the circular curve. The notations followed will be as given in Fig. 2.13 (i) and (ii).

From Eq. (2.10)

$$\text{Length of super-elevation transition } (L_{ST}) = 25.7 + 90 = 115.7\ \text{m}$$

From Eq. (2.11)

$$\text{Depth } Y_{T1} = Y_p = Y_Q = \frac{7.0}{2} \times \frac{2\,\%}{100} = 0.07\ \text{m} = 7\ \text{cm}$$

From Eq. (2.12)

$$\text{Depth } Y_A = \frac{7.0}{2} \times \frac{7\,\%}{100} = 0.245\ \text{m} = 24.5\ \text{cm}$$

As an example, let us take a point at a horizontal distance of 50 m from the point P. The depth $Y_{@50m}$ can be determined by Eq. (2.13).

$$\frac{Y_{@50m}}{0.245\ \text{m}} = \frac{50\ \text{m}}{90\ \text{m}}; \quad Y_{@50m} = 0.136\ \text{m} = 13.6\ \text{cm}$$

Similarly, other depths at required intervals may be determined. Check for maximum allowable longitudinal gradient limit. As the highway is in rolling terrain, the value of $G_{P\&R} \approx 0.67\ \%$. From Eqs. (2.14 a and b)

$$\left(\frac{100}{90\ \text{m}} \times 0.245\ \text{m}\right) \le G_{P\&R};\ 0.272\% \le G_{P\&R} \qquad \text{or}$$

$$\left(\frac{100}{25.7\ \text{m}} \times 0.07\ \text{m}\right) \le G_{P\&R}:\ 0.272\% \le G_{P\&R}$$

For the given conditions, the calculated longitudinal gradient value is less than the maximum allowable limit. Hence, the design is safe.

2.7 WIDENING OF CARRIAGEWAY ON CURVES

There are many reasons why carriageways are widened on curves:

- When a four-wheeled vehicle turns along a horizontal curve, the rear wheels do not pass exactly over the same path traced by the front wheels because the rigid-based axle assembly of the rear wheels tracks along the inner side of the curve. Therefore, the carriageway must have some additional width on curves to accommodate this deviated path of the rear wheels.
- As a safe margin on curves, drivers try to keep an offset distance on both sides of the vehicles. This extra space is maintained while travelling on curves – the extent of the extra space depends on the driver's psychological characteristics.

Let us consider the right-angled triangle ABC in Fig. 2.14. The interior angle at B is 90º.

Let

$d \approx$ BC = wheel base length (m);

R_{out} = AC = radius of path traced by the outer front wheel (m) R_{in} = AB = radius of the path traced by the outer rear wheel (m) $X_m = R_{out} - R_{in}$ = mechanical widening of carriageway (m) and R_c = radius of the curve with respect to the centre-line of the carriageway or mean curve radius (m).

From triangle ABC

$$AB^2 = AC^2 - BC^2$$

$$R_{in}^2 = R_{out}^2 - d^2$$

$$(R_{out} - X_m)^2 = R_{out}^2 - d^2$$

Figure 2.14 Widening of carriageway on curves.

If $R_{out} = R_{in} = R_c$, and we simplify the above equation,

$$X_m = \text{mechanical widening} = 0.5d^2/R_c \text{ (approximately).}$$

If the number of traffic lanes is n, the above equation can be rewritten as

$$X_m = 0.5nd^2/R_c \quad (2.16)$$

In addition to the mechanical widening of curves, it is practical to increase the value of widening to allow margin space during steering and overtaking operations. This extra widening of the carriageway on curves is termed **psychological widening** (X_p). Based on field studies and experience, the IRC has suggested the following formula for determining X_p:

$$X_p = 0.36\, v/\sqrt{R} \quad (2.17)$$

where v = design speed in m/s.

Total widening, X_e (in metres) = Mechanical widening (X_m) + Psychological widening (X_p)

$$X_e = 0.5nd^2/R_c + 0.36\, v/\sqrt{R_c} \quad (2.18)$$

Extra widening of curves is not required if the curve of the radius is longer than 60 m and 300 m for single- and double-lane carriageways, respectively (IRC:73). In the case of hill roads with sharp radii of curves, additional width may be provided along the inner side of the curve. The extra width should be equally provided on both sides of the carriageway. The full width of extra widening is attained gradually – starting from zero at the beginning of the curve and reaching its full value at the middle of the curve; from there it is gradually reduced to zero at the end of the curve.

2.8 DESIGN OF HORIZONTAL ALIGNMENT

Horizontal Curves: Horizontal curves are used to divert highway horizontal alignments from the straight route to any other path. The length and type of curve will be decided based on the location of the site and the horizontal alignment requirements. Horizontal curves are classified based on their geometry as follows:

- Simple circular curves
- Compound curves and multi-compound (three-centred) curves
- Reverse curves
- Transition curves
- Mixed curves

The first three types of curves originate from the fundamental geometric principles of circular arcs. The simple circular arc consists of an arc with a constant finite radius throughout its length. The typical geometry of a circular curve is shown in Fig. 2.15.

The well-known standard relationships for determining various geometric elements of circular curves are given below:

$$\text{Tangent length} = \text{AC} = \text{BC} = R_c \tan\left(\frac{\Delta}{2}\right) \tag{2.19}$$

$$\text{Length of long chord} = \text{ADB} = 2R_c \sin\left(\frac{\Delta}{2}\right) \tag{2.20}$$

$$\text{Mid-ordinate} = \text{DE} = R_c\left(1-\cos\left(\frac{\Delta}{2}\right)\right) \tag{2.21}$$

$$\text{Apex distance} = \text{CE} = R_c\left(\sec\left(\frac{\Delta}{2}\right)-1\right) \tag{2.22}$$

$$\text{Length of circular curve} = \text{AEB} = \frac{\pi R_c \Delta}{180^\circ} \tag{2.23}$$

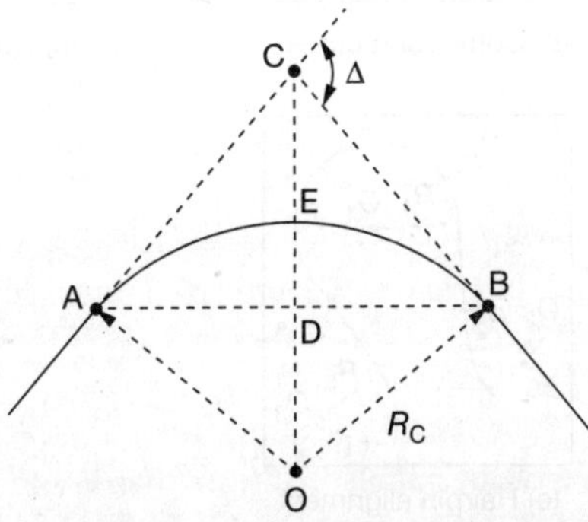

Legend
A: Point of commencement of curve
B: Point of tangency where circular path of curve ends
AEB: Simple circular curve.
AC = BC: Tangents or Straights.
OA = OB = R_C = Circular curve radius
Δ = Deflection angle
ADB: Long chord
DE: Mid-ordinate
EC: Apex distance
Angle AOB = Δ; Angle AOD = Δ/2

Figure 2.15 Elements of a simple circular curve.

Circular curves are generally adopted for slow-moving traffic whereas compound curves are used where deviation of alignment is inevitable. A compound curve consists of a combination of two or more circular arcs having different radii, joined together in the same direction. Their curve centres fall on same side of the curves.

In the case of reverse curves, the circular arcs of different (or same) radii are connected with their centres lying in opposite directions (Fig. 2.16 (d)).

The use of reverse curves in highway alignment is favourable for the alignment of mainline or freeway interchange ramps, where there is minimal need for acquiring land (Dabbour et al., 2004). In the case of sharp curves, a sudden change of direction results in arduous steering where the available sight distance is also insufficient. To satisfy the demand for high driving speeds and automatic drives, transition spiral curves are used in the alignment of high-speed highways such as national highways and expressways.

2.8.1 Sight Distance on Horizontal Curves

When a vehicle travels on a horizontal curve, the driver's line of vision is limited by obstructions such as buildings or any immovable structure, situated at the inside of the curve. Therefore, for horizontal curves, as a safe geometric design criterion, the sight distance provided along the curve should always be longer than the actual distance required for stopping a vehicle or for performing further action. The sight distance availability is governed by the following parameters (Easa, 1993):

- Geometric features of the horizontal curve, such as R_c, Δ, etc.
- Location of the obstacle causing vision obstruction
- Location of the driver with respect to the obstacle and an approaching object considered on the curve

The sight distance also depends on the design speed, characteristics of the vehicle, the driver and the complexity of the location (AASHTO).

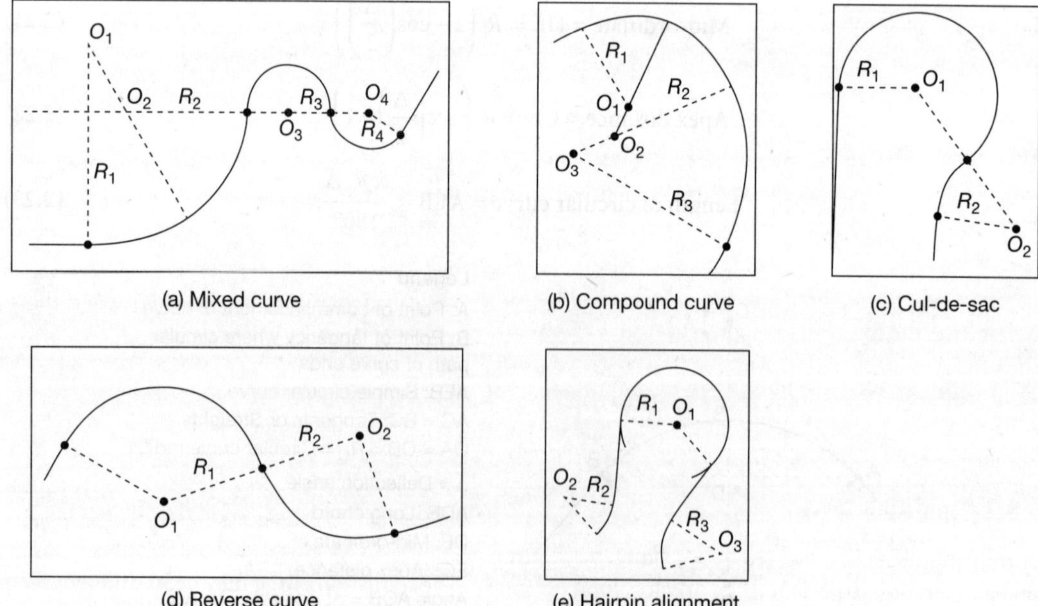

Figure 2.16 Typical horizontal curve alignments.

When an obstacle is located at the mid-point of the long chord of a circular curve on a single-lane carriageway, clear visibility in the lateral direction is obtained by providing a setback distance (DE) between the centre-line of the carriageway and the point of obstruction (Fig. 2.17). IRC: 52-2001 has recommended different setback distances for different values of sight distances (as per Eq. (2.1a)) at different parameters. For the simple case considered in Fig. 2.17, the required setback distance (DE) may be determined by Eq. (2.21).

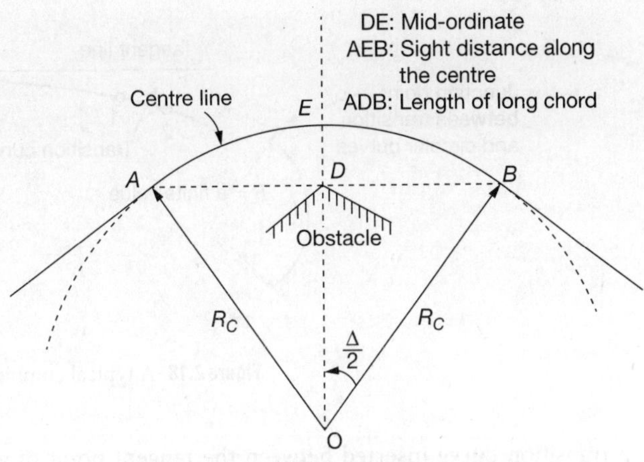

Figure 2.17 Sight distance along a circular curve.

The available sight distances along circular curves for different locations of obstacles may be evaluated according to the AASHTO geometric design guidelines. Available and minimum sight distances on circular curves may also be obtained by exact solutions (Easa, 1991). Sight distances at intersections which are controlled by a stop sign may be determined by using Design Aids – software which help to design and evaluate transportation systems (Easa et al., 2004).

Minimum sight distances on compound horizontal curves having different radii and for obstacles at different positions may be determined using relationships and design tables available in literature (Easa, 1993). In extreme cases, it is preferable to change the alignment, so as to avoid obstructions altogether.

EXAMPLE 2.10

Determine the required setback distance when a two-lane–wide carriageway road has a curve radius of 550 m along its centre-line and minimum stopping sight distance of 180 m.

Answer

With reference to Fig. 2.17, the given values are interpreted as follows

$R_c = 550$ m; sight distance, $S =$ Curve length (AEB) = 180 m; EB = $R_c \times \Delta/2$.

Therefore

$$\Delta/2 = \text{EB}/R_c = 90/550 = 0.1636 \text{ radians} \times (180^\circ/\pi) = 9.373^\circ$$

By Eq. (2.21), the required setback distance from the centre-line of the road

$$= DE = 550(1 - \cos(9.373)) = 7.34 \text{ m}$$

Otherwise, the following equation may be used to calculate the setback distance from the centre-line of the road.

$$DE = R_c\left(1 - \cos\left(\frac{28.647 \times S}{R_c}\right)\right) = 550\left(1 - \cos\left(\frac{28.647 \times 180}{550}\right)\right) = 7.34 \text{ m}$$

2.8.2 Transition Curve

A transition curve is inserted between diverse alignments of highways to smooth the change in centripetal acceleration experienced by drivers along the curve. The transition curve may be in different forms of a mathematical curve, namely, a clothoid, a cubic parabola, a spiral or a lemniscate. As an example,

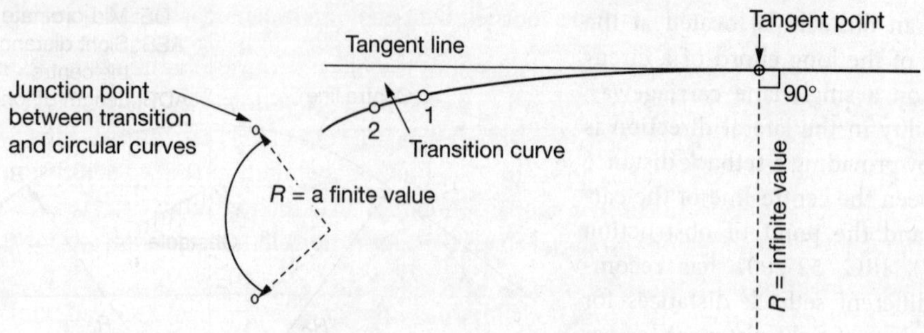

Figure 2.18 A typical combined curve.

a transition curve inserted between the tangent point of a straight alignment and a circular curve is shown in Fig. 2.18. The radius of the transition curve decreases at a uniform rate from an infinite value at the tangent point to a finite value at the junction between the transition and circular curves. At this junction, the radii of the transition and circular curves are equal. Transition curves are used to align the centre-line of highways by assuming the following conditions (Despotopoulos, 1947).

- The essential property of changing the curvature or radius of the transition curve at a uniform rate facilitates comfortable driving at a constant linear velocity (v) within the transition.
- While traversing on a transition curve, the steering wheels and, consequently, the front wheels of the vehicle turn at a constant angular velocity (ω).

By considering points 1 and 2 on the transition curve (Fig. 2.18), the following conditions hold good (Despotopoulos, 1947).

$$R_1 \times \omega \times t_1 = R_2 \times \omega \times t_2 = \text{Constant; and } R_1 \times \frac{L_1}{v} = R_2 \times \frac{L_2}{v} = \text{Constant} \qquad (2.24\text{ a,b})$$

where R_1 and R_2 are the radii of curvature at points 1 and 2, respectively. L_1 and L_2 are the lengths from the beginning of the transition curve corresponding to times t_1 and t_2. However, the above relationships do not satisfy the given condition at the beginning point (or tangent point) of the transition curve.

Transition curves with spirals are preferred in highway alignments for the following reasons (Craus and Polus, 1977; Lipičnik, 1998).

- Super-elevation on the curved path can be increased at a uniform rate, which is the same as the rate of decrease of radius of the transition curve.
- They help avoid the sudden impact of centrifugal forces by providing a smooth change consistent with the vehicle's path.
- The centrifugal force is the reciprocal to the change in alignment curvature. When the speed of vehicles is assumed as constant, non-uniform change in the magnitude of the side pressure takes place due to abrupt changes in curvature. The differential side pressure (or thrust) has a detrimental effect on the surface of the road as well as vehicle performance. In such alignment locations, transition curves should be inserted suitably.
- They facilitate gradual change in the turning angle of steering wheels.
- They provide safe and comfortable travel on curves, even at high speeds.

Following are the possible (or favourable) cases where transition curves can be inserted in the alignment of highways (Lipičnik, 1998):

- Between two straight parts of an alignment
- Between circular curves aligned in the same and opposite directions having the same or different radii
- Between the alignment of a straight part and a circular curve

2.8.3 Length of Transition Curve

Two considerations are important in calculating the required length of a transition curve: the rate of change of radial acceleration and the rate of change of super-elevation. Based on these, the length of the transition curve can be calculated using two methods. The longer length of transition curve is recommended.

By maximum allowable rate of change of radial acceleration (C): Assume that the design speed and rate of increase of centripetal acceleration are constant (Greenstein, 1976). Let the maximum radial acceleration at the end of the transition curve (i.e., at the beginning point of the circular curve) be equal to v^2/R_c, where v is the velocity of the vehicle in m/s. The radial acceleration will be zero at the tangent point when R takes an infinite value.

$$\text{Time taken to travel on a length of transition curve (s)} = t = Lt/v$$

Rate of change of radial acceleration (A_R) = Radial acceleration /Time taken to travel the distance

$$A_R = (v^2/R_c)/(Lt/v)$$

Therefore,

$$L_t \geq v^3/(R_c \times A_R) \tag{2.25}$$

For design purposes, the value of the maximum permissible rate of radial acceleration (A_R) that does not impart any kind of jerk or discomfort while travelling on the transition curve is selected. The value of A_R depends on the speed of the vehicle tracing the transition curve. It can be taken as a maximum of 0.8 m/s^3 and a minimum of 0.5 m/s^3 (IRC: 73). The permissible value of A_R may be calculated using the formula

$$A_R = 80/(75 + 3.6\ v) \tag{2.26}$$

By maximum allowable rate of change of super-elevation (C): To determine the minimum required length of the transition curve (L_t) according to the guidelines of IRC: 73 and as discussed in Section 2.6.6, the following relationships may be used

$$L_t \geq \left(\frac{100}{G_{P\&R}\ \%} \times Y_A \right) \tag{2.27}$$

$$L_t \geq \left(\frac{100}{G_{H\&M}\ \%} \times Y_A \right) \tag{2.28}$$

These notations carry the same meaning as explained before in Section 2.6.6. They hold good when the super-elevation is attained by rotating the surface of the pavement about its centre-line.

EXAMPLE 2.11

A horizontal curve of 270 m radius is to be designed for a speed of 65 kmph. The road is a two-lane road (7 m) and the super-elevation is provided by rotating the pavement surface about its crown. The rate of introduction of super-elevation is 1 in 150. If the maximum wheel base length is 6 m, calculate the required minimum length of the transition curve.

Answer

Design speed, $v = 65$ kmph $= 65 \times 0.278 = 18.07$ m/s

Criterion 1: By rate of change of acceleration (Eqs (2.25) and (2.26)),

$$L_t = \frac{v^3}{A_R \times R_c}$$

Where

$$A_R = \frac{80}{75+3.6v} = \frac{80}{(75+3.6\times 18.07)} = 0.571$$

Take $A_R = 0.571$ $\quad (\because A_{Rmin} = 0.5 \text{ and } A_{Rmax} = 0.8)$

$$L_t = \frac{18.07^3}{0.571\times 270} = 38.27 \text{ m.}$$

Therefore, take $L_t = 39$ m.

Criterion 2: Based on the rate of change of super elevation, by Eq. (2.5)

$$\text{Super elevation, } e = \frac{(0.75\,v)^2}{g\,R_C} = \frac{(0.75 \times 18.07)^2}{9.81\times 270} = 0.0693 \text{ m} < e_{max} = 0.07 \text{ m}$$

$\therefore$ Provide full $e = 6.93\%$

Widening of carriageway on curve: From Eq. (2.18)

$$X_e = \frac{0.5\times 2\times 6^2}{270} + \frac{0.36\times 18.05}{\sqrt{270}} = 0.133 + 0.395 = 0.528 \text{ m}$$

Now, the width of carriageway, $W = W + X_e = 7.0 + 0.528 = 7528$ m

From Eq. (2.12)

$$\text{Depth, } Y_A = \frac{W}{2}\times\frac{\text{full } e\ (\%)}{100} = \frac{7.528}{2}\times\frac{6.93\ \%}{100} = 0.260 \text{ m}$$

From Eq. (2.27), $L_t \geq \left(\frac{100}{G_{P\&R}\ \%}\times Y_A\right)$

$$G_{P\&R}\ \% = \frac{1}{150}\times 100 = 0.67\% \text{ (given)}$$

$$= \left(\frac{100}{0.67\ \%}\times 0.260\right) = 38.8 \text{ m;} \ \therefore L_t \geq 39 \text{ m}$$

Therefore, provide final length of transition curve (L_t) = 40 m.

Note: Always select the higher value from the above criteria.

2.8.4 Elements of the Transition Curve

Generally, a spiral curve is the ideal transition curve because its radius of curvature changes inversely as the curve length increases from the tangent point. Fig. 2.19 presents the basic elements of a combined curve, which has a circular curve adjoined by transition curves on its either end.

Let L_t = length of the transition curve; L_c = length of the circular curve; P and P'= tangent points of transition curves; C and C′ = junction points of transition and circular curves; R = radius of the circular curve and radius of the transition curve at points C and C′; X_Q and X_T = distance along the tangent from point P; Y_c = perpendicular ordinate from point T to point C; α = spiral angle with the line PI by a tangent drawn at C; θ = angle subtended by the circular arc with radius R_c; r = radius of the transition curve at a given point; and ℓ_t = any curved length of transition curve measured from origin P, having radius r.

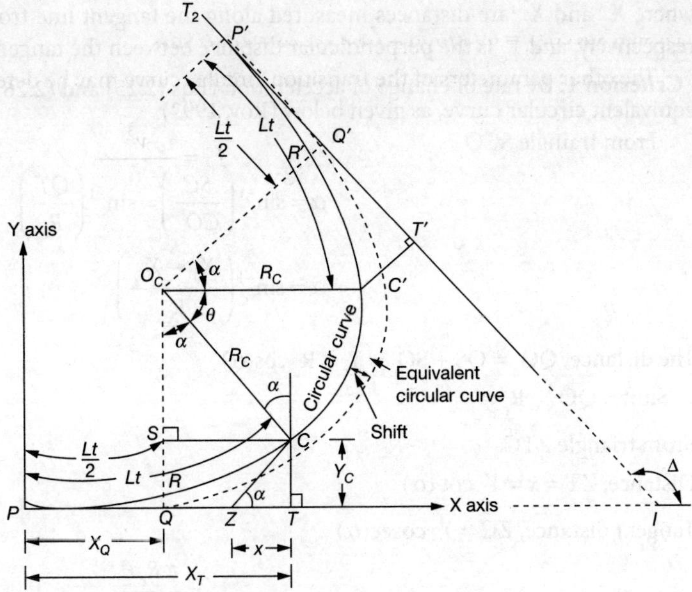

Figure 2.19 Elements of a horizontal combined curve having transition and circular curves (Elaborated; based on Roy, 1992).

From the fundamental principle of a clothoid spiral

$$r \propto \frac{1}{\ell_t};$$

It can also be written as

$$r \times \ell_t = R_c \times L_t = C\text{, where C is a constant} \tag{2.29}$$

The rectangular coordinates of the transition curve at ℓ_t measured from origin P may be written mathematically as (Fig. 2.20)

$$X = \ell_t - \frac{\ell_t^{\,5}}{40C^2} + \frac{\ell_t^{\,9}}{3456C^4} - \frac{\ell_t^{\,13}}{599040C^6} + \ldots \tag{2.30}$$

$$Y = \frac{\ell_t^{\,3}}{6C} - \frac{\ell_t^{\,7}}{336C^3} + \frac{\ell_t^{\,11}}{42240C^5} - \frac{\ell_t^{\,15}}{9676800C^7} + \ldots \tag{2.31}$$

Using the above coordinates, various elements of transition curves as given in Fig. 2.19 may be determined approximately from the following relationships.

$$X_T = L_t - \frac{L_t^{\,3}}{40R_c^{\,2}} \tag{2.32}$$

$$X_Q = \frac{L_t}{2} - \frac{L_t^{\,3}}{1280R_c^{\,2}} \tag{2.33}$$

$$Y_C = \frac{L_t^{\,2}}{6R_c} - \frac{L_t^{\,4}}{336R_c^{\,3}} \tag{2.34}$$

where X_T and X_Q are distances measured along the tangent line from the origin P to points T and Q, respectively, and Y_c is the perpendicular distance between the tangent lines from points T and point C.

The other parameters of the transition circular curve may be determined based upon the concept of equivalent circular curve, as given below (Roy, 1992).

From traingle SCO_c

$$\alpha = \sin^{-1}\left(\frac{SC}{CO_c}\right) = \sin^{-1}\left(\frac{QT}{R_c}\right)$$

$$\alpha = \sin^{-1}\left(\frac{X_T - X_Q}{R_c}\right) \quad (2.35)$$

The distance, $QO_c = QS + SO_c = Y_c + R_c \cos(\alpha)$ (2.36)

$\therefore$ Shift $= QO_c - R_c$ (2.37)

From triangle ZTC:

Distance, $ZT = x = Y_c \cot(\alpha)$ (2.38)

Tangent distance, $ZC = Y_c \operatorname{cosec}(\alpha)$ (2.39)

$$L_c = \frac{\pi R_c \theta}{180°} \quad (2.40)$$

Total length of combined curve $= 2 \times L_t \times L_c$ (2.41)

Total length of tangent, PI = PQ + QI

$$PI = X_Q + (QO_c) \times \tan\left(\frac{\Delta}{2}\right) \quad (2.42)$$

Where

$$\Delta = 2\alpha + \theta \quad (2.43)$$

The design of highway transition curves can also be carried out using methods based on the unit-curve concept (Rao, 1995) and by other methods (García and Conesa, 1997; Lipičnik, 1998).

The design of two- and three-centred compound curves, with and without transition spirals at the points of change of curvature, can be found in literature (AASHTO, 2001 and Roy, 1994).

A generalised solution to determine various elements of hairpin alignments having transition curves is also available (García and Conesa, 1992).

EXAMPLE 2.12

Determine the required length of the transition curve and calculate the salient elements of a combined curve consisting of a circular curve joined with a transition curve to its either end. The required curve design inputs are as follows.

Road class = National Highway
Design speed = 50 kmph
Terrain and climate = Snowbound mountainous terrain
Road type = Undivided two lanes, 7.0 m wide
Radius of the horizontal curve (R_c) = 160 m
Angle of deflection between two tangents of the proposed combined curve (Δ) = 64°
Chainage at the point of intersection (I) = 500.00 m
Permissible rate of change of super-elevation = 1 in 60 ($G_{H\&M} \approx 1.67\%$)
The super-elevation is attained by the rotation of the pavement surface about its centre-line.
Take the maximum permissible super-elevation as 7%.

Answer

This problem is solved in two parts. The first part consists of calculating the length of the transition curve using two criteria. In the second part, elements of the combined curve are determined. The notations used here are with reference to Fig. 2.19.

$$\text{Design speed} = 50 \text{ kmph} = 50 \times 0.278 = 13.9 \text{ m/s}$$

Part I

Criterion 1: By rate of change of radial acceleration and Eqs. (2.25) and (2.26)

Take $A_R = 0.639$, $\quad (\because A_{R\min} = 0.5 \text{ and } A_{R\max} = 0.8)$

$$L_t = \frac{13.9^3}{0.639 \times 160} = \frac{2685.619}{102.24} = 26.27 \text{ m.}$$

Therefore, take $L_t = 27$ m.

$$L_t = \frac{v^3}{A_R \times R_c}$$

Where

$$A_R = \frac{80}{75 + 3.6v} = \frac{80}{(75 + 3.6 \times 13.9)} = 0.639$$

Criterion 2: Based on the rate of change of super elevation by Eq. (2.5)

$$\text{Super-elevation, } e = \frac{(0.75\, v)^2}{g\, R_c} = \frac{(0.75 \times 13.9)^2}{9.81 \times 160} = 0.0692 \text{ m} < e_{max} = 0.07 \text{ m}$$

$\therefore$ Provide full $e = 6.92\%$

Widening of carriageway on curve: From Eq. (2.18)

$$X_T = \frac{0.5 \times 2 \times 6^2}{160} + \frac{0.36 \times 13.9}{\sqrt{160}} = 0.225 + 0.395 = 0.62 \text{ m}$$

Now, the width of the carriageway, $W = W + X_T = 7.0 + 0.62 = 7.62$ m

From Eq. (2.12)

$$\text{Depth, } Y_A = \frac{W}{2} \times \frac{\text{full } e\ (\%)}{100} = \frac{7.62}{2} \times \frac{6.92\%}{100} = 0.263 \text{ m}$$

From Eq. (2.23), $L_t \geq \left(\dfrac{100}{G_{H\&M}\ \%} \times Y_A \right)$

$$G_{P\&R}\ \% = \frac{1}{60} \times 100 = 1.67\% \quad \text{(given)}$$

$$L_t \geq \left(\frac{100}{1.67\ \%} \times 0.263 \right) = 15.7 \text{ m}; \ \therefore L_t \geq 15.7 \text{ m}$$

Note: Select the higher value from the above criteria.
Therefore, provide final length of transition curve (L_t) = 27 m.

Part II

Various geometric elements of the combined curve consisting of a circular curve joined with transition curves on either end are determined below. The notations used here are the same as those used in Fig. 2.19.

From Eq. (2.33),

$$X_Q = \frac{L_t}{2} - \frac{L_t^{\,3}}{1280 R_c^{\,2}} = X_Q = \frac{27}{2} - \frac{27^3}{1280 \times 27^2} = 13.4994 \text{ m}$$

From Eq. (2.32),

$$X_T = 27 - \frac{27^3}{40 \times 160^2} = 26.9808 \text{ m}$$

From Eq. (2.34),

$$Y_C = \frac{L_t^{\,2}}{6R_c} - \frac{L_t^{\,4}}{336 R_c^{\,3}} = \frac{27^2}{6 \times 160} - \frac{27^4}{336 \times 160^3} = 0.75899 \text{ m}$$

From Eq. (2.35),

$$\alpha = \sin^{-1}\left(\frac{X_T - X_Q}{R_c}\right) = \sin^{-1}\left(\frac{26.9808 - 13.4994}{160}\right) = 4.8334°$$

From Eq. (2.36),

$$\text{Distance, } QO_c = QS + SO_c = 0.75899 + 160 \cos(4.8334°) = 160.1900 \text{ m}$$

From Eq. (2.7),

$$\text{Shift} = QCc - Rc = 160.1900 - 160 = 0.1900 \text{ m}$$

From Eq. (2.42),

Tangent length of combined curve (PI)

$$= X_Q + (QO_c) \times \tan\left(\frac{\Delta}{2}\right) = 13.4994 + (160.1900) \tan\left(\frac{64°}{2}\right) = 113.5972 \text{ m}$$

From Eq. (2.43),

$$\theta = \Delta - 2\alpha = 64° - 2 \times 4.8334° = 54.3332°$$

From Eq. (2.40),

$$\text{Length of circular curve } = L_c = \frac{\pi R_c \theta}{180°} = \frac{\pi \times 160 \times 54.3332°}{180°} = 151.7269 \text{ m}$$

From Eq. (2.41),

$$\text{Length of combined curve} = 2 \times L_t + L_c = 2 \times 27 + 151.7269 = 205.7269 \text{ m}$$

Chainage at salient points on the combined curve

Chainage at point I = 500.00 m (given)

Chainage at point P = Chainage at I – PI = 500.00 – 113.5972 = 386.4028 m

Chainage at junction point (C) at the end of the transition curve and the beginning of the circular curve

= Chainage at $P + L_t$

= 386.4028 + 27 = 413.4028 m

Chainage at junction point (C′), at the end point of the circular curve and the beginning of the second transition curve

= Chainage at C′ + L_c

= 413.4028 + 151.7269 = 565.1297 m

Chainage at point, P' = Chainage at C′ + L_t = 565.1297 + 27 = 592.1297 m

2.9 DESIGN OF VERTICAL ALIGNMENT

Longitudinal Gradient

Generally, the natural ground consists of undulations or uneven surfaces. To make a leveled surface, it becomes necessary to fill up or cut the uneven ground surface according to the existing ground profile. Depending on the type of terrain, the magnitude of ground rise or fall is observed. This rate of rise or fall of ground along its length is termed **longitudinal slope** or gradient or grade. Profiles of natural ground and desired pavement surfaces are shown in Fig. 2.20.

The gradient can be expressed in the following forms

1. In percentage: $\text{Grade (\%)} = \dfrac{\text{Level difference}}{\text{Horizontal distance}} \times 100$
2. 1 in N: 1 vertical in N horizontal distance
3. Vertical : Horizontal

Factors affecting gradient

- Existing ground profile and surrounding floor levels of developed areas, especially with reference to surface and sub-surface drainage
- Levels of control, mandatory and obligatory points along the desired alignment
- Type and importance of the highway
- Design speed and type of vehicles anticipated, i.e., bullock cart or motor vehicles
- Operating cost of vehicles and energy saving
- Cost of construction and practical problems with the surrounding locality
- Excessive excavation and high embankments
- Natural sub-soil, water table and artesian conditions
- Earthquake conditions

Types of gradients

- *Ruling gradient*: It is the maximum allowable grade or longitudinal slope considered by a designer during alignment of a highway in the vertical plane. This gradient is also termed the design gradient. The magnitude of the ruling gradient depends on the type of terrain, design speed, curves and altitude.
- *Limiting gradient*: This gradient is slightly steeper than the ruling gradient. Limiting gradient is provided when the natural ground profile is slightly steeper than the ruling gradient. It is necessary to adopt such a steeper gradient to keep the earthwork operation minimum. The limiting gradient is generally adopted to connect plain and rolling terrains or to reach hilly areas.

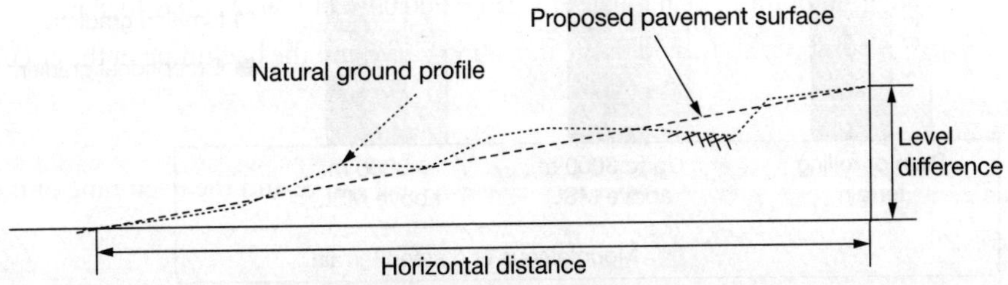

Figure 2.20 Longitudinal gradient.

- *Exception gradient*: This gradient is adopted exceptionally where the ground profile frequently changes its slope and needs an abrupt or a sudden rise or fall in vertical alignment. Exceptional gradient is still steeper in slope than the limiting gradient. Providing this gradient for long distance is inconvenient while ascending as the vehicle would need more tractive effort to overcome the frictional and gravitational forces. Hence, exceptional gradient is provided for a short distance not exceeding 100 m in length at a stretch. Successive sections should be separated by flatter gradients for a distance of at least 100 m (IRC: 52-2001). These gradients are used in steep hilly areas.
- *Minimum gradient*: To drain off surface or rain water in side drains, it is necessary to provide a certain amount of longitudinal slope. A minimum amount of gradient is provided based on the surface characteristics of drains, intensity of rainfall, soil type, type of terrain and existing average floor level of the developed area. A minimum gradient of 0.5% and 1.0% is provided for lined and unlined side drains, respectively (IRC: 73-1980).

The values of the various gradients based on the terrain type are presented in Fig. 2.21 (according to the recommendation by IRC).

2.10 COMPENSATION OF GRADE ON HORIZONTAL CURVES

To facilitate easy driving on horizontal curves in hilly areas, a certain amount of grade is deducted from the design gradient value. Gradients steeper than 4% should be eased. The grade compensation value is determined using the following formula (IRC: 73-1980).

$$\text{Grade compensation} = (30 + R_c)/R_c \tag{2.44}$$

$$\text{The maximum allowable value of grade compensation} = 75/R_c \tag{2.45}$$

The method of determination of grade compensation is explained below.

EXAMPLE 2.13

A ruling gradient of 1 in 18 is to be provided for a hill road having a horizontal curve radius of 70 m. What is the value of grade compensation?

Answer:

Design gradient = (1/18) ×100 = 5.55 % (given)

Radius of horizontal curve, R_c = 70 m (given)

By Eq. (2.44)

$$\text{Grade compensation} = (30 + R_c)/R_c = (30 + 70)/70 = 1.428\%$$

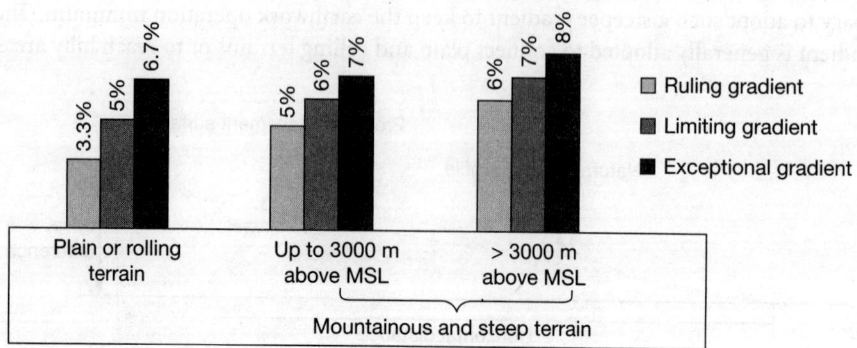

Figure 2.21 Road gradients.

By Eq. 2.45, maximum allowable grade compensation = $75/R_c = 75/70 = 1.071\%$

Now, select the least value from the above.

Grade compensation value = 1.07%.

Therefore, net grade = 5.55 – 1.071 = 4.479%

2.11 VERTICAL CURVES

Proper design of vertical alignment of highways is important to ensure adequate sight distance and to provide uniform change in grade. This will facilitate a safe, smooth and comfortable ride over the road. Circular and parabolic curves are used in the vertical alignment of highways. Parabolic curves are preferred as the rate of change of slope along its length is constant and the sight distance available is more than that available for a circular curve having equal length (Fig. 2.22 (a)). Calculation of reduced levels along the parabolic curve is simple and easy to set out. Therefore, vertical axis parabolas are generally used for the design of vertical alignment of highways.

2.11.1 Types of Vertical Curves

Based on the nature and intensity of steepness of adjacent grade lines, two types of vertical curves are formed (Fig. 2.22 (b)): summit (or crest) curves and sag (or valley) curves. An intersection point of adjacent grade lines is termed the **vertical point of intersection** (I). When the elevation (or reduced

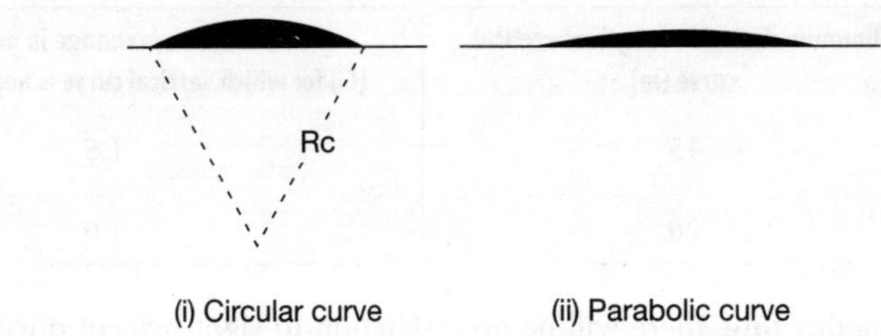

Figure 2.22 (a) Availability of sight distance on a circular and on a parabolic vertical curve.

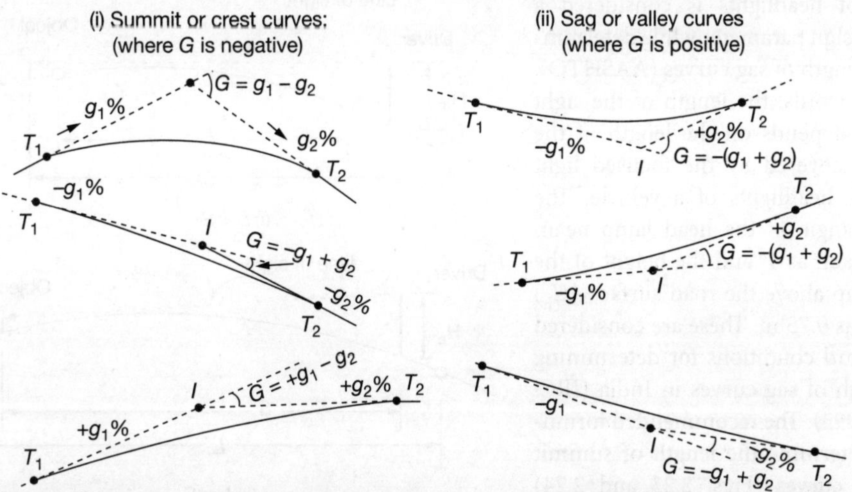

Figure 2.22 (b) Possible cases of forming vertical road curves based on adjacent grade lines.

level, RL) of the point I is higher than or equal to any point on the vertical curve, such curves are termed **summit curves**. When the elevation of the point I is lower than or equal to any point on the vertical curve, such curves are termed **sag curves.**

The point of curvature (which is at the curve beginning) and the point of tangency (which is at the curve end point) are termed T_1 and T_2, respectively. The nature of the gradient (g %) in terms of positive and negative signs indicates the upward and downward slopes, respectively, of the ground. G is the angle of deviation of adjacent grade lines, which is equal to the algebraic difference of the tangent grades $|g_2 - g_1|$ and its value is expressed in decimal form. The above notations are used in the following paragraphs.

2.11.2 Length of Vertical Curves

It is the horizontal distance (L) measured between points T_1 and T_2 with reference to the provision of minimum sight distance (S) along the centre-line of a highway to facilitate safe and efficient operations of vehicles at minimum cost of construction. Hence, determination of the length of the vertical curve is significant while designing the alignment of highways. In the case of lower grade change, vertical curves with minimum required length should be provided for aesthetic reasons (Table 2.10).

Sight distance (S) is a major design parameter used for determining the length of summit (or crest) curves. The main reason is that the driver's visibility (or sight distance) is restricted by the high point of the curve.

Table 2.10 Minimum required length of vertical curves.

Design speed (m/sec)	Minimum desirable length of vertical curve (m)	Maximum change in grade (G) for which vertical curve is not required (%)
≤9.72	15	1.5
13.88	20	1.0

For sag curves, during day time there will be no restriction to sight (except during snow fall) but at night, the sight distance is limited and is equal to the length of the spread of light of the vehicle headlights. Therefore, the length of the spread of headlights is considered a major design parameter while determining the length of sag curves (AASHTO). In other words, the length of the sight distance depends on the length of the roadway covered by the focused light from the headlights of a vehicle. The upward angle of the head lamp beam (Ø) is taken as 1° and the height of the head lamp above the road surface (H_h) is taken as 0.75 m. These are considered as standard conditions for determining the length of sag curves in India (IRC: SP: 23-1993). The recommended formulae to determine the length of summit and sag curves (Figs. 2.23 and 2.24) and the details are given below (IRC: SP: 23-1993).

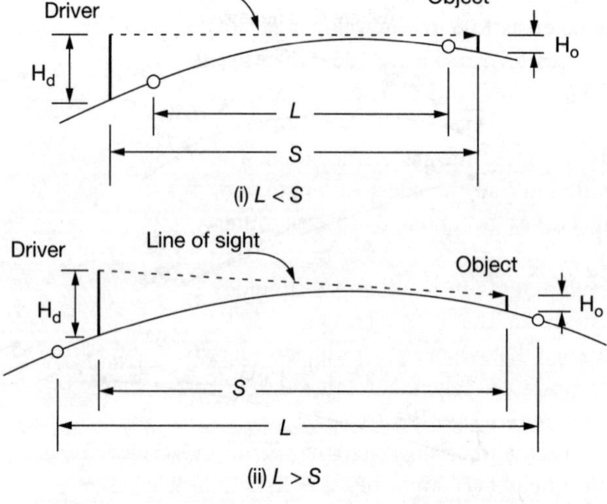

Figure 2.23 Design criteria for summit (or crest) curves.

S. No.	Equation for minimum curve length	For Indian conditions
	For summit curves (Eqs. (2.46) and (2.47))	
1	If $L>S$; $L=\dfrac{\lvert G\rvert S^2}{2\left(\sqrt{H_d}+\sqrt{H_o}\right)^2}$	H_d = Height of the driver's eye level above the road surface H_o = Height of the obstacle above the road surface If S = ISD = OSD, $H_d = H_o = 1.2$ m; If S = SSD, H_d =1.2 m, H_o = 0.15 m; $G = \lvert g_2 - g_1\rvert$, angle of deviation in decimal value. H_h = 0.75 m, height of headlight above the road surface $Ø = 1°$ = Angle of spread of headlight or light cone – focused angle. The values of S may be obtained from design speed relationships or taken according to IRC guidelines.
2	If $L < S$; $L = 2S - \dfrac{2\left(\sqrt{H_d}+\sqrt{H_o}\right)^2}{\lvert G\rvert}$	
	For sag curves (Eqs. (2.48) and (2.49))	
3	If L > S; $L=\dfrac{\lvert G\rvert S^2}{2\ (H_h + S\ \tan Ø)}$	
4	If L < S; $L = 2S - \dfrac{2\ (H_h + S\ \tan Ø)}{\lvert G\rvert}$	

AASHTO also recommends the above equations for determining the length of vertical curves, but with different values. The use of these equations for determining the length of parabolic vertical curves is illustrated with examples.

In the USA, the values of sealed beam headlamp (H_h) and Ø are 0.6 m and 1°, respectively (AASHTO, 2004). The values of the driver's eye height and the height of the approaching vehicle are taken as 1.08 m. The value of H_h was taken as 0.76 m in the 1954 AASHTO design policy, and reduced to 0.6 m in the 1965 AASHTO design policy; it has remained the same since then. This reduction had the effect of lengthening sag curves by about 3–10% at different design speeds. In view of the latest developments in headlamp technology, a study in the USA (Hawkins, 2007) revealed that when H_h = 0.6 m, it is more appropriate to take the Ø value as 0.75° rather than 1°. The effect would lengthen sag curves by 15–30% at different design speeds.

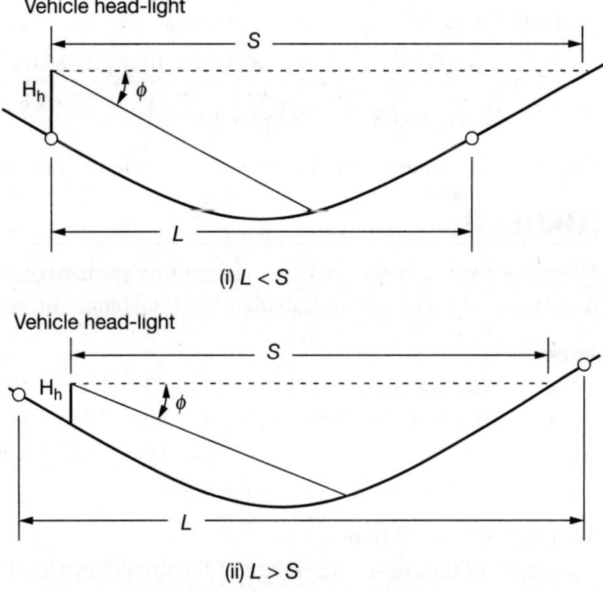

Figure 2.24 Design criteria for sag curves.

Truck and cab-over-engine tractor driver's eye height (H_d) may be taken as up to 2.75 m and the object height (H_o) may be taken as 0.15 or 0.46 m in North America (Easa, 1992).

EXAMPLE 2.14

A vertical summit curve is formed when an ascending gradient of 1 in 30 meets a descending gradient of 1 in 40. The curve is to be designed to provide OSD for a design speed of 80 km/h. Calculate the suitable length of the summit curve. Assume suitable data. **(JNTU, APR/MAY 2005)**

Answer

Ascending gradient $(+g1) = 1$ in $30 = (1/30) \times 100 = 3.33\%$

Descending gradient $(-g2) = 1$ in $40 = (1/40) \times 100 = 2.50\%$

Angle of deviation of adjacent grades

$$G = (-g2) - (g1) = -2.5 - 3.33 = -5.833\% = -0.05833; \quad |G| = 0.05833$$

Assume Indian standard conditions. For the design speed of 80 km/h, assume minimum OSD value as $S = 470$ m.

Assume $L < S$

If $S = \text{OSD}$, $H_d = H_o = 1.2$ m

From Eq. (2.47)

$$L = 2S - \frac{2\left(\sqrt{H_d} + \sqrt{H_o}\right)^2}{|G|} = 2(470) - \frac{2\left(\sqrt{1.2} + \sqrt{1.2}\right)^2}{0.05833} = 940 - \frac{2(4.8)}{0.05833} = 775.42 \text{ m}$$

Therefore, our assumption of $L < S$ is not correct, since $S = 470$ m.

Now, reassume $L > S$ (or $S < L$)

From Eq. (2.46)

$$L = \frac{|G|S^2}{2\left(\sqrt{H_d} + \sqrt{H_o}\right)^2} = \frac{0.05833 \times (470)^2}{2\left(\sqrt{1.2} + \sqrt{1.2}\right)^2} = \frac{12885.097}{2 \times 4.8} = 1342.19 \text{ m} < 470 \text{ m (assumed value)}$$

Therefore, the assumption of $L > S$ is correct. Hence, provide length of the curve = 1343 m.

EXAMPLE 2.15

A summit curve is formed when an ascending gradient of 1 in 20 meets an ascending gradient of 1 in 80. If the value of SSD = 110 m, calculate the total length of summit curve.

Answer

Ascending gradient $(+g1) = 1$ in $20 = (1/20) \times 100 = 5.00\%$ (given)

Ascending gradient $(+g2) = 1$ in $80 = (1/80) \times 100 = 1.25\%$ (given)

$$G = g2 - (g1) = 1.25 - 5.00 = -3.75\% = -0.0375$$

$$|G| = 0.0375$$

Assume, SSD = $S = 110$ m (given)

Length of the curve = L; Height of the driver's eye level above the road surface, $H_d = 1.2$ m;

Height of an object above road surface, $H_o = 0.15$ m

Assume $L < S$

From Eq. (2.47),

$$L = 2S - \frac{2\left(\sqrt{H_d} + \sqrt{H_o}\right)^2}{|G|} = 2 \times 110 - \frac{2\left(\sqrt{1.2} + \sqrt{0.15}\right)^2}{0.0375} = 220 - \frac{2(2.1985)}{0.0375} = 102.74 \text{ m}$$

Our assumption of $L < S$ is correct, since $S = 110$ m. Hence, provide $L = 103$ m.

EXAMPLE 2.16

A valley curve is formed by a descending gradient of 1 in 30 meeting an ascending gradient of 1 in 50. If the design speed is 80 kmph, calculate the total length of the valley curve.

Answer

Descending gradient (–g1) = –1 in 30 = – (1/30) × 100 = –3.33%
Ascending gradient (+g2) = 1 in 50 = (1/50) × 100 = 2.00%
$G = (g2) - (-g1) = 2.00 + 3.33 = 5.833\% = 0.05833; |G| = 0.05833$

For a design speed of 80 km/h, stopping sight distance (SSD) may be assumed as $S = 120$ m (or may be determined by Eq. (2.1)).

Assume $L > S$

From Eq. (2.48),

$$L = \frac{|G|S^2}{2\,(H_h + S\,\tan\text{Ø})} = \frac{0.05833 \times 120^2}{2(0.75 + 120 \times \tan(1°))} = \frac{839.952}{2(0.75 + 120 \times 0.017455)} = 147.64 \text{ m}$$

Therefore, our assumption of $L > S$ is correct, since $S = 120$ m. Hence, provide $L = 148$ m.

Review Questions

1. State the vehicle parameters and dimensions useful for the geometric design of highways, with the relevant design element.
2. State the factors to be considered during the geometric design of highways.
3. Define terrain. What are the different terrains according to IRC: 73?
4. What is the significance of design speed in the geometric design of highways?
5. What is the significance of operating speed in the geometric design of highways?
6. Distinguish between design speed and operating speed.
7. Draw a neat sketch showing the salient cross-sectional details of a highway.
8. Define: (i) control line; (ii) building line; (iii) road land boundary line; (iv) setback distance (SBD); (v) right-of-way and roadway.
9. State the standard width of carriageways based on the number of lanes.
10. Define camber. Why is camber provided to road surfaces?
11. What are the different types of cambers based on the surface profile?
12. State the standard specifications of camber used for the construction of different types of roads based on the materials used.
13. Define and state the different types of sight distances provided in the geometric design of highways with reference to the height of the driver's eye level and an object on the road.
14. Define PIEV theory and SSD with details.
15. State the factors affecting SSD.
16. Derive the formulae for determining the minimum sight distance in the following cases: (i) on a level road surface; (ii) on a sloped road surface.
17. Define overtaking sight distance (OSD or PSD) using a neat sketch with reference to Indian practices.
18. Define OSD. Explain the standard procedure of overtaking according to AASHTO using a neat sketch.
19. State the conditions assumed for determining OSD according to IRC: 66.
20. State the conditions assumed for determining OSD according to AASHTO.
21. State the factors affecting OSD and their influence on minimum OSD.
22. Define intermediate sight distance. How is it determined?
23. Define intersection sight distance using a neat sketch and state the design criterion according to IRC: 66.
24. State the assumption made by AASHTO for determining intersection sight distance.
25. Distinguish between intermediate sight distance and intersection sight distance using diagrams

26. How is the sight distance determined in medians?
27. Define super-elevation and derive an expression based on radius of a curve, design speed, side friction and acceleration due to gravity.
28. What are the factors that affect super-elevation?
29. State the assumptions and formula recommended by IRC: 73 for the determination of the rate of super-elevation.
30. What are the standard values of super-elevation adopted in different countries for different conditions?
31. Define the significance of side friction factor for various design parameters related to horizontal curves.
32. What are the factors affecting side friction factor?
33. Determine the minimum radius of a horizontal curve and restriction of speed recommended by IRC: 73.
34. Define equilibrium cant using a neat sketch.
35. State and explain the ʹmethod of attaining super-elevation by rotating the roadway about the centre-line with a neat sketch.
36. Formulate the relationships used for the design of super-elevation by revolving the roadway about its centre-line.
37. State the limiting values of the rate of change of super-elevation for (i) plain and rolling terrain; (ii) hilly and mountainous terrain.
38. State the reasons for extra widening of the carriageway on horizontal curves.
39. Derive an expression for determining the extra width of carriageway on curves. Also write the formula used for determining the total widening of curves.
40. Draw typical forms of different horizontal curves used for the centre-line alignment of highways.
41. State the assumptions made for the design of transition curves for the geometric design of highways.
42. State the possible cases of insertion of transition curves in the centre-line alignment of highways.
43. Why are transition curves with spirals preferred in highway alignment?
44. How is the length of a transition curve determined?
45. Derive the formulae for determination of the salient elements of a typical combined curve which consists of a circular arc connected with a transition curve on either side. Draw a neat sketch showing the salient features.
46. Derive expressions with neat sketches for the determination of salient elements of (i) twin-centred compound circular curves and (iii) three-centred compound circular curves.
47. Define and describe the different types of longitudinal gradients according to IRC: SP23.
48. What are the factors affecting the selection of appropriate longitudinal gradients in highways?
49. Why is grade compensation necessary for the design of highway curves in hilly areas? What is the formula recommended by IRC: 73?
50. Define summit (or crest) and sag curves with reference to the angle of deviation of adjacent grade lines. Draw figures wherever applicable.
51. State when the design of vertical curves is not required with reference to the maximum change in grade and design speed.
52. State the design criteria according to (i) IRC: SP-23 and (ii) AASHTO standards, used for determining the minimum length of vertical curves with reference to the height of the driver's eye level, the minimum depth of the object and the vehicle's headlight. Draw figures showing the salient details.
53. Write the formulae used to determine the minimum sight distance on a sag curve passing under an overhead structure. Draw figures showing salient details.

Multiple Choice Questions

1. If the design speed is 100 km/hr, perception time of driver is 2.5 sec and coefficient of longitudinal friction is 0.35, the safe stopping sight distance will be
   (a) 200.245 m (b) 182.044 m
   (c) 69.500 m (d) 192.0244 m
2. On highway stretches where the required overtaking sight distance (OSD) cannot be provided,

it is necessary to provide at least the length as equal to ______.
(a) three times the safe stopping sight distance
(b) the intermediate sight distance
(c) safe stopping sight distance
(d) five times the stopping sight distance

3. The value of camber recommended for earthen road surfaces in areas of heavy rainfall is
(a) 1 in 50 (b) 1 in 40
(c) 1 in 33 (d) 1 in 25

4. Width of carriageway for a double-lane divided road is recommended to be
(a) 5.5 m (b) 7.5 m
(c) 7.0 m (d) 3.75 m

5. A parabolic is used to connect a 2.8% upgrade with 1.8% downgrade. What is the length of vertical curve if the rate of change of grade is 0.2% per 20 m chain?
(a) 230 m (b) 360 m
(c) 18.4 m (d) 460 m

6. Coefficient of friction will be greater if the pavement surface is
(a) rough and dry (b) rough and wet
(c) smooth and dry (d) smooth and wet

7. As per IRC, the following equation is used to calculate equilibrium super-elevation
(a) $\frac{v^2}{2gR}$ (b) $\frac{0.75(v)^2}{2gR}$
(c) $\frac{v^2}{gR}$ (d) $\frac{(0.75v)^2}{gR}$

8. The mechanical extra widening required for 7.5m divided carriageway on a horizontal curve radius R is given by ______, where l = length of wheel base of vehicle.
(a) $\frac{l^2}{2R}$ (b) $\frac{l^2}{R}$
(c) $\frac{l^2}{3R}$ (d) $\frac{4l^2}{2R}$

9. The road camber given in the following figure

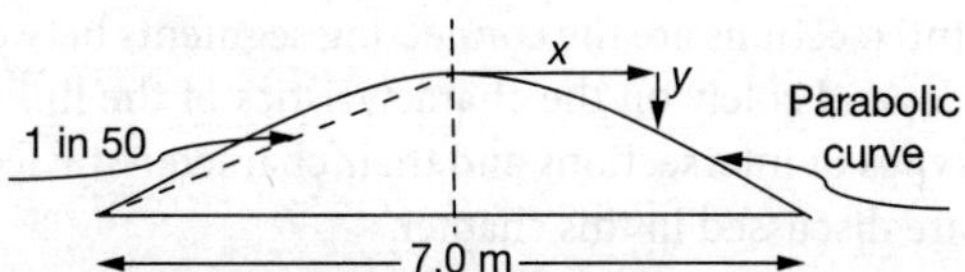

For designing this camber, the equation to be used
(a) $y=\frac{x^2}{175}$ (b) $y=2x^2$
(c) $y=\frac{x^2}{350}$ (d) $y=\frac{x^2}{225}$

10. A traffic stream is moving at an average speed of 100 km/hr with average space headway of 27.8 m. The time headway between the consecutive vehicles is
(a) 0.1 sec (b) 0.2 sec
(c) 1.0 sec (d) None of these

Answers to Multiple Choice Questions

1. (b)
2. (b) On level surface: $\text{SSD} = 0.278\times100\times2.5+\frac{(0.278\times100)^2}{2\times9.81\times0.35} = 182.044 \text{ m}$
3. (d)
4. (b)
5. (d) $G = g_2 - (g_1) = -1.8 - (2.8) = -4.6\ \%$

$$\text{Length of curve} = \frac{|G|}{\text{Permissible rate of change of grade}} = 4.6\% / (0.2\% \text{ per } 20 \text{ m}) = 460 \text{ m}$$

6. (a)
7. (c)
8. (b)
9. (a) Formula used, $y=\frac{2x^2}{nW}=\frac{2x^2}{50\times7}=\frac{2x^2}{350}=\frac{2x^2}{175}$
10. (c) Speed = 0.278 × 100 = 27.8 m/sec time-gap = 27.8 m/27.8 m/sec = 1 sec

CHAPTER 3

Design Guidelines of Intersections

3.0 INTRODUCTION

Intersections are the connecting segments between road links. The quality of any road network does not depend solely on the characteristics of the links, but also on the way the links are interconnected. The types of intersections and their characteristic features of applicability for safe and smooth flow of traffic are discussed in this chapter.

3.1 CLASSIFICATION OF INTERSECTIONS

Based on the grade of intersecting surface

Road intersections can be broadly categorised into two types: 1. At-grade intersections and 2. Grade-separated intersections. The details of each category are listed in Fig. 3.1.

1. **At-grade intersections**: These intersections are formed when the intersecting roads meet at the same level (i.e., on the same plane). Traffic flow manoeuvres such as merging, diverging, crossing and weaving will occur at these intersections. Some of the factors to be considered in the design of at-grade intersections are given below:
   - The area of conflicts should be as small as possible
   - The relative approaching speeds should be small
   - The angle of approach of vehicles entering the intersection area should be small
   - Adequate visibility (with reference to the sight triangle) should be available to the vehicles approaching the intersection
   - Sudden change in lane/path should be avoided by providing adequate flared legs

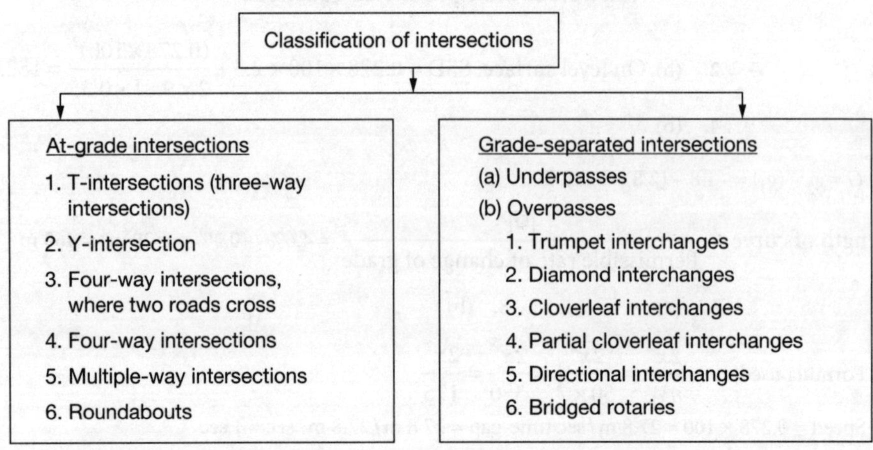

Figure 3.1 Classification of intersections based on the intersecting surface level.

2. **Grade-separated intersections or interchanges**: These intersections are formed when intersecting roads meet at different levels. Traffic flow manoeuvres such as merging, diverging, crossing and weaving are eliminated at these intersections.

Based on the control methods used

At-grade Intersections can also be classified based the control methods used (Fig. 3.2). Details of these are presented in subsequent chapters.

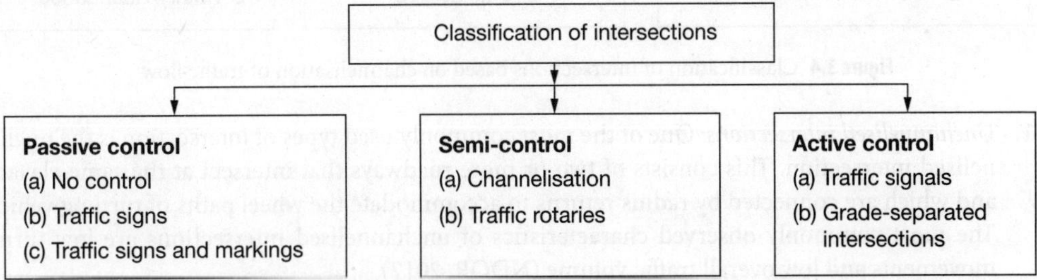

Figure 3.2 Operational classification of intersections (based on control type).

3.2 CLASSIFICATION OF AT-GRADE INTERSECTIONS

At-grade intersections can be classified based on their geometric form and functionality as detailed in the discussion below:

Classification based on geometric configuration of intersection road links

The basic configurations of roads intersecting at-grade with different geometry are shown in Fig. 3.3. The typical range of the angle between the intersecting legs (the angle between the centre lines of the intersection roads) of different intersections may be as follows:

- T-intersection: 75–120°
- Y-intersection or three-legged intersection: 60–75°
- Right-angled intersection: 75–120°
- Oblique four-legged intersection: 75–120°

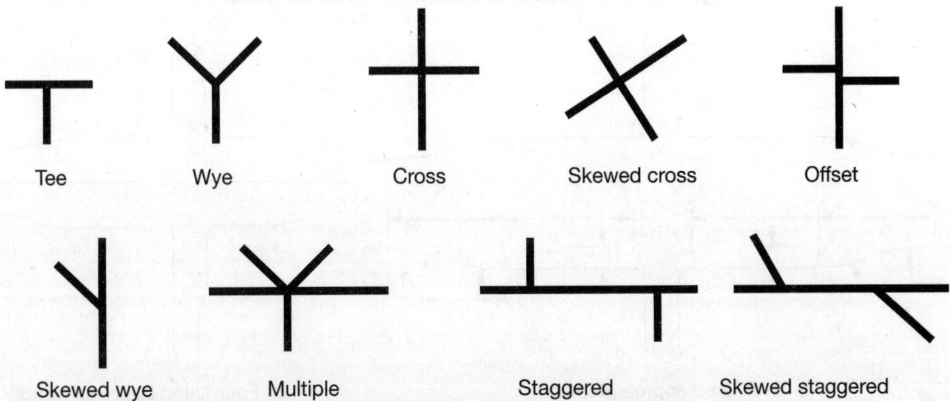

Figure 3.3 Basic configurations of intersections.

Categorisation based on channelisation of traffic flows

At-grade intersections can be categorised based on channelisation of traffic flows to different directions at an intersection area, as shown in Fig. 3.4.

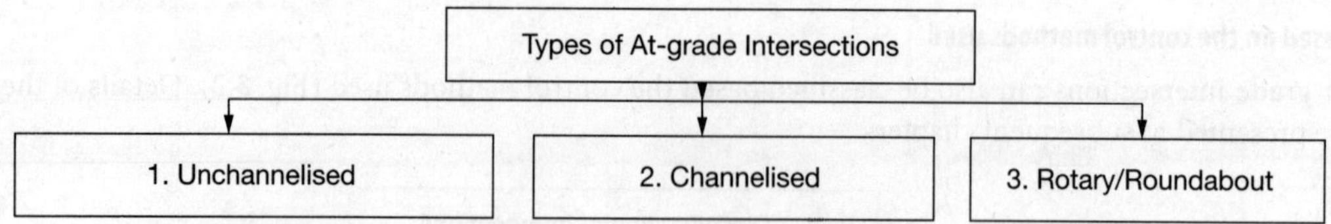

Figure 3.4 Classification of intersections based on channelisation of traffic flow.

1. *Unchannelised intersections*: One of the most commonly used types of intersection is the unchannelised intersection. This consists of two or more roadways that intersect at the same elevation and which are connected by radius returns to accommodate the wheel paths of turning vehicles. The most commonly observed characteristics of unchannelised intersections are low turning movements and low overall traffic volume (NDOR, 2012).

   The intersection area is paved and the paths of the vehicles are not restricted to traversing along their intended direction of travel. When no additional pavement width is provided to facilitate turning movements, such intersections are termed *plain intersections* (Fig. 3.5). When the pavement width is widened at the intersection area by a lane width or more, such intersections are termed *flared intersections* (Fig. 3.6). As the conflict area increases in flared intersections, vehicular flow paths are not restricted. However, while crossing, the vehicles will have to stop while those in the other direction proceed. Flared approaches to intersections can be provided with channelised and rotary intersections as well.

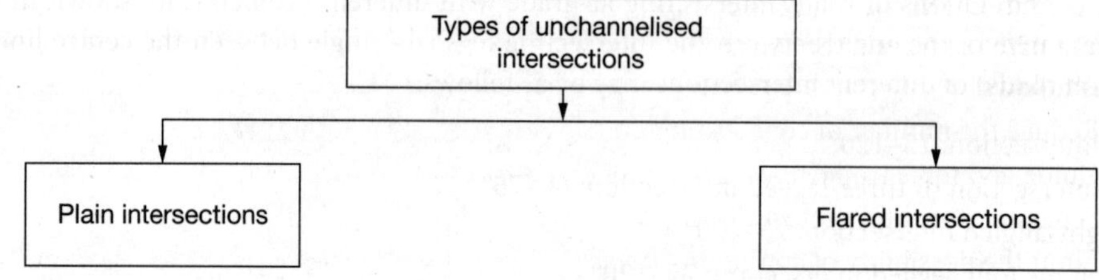

Figure 3.5 Classification of unchannelised intersections.

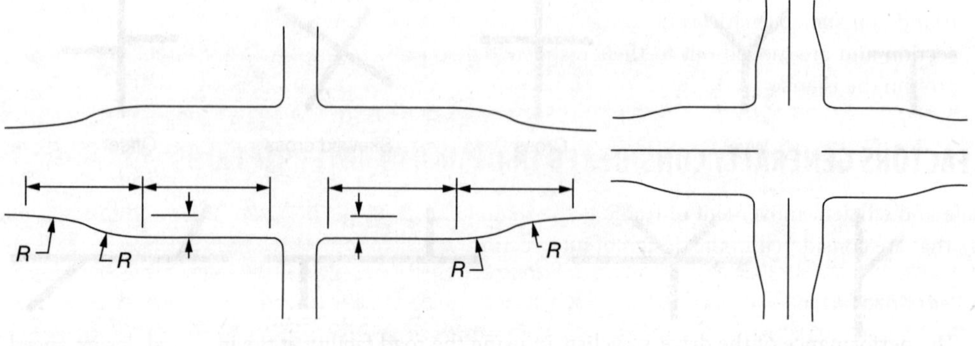

Figure 3.6 Flared intersections.

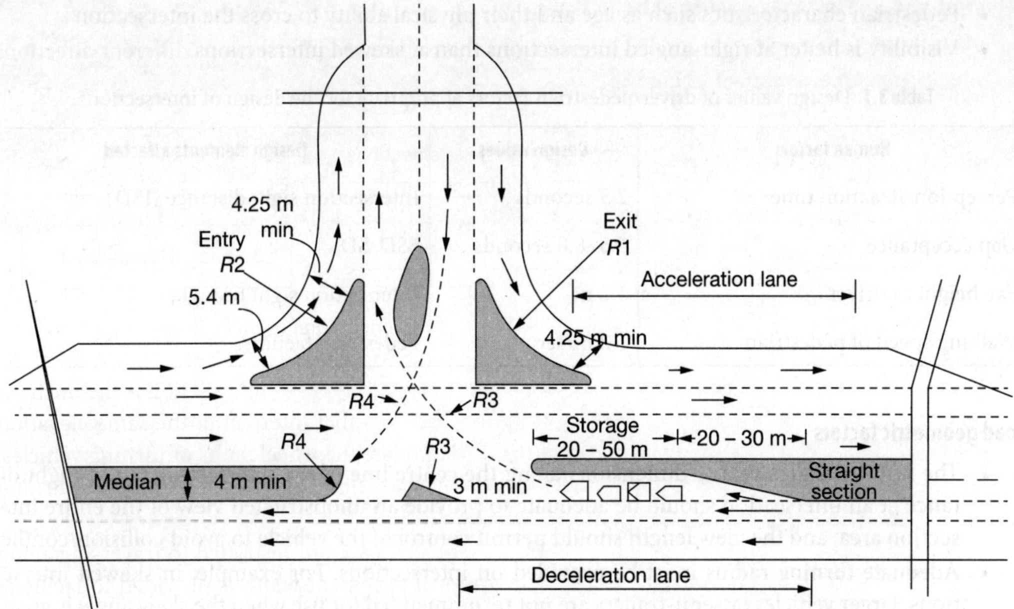

Figure 3.7 Typical layout of a three-legged intersection with channelisation design elements.

2. *Channelised intersections*: The fundamental concept of channelisation is to divide and direct traffic movement. A variety of components and arrangements such as traffic lanes, raised (kerbed) islands, median dividers, signs, signals and markings are used/installed in the intersection area (Fig. 3.7) to channelise the traffic flow paths by separating conflicting traffic movements into definite paths of travel. The functional objectives of the design of channelised road intersections are as follows:
   - Reduce the number of conflicting points in the intersection
   - Guide and force traffic flow to the intended direction and path
   - Separate, store and protect turning and crossing vehicles
   - Limit the possibility of conflicting frequency and severity
   - Provide space in median and islands to install different types of traffic control devices
3. *Rotary intersections*: This is an enlarged intersection area in which a large, centralised and raised island is installed. Vehicles from the various approach roads converge at the entry to the intersection and are moved out to their respective directions in the clockwise direction by weaving around the island.

3.3 FACTORS GENERALLY CONSIDERED IN DESIGN OF INTERSECTIONS

The safe and efficient movement of traffic is the main concern in the design of intersections. The main factors that are considered in the design of intersections are as follows (IRC: SP: 41-1994):

Driver/Pedestrian factors

- The performance of the drivers/cyclists in using the road facility at the intended design speed in terms of skill, perception time and reaction time in understanding the traversing of the intersection (Table 3.1)

- Pedestrian characteristics such as age and their physical ability to cross the intersection
- Visibility is better at right-angled intersections than at skewed intersections.

Table 3.1 Design values of driver/pedestrian factors appropriate for the design of intersections.

Human factors	Design values	Design elements affected
Perception/Reaction time	2.5 seconds	Intersection sight distance (ISD)
Gap acceptance	5.5–8.0 seconds	SSD, SD
Eye height of driver	1.2 m	Intersection sight triangle
Walking speed of pedestrian	1.2 m/sec	Pedestrian facilities

Road geometric factors

- The sight triangle and the dimensions along the centre line of the intersecting roads. Sight distance at an intersection should be adequate to provide an unobstructed view of the entire intersection area, and the view length should permit control of the vehicle to avoid collision/conflict.
- Adequate turning radius must be provided on intersections. For example, in skewed intersections, larger vehicles or semi-trailers are not recommended for use when the skew angle is greater than 200° (Son et al., 2002).

Traffic factors

- Design speed, physical and operating characteristics of vehicles
- Composition of traffic, with the proportion of heavy vehicles
- Capacity of vehicles to stop, accelerate and decelerate
- Height of the driver's eye above the carriageway.

Safety issues

- Long pedestrian crossing distance should be resolved by constructing raised intermediate medians.
- As the intersection area increases with increased value of skew angle, a large conflicting area will form. As a result, the vehicles will be exposed to conflicting traffic for a longer period (Walker, 1981).
- Proximity to on-street parking spaces/transit stops.

Environmental factors

- Drainage facilities such as inclination of the camber, run-off and the location of side drains, proximity to farmland/agricultural areas and mist that affects SSD.

Economic factors

- Land acquisition issues and cost, and the cost of constructing/reconstructing all the required installations.

3.4 DESIGN PRINCIPLES OF AT-GRADE INTERSECTIONS

It is a well-known fact that at-grade intersections provide a lower level of safety than other types of intersections, such as roundabout and interchanges. Due to this reason, at-grade intersections should be designed for maximum safety. To achieve this objective, the following parameters relating to the design of intersection components are considered important.

3.4.1 Design Vehicle

The most commonly used vehicle (85th or 15th percentile use in any particular vehicle class) which is intended to be operated on the proposed roadway intersection is designated as the design vehicle. The wheelbase (WB) and width of the design vehicle are the most important dimensions which determine the minimum turning radius, kerb radii and channelisation design. In India, the most commonly used vehicles (along with their WB) are: passenger car (3.34 m), single unit-truck (6 m), semi-trailer truck (12.2 m and 15.2 m) and truck trailer (18.2 m). The maximum width of vehicles permitted in India is 2.5 m (IRC: 3-1983). Adequate widening of the intersection area at turnings is provided to accommodate the wider turning paths traced between the outer front tyre and the inner rear tyre of the design vehicle. AASHTO (2004)-recommended turning templates may be adopted for the geometry of study intersection area and for the selected design vehicle WB. The kerb radii should be appropriately adjusted to this requirement. A few sample turning templates, as recommended by IRC: 41-1994, are presented in Fig. 3.8.

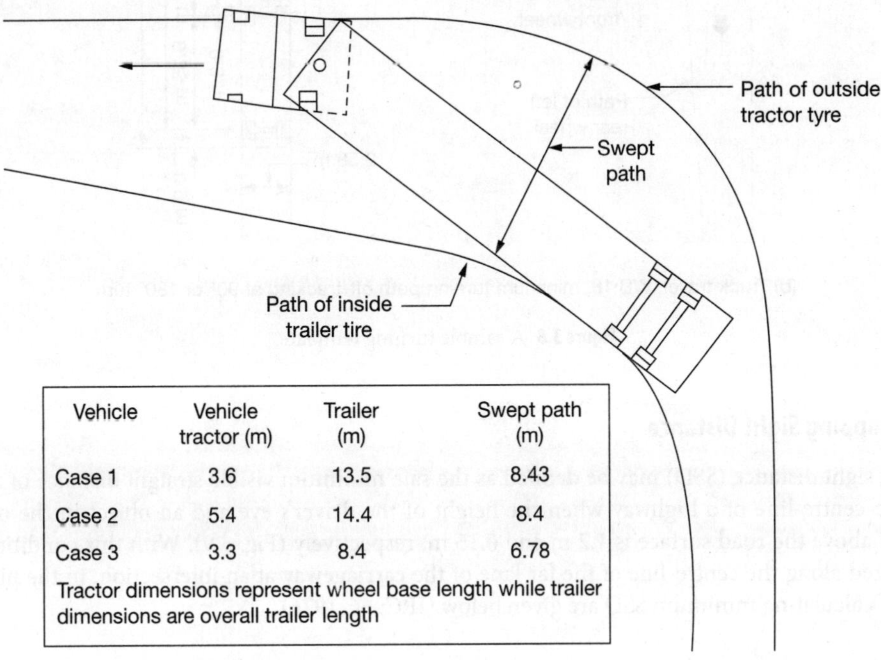

Vehicle	Vehicle tractor (m)	Trailer (m)	Swept path (m)
Case 1	3.6	13.5	8.43
Case 2	5.4	14.4	8.4
Case 3	3.3	8.4	6.78

Tractor dimensions represent wheel base length while trailer dimensions are overall trailer length

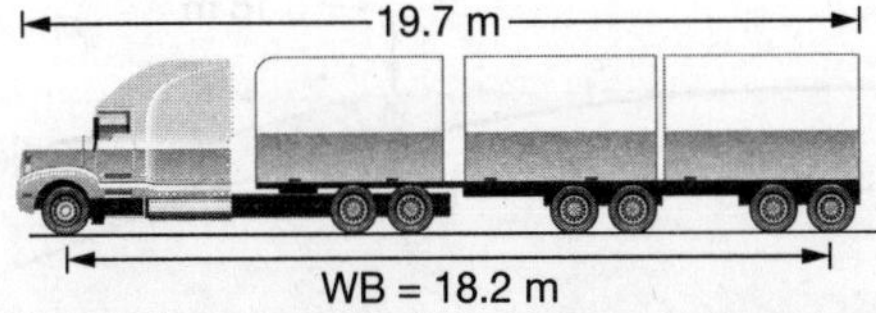

(a) Swept-path width of design vehicle at low-speed off-tracking at 90° turn

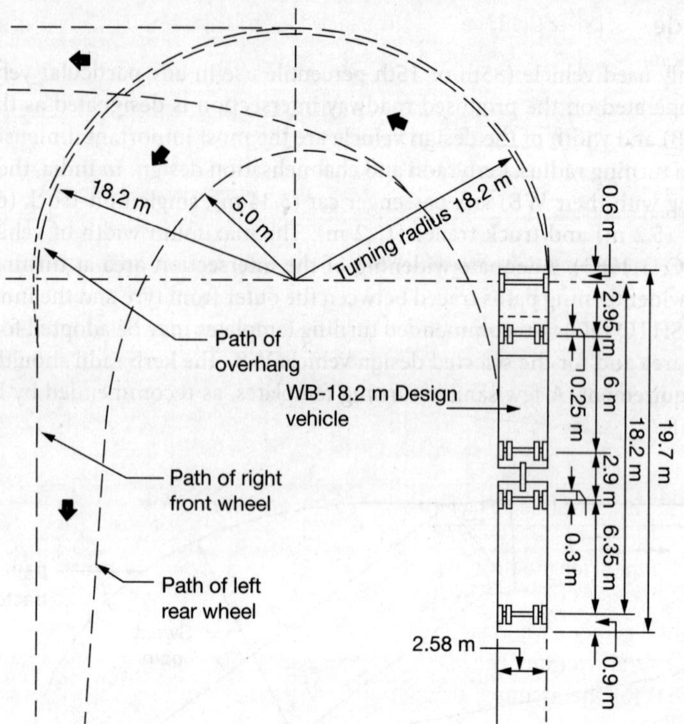

(b) Truck trailer, WB 18, minimum turning path off-tracking at 90° or 180° turn

Figure 3.8 A sample turning template.

3.4.2 Stopping Sight Distance

Stopping sight distance (SSD) may be defined as the safe minimum visible straight distance of a driver along the centre-line of a highway when the height of the driver's eye and an object in the opposite direction above the road surface is 1.2 m and 0.15 m, respectively (Fig. 3.9). With this condition, SSD is measured along the centre-line of the far lane of the carriageway at an intersection, in the plan. The details of calculating minimum SSD are given below (IRC: 66-1976).

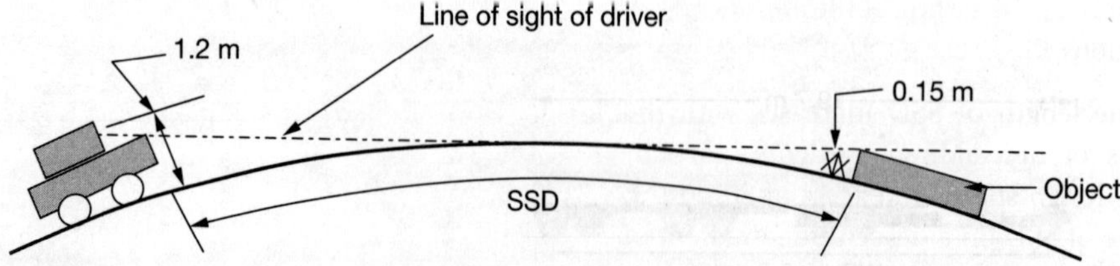

Figure 3.9 (a) Safe stopping sight distance (SSD) on a summit curve across an intersection.

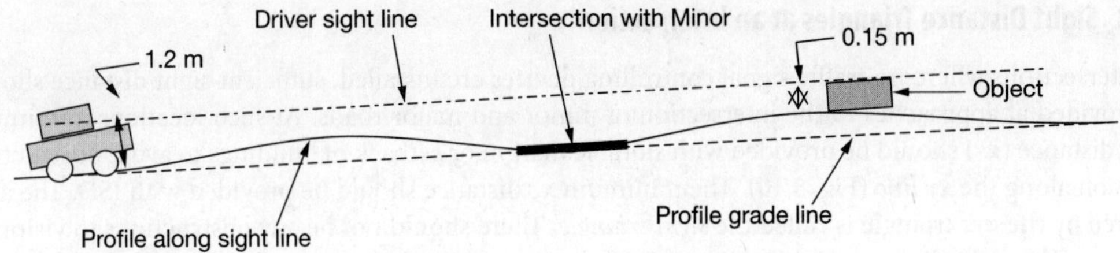

Figure 3.9 (b) Typical longitudinal section and SSD along the summit curve across an intersection.

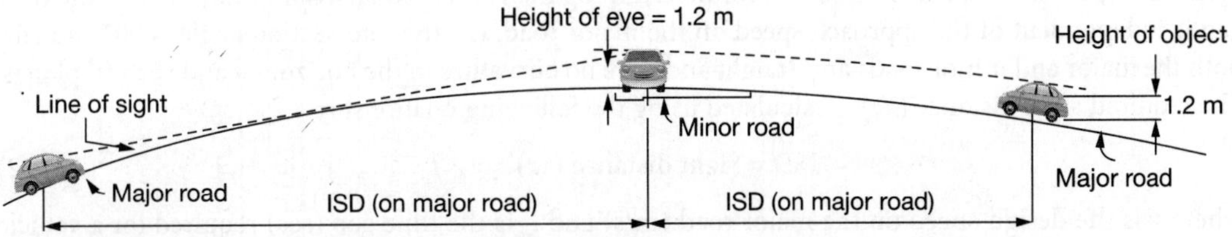

Figure 3.9 (c) Typical longitudinal section and ISD along the summit curve across an intersection.

Minimum sight distance should be provided for the given conditions such as design speed, gradient and coefficient of lateral friction. The sum of lag distance and braking distance is termed the stopping sight distance (SSD) for the assumed driver perception time of 2.5 seconds. The following formula is used to calculate SSD.

Stopping sight distance (SSD) = Lag distance (LD) + Braking distance (BD)

$$\text{SSD} = vt + \frac{v^2}{2g(f_s \pm 0.01\ n)} \tag{3.1}$$

where t = mean total reaction time of the driver (seconds); v = velocity of vehicle (m); g = acceleration due to gravity (m/sec^2); n = ascending or descending gradient (%); and f_s = coefficient of longitudinal surface friction (0.35–0.4 for a speed range of 100–20 km/hr).

The SSD should be provided continuously on all roadways including at the approaches to intersections.

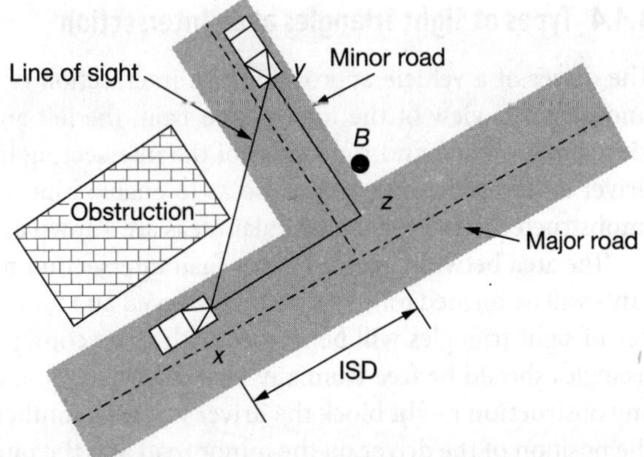

Figure 3.10 Details of intermediate sight distance (ISD)

Note: The length of SSD increases with the steepness of descending gradients, particularly for design speeds in excess of 80 km/hr. The value of SSD on a descending gradient of 35 will be 6% longer than that on a level grade. If this descending gradient doubles, the SSD value will be about 16% longer. In this context, it must be noted that the effect of the descending gradient on the SSD on the minor road is more critical than that on the major road because all the vehicles on the minor road have to stop or yield (NASRAL).

3.4.3 Sight Distance Triangles at an Intersection

At intersections where no traffic signal controlling devices are installed, sufficient sight distance should be provided at approaches to the intersection of minor and major roads. At such locations, minimum sight distance (*xz*) should be provided with suitable margin or setback of buildings to avoid obstruction of vision along the *xy line* (Fig. 3.10). The minimum *xz* distance should be provided with ISD. The area covered by the *xyz* triangle is called the *sight triangle*. There should not be any obstructions to vision in this area. The sight distance (*xz*) is a function of the time gap (t_g) necessary for a stopped vehicle on a minor road to turn or pass through the major road (Harwood et al., 2000). At point B, the total STOP board is provided for vehicles approaching the junction on a minor road. The time gap acceptance method is specified by AASHTO (2001) for determining the ISD based on assumptions that (i) the time gap is independent of the approach speed on the major road, (ii) the intersection angle is 90° and (iii) both the major and minor roads are straight and have no curvature in the horizontal and vertical planes. The required sight distance (*xz*) is calculated using the following relationship.

$$\text{ISD} = \text{Sight distance } (xz) = v \times t_g \tag{3.2}$$

where v is the design speed on the major road (m/s) and t_g is the time gap (sec) required for a vehicle on a minor road to proceed. For different types of vehicles, the value of t_g may be assumed as 7.5 sec. If the approach gradient to the minor road is on an upgrade greater than 3%, increase the ISD value by 10%. More such details may be obtained from *A policy on geometric design of highways and streets*, AASHTO, 2004.

Adequate sight distance should be provided at intersections to view vehicles that are approaching or to stop or reduce speed or cross the main road smoothly without collisions (in the case of uncontrolled intersections). As per Indian practice (IRC: 66-1976), the minimum visibility distance (*yz*) along the minor road is taken to be 15 m and the average time gap to travel the distance (*xz*) along the major road is assumed to be 8 seconds. For example, for a vehicle standing on a minor road 15 m before approaching the major road, the minimum visibility distance required along the major road is $v \times 8$, where v is the design speed in m/s on the major road.

3.4.4 Types of Sight Triangles at an Intersection

The driver of a vehicle approaching an intersection or departing from an intersection should have an unobstructed view of the intersection from the left and right approaches. A sufficient length of clear view along the left and right sides of the intersecting legs of an intersection should be provided to the driver entering the intersection so as to enable him to anticipate and avoid potential collisions. Such unobstructed views form triangular areas are known as sight triangles.

The area between each adjacent road intersection may be termed a quadrant. Three and four quadrants will be formed for a typical three-legged and four-legged intersection, respectively. The same number of sight triangles will be formed with these configuration intersections. The area enclosed in these triangles should be free from any type of obstruction, including traffic controlling installations. If not, any obstruction might block the driver's view of conflicting vehicles, pedestrians and animals. Based on the position of the driver on the minor road and the intended turning movement, the sight triangles are categorised into two types (AASHTO, 2011):

- **Approach sight triangles at uncontrolled or yield-controlled intersections**: These clear triangular spaces are provided for the driver of a vehicle approaching an intersection from the left (Fig. 3.11 (a)) and right (Fig. 3.11 (b)) sides. These triangular spaces should be large enough that vehicle drivers on minor road can see the approaching vehicles/pedestrians on the main road in sufficient time to stop or avoid collision.

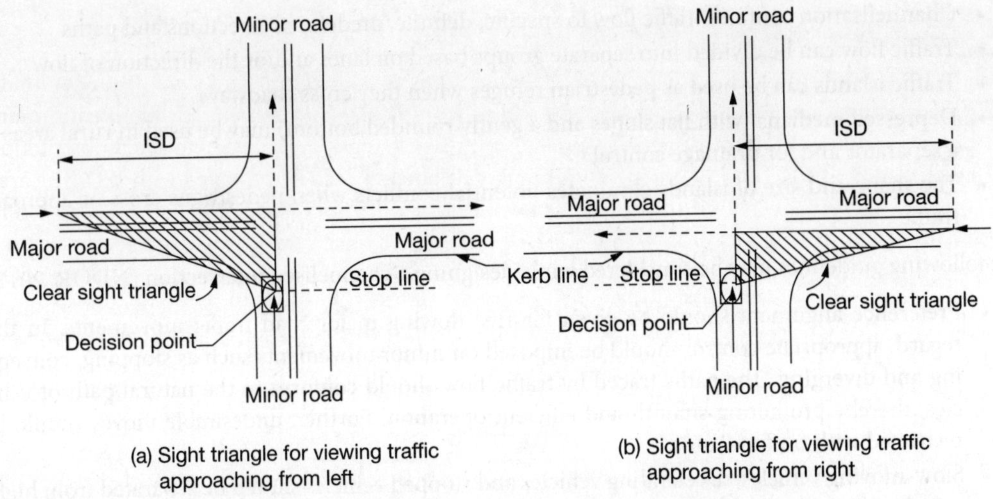

Figure 3.11 Approach clear sight triangles at uncontrolled or yield-controlled intersections on a minor road (Redrawn based on AASHTO, 2011).

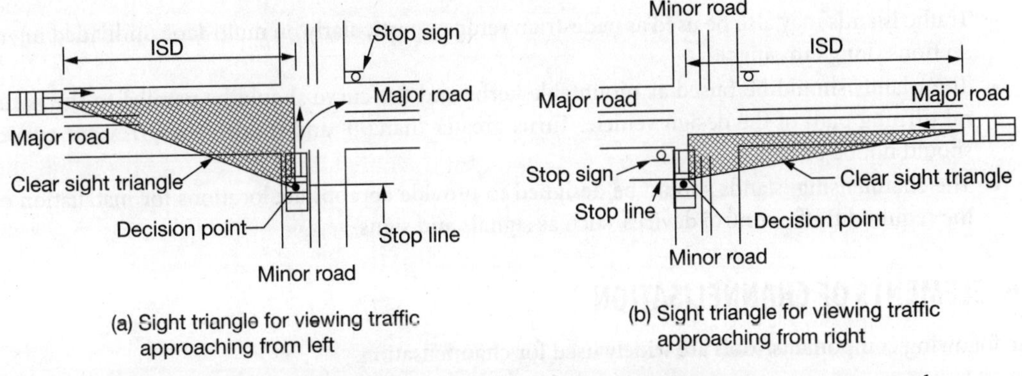

Figure 3.12 Departure clear sight triangles at stop-controlled intersections on a minor road (Redrawn based on AASHTO, 2011).

- **Departure sight triangles at stop-controlled intersections**: These triangular spaces are provided at intersections for viewing traffic on the main road approaching the minor road from the left (Fig. 3.12 (a)) and right (Fig. 3.12 (b)) sides, when the driver of the vehicle stops at the stop line on the minor road. These triangular spaces are smaller than approach triangles because the driver on the minor road stops at the stop line and waits for a favourable chance to cross or enter the major roadway.

3.5 PRINCIPLES OF CHANNELISATION

The purpose of channelisation is to *reduce* areas of conflict by *separating* and *regulating* traffic movement into predefined paths of travel using raised islands and pavement markings. Traffic islands may be composed of paint, kerbs, pavement edges, raised pavement markers, splitter islands or other devices. These islands can be installed at signalised and un-signalised intersection areas, particularly in urban areas, where the traffic speed is not high. Traffic islands serve the following functions:

- Channelisation confines traffic flow to specific, definite, predefined directions and paths
- Traffic flow can be divided into separate groups based on lanes and/or the direction of flow
- Traffic islands can be used as pedestrian refuges when they cross roadways
- Depressed medians (with flat slopes and a gently rounded bottom) may be used in rural areas as a separator and for drainage control
- The shape and size of islands eliminates potential conflicts when pedestrians cross or approach traffic

The following guidelines may be considered while designing a channelised intersection (NDOR, 2012):

- Preference alignment should be given for free flowing major road traffic movements. In this regard, appropriate control should be imposed on minor movements such as stopping, converging and diverging. The paths traced by traffic flow should conform to the natural path of vehicles, thereby promoting smooth and efficient operation. Further, undesirable moves should be restricted with traffic islands.
- Slow-moving vehicles, decelerating vehicles and stopped vehicles should be separated from high-speed vehicles using lanes.
- Points of conflict should be separated and reduced by installing islands/markings of suitable size and shape.
- Traffic islands may also be used as pedestrian refuges, particularly on multi-lane undivided intersections (long crossings).
- The islands should be raised as mountable kerbs and the curve should be parallel to the natural turning path of the design vehicle. Turns greater than 90° and sudden sharp/reverse curves should not be provided.
- The channelising islands should be designed to provide appropriate locations for installation of the required traffic control devices, such as signals and signs.

3.6 ELEMENTS OF CHANNELISATION

The following components/tools are widely used for channelisation.

3.6.1 Alignment of Lanes

Traffic lanes should be properly defined and arranged, with the appropriate width, direction, intersection, weaving length, sight distance and location.

3.6.2 Islands

An island is a defined area between traffic lanes in an intersection used for separating vehicular movements and for pedestrian refuge. Islands may be designed with different sizes and shapes; and they may be flush with the kerb or raised. The shape of the islands varies to facilitate the predefined driving path. In general, their shape may be triangular and elongated. Elongated divisional islands are normally provided with a width of at least 1.2 m and a length of 6 m. In rural areas with high-speed traffic, the minimum and desirable area of triangular islands is 7 m^2 and 9.3 m^2, respectively. Smaller islands (< 9.3 m^2) are acceptable in urban areas with speeds of less than 36 km/hr. However, islands should be least 18.6 m^2 when they contain traffic control devices/luminaires or are used for pedestrian refuge (WSDOT, 2014).

Raised islands provide effective delineation and can also control/restrict undesirable turning movements, as compared to pavement markings. Channelisation with raised islands is preferred in urban areas, particularly where higher pedestrian volume is observed. Island height is usually delineated by pavement markings. The allowable height of the kerb or a raised island is a function of the design speed and may be categorised with reference to the classification of roads in rural/urban areas. The following kerb heights may be adopted (NDOR, 2012):

- On high-speed facilities (≥ 80 km/hr): 7.5–10.0 cm height of concrete slope kerb
- On low-speed facilities (≤ 72 km/hr): 15 cm height of integral concrete kerb

3.6.3 Kerbs

Kerbs are raised (or flat sloping face) elements located adjacent to the extreme sides of the travel way. Kerbs commonly adjoin walkways along roadways, including at intersections. Kerbs are also used as semi-barriers with storm-water drainage channels (or gutters). Kerbing is a widely used practice of channelisation in urban areas, particularly where vehicle speeds are relatively low.

3.6.4 Directional Islands

Islands of different shapes are used to divert traffic to an intended direction through the intersection. In general, triangular islands with an area of not less than 5 m^2 are used with rounded corners (Fig. 3.13). This minimum area criterion is adopted to ensure easy visibility to approaching drivers. The corner radius at the approach ($R = 0.6$ m) should be greater than the exit corner radius ($R = 0.3$ m). The offset between the island and the edge of the travelled way should be 0.6–1.0 m to allow lateral placement of kerbing vehicles. The dimensions and shape of triangular islands dictate the following factors (NASRAL):

- The corner radii of the directional islands and the associated tapers
- Alignment of the islands with reference to the intersection skew angle
- Dimensions of islands appropriate for the turning path of the design vehicle and the alignment of adjacent lanes

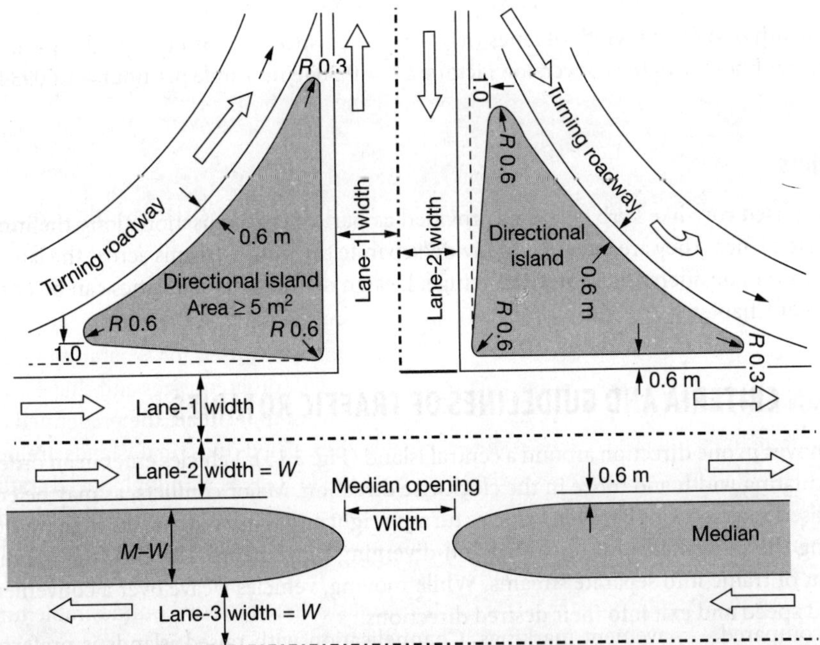

Figure 3.13 Typical layout of a three-legged multi-lane intersection with directional islands and a median opening.

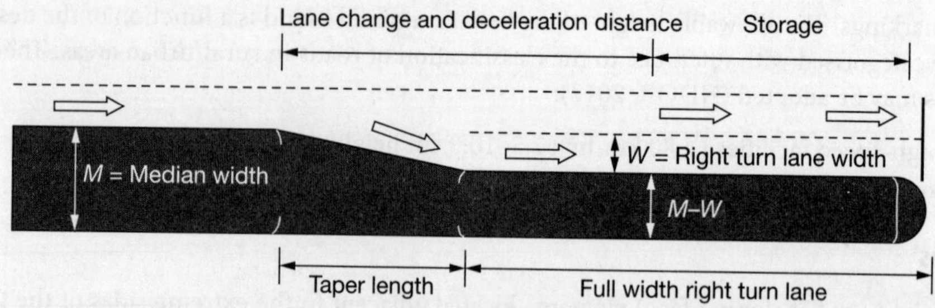

Figure 3.14 Components of a median with a right-turn auxiliary lane (Redrawn from Fitzpatrick et al., 2014)

3.6.5 Median Islands

Medians are used to fully separate opposing traffic flow and clearly define where crossing/turning movements are permitted. Medians may be raised or depressed. Raised medians must have wider islands kerbed along the inner edges of the inside traffic lanes, which also include the central islands and the median shoulders. Raised median islands provide better protection to pedestrians than painted islands. Median widths commonly vary from 1.2 m to over 9 m. It is desirable to provide medians (width 4.2–6.0 m) on multi-lane highways to facilitate storage for separate right turns (3.13).

The transition taper or taper length (Fig. 3.14) value can be calculated as recommended by the FHWA (MUTCD, 2009):

For speeds 45 miles per hour or more: Taper length (ft) = $W \times V$ (3.3)

For speeds under 45 miles per hour: $\text{Taper length (ft)} = \dfrac{WV^2}{60}$ (3.4)

where W = width of offset or width of lanes (ft); V = posted speed or 85th percentile speed or statutory speed (miles per hour) with the conversion factors: 1 ft = 0.3048 m; 1 mile per hour = 1.60934 kilometres per hour.

3.6.6 Dividers

Dividers are raised row-like kerb elements provided as part of channelisation along the inner edges of the inside traffic lanes. They are intended to divide/separate the traffic streams across the lanes. However, dividers are never considered as pedestrian refuge. Design divider characteristics can be obtained from Section 4.3.1 of Chapter 4.

3.7 DESIGN CRITERIA AND GUIDELINES OF TRAFFIC ROTARIES

The traffic moves in one direction around a central island (Fig. 3.15). Vehicles enter in an orderly fashion into the circulating width and move in the clockwise direction. Major conflicts, as may be encountered in un-signalised intersections between vehicles turning right and those moving through traffic, are eased by converting the movements into merging and diverging movements. The end result is division of a single stream of traffic into separate streams. While moving, vehicles weave over a convenient distance with reduced speed and exit into their desired directions.

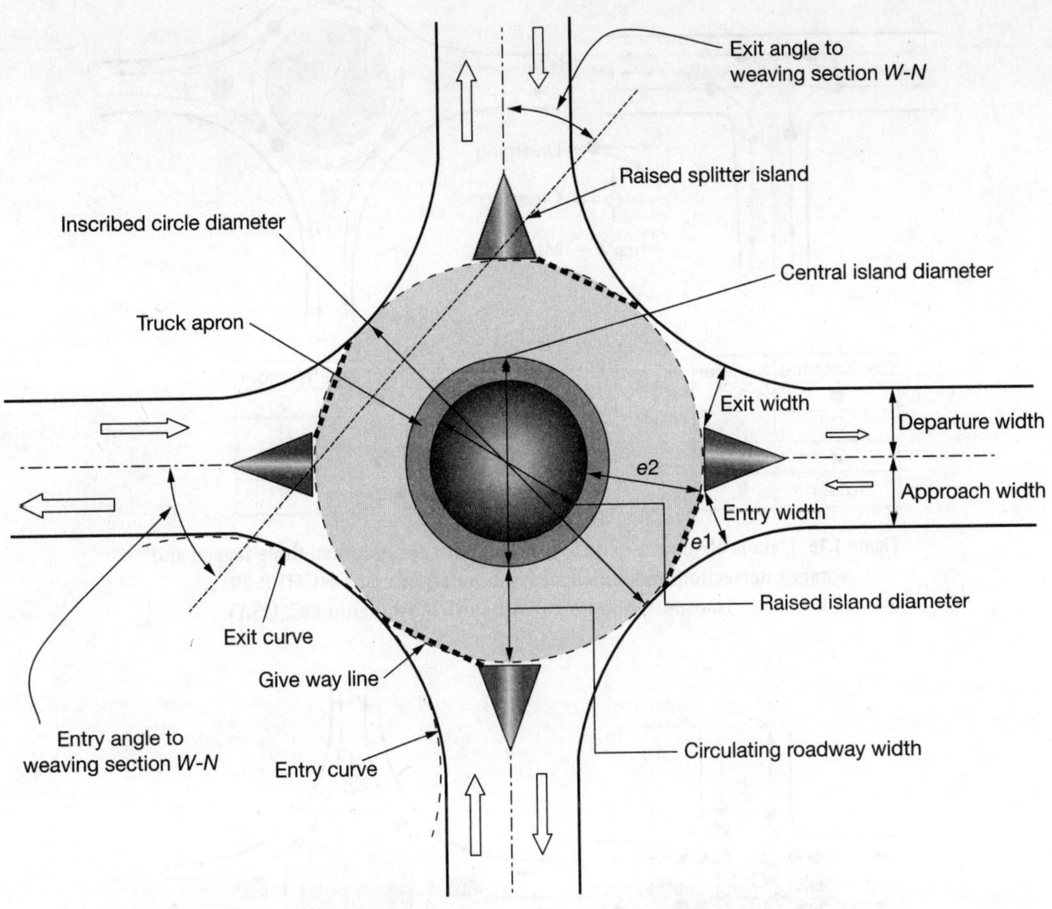

Figure 3.15 Basic geometrics and components of a roundabout (Redrawn based on TRB, 2000).

3.7.1 Advantages of Traffic Rotaries

Traffic rotaries reduce the potential for severe conflicts such as right-angle and right-turn head-on collisions. Examples of the benefits in terms of reduced number of conflicts with single-lane intersections are summarised below.

- Eliminate vehicle–vehicle crossing conflicts and separate vehicle–vehicle conflicts from pedestrian conflicts by using raised splitter islands.
- Benefit of vehicle-vehicle conflicts with T-intersection (from Figs. 3.16, 3.17 and 3.18) $= \frac{(9-6)}{9} \times 100 = 33.33\%$
- Benefit of vehicle-vehicle conflicts with four-legged intersection $= \frac{(32-8)}{32} \times 100 = 75\%$
- Benefit of vehicle pedestrian conflicts with four-legged intersection $= \frac{(16-8)}{16} \times 100 = 50\%$

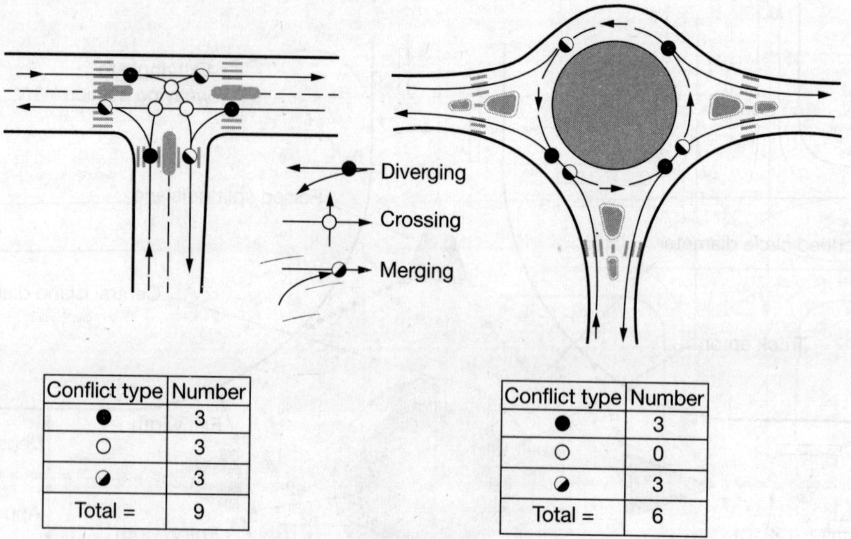

Conflict type	Number
●	3
○	3
◐	3
Total =	9

Conflict type	Number
●	3
○	0
◐	3
Total =	6

Figure 3.16 Details of vehicle–vehicle conflicts with conventional three-legged and rotary intersections (Modified and elaborated based on NCHRP, 2010; *Courtesy*: Transportation Research Board, Washington DC, USA).

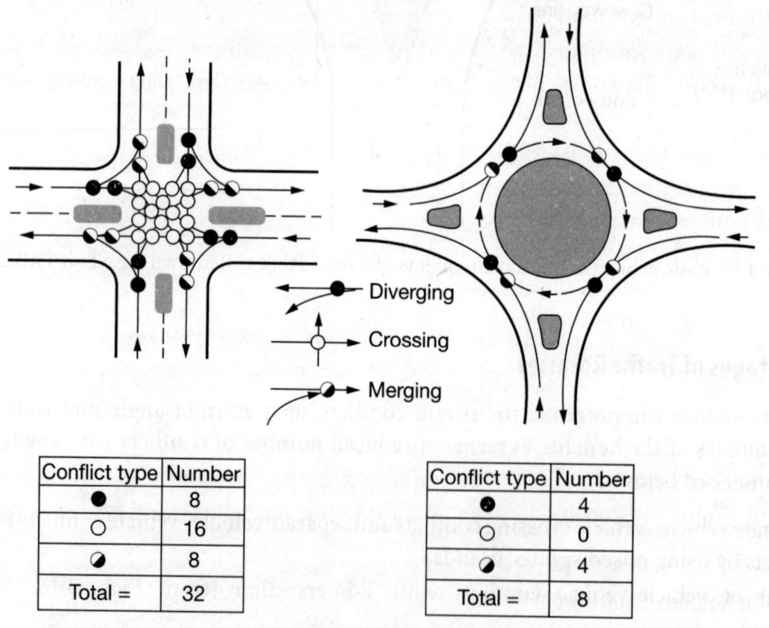

Conflict type	Number
●	8
○	16
◐	8
Total =	32

Conflict type	Number
●	4
○	0
◐	4
Total =	8

Figure 3.17 Details of vehicle–vehicle conflicts with conventional four-legged and rotary intersections (Modified and elaborated based on NCHRP, 2010; *Courtesy*: Transportation Research Board, Washington DC, USA).

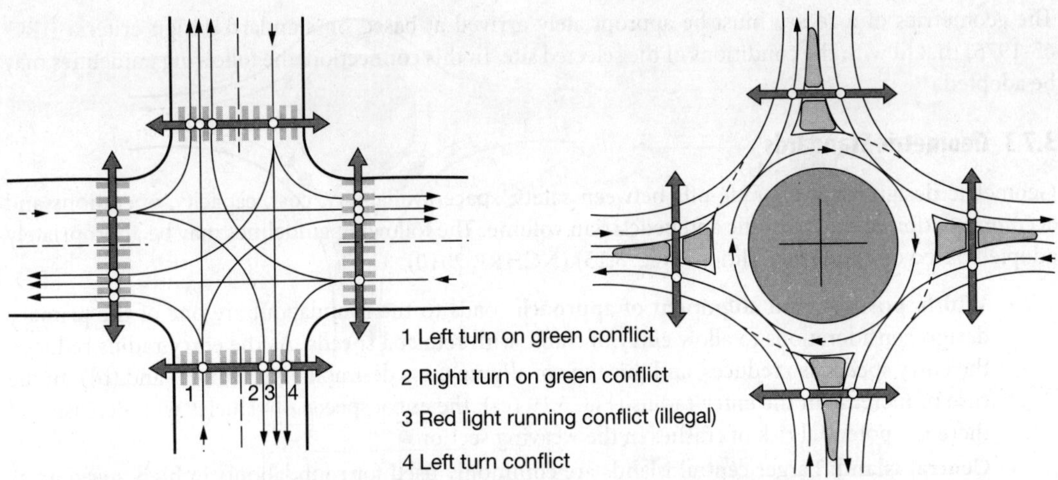

Figure 3.18 Details of vehicle–pedestrian conflicts with conventional four-legged and rotary intersections (Modified and elaborated based on NCHRP, 2010, and ODOT, 2011; *Courtesy*: Oregon Department of Transportation (ODOT) and Transportation Research Board, Washington DC, USA).

3.7.2 Disadvantages of Traffic Rotaries

- Larger area of requirement, initial cost of construction of roundabout with landscape enhancement of the area and its maintenance are some of the primary disadvantages.
- Locations within a coordinated signal network where provision of roundabout would disrupt the platoon (NCHRP, 1998).
- No pedestrian signals to help pedestrians cross the road.
- The right-turning traffic must travel an extra distance around the central island.
- The number of vehicular conflicts may increase with *dual* and *multi-lane* roundabouts due to increased number of lane changes/crossings. This is particularly predominant in the circulating area when (a) vehicles weave to exit by changing lanes and (b) vehicles enter next to another vehicle perceived to be circulating in front of a splitter island corner due to improper left turn brought about by the lane change. Due to these reasons, an increased number of accidents may be seen with multi-lane roundabouts. Hence, specialised geometric design is necessary if two-lane/multi-lane roundabouts are used.
- Acute angles of intersection may result from very closely spaced adjacent approach roads, which may result in the design of inadequate weaving section length.
- If the number of vehicles in a rotary exceeds capacity, the resulting vehicle queue will force entry vehicles to a complete stop and jam the weaving section. Intervention of the traffic police is essential in such cases.

Suitability of Traffic Rotaries/Roundabouts

- The layout of the site consists of plain space with three or more roads intersecting.
- Traffic flow in the intersecting road legs is almost equal in proportion.
- Right-turning traffic at the intersection is predominant and is about equal to or greater than 30% in proportion to through and left-turning traffic.
- The traffic volume is low to moderate (500–3000 vehicles/hour) and enters from all the intersecting roads.

The geometrics of a rotary must be appropriately arrived at based on standard design criteria (IRC: 65-1976) that fit with the conditions of the selected site. In this connection, the following guidelines may be adopted.

3.7.3 Geometric Standards

Geometric design involves trade-offs between safety, space availability, cost, capacity, operations and accommodation of design traffic and pedestrian volume. The following guidelines may be appropriately adopted based on the site conditions (Fig. 3.15) (NCHRP, 2010).

- **Width, position and alignment of approach** roads to the roundabout are one of the primary design considerations to allow entry of traffic with reduced speeds. As the entry radius reduces, the entry speed also reduces, and this type of alignment is desirable (Fig. 3.19 (a) and (b)). In the case of increase in the entry radius (Fig. 3.19 (c)), the entry speeds of vehicles will increase and there is a potential risk of crashes in the weaving section.
- **Central island:** Larger central islands are commonly used for roundabouts in high-speed areas and at intersections with more than four legs. These larger islands can be non-circular to suit the specific intersecting legs, with the long side of the island oriented for major flow. However, islands of less than the nominal 5 m radius should be circular.
- **Inscribed circle diameter**: It is the sum of the diameter of the central island and twice the width of the circulatory roadway. The dimension must be large enough accommodate the turning radius of the longest vehicle (design vehicle) and the design speed. In general, the diameter should be at least 32 m to accommodate bus or single-unit trucks as design vehicles. The diameter is typically in the range of 40–46 m, if the design vehicle is longer than a bus. Typical ranges of inscribed circle diameters for two-lane and three-lane roundabouts require at least 50–67 m and 61–91 m, respectively.
- **Entry approach angle:** It is selected to provide adequate sight distance and visibility, and its value ranges between 20° and 40°.
- **Entry and exit angles:** The entry angle should be larger than the exit angle. Ideally, the entry angle should be 60° with an exit angle of 30°. Staggering of the approach legs is recommended to achieve an entry angle that is close to 60°.

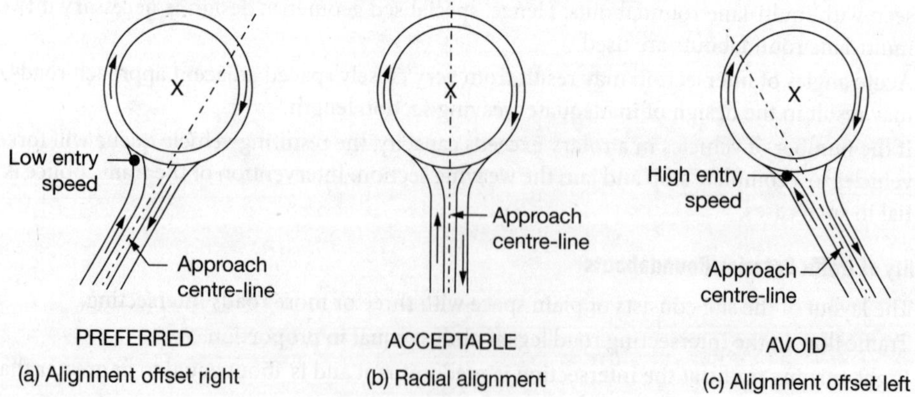

Figure 3.19 Typical alignment of approach roads to a roundabout (Elaborated based on NDOR, 2012; *Courtesy*: Nebraska Department of Roads, USA).

- **Exit curb radii** should not be less than 15 m (commonly between 30 m and 60 m). Further, larger exit radii may be desirable in areas with high truck volumes to improve ease of exit and reduce the risk of trailers tracking over the outside kerb.
- **Entry radii:** Larger entry radii are often not provided in order to discourage high entry speeds. A low-speed value is commonly recommended in urban areas. Entry radii at the kerb line of single-lane roundabouts typically range 15–30m. With reference to design speeds of 40 km/hr and 30 km/hr, the recommended radius of entry may be in the range of 20–35 m and 15–25 m, respectively (IRC: 65-1976).
- **Truck aprons** provide additional paved width around the central island which allows over-tracking of large semi-trailers without compromising the deflection for smaller vehicles. The width of the truck apron is decided based on the swept path of the longest vehicle permitted. Its width is generally provided between 1 m and 4.6 m, with a cross slope of 1% to 2% away from the central island. The outer edge of the apron is raised approximately 50–75 mm above the circulatory roadway surface to discourage use by passenger vehicles.
- **Entry/Exit width** of a single-lane entrance ranges from 4.2 m to 5.5 m, with a flared approach. In dual lanes (7 m), the minimum width is taken as 6.5 m and 7 m with an entry radius of 25–35 m and 15 25 m, respectively. In multi-lane road approaches, the standard values may be suitably selected from IRC: 65-1976.
- **Width of the weaving section:** Circulating width depends on the swept paths, layout, and width of exits and entries. In general, a consistent width throughout the roadway is preferable. For single-lane roundabouts, the width typically ranges from 4.8 m to 6.0 m. In India, it is wider than the mean entry width by 3.5 m. California DOT recommends that the width should not exceed 15 m (Taekratok, 1998).
- **Weaving length:** Adequate weaving length is provided to accommodate the design volume of traffic with reference to the number of lanes, weaving width and proportion of weaving traffic to total traffic. As the design speed increases, the required weaving length increases. As a thumb rule, the length of the weaving section must be equal to four times its width. With reference to design speeds of 40 km/hr and 30 km/hr, the recommended minimum weaving length may be taken as 45 m and 30 m, respectively. As a common guideline, weaving section length is longer by 33–50% to avoid decrease in design speed while weaving. However, over-speeding of traffic in the weaving section is checked by restricting its length to twice the prescribed values (IRC: 65-1976).
- **Grade:** According to French guidelines, any grade lower than 3% is generally considered acceptable; but on roadways, any grade greater than 6% is generally considered unsuitable for the installation of roundabouts (Taekratok, 1998).
- **Super-elevation**: Super-elevation on curved circulating width through the roundabout is generally not necessary as the operating speeds are constrained (up to 50 km/hr) and the drivers can tolerate higher values of the coefficient of sideways friction within the intersection (MDOT, 1995).

3.7.4 Operating Standards

Achieving appropriate operating speeds for entering, circulating and exiting the roundabout is a critical design objective which impacts on the safety and comfort of the road users. Hence, operating speeds at entry and through the roundabout are managed with geometric design (Tian et al., 2007).

The maximum permissible design speeds at *entry* may be considered in the range 32–40 km/hr and 40–48 km/hr, for single-lane and multi-lane roundabouts (NCHRP, 2010).

3.7.5 Design Guidelines

Operational Performance Evaluation

The operational performance of existing roundabouts can be evaluated by measuring the control delay. This can be estimated by measuring the average time taken by vehicles to travel between any study entry and exit points of a roundabout under constrained conditions and ideal conditions.

The design of the geometric components of a roundabout such as entry radius, exit radius and circulating radius with sight distances can be estimated based on travel time delays along a corridor.

Capacity Analysis

The capacity of a rotary is determined by the capacity of its weaving section (Fig. 3.20). The following TRL-recommended formulae are used in India to determine the capacity (in Passenger Car Units per hour – PCU/hr) of the weaving section of a rotary (IRC: 65-1976).

$$\text{Capacity of a weaving section of roundabout} = Q = \frac{280w\left(1+\frac{e}{w}\right)\left(1-\frac{p}{3}\right)}{\left(1+\frac{w}{L}\right)} \quad (3.5)$$

where,

e = average of entry widths $= \left(\frac{e1+e2}{2}\right)$ (recommended range of e: $0.4w – 1.0w$ m)

$e1$ = entry width, which is measured/set across the width between the exit entry corner of the raised splitter island and the adjacent kerb (Fig. 3.20);

$e2$ = width of the non-weaving section, which is measured/set across the width between the exit end corner of the raised splitter island and the central island;

L = length of the weaving section (recommended range of L: $2.5w$–$8.33w$ m)

$$w = \text{weaving width} = \left\{\left(\frac{e1+e2}{2}\right)+3.5\text{ m}\right\} = \{e+3.5\text{ m}\}$$

(Recommended range of w: 6 – 18 m)

p = proportion of weaving traffic to total traffic in the weaving section, (range: 0.4–1.0).

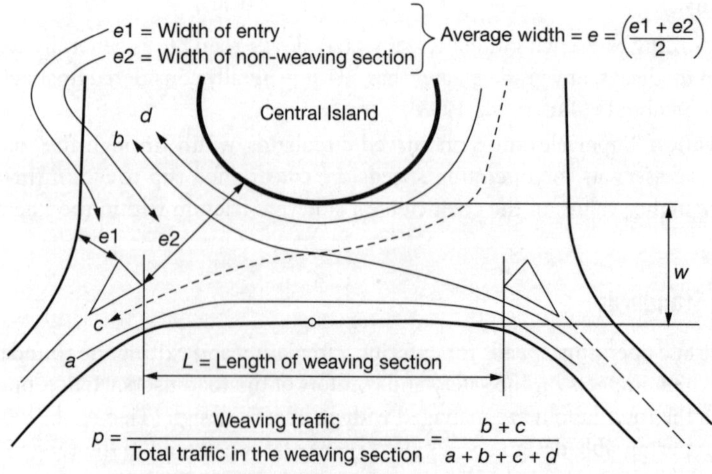

Figure 3.20 Definition of average entry width, weaving traffic and total traffic in a weaving section.

Adjustments: The calculated capacity values are adjusted for the following site conditions (IRC: 65-1976):

- Deduct 5% from the calculated capacity of the weaving section if the entry angle is between 0° and 15°.
- Deduct 2.5% from the calculated capacity of the weaving section if the entry angle is between 15° and 30°.
- Deduct 2.5% from the calculated capacity of the weaving section if the exit angle is between 60° and 75°.
- Deduct 5% from the calculated capacity othe weaving section if the exit angle is greater than 75°.
- Deduct 5% from the calculated capacity of the weaving section if the internal angle of the approach roads is greater than 95°.
- Deduct 1/6th of the calculated capacity of the weaving section if pedestrian flow at the exit from the roundabout exceeds 300 per hour.

Remarks: The PCU values considered for the calculation of capacity of the weaving section is 2.8 for buses and heavy commercial vehicles and 4–6 for animal-drawn vehicles. The necessary adjustment factors may be derived from the prevailing site conditions if these values of PCU differ.

Note:

- The capacity of a rotary is determined based on the highest value of p, which is obtained by considering the directions of traffic flow. This criterion holds well only if the rotary geometrics (w, e and L) are same for all the intersecting approach roads.
- If the rotary geometrics are not same for all the intersecting approaches, the lowest value of the capacity calculated is said to be the *capacity of the rotary* intersection.
- If the calculated capacity value is greater than the standard limiting value of a single lane, then an additional lane or a multilane rotary may be designed.

EXAMPLE 3.1

Determine the capacity of a roundabout for the following values:

Average entry width = 8.5 m; Width of the weaving section = 15 m; Length of the weaving section between channelising islands = 40 m; Conflicting traffic volume observed in the weaving section = 1200 PCU/hr; Total traffic volume observed in the weaving section = 2500 PCU/hr.

Answer

The given values are rewritten as

$$e = \text{Average entry width} = 8.5 \text{ m}$$

$$w = \text{Width of the weaving section} = 15 \text{ m}$$

$$L = \text{Length of the weaving section between channelising islands} = 40 \text{ m}$$

$$b + c = \text{Conflicting traffic volume observed in the weaving section} = 1200 \text{ PCU/hr}$$

$$a + b + c + d = \text{Total traffic volume observed in the weaving section} = 2400 \text{ PCU/hr}$$

The following equation is used to determine the capacity of the weaving section. Substituting, we get

$$\text{Capacity of weaving section} = \frac{280w\left(1+\frac{e}{w}\right)\left(1-\frac{p}{3}\right)}{\left(1+\frac{w}{L}\right)}$$

$$p_{E\to S} = \frac{b+c}{a+b+c+d} = \frac{1200}{2400} = 0.5$$

$$Q_{N\to E} = \frac{280\times 15\times\left(1+\frac{8.5}{15}\right)\times\left(1-\frac{0.5}{3}\right)}{\left(1+\frac{15}{40}\right)} = \frac{4200\times(1.566)\times(0.833)}{(1.375)} = 3987.8 \text{ PCU/hr}$$

Note: For more example and solved problems, readers are referred to see the book titled "Introduction to Traffic Engineering" by R. Srinivasa Kumar, Universities Press.

3.8 DESIGN GUIDELINES FOR GRADE-SEPARATED INTERSECTIONS

3.8.1 Classification of Grade-separated Structures

Grade-separated structures (without ramps) are broadly classified into two types: underpass and overpass (Fig. 3.21). The choice of these structures depends on several factors such as topography, slope of intersecting grades, vertical alignment, drainage, aesthetic features, preferential aspects relating to the habits of local people/users and the cost of construction. A brief description of these structures is as follows.

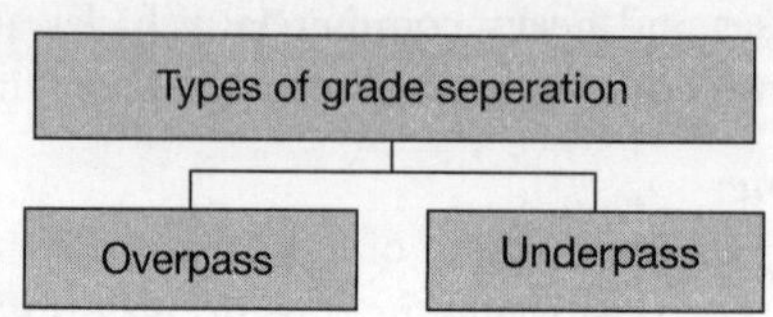

Figure 3.21 Types of grade-separated structures.

Overpass or Bridge

An overpass or flyover is an elevated road structure on which major traffic travels *over a point* of an intersecting road, river or railway line. Both the terms are *synonymous* and commonly referred to as a bridge structure in the UK and most commonwealth countries (Fig. 3.22). These structures are also constructed to facilitate crossing of pedestrians and bicyclists.

Advantages

- Drainage-related problems can be reduced by aligning the major highway above the cross road, instead of providing it as an underpass.
- Overpass of major highways is aesthetically preferred over an underpass as it is associated with a lower sense of restriction and confinement and greater visibility.
- Future widening of existing overpasses is simpler than that of underpasses because sub-structure disturbances may occur during widening.

Disadvantages

- Provision of a steeper (or rolling) gradient of the major highway ramp may be required in case it passes through rolling terrain.
- Provision of a longer embankment may become necessary to attain adequate sight distance for the design speed. This may increase the total cost of construction (land cost + construction cost).
- There may be a reduction is design speed due to the steeper gradient of the interchange ramp. Heavy vehicles need more tractive energy to overcome this steeper gradient.

Underpass

An underpass is a grade crossing where the traffic flows *under* another structure at a point such as a road, river or railway line. In some cases, the underpass may be constructed as a tunnel through mountains.

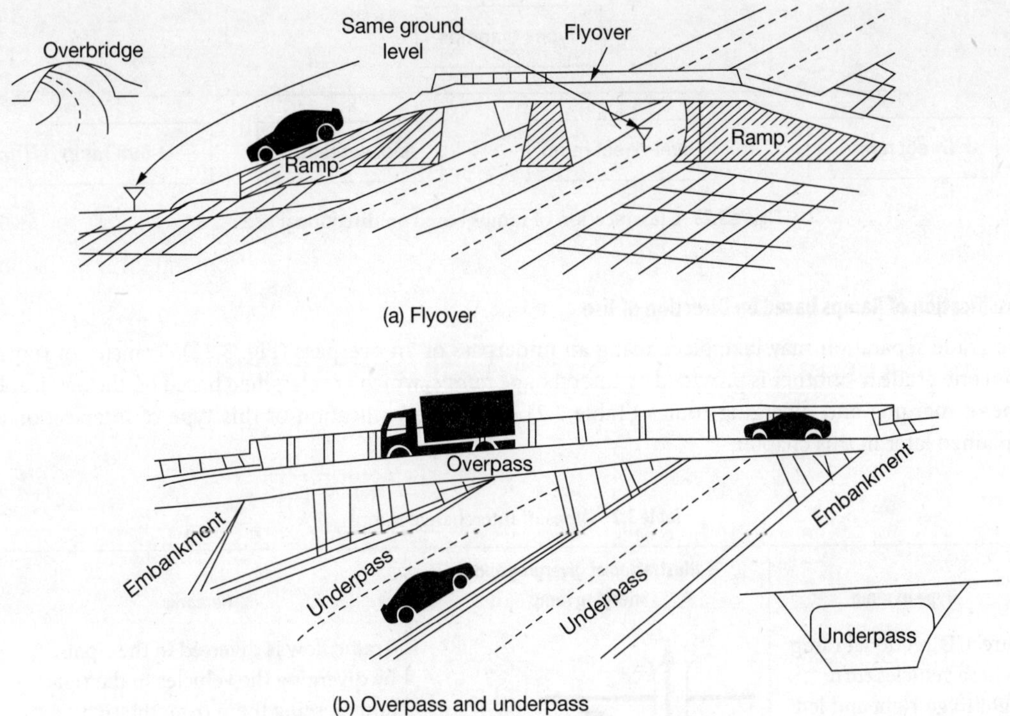

Figure 3.22 Flyover, overpass and underpass structures.

The underpass created to facilitate the safe crossing of pedestrians, wild-life and bicyclists under a railway line or roadway is often termed a *subway*.

Advantages

- Underpasses are useful when the main highway needs to be constructed on the existing grade, without any change in its vertical alignment.
- Underpasses are constructed when the crossing site conditions demand such alignment of the main highway.

Disadvantages

- Rainwater may stagnate during the monsoon season and there may also be a rise in ground water levels in agricultural areas nearby.
- The sight distance through the underpass may be inadequate if a sag curve passes under an overhead structure. The structure may restrict the sight distance.
- There may be a feeling of restriction to traffic at the sides while passing along the underpass and this may affect the capacity of the intersection.
- No changes can be made during the construction of underpasses when there is a paucity of funds.

3.8.2 Classification of Ramps

A ramp is a short section of a road length that connects intersecting roads, from one level to another. Ramps are classified based on their function and intended use (Fig. 3.23).

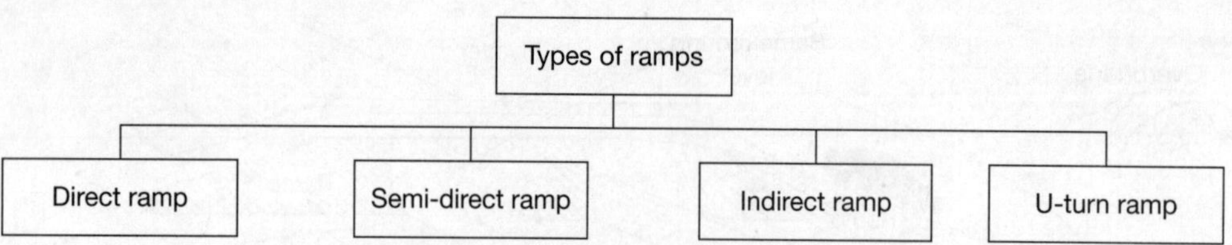

Figure 3.23 Classification of ramps based on directional use.

Classification of Ramps based on Direction of Use

The grade separation may take place using an underpass or an overpass (Fig. 3.22). Transfer of routes from one grade to another is provided by interchange ramps, which are classified based on the applicable type of merging and diverging routes (Table 3.2). Further classification of this type of intersection is explained later in this chapter.

Table 3.2 Types of interchange ramps.

Type of ramp	Illustration of diverging and merging ramp	Remarks
Direct/Directional ramp (where vehicles turn right to go right and left to go left)	Ramp	Traffic flow is diverted to the right by diverging the vehicles to the right and merging them from the right. As per Indian traffic rules, this may be unsafe and hazardous because both the diverging and the merging of traffic flow takes place on the right. However, the distance to be traversed on direct interchange ramps will be less. The capacity and speed on these ramps will be high.
Semi-direct/Semi-directional/Jug-handle ramp (where vehicles must first manoeuvre left to make a right-turn and vice versa for a left turn)	Jug-handle ramp	Traffic flow is diverted to the left by diverging the vehicles to the left and merging them from the right. This is a ramp provided on the left-hand side of the road, and used for making right turns.
Indirect/ Non-directional ramp	Loop ramp	Traffic is diverted to the left by diverging the vehicles to the left and merging them from the left. As per Indian traffic flow rules, this is considered safe and less hazardous because the diverging and merging of traffic flow takes place on the left. However, the distance to be traversed on will be greater.

(Continued)

Table 3.2 (*Continued*)

Type of ramp	Illustration of diverging and merging ramp	Remarks
4. U-turn ramp (where a vehicle must drive on a 360° curve to make a U-turn)	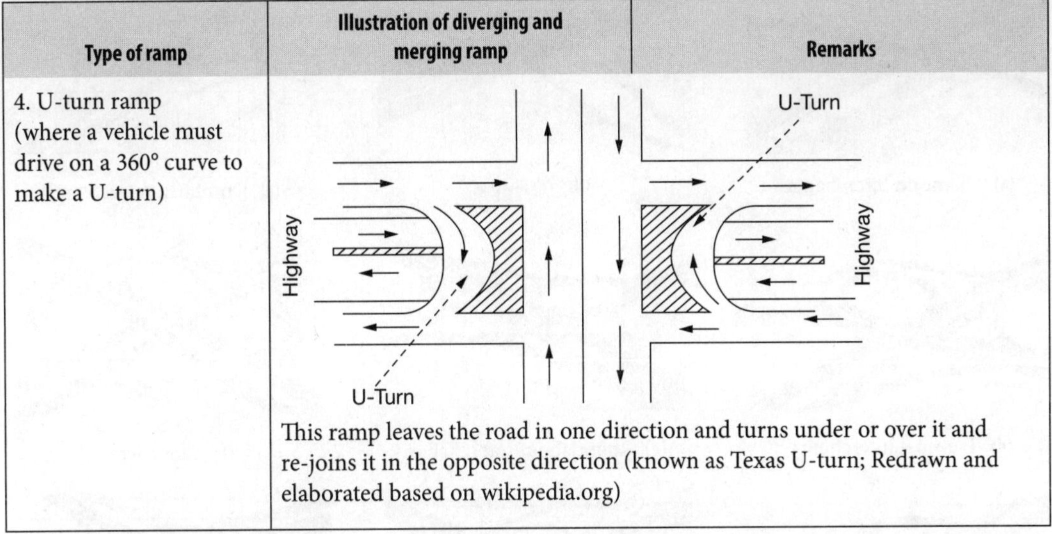 This ramp leaves the road in one direction and turns under or over it and re-joins it in the opposite direction (known as Texas U-turn; Redrawn and elaborated based on wikipedia.org)	

These types of ramps can be used in combination for a specific flow direction These combinations are useful when designers need to sketch a suitable configuration for an interchange in a study area, during the preliminary design process.

3.9 CLASSIFICATION OF INTERCHANGES

Interchanges can be classified based on different parameters. These classifications may be useful as a guide to prepare a list of alternative candidate interchanges for evaluation, and finally, for the selection of the most suitable configuration for a design situation.

3.9.1 Classification Based on the Number of Approach Legs

(a) **Three-way interchanges** (with three approach roads)
- Trumpets (all variants) (Fig. 3.24 (d))
- Directional Y (Fig. 3.24 (b)) or T (Fig. 3.24 (g))
- Half-diamond
- Hybrids*

(b) **Four-way interchanges** (with four approach roads)
- Full cloverleaf with C–D roads ((Fig. 3.24 (f)))
- Parclo or *par*tial *clo*verleaf (all variants) ((Fig. 3.24 (h))
- Diamond (with variants) (Fig.3.24 (a) and (c))
- Directional with loop ramps (Fig. 3.24 (j)) or without loop ramps
- Hybrids*

(c) **Five-way or more interchanges** (with five or more approach roads)
- Rotary (Fig. 3.24 (e))
- Directional with loop ramps or without loop ramps (Fig. 3.24 (j) and (k))
- Hybrids*

*Hybrid interchange configurations do not exactly fit into any one of the standard geometrical configurations. They are interchanges modified to meet site constraints.

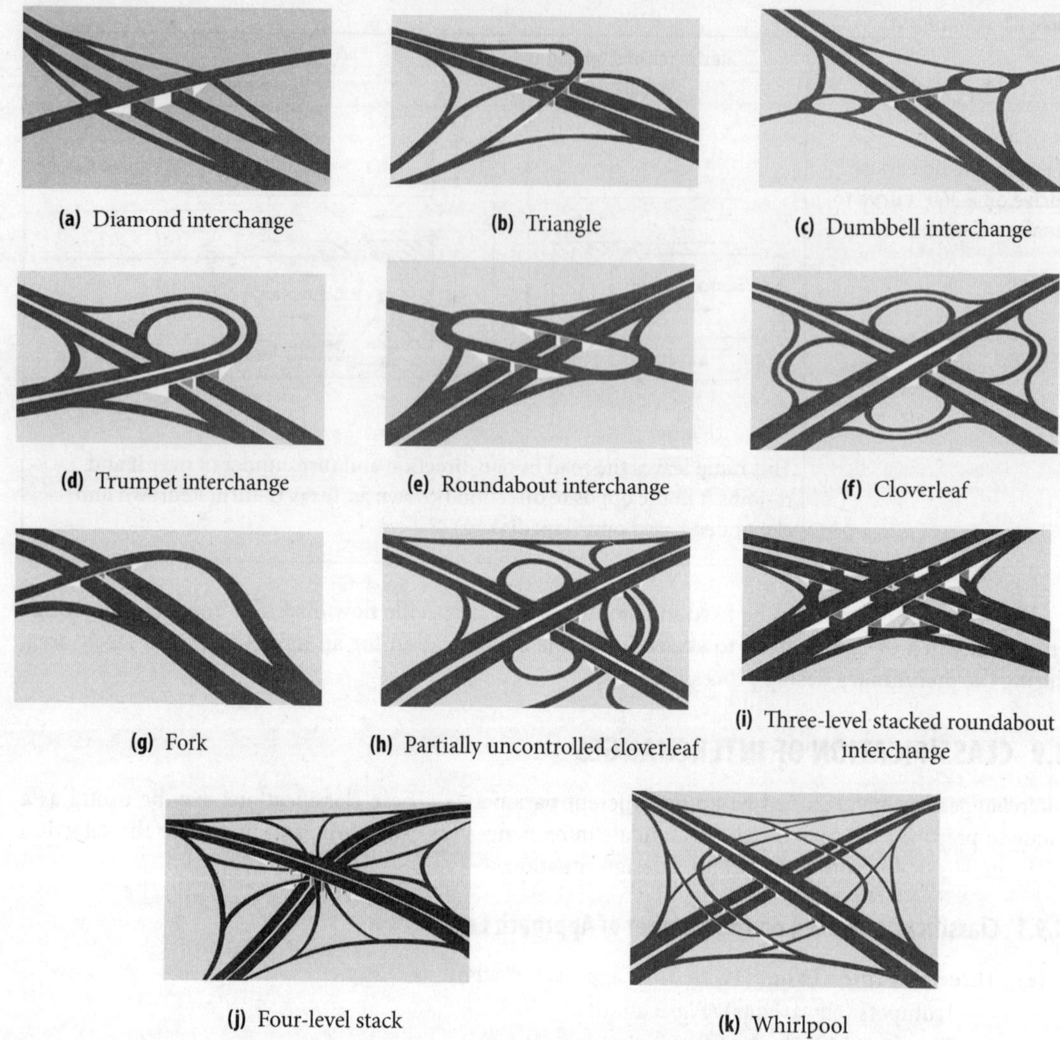

Figure 3.24 Perspective views of a few interchange configurations.
(*Source*: www.cbrd.co.uk; *Courtesy*: Mr Chris Marshall)

3.10 INTERCHANGE CONFIGURATIONS

The types of interchange configuration based on the number of intersecting roadways are presented in the following sections.

3.10.1 Three-leg Interchanges

Three-leg intersections can be configured in different ways. The type of configuration fit for a particular location of an intersection is determined mainly by the alignment of the intersecting roads, the topography and the directional distribution of traffic. It is important to take into consideration that it is very difficult to include a fourth leg in any three-leg intersection, as part of future expansion. To plan for such expansion, any suitable configuration of four-leg intersections which can fit into the current requirement for a three-leg intersection must be selected.

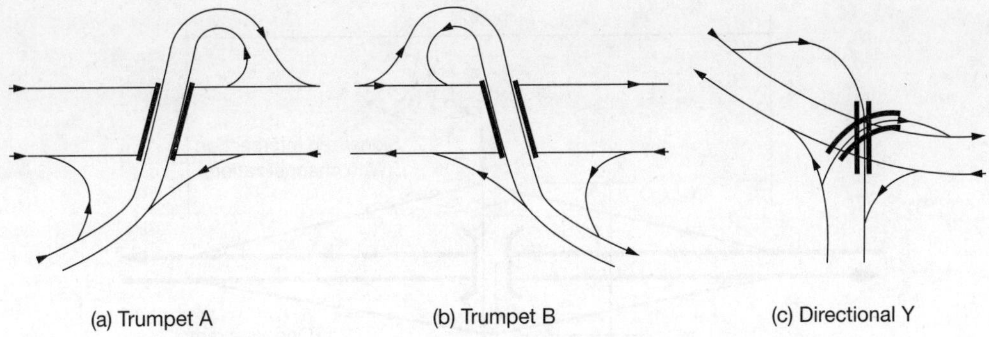

Figure 3.25 Three-leg interchanges (*Note*: Changed for left-hand traffic).
(*Source*: CDOT, 2005; *Courtesy*: www.codot.gov)

Trumpets are the most widely used configuration for three-leg intersections (Fig. 3.24 (d) Fig. 3.25 (a) and (b)). As a thumb rule, preference is given to major turning movements on directional/semi-directional ramps and the lowest traffic movements should be facilitated by a loop ramp.

A composite listing of the advantages and disadvantages of trumpet interchanges is presented in Table 3.3.

Table 3.3 Advantage and disadvantages of trumpet interchanges.

Advantages	Disadvantages
✓ Loop serves as off ramp to the highway ✓ Fits in T-intersection traffic flow pattern ✓ Used for highway-to-arterial connection ✓ Can handle large volumes	✗ Difficult to extend truncated approach road if demand increases ✗ Indirection in movement ✗ Not used for highway-to–highway connections

3.10.2 Diamond Interchanges

These are one of the most commonly used interchanges in urban and rural areas, where highways (major road) and arterial roads (minor road) intersect (Fig. 3.26). Usually, the minor road is provided as an overpass and the turning movements are facilitated by one-way diagonal ramps as interchanges, in each quadrant (Fig. 3.27 (a)). The design allows free flow of operation on the major highway but creates two at-grade intersections on the minor road with ramps. These two at-grade intersections are often signalised if the traffic volume is at medium level or just operated with a Stop sign, when the turning traffic is very small.

An unsignalised system of diamond interchange can be operated with an acceptable level of service (LOS) when the entering volume is up to 1500 veh/hr. The diamonds can operate at LOS D or higher until entering volume reaches 4500 veh/hr (Garber and Fontaine, 1999).

Depending on the distance between these intersections, diamond interchanges can be classified into three types (Chaudhary and Chu, 2000):

- *Conventional diamond interchange*: The distance between the two at-grade intersections (Y) is greater than 244 m (800 ft). These intersections are controlled by Stop signs in rural areas (Fig. 3.26).

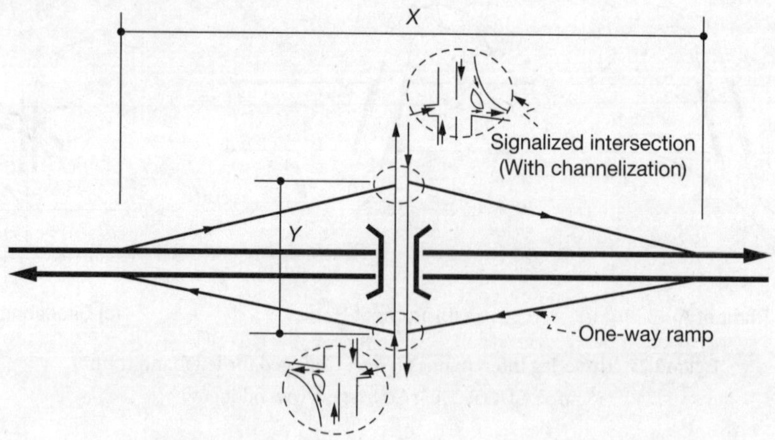

Figure 3.26 Conventional diamond interchange.

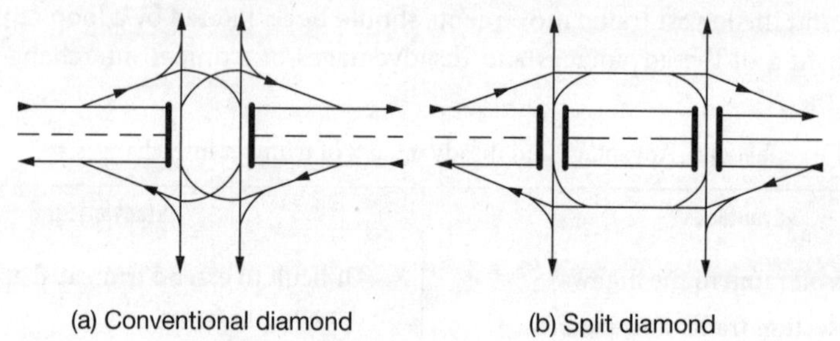

Figure 3.27 Typical directions of traffic flow permitted in diamond interchange variants (*Note*: Changed for left-hand traffic).
(*Source*: CDOT, 2005; *Courtesy*: www.codot.gov)

- *Compressed diamond interchange*: The distance between the two at-grade intersections (Y) is between 122 m (400 ft) and 244 m (800 ft). Both the intersections of the interchange are controlled by signals with or without interconnection.
- *Tight diamond interchange* (TDI*) or tight urban diamond interchange* (TUDI): The distance between these two at-grade intersections (Y) is between 61 m and 122 m (200–400 ft). Because of the close proximity, *two* signals at these intersections are always controlled and operated as *one* signal system (Fig. 3.28). Optimum signal timing is developed to minimise storage of vehicles between the ramp intersections and also to provide coordinated movement of traffic along the street and ramps. TDIs are commonly designed for developed urban areas, where land acquisition is difficult and costly.

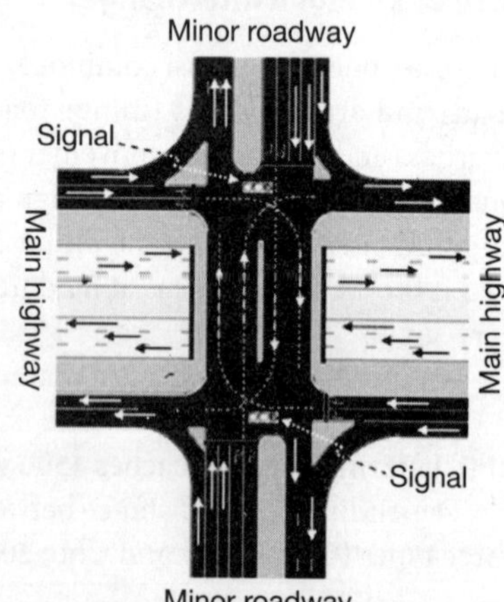

Figure 3.28 Tight diamond interchange (TDI) or tight urban diamond interchange (TUDI).

A composite listing of a few advantages and disadvantages of TDI is presented in Table 3.4.

Table 3.4 Advantages and disadvantages of TDI.

Advantages	Disadvantages
✓ Minimum property required outside of the usual right of way (RoW) along the major highway ✓ No weaving on the through road ✓ Economical, with one grade-separated structure ✓ Flared ramp terminals can improve capacity ✓ Single exit from highway at high speeds ✓ Suitable in rural and urban areas as local road intersection treatment ✓ Fits service road pattern in case of urban areas ✓ Pedestrian and bicycle movements can be better accommodated than in full cloverleaf ✓ Diamonds are preferred over single-point urban interchanges (SPUI) when frontage roads are present.	✗ The operating capacity is limited due to the at-grade intersections on the minor road ✗ There would be inadequate sight distance at ramp terminals on the at-grade intersections ✗ Need to stop on crossroads for all turn movements on \the minor road ✗ Possibility of wrong way movement at the at-grade intersections ✗ Difficult to accommodate high right-turn volumes ✗ Three-phase signal control with signal coordination is often required to operate the traffic on the minor road of at-grade intersections.

3.10.3 Variants of Diamond Interchanges

- *Single-point urban interchange* (SPUI) (Figs. 3.28(a) and (b)): Compresses the two intersections into one single signalised intersection. SPUI is preferred over TDI because it can operate with less delay and handle heavier traffic volumes as vehicles need to cross only one signalised intersection.

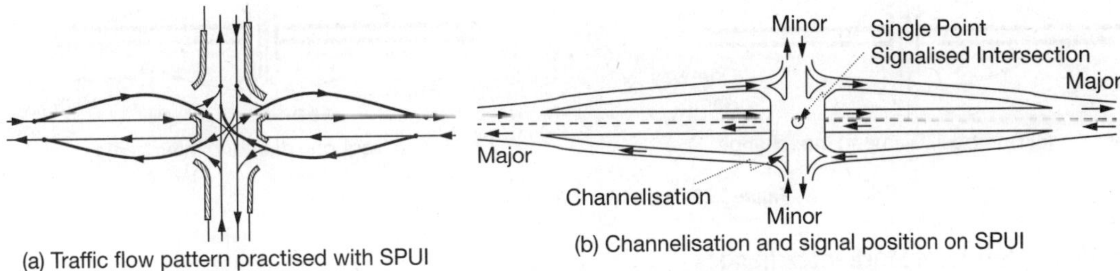

Figure 3.28 Schematic representations of single-point urban interchange (SPUI)

- *Spread diamond interchange* (SDI) (Fig. 3.29 (c)): The ramps of the diamond interchange are spread wider away from the through mainline. There will be flexibility to provide good sight distance at the ramp terminals of the at-grade intersections. The disadvantages are the same as listed for TDI.
- *Split diamond interchange*: This (Fig. 3.29 (a)) provides good bypass in case of accidents as its configuration provides alternatives for entering and leaving the highway (Fig. 3.29 (b)). There would be a capacity problem due to queuing, which is because of closely spaced at-grade intersections. The split diamond (Fig. 3.29 (d)) can be operated with two-phase signals. It requires two structures. The turning movements are simplified but are not convenient for right turns.
- *Diamonds with right-turn structures* or *dumbbell diamond* (Fig. 3.29 (e)): This is similar to the diamond interchange, but uses a pair of mini-roundabouts to join the ramps with the non-highway

street. Dumbbell diamonds increase the efficiency of the interchange when compared to a conventional diamond interchange.

- *Diamonds with U-turn structures* or *dogbone diamond* (Fig. 3.29 (f)): This is a variation of the dumbbell roundabout and diamond interchange. The central island is shaped like a toy dog bone, in the plan view. The dog bone is formed by two interconnected mini-roundabouts that look like raindrop- or teardrop-shaped islands at either ends. These two roundabouts act as traffic calming devices. There will be no chance of wrong-way entry at the ramps and such conflicts are reduced. For these reasons, it has been observed that these configurations increase capacity by reducing queuing and delays, as compared to the dumbbell roundabout.

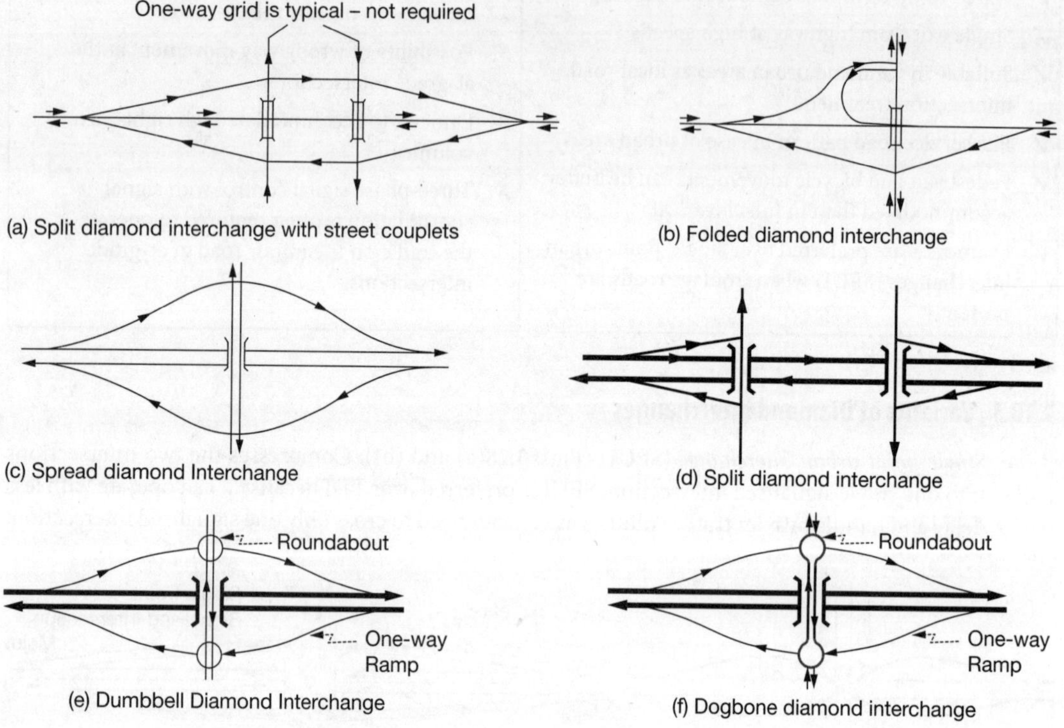

Figure 3.29 Variants of diamond interchanges.

3.10.4 Multi-level Stack Interchanges

Stack interchanges are designed as free-flowing grade-separated intersections. In general, any stack interchange consists of two pairs of curved right-turning directional ramps, which may be stacked at the geometric centre of the interchange (Fig. 3.24 (j)). The right-turning ramps originating from each leg are configured at grade or above or below the two interchanging highways. If these four right-turning ramps are designed to merge at a common level, such an interchange is termed a three-level stacked diamond interchange or three-level stack (Fig. 3.30).

In other cases, if any pair of right-turning directional ramps is configured one above the other, such an interchange is termed a four-level stacked diamond interchange or four-level stack (Fig. 3.24(j)). As the two pairs of ramps are grade-separated, there is no need of signal control. A common operational feature of three-level and four-level stacks is that traffic flows on each pair of curved ramps at their respective levels are reversed.

Five-level and six-level stacks also have a similar configuration, with additional frontage roads or elevated highways adding the levels. Intersections of such additional frontage roads are typically at grade or nearer to the through road.

All variants of stack interchanges can serve high volumes of traffic at highway speeds, except at signalised intersections of three-level stacks. They occupy smaller land area. Stacks eliminate the problem of weaving and can provide the highest vehicle capacity among similar types of four-way interchanges. Their ramps are directional (or sometimes semi-directional), and hence, their use is very comfortable, with adequate sight distance. Despite these advantages, the construction of raised stack ramps and maintenance costs are higher; they do not usually permit U-turns.

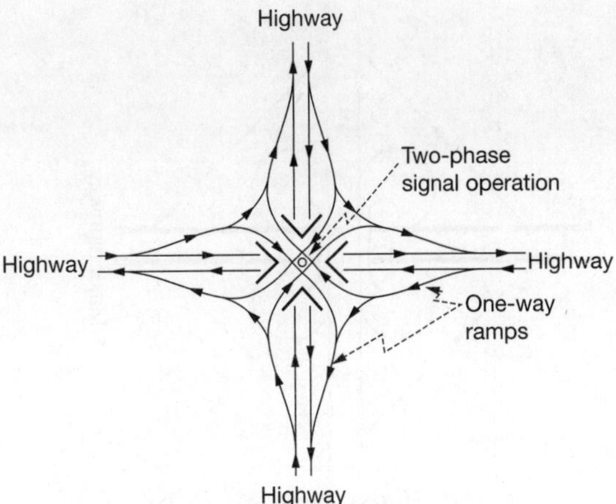

Figure 3.30 Three-level stack diamond interchange.

3.10.5 Cloverleaf Interchanges

A four-quadrant, full cloverleaf has eight ramps (four direct ramps and four loop ramps) which are used to eliminate right-turning conflicts through the construction of a two-level interchange. Cloverleaf interchanges are better suited to rural areas with low turning volumes (Fig. 3.31).

A full cloverleaf *without* C–D roads is used only when weaving volumes on through roads are small. It was observed in a study that as the number of weaving vehicles approached 1000 veh/hr, a significant reduction in speed took place, and hence, collector–distributor (C–D) roads (Fig. 3.32) should be used (Garber and Fontaine, 1999).

A few advantages and disadvantages of full cloverleaf are listed in Table 3.5.

Table 3.5 Advantages and disadvantages of full cloverleaf.

Advantages	Disadvantages
✓ All movements are free flow ✓ No right-turn movements on ramps or cross roads ✓ Weaving problem can be solved by providing C–D roads ✓ Provide higher capacity than most diamonds since movements at the ramp terminal are usually free flow by indirect loop ramps ✓ Considered more appropriate than diamonds where traffic volumes are relatively high	✗ Large area of right-of-way required ✗ Free flowing exits may pose problems at adjacent limited access control at-grade intersections ✗ All are indirect loop ramps and may offer limited capacity/restricted capacity in the case of significant truck traffic ✗ Indirection in movement ✗ There will be a change of tight-weaving on the mainline if there are no C–D roads ✗ Additional cost of C–D roads ✗ Bicycle and pedestrian movement along crossings is difficult ✗ Two entrances and two exits ✗ Higher construction cost for right-turning vehicles than diamonds ✗ The travel distance may increase

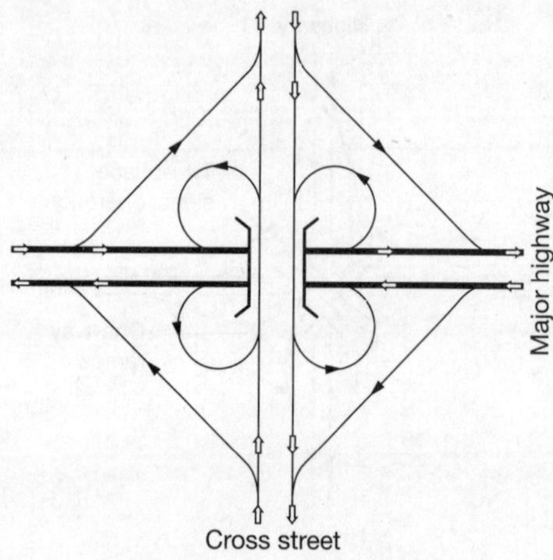

Figure 3.31 Cloverleaf or full cloverleaf (Modified based on Pinnell et al., 1963).

Figure 3.31 (a) A four-leaf clover: An inspiration for interchange configuration. (*Source:* http://desktopbackgroundshq.com)

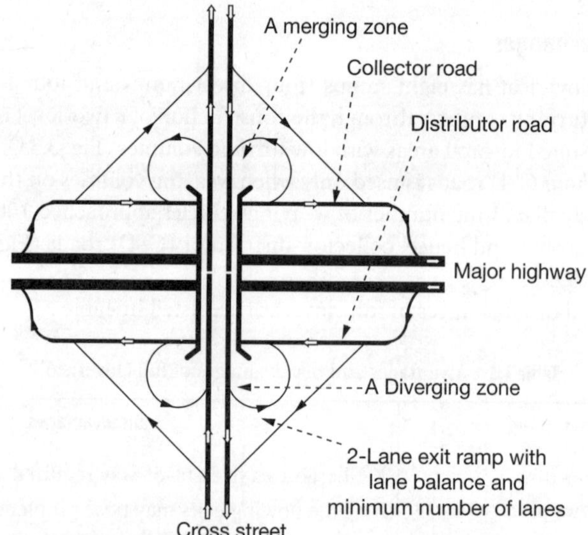

Figure 3.32 Full cloverleaf with collector–distributor roads (Modified and elaborated based on Pinnell et al., 1963).

3.10.6 Parclo (Partial Cloverleaf Interchange)

Cloverleaf interchanges that provide a *limited* number of movements with loops in one, two or three quadrants are referred to as partial interchanges or partially-directional interchanges or parclo. These types of interchanges are used when unavoidable property/right-of-way restrictions preclude ramps in one or more quadrants, and where right-turn/left-turn volumes are *low* or there is a significant impact on the environment due to artificial disturbances. Several examples of Parclos are illustrated in Fig. 3.24 (h), Fig. 3.33 and Fig. 3.34.

Signalised delays observed with parclo will be less than with diamonds or SPUIs for all entering volumes. This may be because large right-turning traffic flow is handled by ramps and therefore not influenced by phase/signal timings (Garber and Fontaine, 1999).

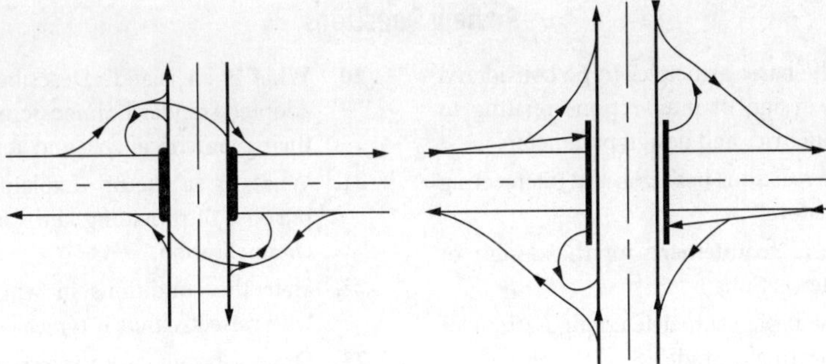

Figure 3.33 Variants of partial cloverleaf (*Note*: Changed for left-hand traffic). (*Source*: CDOT, 2005; *Courtesy*: www.codot.gov)

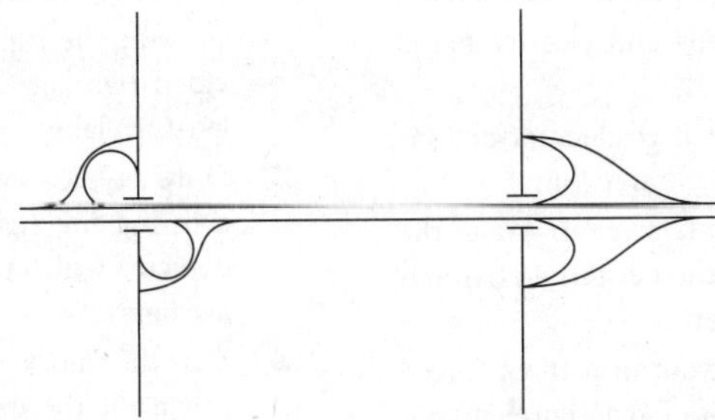

Figure 3.34 Partial cloverleaf. (*Note*: Changed for left-hand traffic). (*Source*: WSDOT, 2014; *Courtesy*: www.wsdot.wa.gov)

It is preferred that the movements provided in a partial interchange should have the facility to provide the corresponding return movement also, because the driver may expect to *retrace* his route in the return direction on a particular trip.

A composite listing of a few advantages and disadvantages of parclo is presented in Table 3.6.

Table 3.6 Advantages and disadvantages of parclo.

Advantages	Disadvantages
✓ Appropriate when one or more quadrants pose adverse terrain/right-of-way problems that preclude the use of partial diamond or full diamond or full cloverleaf interchanges ✓ Good design for unbalanced traffic movements ✓ Can eliminate or increase weaving distances on major roads, as with full cloverleaf with C–D roads ✓ The loops in the opposite quadrants can eliminate the weaving problem associated with full clover ✓ Can accommodate heavy right-turn traffic by indirect ramp loops and thereby improve operational and safety conditions ✓ Overall width of the interchange structure can be reduced in terms of the number and size of the loops ✓ Easily convertible to full cloverleaf or compliant for upgradation	✗ Right turns are made off the cross-road ✗ Channelisation is required at the cross-road ✗ At-grade intersections may require three-phase signals ✗ Chances of wrong-way movements at these cross-roads

Review Questions

1. What are the basic principles to be considered during the design of intersections relating to general geometrics and design parameters?
2. Classify intersections based on the intersecting road surface level.
3. List the basic requirements of the design of at-grade intersections.
4. Describe the basic geometric configurations of roads intersecting at grade.
5. What is skew intersection? Draw typical sketches showing left- and right-skewed intersections with negative and positive skew angles.
6. Define flared intersections and classify flared at-grade intersections.
7. Write the classification of at-grade intersections based on channelisation of traffic flow.
8. What is a channelised intersection? Write the functional objectives of the design of channelisation of road intersections.
9. Draw a typical sketch layout of a three-legged intersection with channelisation design elements.
10. What are the factors considered for the design of intersections relating to (a) Driver/Pedestrian factors (b) Road geometric factors (c) Traffic and vehicular factors (d) Safety issues (e) Environmental factors (f) Economic factors?
11. Explain the design principles of at-grade intersections relating to (a) Design vehicle (b) Stopping sight distance (c) Sight distance triangles at intersections.
12. What is a sight triangle? Describe the different types of sight triangles at an intersection.
13. Draw typical layouts showing the approach clear sight triangles at intersections with uncontrolled or yield-controlled minor roads.
14. Draw typical layouts showing the departure clear sight triangles at intersections with stop-controlled minor roads.
15. Explain how sight distance at medians is determined, with a neat sketch.
16. Write the basic functionalities of providing islands as channelisation guides.
17. Write the principles of channelisation.
18. Write the guidelines considered while designing a channelised intersection.
19. What are the elements of channelisation? State the standards that are most widely adopted.
20. What is an island? Describe the most widely adopted standard dimensions with reference to their geometrical area and kerb height.
21. What is a kerb? Explain the objectives/purpose of providing kerbs as a component of channelisation.
22. State the conditions in which channelisation with raised islands is typically omitted.
23. Draw a layout of an intersection showing the elements of median openings.
24. What are tapers and transitions and where are they used? Also draw a typical layout sketch showing the right-turn lane added to an undivided two-lane highway with fully shadowed left-turn lane.
25. Draw a typical layout of an at-grade intersection showing the right-turn lane within a median depicted with a taper bay, deceleration and storage lanes.
26. State the values of human factors that are appropriate for the design of intersections. Explain how the effect of a descending gradient on the SSD of a minor road is more critical than that on a major road intersection.
27. State and explain the three main functions of channelising islands. (*Hint*: Direction. Division and Refuge)
28. Define traffic island and state the functions of traffic islands.
29. State the situations where grade-separated interchanges are appropriate.
30. State the differences between grade separation and interchange. (*Hint*: Traffic may be interrupted in the case of grade separation but never with an interchange)
31. State in what situations it is appropriate to provide interchanges.
32. State the basic requirements of grade-separated intersections.
33. Write the advantages and disadvantages of grade-separated intersection.
34. What are the factors affecting the selection of an appropriate configuration of interchange?
35. Write the advantages and disadvantages of overpass structures.
36. Write the advantages and disadvantages of underpass structures.

37. Classify and detail the constraints considered for the selection of a suitable interchange configuration.
38. Explain the cardinal principles of interchange design. Write with suitable sketches.
39. Explain the secondary principles of interchange design. Write with suitable sketches.
40. Describe the classification of interchanges based on the directional type. Draw a few examples of directional and non-directional interchange layouts.
41. Classify the ramps based on their configuration with sketches for each case. Also explain their suitability and limitations with reference to the Indian practice of left-side driving?
42. What is a trumpet? Draw sketches showing configurations of three-leg intersections. Also state the advantages and advantages of trumpet interchanges.
43. What is meant by directional three-leg interchanges? Draw any three configuration variants of Y-interchanges with directional ramps and T-interchanges with semi-directional and directional ramps.
44. Write the advantages and disadvantages of diamond interchanges.
45. What is a three-level diamond interchange? Describe it with a configuration.
46. List any five variants of diamond interchanges and describe their operational features with an appropriate configuration.
47. Draw the configurations of these diamond interchange variants: (a) Split single-point/ Split tight-diamond interchange (b) Folded diamond interchange (c) Split-diamond interchange (d) Dogbone diamond interchange (e) Diamond with staggered intersections
48. Draw a sketch showing the configuration of a three-level stack diamond interchange and write its operational advantage.
49. Describe the various forms of three-leg interchanges and state the advantages and disadvantages of trumpet interchange.
50. Describe the various forms of Y-interchanges and T-interchanges that are commonly in use.
51. Classify diamond interchanges based on the distance between the intersections. Draw their sketches and explain their operating principles.
52. Describe with a sketch how C–D roads can be considered as a basic design element of a clover leaf interchange.
53. Write the advantages and disadvantages of each basic type of the following interchanges:
    (a) Diamond (b) Parclo
    (c) Cloverleaf (d) Trumpet
    (e) Directional
54. Draw a configuration showing a full cloverleaf with C–D roads and label their component road parts.
55. Draw the configuration of any two variants of partial clover leafs and state their uses.

Multiple Choice Questions

1. As per IRC: 3, the maximum width of vehicles permitted in India is ________ m.
   (a) 2.4 (b) 2.5
   (c) 3.0 (d) 3.2
2. Match List-I (Unit) with List-II (Purpose) and select the correct answer using the codes given below the lists:

List-I	List-II
A. Trumpet	1. information sign
B. Traffic studies	2. Traffic regulatory sign
C. Rotary island	3. Traffic flow characteristics
D. Road intersection approach sign	4. Grade separation

	A	B	C	D
(a)	1	4	3	2
(b)	2	3	1	2
(c)	3	1	2	4
(d)	4	3	2	1

3. The interchange having with the following ramp is safe and less hazardous for the Indian traffic rules:
   (a) Direct ramp
   (b) Indirect ramp
   (c) Semi-direct ramp
   (d) Indirect ramp and Semi-direct ramp.
4. The appropriate value of entry approach angle should be provided to rotary intersection so as to ensure adequate sight distance and visibility

(a) between 20^0 and 40^0
(b) $< 20^0$
(c) $>100^0$
(d) between 60^0 and 90^0

5. The following condition is favourable to design roads intersecting at grade as a roundabout/ rotary:
(a) The number of intersecting roads is between 4 and 7
(b) Traffic volume in the range: 4000–5000 veh/hr
(c) At intersections with less delays
(d) Traffic volume is less than 400 veh/hr.

6. What would be the capacity of the weaving section of a rotary having length 45 m, width 13.9 m, average entry width 10.4 m and the proportion of weaving section to the total traffic 0.653.
(a) 4067.3 PCU/hr
(b) 4263.5 PCU/hr
(c) 4380.3 PCU/hr
(d) 4277.2 PCU/hr

7. Consider the following parameters related to a roundabout intersection:
A. Width of weaving section
B. Length of the weaving section
C. Proportion of weaving traffic
D. Weaving angle
E. Width of carriageway at entry.
Which of the above parameters is not considered for calculation of capacity of a roundabout?
(a) A (b) B (c) D (d) E

8. As for Indian traffic-flow rules, this ramp is safe and less hazardous:
(a) Semi-direct ramp
(b) Direct ramp
(c) Indirect ramp
(d) U-turn ramp

9. As per IRC: SP: 90, it appropriate to provide flyovers when the intersection traffic volumes approach ______ PCU/hr.
(a) 5000 (b) 7000
(c) 8000 (d) 10000

10. As a common guideline of route continuity, the design speed differential between the main lines and the interchange ramp should be smaller than ______ of the main line design speed.
(a) 0.3 times (b) 0.1 times
(c) 0.01 times (d) 1.0 times

Answers to Multiple Choice Questions

1. (b) 2. (d) 3. (b) 4. (a) 5. (a)

6. (a) $$Q_{N\to E} = \frac{280w\left(1+\frac{e}{w}\right)\left(1-\frac{p}{3}\right)}{\left(1+\frac{w}{L}\right)} = \frac{280\times13.9\times\left(1+\frac{10.4}{13.9}\right)\times\left(1-\frac{0.653}{3}\right)}{\left(1+\frac{13.9}{45}\right)} = 4067.3 \text{ PCU/hr}$$

7. (c) 8. (c) 9. (d) 10. (a)

CHAPTER 4

Traffic Engineering and Control

4.0 INTRODUCTION

Traffic is defined as pedestrians, bicyclists, ridden or herded animals, all types of motorised vehicles, tramcars and other conveyances, either singly or together, using any highway for the purpose of travel (FHWA, 2003).

Traffic Engineering is a part of Transportation Engineering that uses engineering techniques to promote safety and efficiency in the *orderly movement* of all road users (i.e., vehicular traffic and pedestrians). The objective of traffic engineering is to provide efficient and safe flow of traffic in all weather conditions on different terrains that have well-designed roads, geometric components and their appurtenances.

4.1 TRAFFIC PROBLEMS

As the volume of traffic continues to increase, the resulting traffic demand of the present and future years should be adequately accommodated on existing as well as new highway systems – a challenging task for traffic engineers and transport planners. Provision for such timely traffic facilities is key to the success of any traffic operation and control system. The following possibilities are considered under the broad scope of road traffic engineering and may be explored to ensure systematic development of traffic facilities and infrastructure needs:

- Traffic control and road management measures to cater to present and future needs
- Planning of new roadway and highway development projects to meet the forecasted traffic needs
- Allocation of budgetary and non-budgetary funds, with appropriate public–private partnership (PPP) models, for sustainable highway development
- Enforcement, control, management, administration and safety measures of traffic, including vehicles and road users, for healthy and eco-friendly traffic flow condition
- Establishment of dedicated and non-profit traffic research institutions at local, regional and national levels for updating and revising the knowledge on related technological aspects

4.2 ROAD USER AND VEHICLE CHARACTERISTICS

The main objective of the design of traffic facilities is to ensure orderly and systematic movement of traffic within safety and economic considerations. To achieve this objective, there should be coordination among the components involved. The **main components** of traffic engineering are listed below:

- Drivers
- Pedestrians
- Vehicles
- Roads
- Traffic control devices
- The general environment

The characteristics of each of the above components affect the performance of any transportation system, and the traffic engineer should have an understanding of the local/regional problems and a sound knowledge of how these factors interact with each other.

We will now examine in detail the characteristics of these components and their relevance with respect to the performance of traffic operations.

4.3 ROAD USER

Road users can be broadly classified into three types: drivers, pedestrians and passengers (Fig. 4.1). Regardless of their type, they come together to form a complex human element – one that displays large variations in performance while interacting with the transport environment. This is of crucial concern to traffic engineers in their quest for optimal design of facilities.

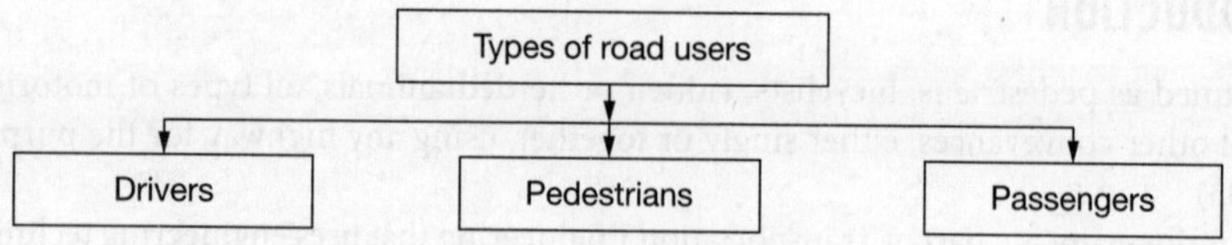

Figure 4.1 Types of road users (humans).

According to a survey (www.visualexpert.com), over 90% of road accidents are caused fully or in part by human error. Of these, 90% of accidents are caused by perception error and the remaining 10% by response error. Accidents are also caused due to inadequate highway visibility conditions. Therefore, knowledge of human performance and their capabilities and behavioural characteristics in traffic flow operations is a vital input for the design of the various geometric elements of roadways.

The internal and external factors that affect general driver behaviour are presented in Fig. 4.2.

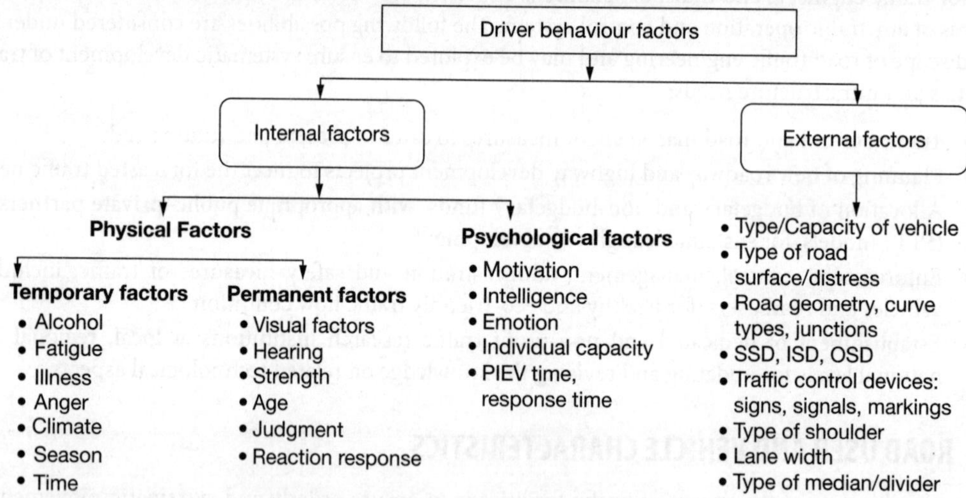

Figure 4.2 Behaviour factors affecting driver performance.

4.3.1 Road User Characteristics

The factors affecting road user characteristics may broadly be classified under five heads:

- Physical characteristics
- Mental characteristics
- Psychological factors
- Environmental factors
- Pedestrian characteristics

These factors are briefed below:

Physical characteristics

The physical characteristics that are required of a road user are vision, sense of touch, hearing and strength.

1. **Vision:** Visual factors affect the ability of a road user to detect objects, receive information, and process and identify objects. Vision depends on the following abilities of the eyes.
   - **Visual acuity or clearness of vision:** It is the ability of the eyes to see the fine details of an object under average illumination conditions. Visual acuity is the one of the important factors in the placement of traffic information devices such as roadside signs. Visual acuity is of two types: static and dynamic. The static acuity relates to stationary objects and the dynamic acuity relates to moving objects. Clarity of vision depends on the cone angle of vision (Fig. 4.3) and the speed of the vehicle-driver under the given conditions of illumination. Some cone angles of vision and their corresponding clarity of vision level are presented in Table 4.1.

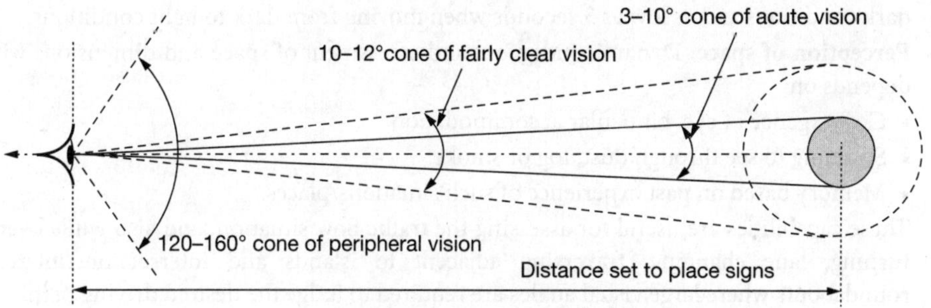

Figure 4.3 Field of vision.

Table 4.1 Cone angles and clarity of vision.

Cone angle of vision	Clarity of vision
3–5°	Very clear vision; legend can be read within this cone
5–7°	Clear and good vision
10–12°	Satisfactory vision; object shapes/colour can be identified
>12°	Blurred vision starts; movement of objects can be detected

   - **Peripheral vision:** This is the field of view beyond the cone of clearest vision (Fig. 4.3). In this field of view, an individual can only see objects without clarity of details or colour. The cone of vision can provide a warning to the driver and act as a safety measure. In static conditions, the cone angle of peripheral vision ranges between 120° and 160°; the cone angle reduces with increase in speed. For example, the peripheral vision cone angle may reduce from 100° to 40° as the driving speed increases from 30 km/hr to 90 km/hr.
   - **Eye movement and visual concentration:** The movement of the eyes and the time taken to focus on different objects may be categorised, as shown in Table 4.2.

Table 4.2 Eye movement and time lapsed.

Eye movement	Time taken
Fixing of eyes on objects	0.10–0.30 seconds
Jumping of eyes	0.12–0.23 seconds
Shifting of eyes (converging and diverging)	0.30–0.50 seconds

- **Colour vision:** As we know, different colours have different wavelengths and thus relative visibility. At very low levels of illumination, colour sensation may disappear altogether. Colour blindness is another factor which may be due to defective vision.
- **Glare vision and recovery:** Glare vision results in decrease in the seeing ability of a driver and can thus cause discomfort to the driver. This depends on the eye's adaptability to changes in intensity of light. Glare recovery is the time duration in which a driver recovers from the effect of glare, just after passing a light source. It may as high as 6 seconds when moving from light to dark conditions and as low as 3 seconds when moving from dark to light conditions.
- **Perception of space:** Dynamic judgment and assessment of space and dimensions with time depends on:
  - Convergence of eye, binocular accommodation
  - Straining to see through dust, fog or smoke
  - Memory based on past experience of such situations/places.

  These capabilities are useful for assessing the traffic flow situation, and also while overtaking, turning, lane changing, traversing adjacent to islands and intersections/interchanges/roundabouts where large visual angles are required to judge the desired driving path.

2. **Hearing:** Sound is the most important factor by which the drivers and pedestrians move away from a situation, with the horn (or sounds) acting as a warning. In addition, hearing is also important as it lets the road user know when to move away from emergency vehicles, which indicate their presence with a siren.
3. **Vibrations:** Raised markers on pavements are used to indicate/warn the drivers about the prohibited path along the edge of a carriageway or just before the crossing of an uncontrolled intersection, so that drivers reduce their speed and take the necessary action.
4. **Strength:** Lack of strength may affect the prompt functioning of the sense organs and also affect parking manoeuvres, especially while driving heavy trucks. A strong pedestrian can walk fast, depending on the situation, whereas a weak pedestrian may not be able to do so. The overall health condition of road user is also a consideration.

Mental characteristics

Level of literacy, knowledge of vehicle characteristics and traffic rules/signs, driving/walking skills and the ability to understand the surroundings based on intelligent analysis of each situation on the road can influence the performance of the road user.

Psychological/Behavioural factors

Factors such as attentiveness, impatience, anxiety, anger, fear and the attitude of the road user may affect traffic performance to great extent. The willingness to allow other drivers to overtake and to give way to other vehicles and pedestrians are some of the behavioural factors of road users in a traffic stream.

Environmental factors

Rain, fog, snow, mist, dust, sun radiation, cold weather, wind intensity, night visibility and surface water drainage conditions also affect the performance of road users.

Pedestrian characteristics

Pedestrians cross the streets at mid-block (i.e., across a straight section length of a road) locations and at intersections. Walking speed is the most critical input parameter in the design of pedestrian crossings at intersections. The intersections may be signalised or un-signalised. The 85th percentile of walking speed is used for the design of signal timing at signalised intersections. The gap acceptance and time-gap accepting/rejecting behaviour of pedestrians is considered in the design of un-signalised intersections. A study by the Institute of Transportation Engineers (ITE) found that the 85th percentile space gap was 38 m.

As large variations exist in the walking speed of pedestrians with reference to age, gender, physical fitness and location, the 85th percentile speed of the sampled pedestrian walking speed in the locality is used as the pedestrian walking speed for design purposes. In the absence of such data, a walking speed of 1.2 m/sec may be assumed for (a) design of timing signal and (b) location planning of off-street parking facilities which are adjacent to the central business district (CBD). The 15th percentile speed of 0.95 m/sec is used for the design of pedestrian crossing facilities in India (IRC: 103-2013).

4.3.2 Reaction Process

Perception–Reaction Time (PRT)

It is the time taken by a driver to initiate a physical response on detection of an event. The driver performs four distinct processes in sequence: detection, identification, decision and response. The factors affecting PRT are age, fatigue level, complexity of the situation and the presence of alcohol/drugs in the bloodstream of the road user.

The average PRT value commonly suggested for sight distance design purposes is 2.5 seconds (based on the 90th percentile criterion by the American Association of State Highway and Transportation Officials [AASHTO]) and the PRT value for driver response to signal timing purpose is 1.0 seconds (based on the 85th percentile criterion by the Institute of Transportation Engineers).The 85th percentile represents the value that 85% of the population can meet or exceed.

Perception, identification, emotion and volition (PIEV)

The time required for this process depends on the complexity of the task, the level of expectancy and variability in the driver's capabilities. The PIEV process is as follows:

1. **Perception:** Stimulus of senses (eyes, ears) take place while perceiving (observing) the situation ahead.
2. **Identification:** Identification and understanding of the stimulus with speed alternatives develops.
3. **Emotion:** A decision/judgment is made about the best course of action.
4. **Volition:** Reaction based on the decision.

The PIEV time depends on the driver's physical and mental capabilities, fatigue level and the influence of alcohol or drugs. The PIEV distance is determined using the formula

$$\text{PIEV distance} = \text{Reaction distance} = \text{Speed (m/sec)} \times \text{PRT (sec)} \tag{4.1}$$

For example, assume that a driver is travelling on a curve at a speed of 40 km/hr and suddenly sees an animal crossing the roadway. The distance travelled by the vehicle before the driver's foot presses the brake is (0.278 × 40) m/sec × 2.5 sec = 27.8 m.

The concept of PIEV is useful for determining the following:

- Overtaking sight distance
- Stopping sight distance
- Decision sight distance
- Safe approach speeds at intersections and interchanges
- Design of signal timing
- Traffic signal change intervals

Note: The designer must consider two types of situations on any roadway operation, namely, the expected and the unexpected. The expectancies can be listed based on the situation, such as continuity and temporal structure of event. Average design parameters (say, the 85th percentile) or any site-specific input parameter value are selected based on the sample data collected. The traffic engineer must avoid designing unexpected events into roadways geometrics and traffic controls.

Design driver

Following are the indicative and average driver characteristics/capabilities chosen by the traffic designer, and such a driver is referred to as the design driver.

- PRT = 2.5 seconds (AASHTO)
- Height of the driver's eye level above the road surface = 1.2 m
- Comfortable deceleration rate = 3.0 m/sec^2 (ITE) or 3.4 m/sec^2 (AASHTO)
- Allowable jerk = 0.7 m/sec^3; standard values of lateral jerk range from 0.3 m/s^3 to 0.9 m/s^3 (AASHTO)

Note: The rate of change of acceleration, i.e., the third derivative of displacement, is known as jerk.

The critical gap value should be maintained between 4.0 seconds and 7.5 seconds, depending on the complexity of the manoeuvre, speed of opposing streams and the situation.

4.4 VEHICLE

The vehicles that operate on roadways can be broadly classified into two types: motorised and non-motorised (Fig. 4.4).

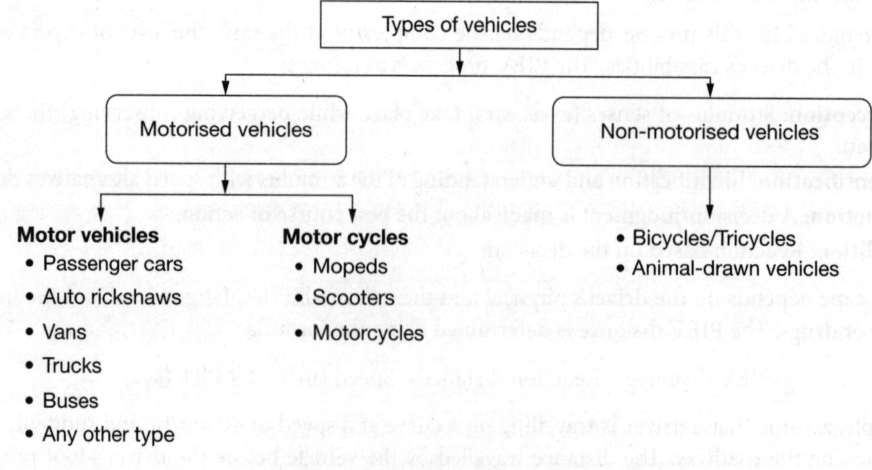

Figure 4.4 Classification of vehicles.

4.4.1 Vehicle Characteristics

Traffic engineers should have knowledge of vehicle performance characteristics because the geometric design of roadways must take into account and accommodate such vehicle-related inputs. The road design should cater to the needs of present and future anticipated vehicles.

The geometric design of highways is carried out by considering three types of characteristics: static, kinematic and dynamic. The physical and performance features of each are discussed below.

Static characteristics

The physical features and maximum permissible loads for different type of vehicles in India are given in Table 4.3.

Table 4.3 Maximum permissible gross weight of vehicles in India (from IRC: 3, 1983).

S. No.	Vehicle class	Vehicle type	Maximum gross weight (Tonnes)
1	Type 2	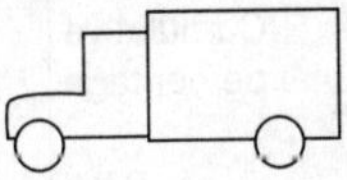	Both single tyres: 12 Front single tyre; rear dual tyre: 16.2
2	Type 3	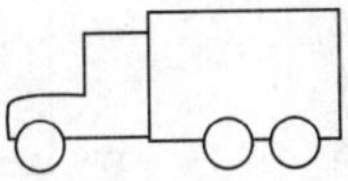	24
3	Type 2-S1	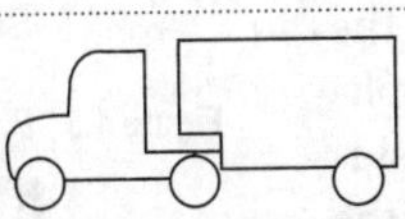	26.4
4	Type 2-S2	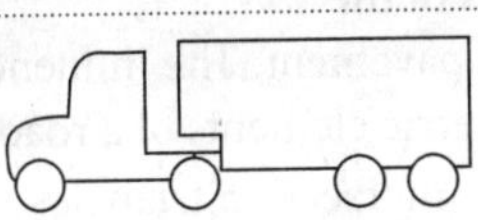	34.2
5	Type 3-S1	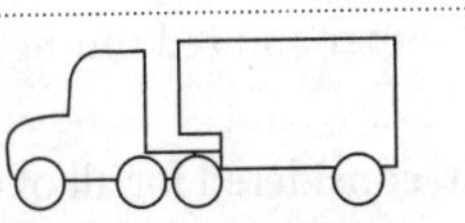	
6	Type 3-S2	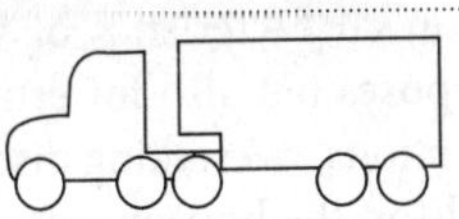	42
7	Type 2-2	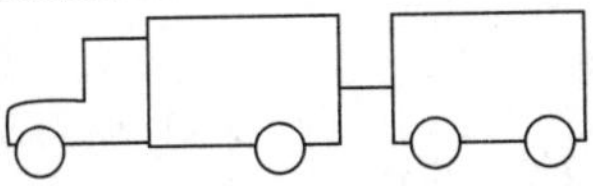	36.6
8	Type 3-2	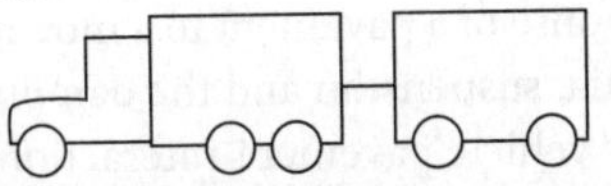	44.4
9	Type 2-3	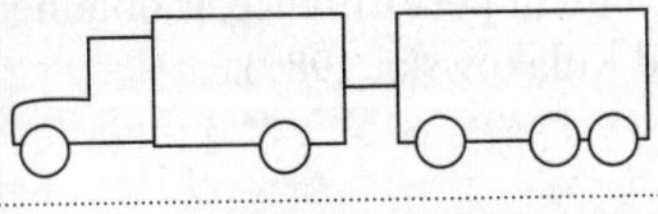	
10	Type 3-3	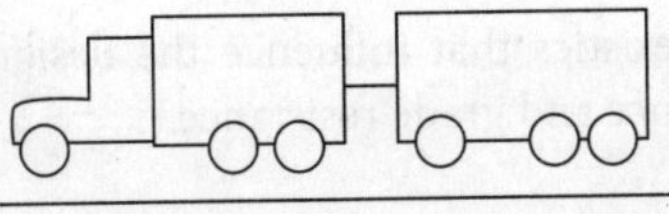	52.2

Vehicle dimensions

The maximum values of width, height, wheel-base distance, overhang and length of vehicles influence the selection of design criteria such as limiting of lane width, curb/curve radii with extra widening of carriageway, operating speeds over a road-section length, vertical clearance under a bridge, tunnel and under pass, provision of standard size of parking bays with turning radius and other such dimensions. The details of such factors and the permissible values which are applicable as per IRC specifications are presented in the Chapter 2.

Factors related to axle loads and vehicle gross loads: These details are discussed in Chapter 9.

Kinematic characteristics

1. **Speed limit:** Variability in the performance of drivers is the main factor in the operational functionality of any road facility. To take account of this variability, the speed limit of any road section is often fixed as the 85th percentile. This indicates that at least 85% of the user population is safely covered (Fig. 4.5).

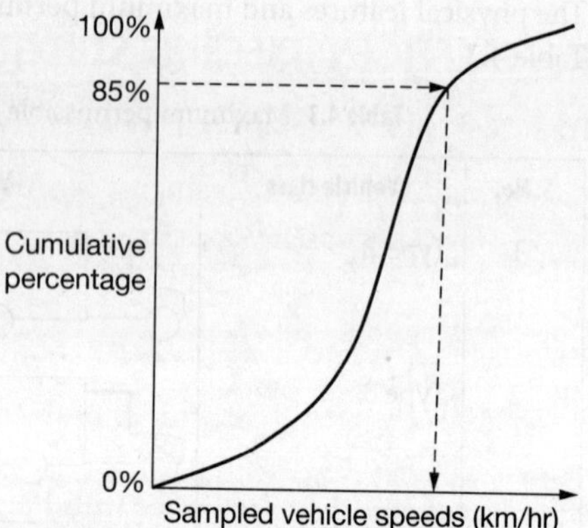

Figure 4.5 Typical presentation of traffic speed data.

2. **Braking performance:** The braking performance and deceleration characteristics of vehicles are very important factors in the safety evaluation of vehicles with roadway geometric design considerations. The time and distance travelled by vehicles after application of the brake is an important design consideration, which depends on the characteristics of the vehicle and the surface characteristics of the pavement. The influence of braking performance is considered for designing the following geometric elements of a roadway:
   - Safe stopping sight distance, including at intersection triangles
   - Safe clearance between vehicles, between amber and red timing
   - Placement of signs along a roadway

   Perception–response time (PRT) is, by default, considered for all of the above cases.
3. **Acceleration characteristics:** The acceleration characteristics of vehicles are not only useful for accident reconstruction/investigation purposes but also for estimating the time required to cross an intersection area at signalised intersections, overtaking distance and for gap acceptance purposes. The acceleration of vehicles depends on the horsepower (hp) and speed of the vehicle and also on grades.

Dynamic characteristics

Transient pavement response defines the response of a pavement to a moving tyre load input. The effects of operating vehicle speed, vehicle mass inertia, suspension and the damping system are considered for modelling/designing vehicle components for vehicle pavement interaction. Vehicle dynamics, indicating the influence of each component on pavement performance, is obtained by analytical, experimental and computer-based simulations (Todd and Kulakowski, 1989).

Other characteristics

Other factors relating to vehicle characteristics that influence the design elements of roadways are inertial force, air resistance, rolling resistance and grade resistance.

4.4.2 Lighting

The lighting characteristics of vehicle headlamps are useful for determining the minimum length of sag curves, and a few important standards are given below (IRC: SP: 23 1993).

- Height of headlamp above the road surface = 0.75 m
- Height of the driver's eye level above the road surface = 1.2 m
- Angle of spread of headlight or light cone focus angle = 1°
- Minimum height of the obstacle above the road surface = 0.15 m

4.5 TRAFFIC ENGINEERING STUDIES

Traffic data in different forms is collected for the following reasons:

- Evaluating and managing an existing road facility or traffic flow
- Upgrading of any existing road to a National Highway or widening it to two or multi-lane roadway
- Determining the rate of road distress due to traffic volume and for assessment of future needs of budget
- Assessing traffic delays and for determining level of service of existing traffic facilities such as at signalised intersection
- Assessing the effectiveness of improvements using before-and-after studies of traffic data. For example, the number of road accidents are often compared and justified with the investment
- Assessing traffic impact due to new facilities and future scope of development

4.5.1 Types of Studies

For the reasons and purposes mentioned above, the following types of studies are conducted in any selected area:

- **Traffic volume studies:** This is the most basic traffic study and the data collected is used for many purposes.
- **Speed studies:** The speed of vehicles is measured for the implementation of safe operation of traffic at all locations and in any climate.
- **Travel-time studies:** Journey speed with delays along any selected road segment is used to measure the quality of traffic flow service.
- **Delay studies:** These are considered a part of travel-time studies and deal with factors that the user finds particularly annoying or which cause delay/discomfort.
- **Density studies:** These studies are mostly carried out by indirect measurement.
- **Accident studies:** These involve investigation of accident characteristics, causal factors, specific locations, grey/black spots and preventive measures.
- **Parking studies:** These involves with the collection of data about inventory, parking supply/ demand and required parking facilities.
- **Goods movement and transit studies:** These involve the collection of data about truck volume, truck loading/unloading facilities and transit systems.

A list of traffic surveys that are widely carried out for the improvement/development of road projects are presented in Table 4.4.

Table 4.4 Summary of the main types of traffic surveys, with intended output.

Survey type	Required information/Data	Methods of survey*	Output
Road inventory	Road and network geometric characteristics	• Observation • Survey of India maps • Google maps	Geometric details, land use and road traffic furniture
Traffic volume	Traffic demand	• Manual counts • Automatic counts ▪ Intrusive ▪ Non-intrusive	Pedestrian and vehicular traffic volume, flows on links and across junctions, annual average daily traffic (AADT), traffic growth rate, peak hour factor (PHF) and other traffic factors
Spot-speed	Speed control, safety measures and vehicle speed performance on road links	• Manual measurement by short-base method • Radar measurement	85th percentile speed, space mean speed, median speed, modal speed, pace and percent values in pace
Road network speed and delay	Route performance	• Floating car method	Average and running speed on routes/networks, delay and congestion at points
Road junction speed and delay	Junction performance	• Stopped vehicle count • Elevated observer method	Approach delay, total delay, distribution of delay by turning movements and causes of delay
Saturation flow	Capacity of junction and design of signal time with phases	▪ Flow profile method ▪ Saturated period count	Saturation flow and capacity of junction
Origin and destination	Traffic demand forecasting	▪ Registration number methods ▪ Postcard method ▪ Home interview	Trip matrix, route choice and through-traffic travel time
Parking use	Parking supply and demand for space	▪ Parking patrol survey	Types of parking in use, space usage for parking and occupancy time
Accident report	Accident reduction/ Preventive measures	▪ Field investigations ▪ Statistical data collection	Causes and reasons for accidents, block spots

* Only a few methods are listed here. A detailed discussion is presented in subsequent chapters.

These road traffic studies and their objective details are briefly described in subsequent sections.

4.6 TRAFFIC VOLUME STUDIES

Traffic volume data is considered the most important input parameter of traffic streams. It is useful for ascertaining the importance of a route with respect to other routes, based on hourly variation, seasonal variation of traffic flow rate, distribution of traffic flow in a road network and its composition by vehicle type. Traffic flow data is collected for planning, developing and managing road networks for the following objectives (MWT 2004):

- To ascertain the necessity of road-widening and planning of new road network development
- To determine the general needs of structural and functional improvement, i.e., for overlay thickness design and thin overlays to meet roughness reduction requirements
- To ascertain the efficiency of the road network by comparing current traffic flow parameters with the level of service or with the calculated capacity
- To design signal systems, in particular, coordinated signal systems based on the directional movement of traffic at road junctions
- To enable prioritisation of maintenance work based on traffic volume, fund allocation for improvement schemes, designation of a road as a national highway and expressway with their related facilities
- To assess the economic benefits arising from road improvements
- To estimate the needs of parking facilities
- To estimate the possible demands of traffic produced due to planning of new roads in a new town and the associated land use
- To assess the impact of traffic due to proposed new shopping malls or any such traffic-attracting and -generating service stations
- To determine traffic warrants for traffic control and traffic management techniques, such as no-entry, one-way, stop signs, etc.
- To project the future traffic growth for a given period and for deciding the optimal time for maintenance activities
- To classify roads/road networks on their functional basis
- To relate road accidents with the traffic volume and predominant vehicle
- To evaluate performance of pavements based on traffic volume variability with seasonal and weather factors
- To establish thc cconomic and social implications of design of infrastructure facilities and, as part of their feasibility studies, of all development projects countrywide or region wide.
- To use weigh-in-motion (WIM) data relating to total tonnage moved on roads to estimate the freight travelling on each roadway. Such input is useful for determining the costs of congestion and the resulting benefits due to new construction and proposed operating strategies (FHWA 2012).

4.6.1 Types of Traffic Counts

As a broad classification, the traffic flow on roads may be categorised as vehicular traffic and pedestrian traffic (Fig. 4.6). The summary of data obtained from the respective traffic volume counts are used for specific projects or for the overall development/improvement of a road transportation facility.

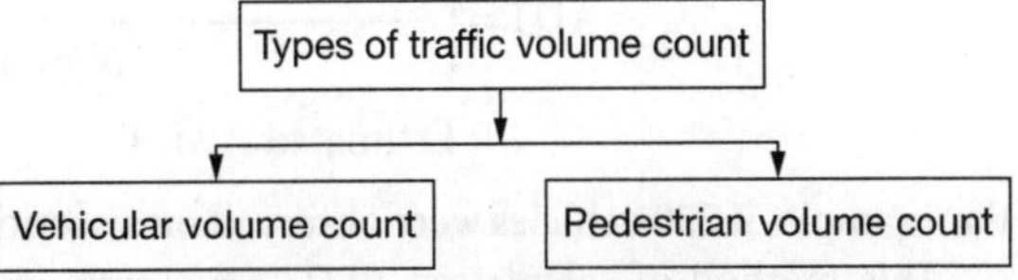

Figure 4.6 Types of traffic volume counts.

Vehicular Traffic Volume (or Flow) Survey

Traffic volume (VOL) or flow (Q) is defined as the number of vehicles (n) passing across a section of a road, or a lane (Fig. 4.7), or a direction of a two-way highway, in a specified time interval (t), say, 1 hour or 15 minutes or 24 hours. In general, the volume (Q) is expressed as vehicles/hour (veh/h) or veh/day or veh/yr. This concept is also applicable for the counting of non-motorised vehicles.

$$Q = VOL = \frac{n}{t} \tag{4.1}$$

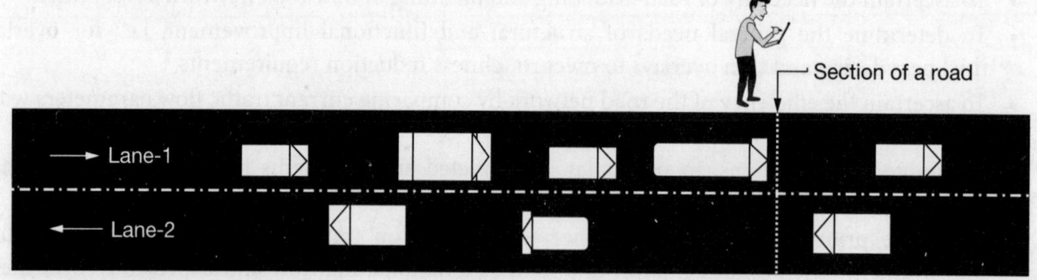

Figure 4.7 Traffic volume measurement: Vehicles crossing over a section of a road per lane.

The most widely used parameters that can be derived from traffic count data are described below:

Average Annual Daily Traffic (AADT) The average of 24-hour traffic volume observed over a full year (i.e., 365 days) at a location is considered. It is an estimate of the mean daily traffic on a road segment on all the days of the week (i.e., Monday through Sunday) over a period of one year. It is the most widely and frequently used traffic data statistic. AADT data is based on site-specific counts. It is the rate flow that indicates how busy a road is. It is expressed as the total number of vehicles passing a section of a road in a year divided by 365. It is a simple average volume of 365 days, except for leap years which have 366 days.

$$\text{AADT} = \left(\frac{\text{Total of average daily volume of traffic in a year}}{365\text{ days}}\right) \tag{4.2}$$

or

$$\text{AADT} = \frac{\sum_{i=1}^{365} VOL_{24\text{hour}}}{365} \tag{4.3}$$

Average Daily Traffic (ADT) The average 24-hour traffic volume observed over a period of less than a year (say N number of days) at a location is considered. It is obtained by a short-duration traffic count, and the data is based on site-specific counts. The traffic on a section of road measured during the N number of days may be considered with reference to a season, festival days, months, a week or say 3 days. The calculated value is reported as the ADT for any such period of the year. For this reason, the ADT is not considered as the standard measure of vehicular traffic flow on a section of road and is not the basis for most decisions related to transport projects.

$$\text{ADT} = \left(\frac{\text{Total daily volume of traffic in } N \text{ number of days}}{N \text{ number of days}}\right) \tag{4.4}$$

$$= \text{Estimated AADT}$$

For example, if 750 vehicles were observed on a section of road over three days, the ADT = (750/3) = 250.

This method of calculating ADT is a simple average of all the days measured. The disadvantage of this method is that the missing data can cause inaccuracies in the determination of true AADT. The resulting bias in the AADT will be large when an unequal number of week day or weekend days are missing or a considerable number of days is missing from monthly or yearly traffic volume data.

To overcome this effect, AASHTO adopted a different approach for calculating AADT. In this approach, first the average monthly days of the week are computed and these 84 values (i.e., 7 weeks × 12 months = 84) are then averaged to yield the AADT. The AASHTO (1992) recommended equation is as follows:

$$\text{AADT} = \frac{1}{7}\sum_{i=1}^{7}\left\{\frac{1}{12}\sum_{j=1}^{12}\left(\frac{1}{n}\sum_{k=1}^{n}\text{VOL}_{\text{ijk}}\right)\right\} \tag{4.5}$$

where

VOL = daily traffic for day k, of week i and month j
i = day of the week
j = month of the year
k = 1, when the day is the first occurrence of that day of the week in a month; 4 when it is the fourth day of the week
n = number of days of that day of the week in month j

The vehicle flow across a road section is random. Due to this reason, fluctuations in traffic volume are observed on any section of road with time. The fluctuations in flow can be observed with reference to seasons, days or months. The maximum traffic volume observed in an hour of a day is termed **peak hour traffic,** which is usually up to 10% of the total daily traffic flow, or 2–3 times the average hourly volume observed on that day.

Peak Hour Factor (PHF) It is a ratio of the average traffic volume during the peak-hour period to four times the average volume during the peak 15-minute period.

PHF represents traffic flow variation within the peak hour. The maximum value of PHF is 1.00 and the minimum value is 0.25. The minimum value would occur if the total volume observed in the peak hour is found in only one of the 15-minute intervals. Typically, the PHF value will fall in a range of 0.70 to 0.98. The higher the value of PHF, the lower will be the degree of variation in volume during the peak hour, and vice-versa. As the value of PHF is ≤0.50, it represents a non-homogeneous traffic flow condition within the peak hour. The following expression may be used to calculate PHF:

$$\text{PHF} = \left(\frac{\text{Average volume during peak hour}}{4 \times \text{Average volume during peak 15-minute period}}\right) \tag{4.6}$$

For example, the traffic volume data obtained in a peak hour at a location is given below. Calculate the PHF.

Time duration within an hour	Traffic volume (number of vehicles)
10:00–10:15	950
10:15–10:30	1010
10:30–10:45	1050
10:45–11:00	990
Total =	4000

$$\text{PHF} = \left(\frac{4000}{4 \times 1050}\right) = 0.95$$

Design Hour Volume (DHV) It is the traffic volume that is estimated to use a highway segment during the design hour (DH) of the design year. It is an adopted volume of traffic which is obtained below peak and at nth highest hour (Fig. 4.10). Usually, the 30th highest hourly volume is adopted as the DHV because the design of facilities for maximum volume of traffic observed in a year is uneconomical. It means this hourly volume will be exceeded only 29 times in a year. DHV can be obtained using the following equation:

$$DHV = AADT \times K \text{ for the design year} \tag{4.7}$$

For example, it can be observed from Fig. 4.8 that the variations in hourly volume up to the 50th highest hourly volumes in a year are less on a national highway than on a recreational rural route, where seasonal variations are predominant.

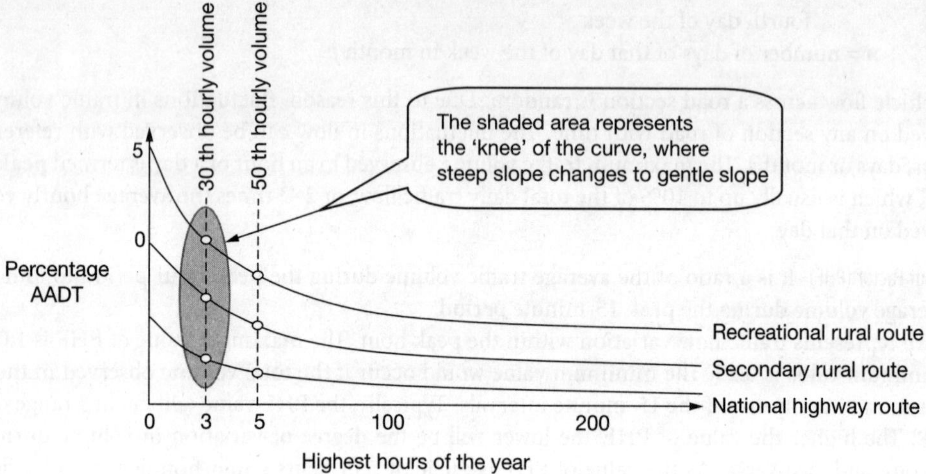

Figure 4.8 Typical relationships between hourly volume and AADT.

K-**Factor or Design Hour Factor** It is the ratio between traffic volume in the study hour and AADT. The K-factor represents the proportion of AADT that is expected to occur in the design hour. The routes that are subjected to occasional extreme traffic volumes generally exhibit the highest K-factors. The K-factor is useful for corridor traffic forecasting and project traffic forecasting projects because it defines the volume of traffic that the road is designed to handle (FDOT, 2012).

Design hour volume (DHV) may be considered as the traffic volume in the study hour, and the calculated factor is termed the design hour factor *K*. The relationship between DHV, AADT and *K* is expressed below

$$K = \frac{\text{DHV}}{\text{AADT}}$$

Pedestrian traffic volume survey

The number of persons passing across a section of a road during a specified period of time is often reported as **pedestrian traffic volume**. In general, pedestrian traffic volume is expressed as Ped/min, Ped/hr, Ped/day, or Ped/yr. It is known that pedestrian traffic volume changes depending on several factors, such as time of day, holidays, accidents, occasional events, road closure and weather condition.

Types of traffic counting methods

Counting of traffic flow across a given section of road can be carried out using different methods, which may be broadly categorised as traditional (using intrusive devices) and non-traditional (using non-intrusive devices) (Fig. 4.9). These methods are explained briefly below.

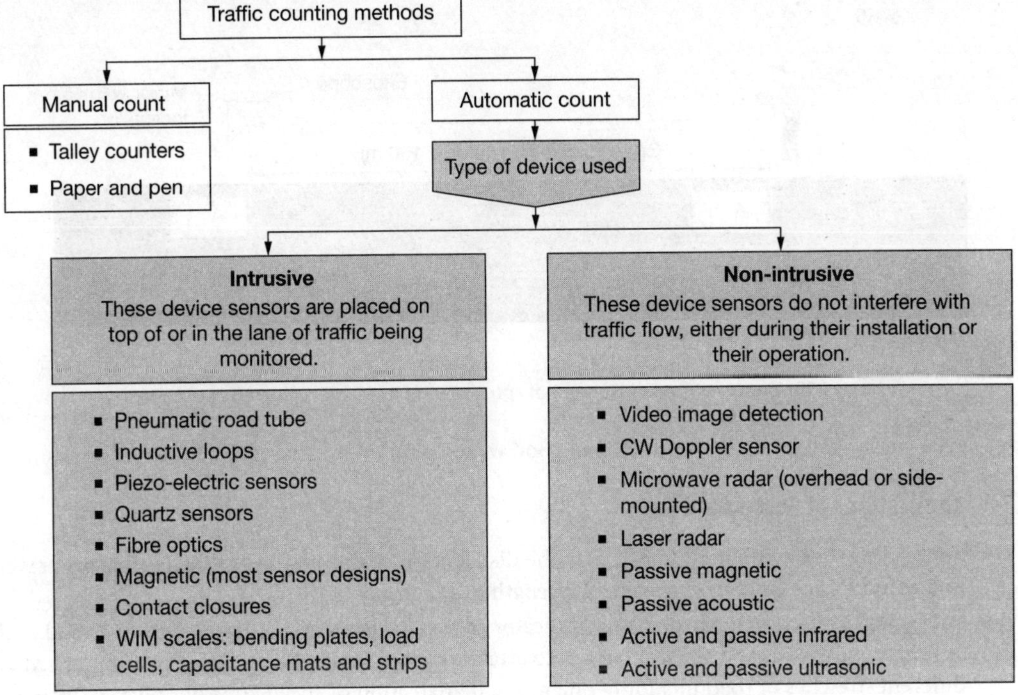

Figure 4.9 Traffic counting methods and the types of devices used.

4.7 SPOT-SPEED STUDY

Speed is the rate of motion of a vehicle which is expressed as the distance travelled in metres per unit time. Speed or velocity (V) is usually denoted in terms of kilometres per hour (km/h) or metres per second (m/s). The speed of vehicles is variable and is subject to *time* and *space*. The different types of speeds used to represent these variations are discussed below.

Spot-speed is the *instantaneous speed* of a vehicle measured over a specific distance (often <100 m) along a road. Spot-speed can be measured using the following methods:

- An enoscope and a stopwatch timing procedure (Fig. 4.10)
- Intrusive sensors
- A radar speedometer
- Time-lapse photographic or videographic methods
- A graphic recorder

The basic principle involved in these methods is to measure the time taken by a vehicle to travel between any two predetermined markings made along the road. The spot-speed is then determined by dividing the measured distance between the markings by the elapsed time measured. It is expressed in terms of metres per second (m/s) or kilometres per hour (km/h). The distance between such markings along a carriageway may be generally taken as a plain and straight length between 30 and 100 m. According to a TRL (1993) guideline, the following distances may be considered with the average speed of traffic:

Average speed of traffic (km/hr)	Distance between the markings (m)
< 40	25
40–65	50
> 65	75

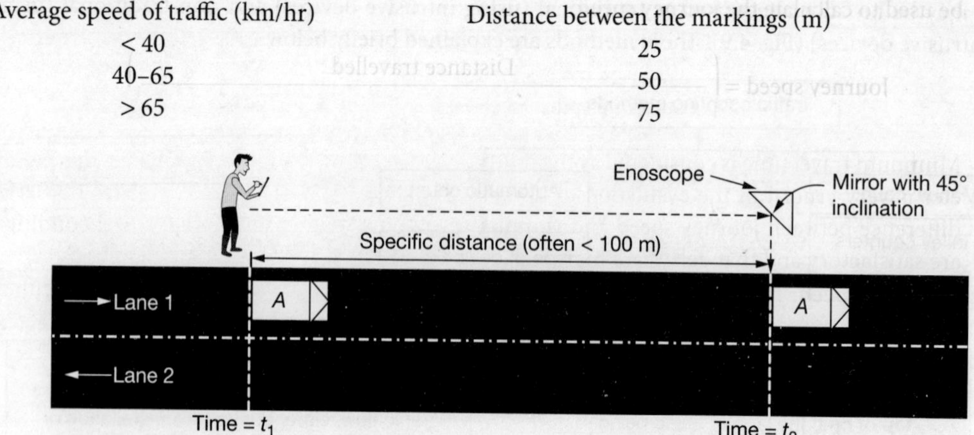

Figure 4.10 Concept of measurement of spot speed of a vehicle, using an enoscope.

Spot-speed studies are generally conducted in good weather and in normal traffic flow conditions.

4.7.1 Applications of Spot-speed Data

- Spot-speed studies are used to estimate the distribution of the speeds of vehicles that are travelling in a stream of traffic on a particular length of a highway.
- Spot-speed is used as a performance indicator of specific geometric components of a road such as horizontal curves and vertical curves, to establish speed limits/trends, safe operating speed on different stretches of road including on curves, demarcation of speed zone boundaries, intersection sight distance, location of signboards and their visibility, design of signals, delay modelling and for accident analysis as well as before-and-after studies.

4.7.2 Other Definitions of Speed

The terminology used to characterise the speed of traffic using different parameters is explained below.

- **Operating speed:** If the design is according to British and German design practices, estimates of 85th percentile operating speeds should be adopted (Kanellaidis, 1996). This means that on any given segment of roadway, the design speed should not be less than the estimated 85th percentile operating speed (Nicholson, 1998). To design horizontal curves, the operating speed at the 85th percentile is generally used as a parameter for controlling the speed (Bonneson, 1999). If the proposed design speed is to be assumed, it should be a logical one, determined primarily by considering the natural terrain, land use and highway classification (AASHTO, 1990).
- **Project/Design speed:** Design speed may be defined as the maximum average safe operating speed of vehicles that is either specified or maintained by most drivers travelling through all the features of a roadway. It is one of the prime parameters that dictate the overall geometry of highways in plain as well as hilly areas. According to the IRC (IRC: 73-1980), the ruling design speed is generally considered as a design criterion and taken as a primary input parameter for the design of various geometric elements of highways.

  This value of speed is used to design geometric road elements such as the radius of curves and super-elevation on curves.

- **Journey/Travel speed:** It is the average speed of a vehicle travelling between any two points, taking into account all possible delays in the course of the journey. The following expression can be used to calculate the journey speed.

$$\text{Journey speed} = \left(\frac{\text{Distance travelled}}{\text{Total time spent to cover the distance, including delays}}\right)$$

Minimum travel time is considered as the criterion for selecting the route, and hence, this parameter is very crucial in the evaluation of a highway system (NCUT, 1958). A smaller numerical difference between journey speed and running speed on a road indicates that travel conditions are satisfactory and that delays are marginal.

- **Running speed:** It is the average speed of a vehicle travelling between any two points without considering any delays in the course of the journey. The following expression can be used to calculate the running speed.

$$\text{Running speed} = \left(\frac{\text{Distance travelled}}{\text{Journey time} - \text{Time for delays and stoppages}}\right)$$

Note: The quality of traffic flow is best judged on the basis of the journey speed and running speed of vehicles. The ratio between running speed and journey speed describes the quality of flow. The higher the ratio, the better the quality of flow, and vice versa (Jalihal et al., 2005).

- **Time mean speed:** It is the *arithmetic mean speed* of all instantaneous vehicle speeds at a specified location on a road length. The time mean speed can be calculated as follows:

$$\text{Time mean speed, } V_{\text{mean}} = \frac{\sum_{i=1}^{n} V_i}{n} \tag{4.8}$$

where V_i = individual velocities of different vehicles in a group and
n = total number of vehicles considered in the group.
In the case of grouped vehicular data, the following equation is used:

$$V_{\text{mean}} = \frac{\sum_{i=1}^{n} f_i \times V_i}{\sum_{i=1}^{n} f_i} \tag{4.9}$$

where f_i = frequency of the ith vehicle.

- **Space mean speed:** It is the mean travel speed of all the vehicles traversing a specified section-length of a road, over a specified time period. The space mean speed can be calculated as follows:

$$\text{Space mean speed} = \left(\frac{\text{Distance}}{\text{Average of the time spent to travel the distance}}\right)$$

$$\text{Space mean speed} = \frac{L}{\dfrac{\sum_{i=1}^{n} t_i}{n}} = \frac{nL}{\sum_{i=1}^{n} t_i} \tag{4.10}$$

where L = distance covered by the group of vehicles considered;
t_i = individual time spent by the ith vehicle to cover the distance L; and
n = total number of vehicles considered in the group.

Space mean speed is considered the harmonic mean of vehicles and is useful in traffic applications. The space mean speed is always less than the time mean speed.

- **85th percentile speed:** It is the speed of vehicles that is only exceeded by 15% of vehicles. This speed is usually 20% higher than the average speed of the vehicles.
- **Median speed:** It is the value of speed that represents a middle value among a series of speed values arranged in ascending order. It is the 50th percentile speed which demarcates the point at which one half of the values of speed will be above median speed and the remaining half will be below it.
- **Modal speed:** It is the value of speed that is found to have the highest frequency of occurrence. In other words, it is the speed observed most frequently at a study site.
- **Pace:** It is an interval or range of speed; usually a 10-mi/hr interval, found in the greatest number of speed observations.

A typical presentation of spot-speed observations by frequency curve and cumulative frequency curve are plotted in Fig. 4.11. The typical values of median, mode, pace and the percent of vehicles at pace are shown on the respective curves.

EXAMPLE 4.1

This example describes the procedure of calculating time mean speed and space mean speed from the following data, obtained from a field survey.

Answer

Vehicle number	Time (in seconds) taken to travel the distance $L = 50$ m	Speed (km/h)
1	4	$= 3.6 \times (50\text{m}/4\text{s}) = 45$
2	3	$= 3.6 \times (50\text{m}/3\text{s}) = 60$
3	4	$= 3.6 \times (50\text{m}/4\text{s}) = 45$
4	4	$= 3.6 \times (50\text{m}/4\text{s}) = 45$
5	5	$= 3.6 \times (50\text{m}/5\text{s}) = 36$
6	6	$= 3.6 \times (50\text{m}/6\text{s}) = 30$
7	3	$= 3.6 \times (50\text{m}/3\text{s}) = 60$
8	5	$= 3.6 \times (50\text{m}/5\text{s}) = 36$
9	6	$= 3.6 \times (50\text{m}/6\text{s}) = 30$
10	6	$= 3.6 \times (50\text{m}/6\text{s}) = 30$
$n = 10$	Total = 46 seconds	Total = 417 km/h

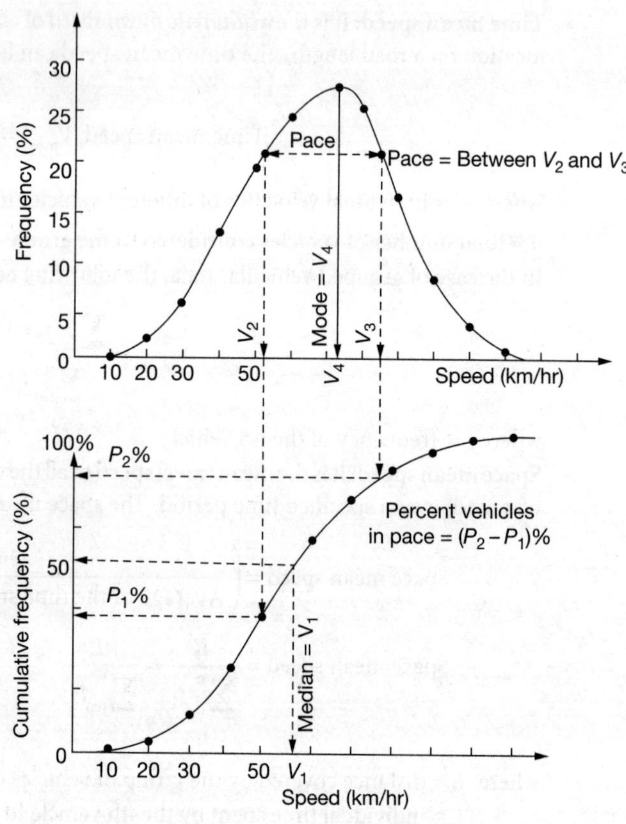

Figure 4.11 A typical presentation of spot-speed observations.

$$\text{Time mean speed} = \frac{\sum_{i=1}^{n} V_i}{n} = \frac{417}{10} = 41.7 \text{ km/h}$$

$$\text{Time mean speed} = \frac{nL}{\sum_{i=1}^{n} t_i} = \frac{10 \times 50 \text{ m}}{46 \text{ s}} \approx 10.87 \text{ m/s} = 10.87 \times 3.6 = 39.132 \text{ km/h}$$

EXAMPLE 4.2

Calculate the time mean speed and space mean speed of three vehicles travelling over a 1-km length in 1.2 minutes, 2.0 minutes and 2.9 minutes, respectively.

Answer

The speeds of vehicles are calculated as

$$V1 = 1 \div (1.2 \div 60) = (1 \times 60 \div 1.2) = 50.00 \text{ km/hr}$$

$$V2 = 1 \div (2.0 \div 60) = (1 \times 60 \div 2.0) = 30.00 \text{ km/hr}$$

$$V3 = 1 \div (2.9 \div 60) = (1 \times 60 \div 2.9) = 20.68 \text{ km/hr}$$

Time mean speed = Arithmetic average of speeds = (50.00 + 30.00 + 20.68)/3 = 33.56 km/hr.

Average time = (1.2 + 2.0 + 2.9) ÷ 3 = 2.03 minutes

Space mean speed = (1 km) ÷ (2.03÷60) hr = (1 × 60 ÷ 2.03) = 29.55 km/hr.

or

$$\text{Space mean speed} = \frac{3}{\left(\frac{1}{50} + \frac{1}{30} + \frac{1}{20.68}\right)} = 29.5 \text{ km/hr}$$

EXAMPLE 4.3

Determine the running speed and journey speed of a vehicle travelling from A to G. Use the following field data:

Average total travel time including delays due to traffic at junctions = 36 minutes
Average delays at junction B = 2 minutes, C = 3 minutes, D = 6 minutes, E = 4 minutes, F = 5 minutes.

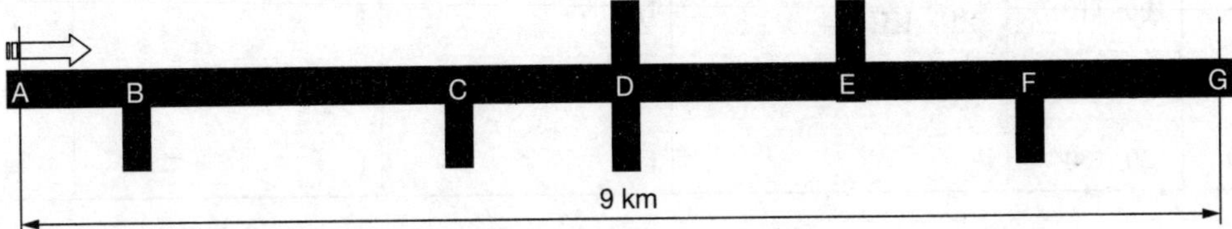

Answer

$$\text{Average journey speed between A and G} = \left(\frac{9 \text{ km}}{\left(\frac{36}{60}\right)\text{hr}}\right) = \left(\frac{9 \text{ km}}{0.6\text{hr}}\right) = 15 \text{ km/hr}$$

Total delay = (2 + 3 + 6 + 4 + 5) = 20 minutes

Average running journey speed between A and G

$$= \left(\frac{9\text{ km}}{\left(\dfrac{36-20}{60}\right)\text{hr}}\right) = \left(\frac{9\text{ km}}{\left(\dfrac{16}{60}\right)\text{hr}}\right) = \left(\frac{9\text{ km}}{0.267\text{ hr}}\right) \approx 33.7\text{ km/hr}$$

$$\text{Per cent difference in speed} = \left(\frac{33.7-15}{15}\right)\times 100 = 124.76\%$$

As the journey speed is 124.67% less than the running speed, this indicates that an inefficient traffic control system exists at the junctions, and hence, the system requires improvement.

EXAMPLE 4.4

The speed data collected for a segment of a national highway is given below.

S. No.	Speed class	Vehicle type			Total
		Cars	Trucks/Buses	Others	
1	30–40	~~IIII~~ ~~IIII~~	~~IIII~~ ~~IIII~~	IIII	24
2	40–50	~~IIII~~ ~~IIII~~ III	~~IIII~~ ~~IIII~~ IIII	~~IIII~~ ~~IIII~~ I	38
3	50–60	~~IIII~~ ~~IIII~~ ~~IIII~~ ~~IIII~~ I	~~IIII~~ ~~IIII~~ ~~IIII~~ III	~~IIII~~ ~~IIII~~ ~~IIII~~ IIII	58
4	60–70	~~IIII~~ ~~IIII~~ ~~IIII~~ ~~IIII~~ ~~IIII~~ II	~~IIII~~ ~~IIII~~ ~~IIII~~ IIII	~~IIII~~ ~~IIII~~ ~~IIII~~ ~~IIII~~ ~~IIII~~ ~~IIII~~	76
5	70–80	~~IIII~~ ~~IIII~~ IIII	~~IIII~~ ~~IIII~~ IIII	~~IIII~~ ~~IIII~~ IIII	42
6	80–90	~~IIII~~ ~~IIII~~	~~IIII~~ ~~IIII~~ I	~~IIII~~ ~~IIII~~ I	32
7	90–100	~~IIII~~ ~~IIII~~	~~IIII~~ ~~IIII~~	II	22
8	100–110	~~IIII~~ ~~IIII~~	IIII	I	15
9	110–120	~~IIII~~	III	I	9
10	120–130	II	I	I	4

(a) Draw a histogram of the speeds.

(b) Draw the cumulative frequency curve and find the values of (i) median speed, (ii) modal speed, and (iii) 15th, 85th and 98th percentile speeds.

Also determine the (c) mean speed, (d) standard deviation of the speed values, (e) coefficient of variation (CoV) of the speeds, (f) standard error (SE) of speed values, (g) range of speed at which the mean speed lies within 95% confidence, and (h) the size of the sample at which the SE is restricted to 0.5 km/hr.

Answer

The following calculations are performed on the given data for the initial process of analysis.

S. No. [0]	Speed class [1]	Mid class (V_i) [2]	Observations or Frequency (f_i) [3]	% of f_i [4]	% Cum. of f_i [5]	$V_i \times f_i$ [6] = 2 × 3	$(V_i - V_m)^2$ [7]	$f_i \times (V_i - V_m)^2$ [8]
1	30–40	35	24	7.50	7.5	840	1086.938477	26086.52
2	40–50	45	38	11.88	19.375	1710	527.5634766	20047.41
3	50–60	55	58	18.13	37.5	3190	168.1884766	9754.932
4	60–70	65	76	23.75	61.25	4940	8.813476563	669.8242
5	70–80	75	42	13.13	74.375	3150	49.43847656	2076.416
6	80–90	85	32	10.00	84.375	2720	290.0634766	9282.031
7	90–100	95	22	6.88	91.25	2090	730.6884766	16075.15
8	100–110	105	15	4.69	95.9375	1575	1371.313477	20569.7
9	110–120	115	9	2.81	98.75	1035	2211.938477	19907.45
10	120–130	125	4	1.25	100	500	3252.563477	13010.25
		$\sum V_i = 802$	$\sum f_i = 320$			$\sum = 21750$		$\sum = 137487.7$

(a) The histogram of the speeds is drawn (see Fig. 4.12) by plotting the mid-speed values on the y-axis and frequency values on the x-axis.

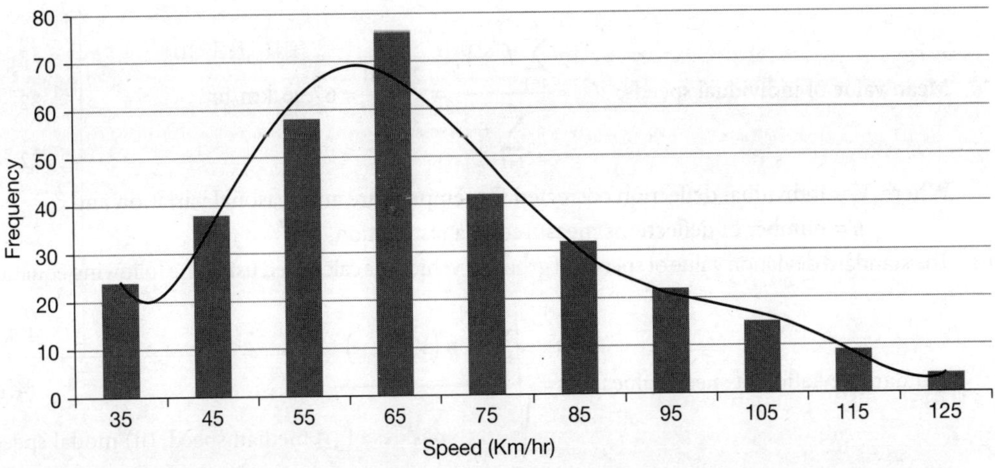

Figure 4.12 Example 4.4 - Histogram/Frequency curve presentation of speed data.

(b) The cumulative frequency curve is drawn (see Fig. 4.13) by plotting the % of cumulative frequency values on the y-axis and the mid-speed values on the x-axis.

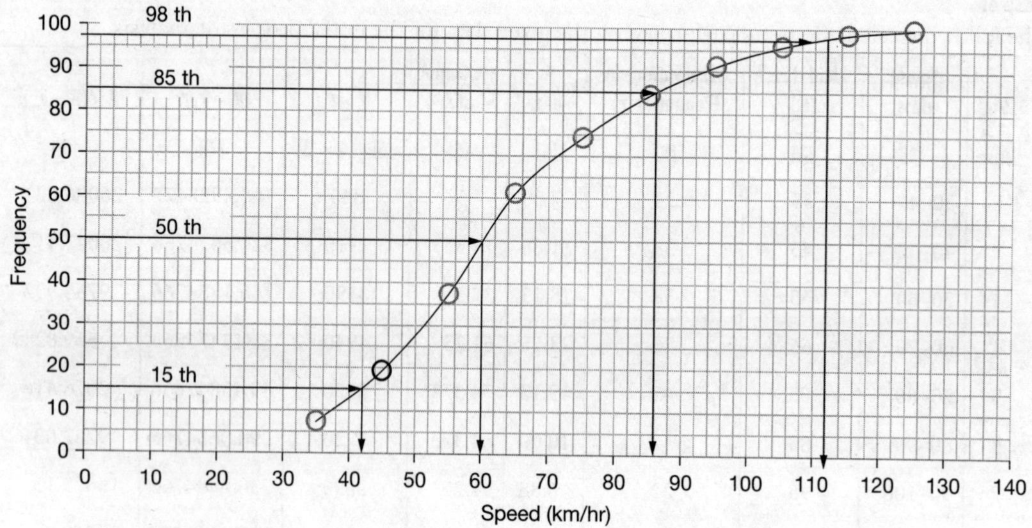

Figure 4.13 Example 4.4 Cumulative frequency curve presentation of speed data.

The following values are obtained from Fig. 4.13

(i) Median speed (i.e., 50th percentile speed) = 60 km/hr

(ii) Modal speed (i.e., peak point on frequency distribution curve, Fig. 4.12) = 65 km/hr

(iii) 15th percentile speed = 42 km/hr

(iv) 85th percentile speed (used for traffic regulation) = 86 km/hr

(v) 98th percentile speed (used for geometric design) = 110.2 km/hr

(c) Mean speed is calculated using the following equation:

$$\text{Mean value of individual speeds, } V_m = \frac{\sum_{i=1}^{n} f_i \times V_i}{\sum_{i=1}^{n} f_i} = \frac{21750}{320} \approx 67.96 \text{ km/hr}$$

Where V_i = individual deflection corrected for temperature and seasonal variation and n = number of deflections measured in a test section.

(d) The standard deviation value of speeds of grouped vehicles is calculated using the following equation:

$$\text{Standard deviation of speed values} = s = \sqrt{\frac{\sum_{i=1}^{n} f_i \times (V_i - V_m)^2}{\sum_{i=1}^{n} f_i}} \tag{4.11}$$

$$s = \sqrt{\frac{137487.7}{320}} = 20.72 \text{ km/hr}$$

(e) $$\text{CoV} = \frac{\text{Standard deviation of the speed values } (s)}{\text{Mean speed value } (V_m)} \times 100 \tag{4.12}$$

$$CoV = \frac{20.72}{67.96} \times 100 = 30.49\%$$

(f) Standard error (SE) of speed values is calculated using the following equation:

$$SE = \frac{s}{\sqrt{n}} \tag{4.13}$$

$$SE = \frac{20.72}{\sqrt{320}} \approx 1.15 \text{ km/hr}$$

(g) Range of speed at which the mean speed lies with 95% confidence is
$V_m \pm SE$ = (67.96 ± 1.15) = between (67.96 + 1.15) and (67.96 – 1.15) = between 69, 11 and 66.81 km/hr.

(h) Size of sample at which the SE is restricted to 0.5 km/hr:
Using the standard equation,

$$SE = \frac{s}{\sqrt{n}}, \quad 0.5 = \frac{20.72}{\sqrt{n}}, \quad n \approx 1717$$

This means, 1717 vehicle speeds need to be measured in the speed study to restrict the SE to 0.5 km/hr. Then, the range of the speed calculated above will be narrowed to (67.96 ± 0.5) km/hr.

EXAMPLE 4.5

A spot speed study was conducted on a stretch of a highway and the data collected was consolidated as given below:

Speed range, km/hr	0–10	10–20	20–30	30–40	40–50	50–60	60–70	70–80	80–90	90–100
No. of vehicles observed	15	35	40	60	150	260	290	60	70	20

Determine (i) The upper and lower limit of speed for speed regulation of mixed traffic
(ii) The design speed for checking the geometric design of the highway
(iii) The median speed observed

Answer

Speed Range, Km/hr	Mid-speed, (Values plotted on X-axis)	No. of vehicles observed, *f*	Percentage frequency %*f*	Cumulative frequency, % (Values plotted on Y-axis)
0–10	5	15	1.5	1.5
10–20	15	35	3.5	5
20–30	25	40	4	9
30–40	35	60	6	15

(Continued)

Speed Range, Km/hr	Mid-speed, (Values plotted on X-axis)	No. of vehicles observed , *f*	Percentage frequency % *f*	Cumulative frequency, % (Values plotted on Y-axis)
40–50	45	150	15	30
50–60	55	260	26	56
60–70	65	290	29	85
70–80	75	60	6	91
80–90	85	70	7	98
90–100	95	20	2	100
	Sum =	1000	100	

Using the mid-speed and cumulative frequency values, the cumulative frequency distribution curve is plotted (Fig. 4.14) and the following results are obtained.

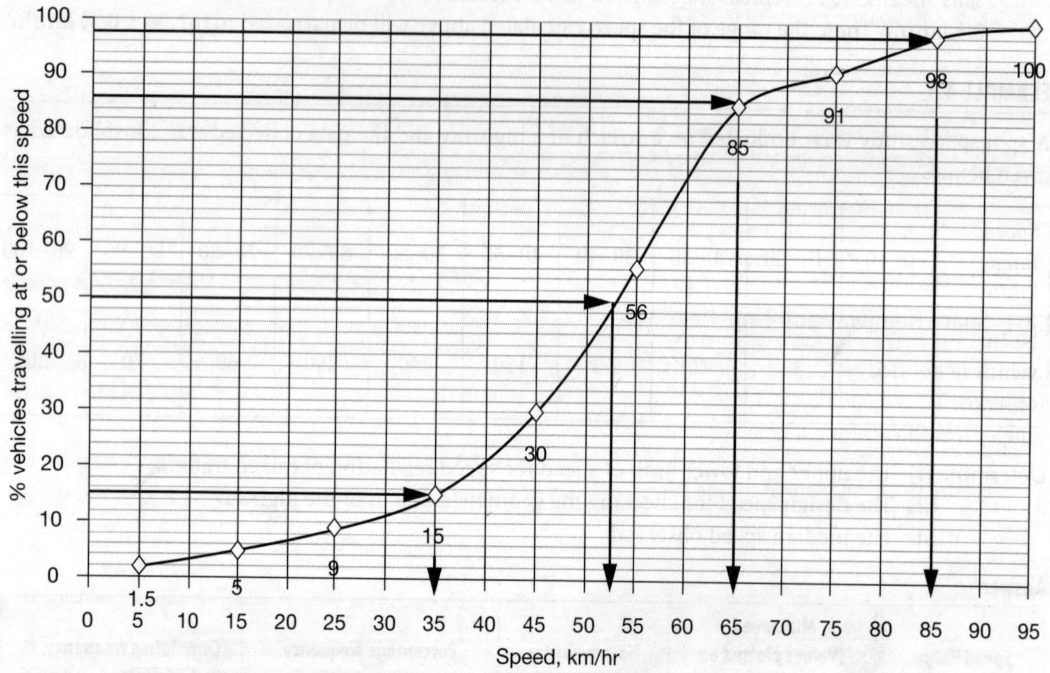

Figure 4.14 Example 4.5 – Cumulative speed distribution.

(i) The upper and lower speed for speed regulation of mixed traffic:
The upper speed limit for speed regulation = 85th percentile speed = 65 km/hr.
The lower speed limit for speed regulation = 15th percentile speed = 35 km/hr.

(ii) The design speed for checking the geometric design of the highway:
The speed to check for design elements = 98th percentile speed = 85 km/hr.

(iii) The median speed observed = 53 km/hr.

4.8 TRAFFIC FLOW CHARACTERISTICS

Based on the type of vehicular flow along a selected length of a road, the traffic stream parameters may be categorised as follows (Fig. 4.15):

- Macroscopic parameters
- Microscopic parameters

Parameters such as rate of flow (*Q*), average speed (*V*) of vehicles in a stream of traffic and their density (*K*) are considered to belong to the macroscopic category as they describe the traffic stream as a whole (i.e., considered as a group of vehicles). Parameters such as speed (*V*), spacing and time headway are considered to belong to the microscopic category and they describe the individual behaviour of vehicles or pairs of vehicles within a stream of traffic.

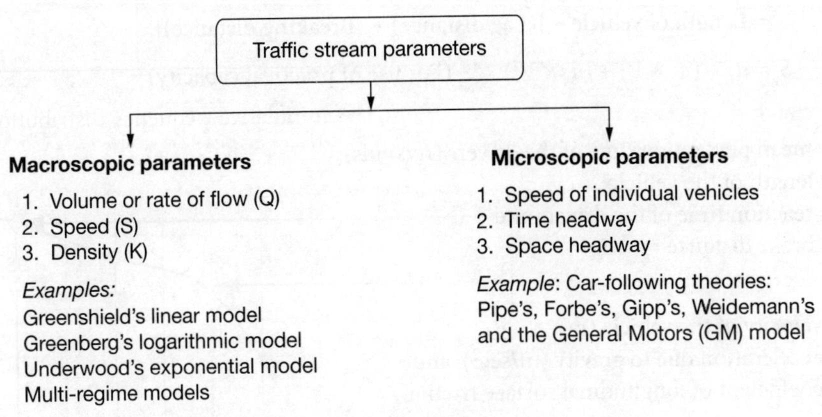

Figure 4.15 Classification of traffic stream parameters.

4.8.1 Space Headway and Time Headway

Spacing is defined as the distance (m) between successive vehicles in a traffic lane, at a given point (Fig.4.16). Time headway is defined as the time gap (seconds) to pass the space gap between successive vehicles travelling along a lane.

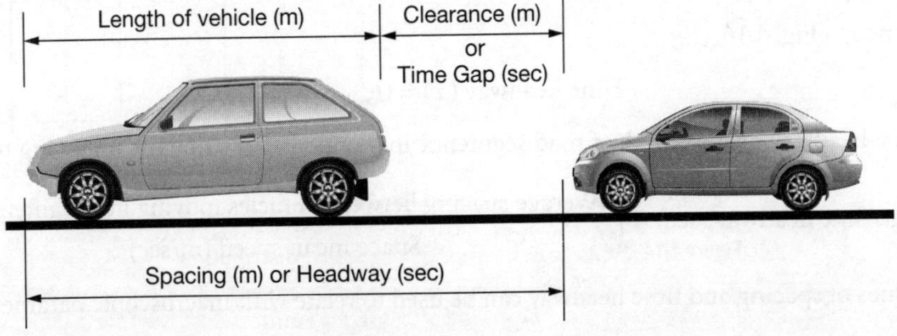

Figure 4.16 Measurement of spacing between a pair of vehicles travelling at a point.

Space headway and time headway are considered microscopic measures of traffic flow as they apply to pairs of vehicles travelling in a traffic stream. These parameters are useful for determining the number of gaps in a traffic stream or with pedestrians to cross a roadway and for the design of pedestrian signals at junctions.

The following additional parameters can be determined from field measurements:

Clearance (m) = Spacing (m) – Average vehicle length (m)

Gap (seconds) = Time headway (seconds) –
Time equivalence of the average vehicle length (seconds)

Space headway (S_s) can be calculated using any one of the following equations:

Space head way = Length of rear vehicle + Space gap between the vehicles

$$S_S = 6\,\text{metre} + \left(\frac{V}{3.6}\right) \times t_p \quad \text{(For use of theoretical capacity, assume } t_p = 0.7 \text{ sec)} \tag{4.14}$$

S_s = Length of vehicle + [Safe stopping distance, SSD]

S_s = Length of vehicle + [(Lag distance) + (Breaking distance)]

$$S_s = a_1 + (a_2 \times V) + (a_3 \times V^2) \quad \text{(For use of practical capacity)} \tag{4.15}$$

where,

t_p = mean perception time of the drivers (seconds);
a_1 = length of the vehicle;
a_2 = reaction time of the drivers and
a_3 = brake distance = $v^2/2gf$

where,

v = velocity of the vehicle (m);
g = acceleration due to gravity (m/sec²); and
f = coefficient of longitudinal surface friction.

If the S_s value is expressed in terms of time gap in seconds, it is termed the *time headway* (T) and it can be calculated as follows:

$$\text{Time headway } (T) = \left(\frac{\text{Space gap between the front of one vehicle to the front of the other vehicle considered}}{\text{Average velocity of the vehicles}}\right) \tag{4.16}$$

With reference to Fig. 4.16,

$$\text{Time headway } (T) = (t_2 - t_1) \text{ sec.} \tag{4.17}$$

Average time headway over a length of road segment can be calculated using the following relationship:

$$\text{Average time headway (sec)} = \left(\frac{\text{Average spacing between vehicles moving in a segment (m)}}{\text{Space mean speed (m/sec)}}\right)$$

Average values of spacing and time headway can be used to relate with macroscopic parameters:

$$\text{Density} = K \text{ (Veh/km per lane)} = \frac{1000 \text{ m}}{\text{Average spacing in m}} \tag{4.18}$$

$$\text{Volume} = Q \text{ (veh/hr per lane)} = \frac{3600 \text{ sec}}{\text{Average time headway in sec}} \tag{4.19}$$

$$\text{Speed} = V(\text{m/sec}) = \frac{\text{Average spacing in m}}{\text{Average time headway in sec}} \tag{4.20}$$

Headway data is useful for the design, operation, control and evaluation of road sections, particularly the overtaking zones, lane changing transitions, intersection operations and acceleration/deceleration lanes.

4.8.2 Density

It is defined as the number of vehicles (n) travelling in unit length (l) of a road, per lane, at an instance (Fig. 4.17). The density or concentration (K) value is measured in vehicles/kilometre (veh/km). If there is no space between the vehicles at an instance, the value of density will be maximum. Density is a function of mean vehicle dimension, and this parameter can be used to describe road performance.

$$K = \frac{\text{Number of vehicles}}{\text{Distance}} = \frac{n}{\ell} = \frac{Q}{V} \tag{4.21}$$

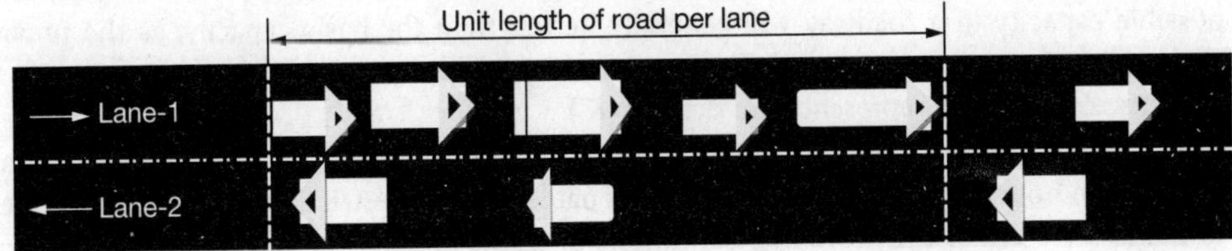

Figure 4.17 Density of vehicles.

This equation may be re-written as given below, to assess the impact of a development project by determining the Level of Service (LoS) of a road segment (SCVTA, 2003):

$$\text{Density} = K \text{ (veh/hr per lane)} = \frac{Q}{N \times V} \tag{4.22}$$

Where Q = peak hour traffic volume (i.e., flow rate in veh/hr); V = average speed of vehicles(km/hr); and N = number of travel lanes.

4.8.3 Lane Capacity

Capacity is a measure that represents the ability of a roadway to accommodate traffic volume. Theoretical maximum number of vehicles that can *pass* a given point on a lane per unit time under ideal roadway, traffic and control conditions is often termed the *basic capacity of a lane* (C). The capacity of a lane can be calculated from the following equation.

$$C = \frac{V}{S_s} \times 1000 \tag{4.23}$$

where C = basic capacity of a single lane (veh/hr per lane)
V = average speed or velocity of the vehicles (km/hr) and
S_s = safety mean spacing (m) between the front of two consecutive vehicles that are travelling on the same lane at a given speed (Fig. 4.18). This parameter is also termed *space headway* or *spacing* (by Eq. 4.14).

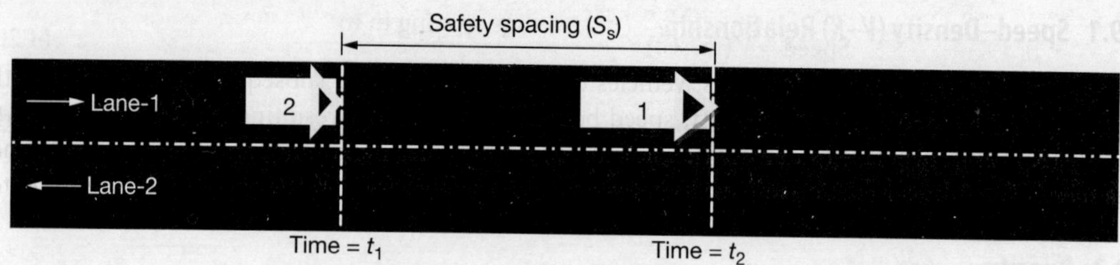

Figure 4.18 Concept of determination of safety spacing.

Types of traffic capacity

- **Basic capacity:** It is the maximum possible number of vehicles that can *pass* a given point on a lane in an hour, under ideal conditions of traffic flow on the roadway.
- **Possible capacity:** It is the maximum number of vehicles that can *pass* a given point on a lane in an hour, under prevailing conditions of traffic flow on the roadway. It must be noted that the possible capacity of a roadway will generally be *less* than the basic capacity, as the prevailing traffic conditions are seldom ideal. This means that possible capacity will vary from zero to basic capacity. Zero density represents jam density (K_j).
- **Practical/Design capacity:** It is the maximum number of vehicles that can *pass* a given point on a lane in an hour, without causing any unreasonable delay or restrictions to the driver's freedom to manoeuvre under the prevailing conditions of traffic flow on the roadway. It must be noted that roadway width and the number of lanes is designed in such a way that the measured traffic volume is less than or equal to its lane design capacity. For this reason, practical capacity is also termed design capacity.

Factors Affecting Practical Capacity: The capacity of a roadway depends on several factors relating to road geometrics and mixed traffic condition (i.e., with the proportion of vehicle types).

- **Space headway:** As the space gap between vehicles increases, there will be a decrease in the value of capacity. If the vehicles are in a jammed condition, the density will be maximum, and it is termed jamming density. In addition, all the factors affecting space headway may also be considered under this factor.
- **Lane width:** As the lane width increases (i.e., the number of lanes increases), the capacity to pass vehicles per hour also increases, and vice versa.
- **Shoulder width:** As the width increases, the capacity value also increases, and vice versa.
- **Proportion of commercial vehicles:** As the proportion of commercial vehicles (large dimensions) increases, the capacity value decreases.
- **Curve alignment:** In case of horizontal or vertical curves, there is inadequate sight distance. So there may be a reduction in speed, and hence, the capacity decreases.

4.9 FUNDAMENTAL RELATIONSHIPS BETWEEN TRAFFIC FLOW PARAMETERS

The fundamental traffic flow characteristics and their relationships which pertain to uninterrupted traffic flow condition are discussed below.

4.9.1 Speed–Density (*V*–*K*) Relationship

As spacing between vehicles decreases, vehicles density increases, and subsequently, the speed of the vehicles decreases. At very high density, speed becomes zero and the resulting density is termed *jam density* (K_j). With decrease in density (i.e., $K = 0$), speed becomes maximum, and this speed is termed *free speed* (V_f). A linear relationship exists between speed and density, and the correlation is represented in Fig. 4.19.

$$\text{Critical density } (K_c) = \left(\frac{\text{Jam density}}{2}\right) = \left(\frac{K_j}{2}\right) \tag{4.24}$$

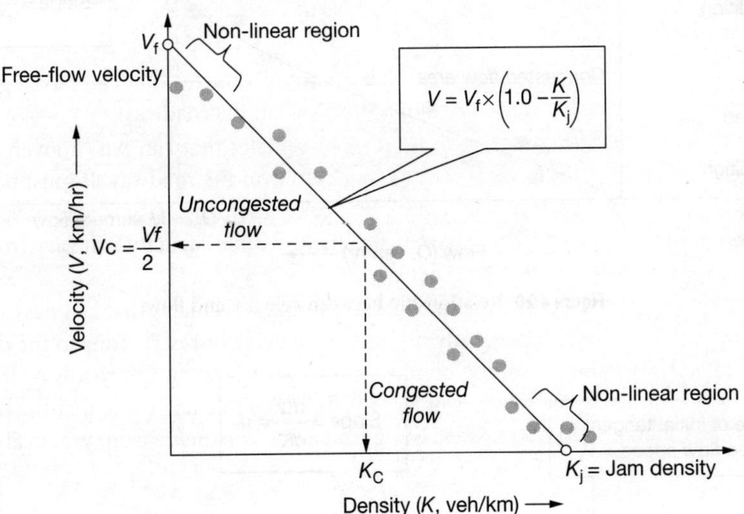

Figure 4.19 Relationship between velocity and density.

4.9.2 Speed–Flow (*V*–*Q*) Relationship

As vehicles speed decreases (from free-flow condition), the flow rate reduces. As vehicle speed increases, the initial flow rate also increases, up to certain maximum limit, with decreased space headway. However, as speed increases further, the space headway increases and the flow rate of the vehicles decreases (Fig.4.20). The speed at which the flow rate is maximum (Q_m) is termed optimum speed or *critical speed* (V_c).

$$\text{Critical speed } (V_c) = \left(\frac{\text{Free-flow velocity}}{2}\right) = \left(\frac{V_f}{2}\right) \tag{4.25}$$

$$\text{Capacity} = \text{Maximum flow } (Q_m) = V_c \times K_c = \left(\frac{V_f}{2}\right) \times \left(\frac{K_j}{2}\right) = \left(\frac{V_f \times K_j}{4}\right) \tag{4.26}$$

4.9.3 Flow–Density (*Q*–*K*) Relationship

As density increases from zero density (Fig. 4.21), the vehicle flow rate increases, up to the point of critical density (K_c), which corresponds to the maximum flow rate (Q_m). Further, as the density reaches maximum – termed jam density (K_j) – the vehicles are stopped and flow rate becomes zero.

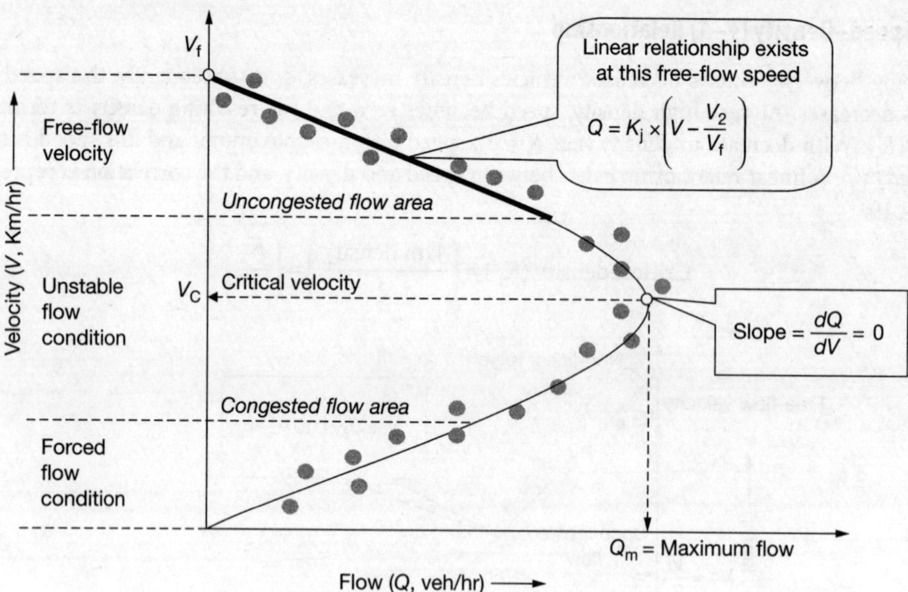

Figure 4.20 Relationship between velocity and flow.

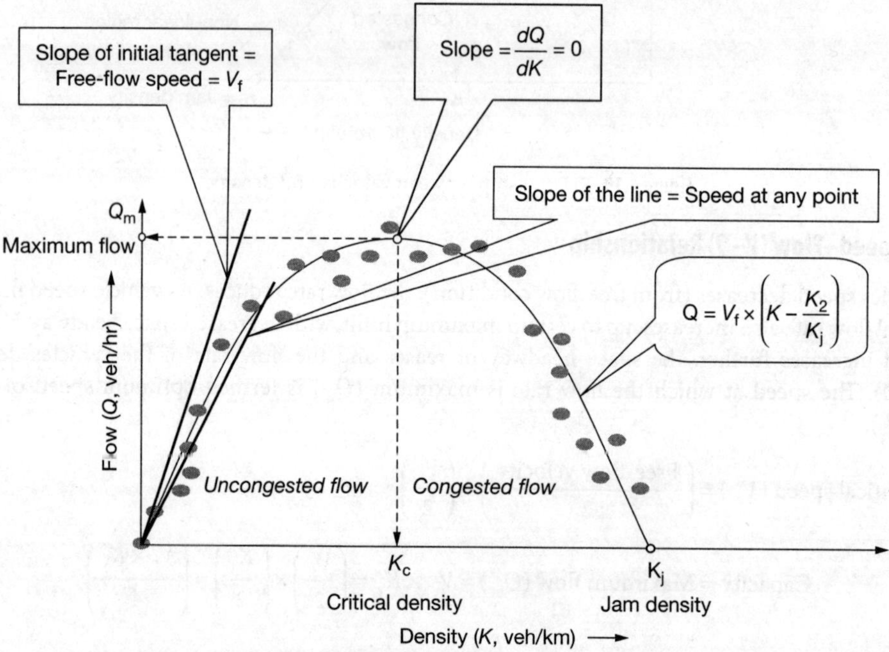

Figure 4.21 Relationship between flow and density.

The following relationship may be used to determine the density or any unknown parameter from the other known values using the standard trend curves (Fig. 4.22).

Traffic flow (Q) = Space mean speed (V_s) × Traffic density (K)

$$Q = V_s \times K \tag{4.27}$$

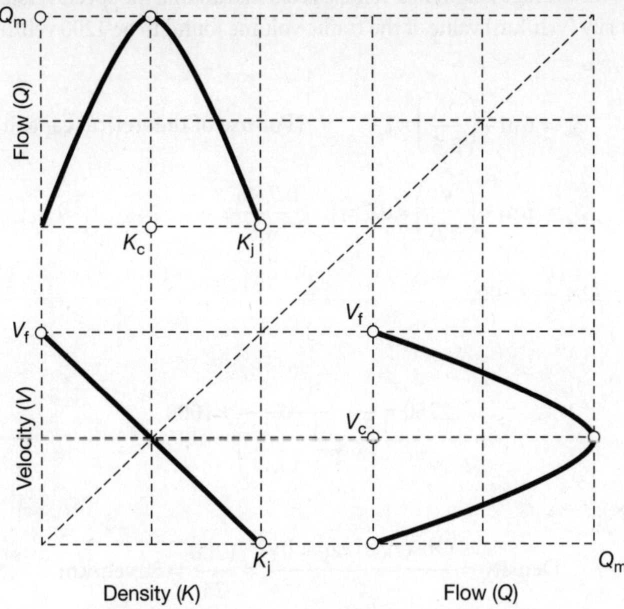

Figure 4.22 Velocity–Flow–Density relationship.

Derivation of the parabolic flow–density model

The fundamental relationships of traffic flow models are derived as shown below:

$$K = \left(\frac{K_j}{2}\right) \tag{4.28}$$

Maximum flow or volume occurs when the density is half the jam density. In other words, the density at maximum flow = critical density (K_c) = $K_j/2$.

The critical speed (i.e., space mean speed, V_s) is half the free-flow speed at maximum flow (Fig. 4.20).

i.e., $V_c = \dfrac{V_f}{2}$

Using Eq. (10.14), $Q = V_s \times K$

$$Q_m = \frac{V_f}{2} \times \frac{K_j}{2}$$

$$Q_m = \frac{V_f \times K_j}{4} \tag{4.29}$$

4.9.4 Density–Spacing (K–S_s) Relationship

As the spacing between the vehicles travelling on a lane decreases, their density increases. The density value can be determined by measuring the flow rate and average speed of the vehicles travelling on a length of road.

EXAMPLE 4.6

A two-lane urban road with one-way traffic flow control was designed for a maximum capacity of 2250 veh/hr. Under jam conditions, the average length of a vehicle is 6.0 m. Assume the speed versus density relationship is linear. What is the density (veh/km) value, if the traffic volume found to be 1200 veh/hr?

Answer

$$S_s = 6\text{m} + \left(\frac{V}{3.6}\right) \times t_p \qquad \text{(For use of theoretical capacity assume } t_p = 0.7 \text{ sec)}$$

$$S_s = 6\text{m} + \left(\frac{V}{3.6}\right) \times 0.7 = 6 + \frac{0.7 \times V}{3.6}$$

We know, Capacity $= C = \frac{V}{S_s} \times 1000$

Substituting, we get

$$2250 = \frac{V}{\left(6 + \frac{0.7 \times V}{3.6}\right)} \times 1000$$

Solving, $V = 24$ m/sec

$$\text{Density} = \frac{\text{Observed capacity}}{V} = \frac{1200}{24} = 50 \text{ veh/km}$$

EXAMPLE 4.7

Based on a traffic survey conducted on a National Highway (NH), it was found that free speed was 110 km/hr and jam density was 75 veh/km. Calculate the maximum flow rate (Q_m) that could be expected on this NH and the corresponding speed (V_c).

Answer

$$\text{Critical speed } (V_c) = \left(\frac{\text{Free flow velocity}}{2}\right) = \left(\frac{V_f}{2}\right) = \left(\frac{110}{2}\right) = 55 \text{km/hr.}$$

$$\text{Capacity} = \text{Maximum flow}(Q_m) = V_c \times K_c = \left(\frac{V_f \times K_j}{4}\right) = \left(\frac{110 \times 75}{4}\right) = 2062.5 \text{veh/hr.}$$

EXAMPLE 4.8

The free-flow speed found in a study section of road is 82 km/hr and the jam density is 98 vehicles per km length. Determine the maximum flow expected on the study section and the speed at maximum flow. Assume a linear relationship between speed and density on the study sections.

Answer

By Eq. 4.26,

$$\text{Capacity} = \text{Maximum flow}(Q_m) = V_c \times K_c = \left(\frac{V_f}{2}\right) \times \left(\frac{K_j}{2}\right) = \left(\frac{V_f \times K_j}{4}\right)$$

$$\text{Maximum flow } (Q_m) = V_c \times K_c = \left(\frac{82}{2}\right) \times \left(\frac{98}{2}\right) = \left(\frac{8036}{4}\right) = 2009 \text{ veh/hr}$$

Speed at the above maximum flow = Critical speed = $V_c = 82/2 = 41$ km/hr

EXAMPLE 4.9

The details of traffic data obtained on a road segment are given below.

Speed, V (km/hr)	59	51	46	41	34	29	22	15	10	4
Density, K (veh/km)	11	21	32	42	52	63	70	81	91	100

Calculate the volume data and the corresponding time headway and space headway values.

Answer

The following relationships are used to calculate the values:

$$\text{Volume } (Q) = \text{Speed } (V) \times \text{Density } (K)$$

$$\text{Time headway (sec)} = 3600 \text{ sec/Volume} = 3600/Q$$

For example, Time headway = 3600/649 = 5.55 sec

$$\text{Space headway (m)} = 1000 \text{ m/Density} = 1000/K$$

For example, Space headway = 1000/11 = 90.91

By Eq.4.26 $Q = V \times K$ Volume (veh/hr)	649	1071	1472	1722	1768	1827	1540	1215	910	400
Time headway, T (sec) = (3600/Q)	5.55	3.36	2.45	2.09	2.04	1.97	2.34	2.96	3.96	9.00
Space headway, S (m) = (1000/K)	90.91	47.62	31.25	23.81	19.23	15.87	14.29	12.35	10.99	10.00

EXAMPLE 4.10

Vehicles in a traffic stream move at an average time headway (T) of 2 sec. The average speed of the vehicles was found to be 75 km/h. Estimate the flow rate and density of the traffic stream.

Answer

We know that the flow rate (Q) is equal to the reciprocal of time headway (T), and it can be written as

$$Q(\text{veh/hr}) = \frac{1}{(T)} = \frac{1}{\left(\frac{2\text{seconds}}{3600}\right)} = 1800 \text{ veh/hr.}$$

By Eq.10.14, $Q = KV$, can be rewritten as

$$\text{Density} = K = Q/V = 1800/75 = 24 \text{ veh/km.}$$

4.10 LEVEL OF SERVICE

Level of service (LoS) is a qualitative measure of traffic flow conditions which describes the operational quality within a traffic stream, as experienced by drivers/passengers. It represents the quality of service of a roadway in the prevailing traffic and roadway geometric conditions. The quality of performance is measured on an A–F scale (Fig. 4.23), with LoS-A representing the *best* operating conditions and LoS-F representing the *worst* operating conditions, from the traveller's perspective (HCM, 2010).

LoS is commonly determined using travel projections twenty to thirty years in the future.

The general operating conditions for the six levels of service are defined in Table 4.5.

The LoS of a roadway along a selected length measures the effect of the following factors (IRC: 106-1990):

- Speed and travel time
- Road user freedom to manoeuvre
- Traffic interruptions, delays and hindrances
- Driving comfort and convenience
- Safety and operating costs

The following factors affect the LoS on urban roads (IRC: 106-1990):

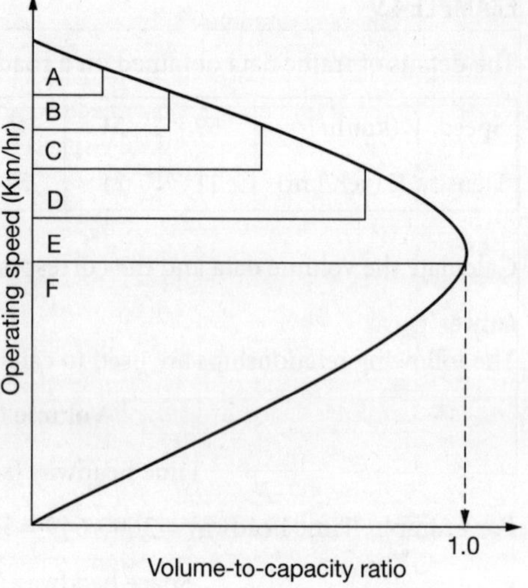

Figure 4.23 Level of Service A to F.

- Heterogeneity of traffic: The proportion of slow-moving vehicles and fast-moving vehicles, and the size and tractive energy of vehicles. Traffic service quality or traffic flow on the road may vary with time and day.

Table 4.5 General operating conditions for levels A–F.

LOS	Volume-to-capacity ratio limit	Condition	Remarks	Typical illustration
A	0.35	Free flow: Average speed = 90% of free-flow speed	With low volume and density, and high speed. Speed is controlled as driver desires. No delay.	
B	0.54	Stable flow: Average speed = 70% of free-flow speed; flow <55% of capacity	Speeds beginning to be restricted due to traffic conditions	
C	0.77	Stable flow: Average speed = 50% of free-flow speed; flow <75% of capacity	Speeds are restricted	
D	0.93	High density: Average speed = 40% of free-flow speed; flow up to 89% of capacity	Drivers have little freedom to manoeuvre	

(*Continued*)

LOS	Volume-to-capacity ratio limit	Condition	Remarks	Typical illustration
E	1.0	Unstable flow; Average speed = 33% of free-flow speed; flow at 100% of capacity	With short stoppages	
F	>1.0	Breakdown: Forced flow; Average speed at 25–33% of free flow speed	Congested flow	

- Speed regulation/control: One-way, two-way, signalisation, unsignalised intersection
- Frequency of intersections and spacing
- Presence of on-street bus stoppages
- On-street parking
- Intensity of roadside commercial activities and land use
- Speed breaker
- Pedestrian volume
- Roadway geometrics: Number of lanes, lane width, sight distance, curvature
- Access road alignment (suburban and rural areas)
- Weaving length, weaving angle and space headway

To determine the performance of a road, LoS is an important factor. The following six factors are used to determine LoS and such studies will be helpful in the future development of a road facility:

- Volume-to-capacity ratio
- Density
- Speed
- Travel time
- Delay
- Control delay (sec/veh)

The Highway Capacity Manual (HCM) recommends the LoS of intersections based on measured delays (Table 4.6).

Table 4.6 HCM-recommended criteria for LoS of an intersection.

LoS	Signalised intersection	Unsignalised intersection
	Average control delay (sec/veh)	Delay (sec/veh)
A	≤10	≤10
B	10–20	10–15
C	20–35	15–25

(Continued)

LoS	Signalised intersection	Unsignalised intersection
	Average control delay (sec/veh)	Delay (sec/veh)
D	35–55	25–35
E	55–80	35–50
F	>80	>50

4.11 EQUIVALENCY FACTORS

The *interaction* and *influence* of slow-moving vehicles on fast-moving vehicles and the *effect* of commercial vehicles (i.e., larger dimensions) on other types of vehicular traffic (in a heterogeneous mix) is one of the major important considerations in a study of the performance characteristics of non-uniformity in the traffic stream. Due to the heterogeneous traffic mix, one class of vehicles cannot be considered equal to any other class. Interactions among various classes of vehicles travelling in a stream are accounted for in prevailing roadway and traffic condition using a common factor, termed passenger car unit (PCU). PCU is a multiplication factor (Fig. 4.24 and Fig. 4.25) used to convert different types of vehicles into the *equivalent* passenger car units, based on their relative interference.

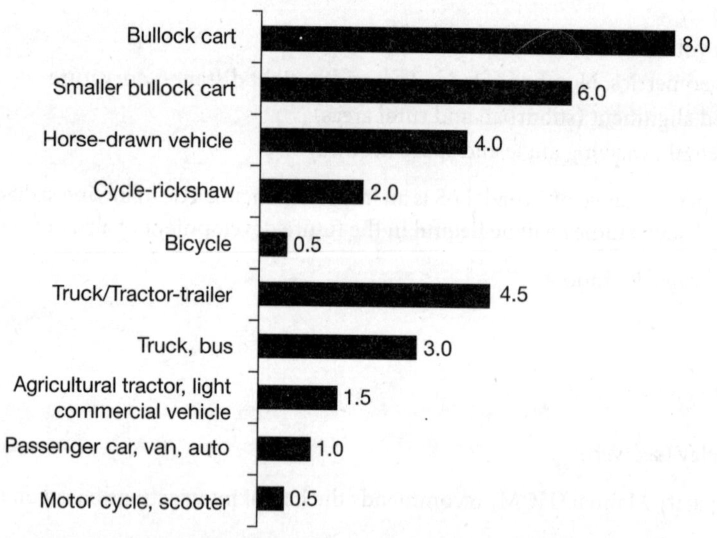

Figure 4.24 IRC recommended PCU factors applicable on rural roads (not be applied to road intersections) (Modified from IRC: 64-1990).

PCU, also called passenger car equivalent (PCE), was initially introduced in the 1965 Highway Capacity Manual to account for the effect of heavy vehicles in the traffic stream (HRB, 1965). Of several methods (Walker's, headway, multiple linear regression, simulation and density), the most suitable one for non-uniform mixed traffic flow conditions in India is the one proposed by Chandra et al. (1995). According to this method, the PCU value is directly proportional to the speed ratio and inversely proportional to the space occupancy ratio with respect to a standard design vehicle, i.e., car. The equation used for estimating the PCU values for a study area is given below (Chandra and Kumar, 2003).

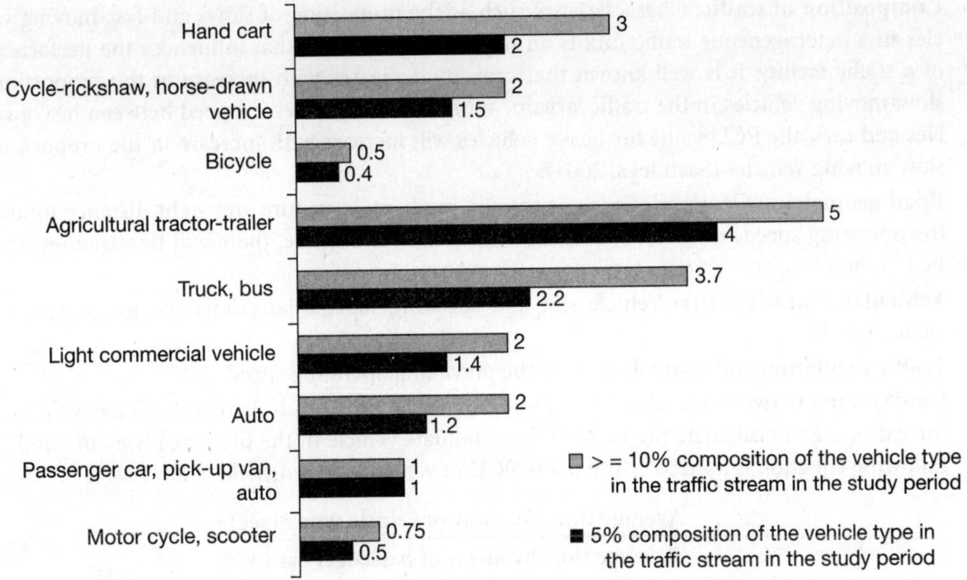

Figure 4.25 IRC recommended PCU factors applicable on urban roads in plains (Modified from IRC: 106-1990).

$$\text{PCU}_\text{i} = \frac{(\text{Speed ratio of the car to the } i\text{th vehicle})}{(\text{Space ratio of the car to the } i\text{th vehicle})} = \frac{\left(\dfrac{\text{Velocity of the car}}{\text{Velocity of the } i\text{th vehicle}}\right)}{\left(\dfrac{\text{Space occupied by the car}}{\text{Space occupied by the } i\text{th vehicle}}\right)} \tag{4.30}$$

4.11.1 Factors Affecting Equivalency Factors (PCU Values)

The PCU conversion factors are subject to *variation,* which depends on the following factors:

- **Width of the pavement:** It was observed that the mean free speed of cars increases linearly with the width of the carriageway over a certain range of width—5.2 11.3 m (Leong, 1968; Farouki and Nixon, 1976). As the freedom of movement increases with the width, drivers will have more opportunities to increase their speed on the roads. This results in higher PCU values for buses and trucks.
- **Condition of the shoulder:** The PCU value of a vehicle on a road increases with increase in the quality of the shoulder. A better shoulder increases the effective width of the carriageway and results in higher PCU values. This is because of the higher speed differential between cars and heavy vehicles (Sachdeva, 2004).
- **Surface roughness/unevenness characteristics:** It is widely known that the free-flow speed of a vehicle decreases with increased roughness of the road surface. As speed decreases, the PCU value also decreases.
- **Directional split of traffic:** In two-lane roads, the PCU value decreases as the directional split of traffic deviates from the 50/50 proportion. If the traffic is not balanced in both directions, overtaking opportunities will reduce and the vehicles in the stream will be forced to travel at low speed. Hence, the PCU values for heavy vehicles will be reduced (Sachdeva, 2004).

- **Composition of traffic:** Characteristics such as the proportion of slow- and fast-moving vehicles in a heterogeneous traffic mix is an important parameter that influences the performance of a traffic facility. It is well known that capacity decreases with increase in the proportion of slow-moving vehicles in the traffic stream. With large differences in speed between heavy vehicles and cars, the PCU value for heavy vehicles will increase with increase in the proportion of slow-moving vehicles (Sachdeva, 2004).
- **Road geometrics:** Geometrics such as terrain, gradient, curvature and sight distance influence the operating speeds at different locations in an area, and hence, there will be variations in the PCU values.
- **Vehicular characteristics:** Vehicle size and speeding capabilities contribute to change in the value of PCU.
- **Traffic regulation and control:** As with the prevailing operating speed
- **Gap/Spacing between vehicles:** The ratio between the time headway of a candidate vehicle and the car is used to calculate the PCU of the candidate vehicle in the basic headway method. The following equation is used to estimate the PCU of vehicles (Greenshield et al., 1947):

$$PCU_i = \frac{\text{Average time headway of vehicle type } i \text{ (sec)}}{\text{Average time headway of passenger car (sec)}} \tag{4.31}$$

Several road user cost studies have been conducted in India, and separate sets of PCU values have been recommended for use in India *for specific traffic conditions*:

- Roads in rural areas (IRC: 64-1990)
- Urban roads (IRC: 106-1990)

4.12 ORIGIN AND DESTINATION STUDIES

Origin and destination (OD) studies are conducted to understand traffic movements across and within a selected study area in a period of time. The survey may be related to specific trip studies pertaining to vehicles, pedestrians, cyclists and goods in a small residential area or a large urban area or any geographical region or crossing key screen lines within a study area.

While compiling the OD data, the percentage of the sample interviewed should be converted to represent a 24-hour day and consolidated under the following heads:

- Date and time
- Place
- Trip origin (specify the place on the street)
- Trip destination (specify the place on the street)
- Purpose of the trip (business or non-business)
- Trip frequency
- Type of vehicle used, with occupancy
- Vehicle ownership
- Type of fuel used (for air quality planning)
- Vehicle registration number (for private transport vehicles)
- Total average time of travel, from origin to destination
- Weather conditions during the journey
- Details of the delays that occurred (with place and duration)

The OD study area

Initially, the transportation engineer must identify a study area with its boundary based on an objective. An imaginary line, considered the boundary line of the study area, is termed the *cordon line*. The political borders of an area such as a town, district, mandal, village or a group of adjacent colonies with a similar traffic flow pattern are often considered as cordon lines for OD surveys. Direction-wise (i.e., entry and exit) vehicle flow data across the cordon line is recorded using any one or a combination of methods, discussed in the next section.

The study area is further divided into a suitable number of zones based on similar traffic flow pattern. These zones are often termed traffic analysis zones (TAZs). Each TAZ may have one main and other minor internal traffic centres. The main traffic movements across the cordon line (i.e., boundary line) are identified as Entry/Exit nodes. A sample demarcation of a study area with 14 TAZs and 4 Entry/Exit nodes are presented in Fig. 4.26.

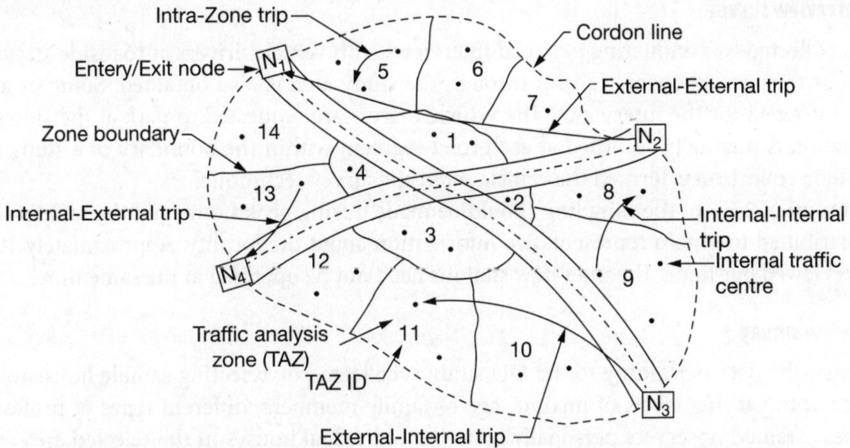

Figure 4.26 A sample OD study area with traffic analysis zones (TAZ) and types of trips (modified and elaborated, based on Guy and Fricker, 2005).

After conducting the OD survey and collecting the data, the trip distribution in a study area is presented as an OD matrix. The OD matrix is represented as an $n \times n$ matrix; where n = (Number of TAZs + Number of Entry/Exit nodes) considered in a study area. Based on the origins and destinations in the study area, the trips can be classified into four types:

I–I = Internal–Internal trips

I–E = Internal–External trips

E–I = External– Internal trips

E–E = External–External (through) trips

The methods of collecting OD data may be categorised as three types: questionnaire mail-back-based methods, (b) interview-based methods and (c) camera-based detection methods. The camera-based detection methods are suitable in situations where most of the respondents are unwilling to share information because of concerns about privacy. The details of conducting OD surveys are discussed below.

Registration questionnaire survey

A list of the addresses of vehicle owners, arranged according to their licence plate number within a study area, is obtained from the concerned Motor Vehicle Registration Department. Each questionnaire is printed with the full licence plate number. Each vehicle owner is sent the questionnaire to complete and is requested to mail it back within a certain date. Low response rates are usually expected, generally 15–30%. The cost of printing and mailing the survey questionnaire forms will be additional. Alternatively, internet-based online forms may be explored, if possible.

Mail-back questionnaire or self-completion postcard survey

A prepaid return postcard printed with the questionnaire is distributed at a given time to all the drivers entering roadways into the boundary of a study area. However, the cards may also be distributed to drivers who have already stopped/halted at rest areas, toll booths and at specific border crossings. This method is less expensive than the licence plate survey. Traffic disruption is inevitable during distribution of the postcards. This method of surveying is less disruptive than the roadside interview method. However, the challenges associated with the registration questionnaire method hold good for this as well.

Roadside interview survey

OD data is collected by conducting personal interviews with vehicle drivers at roadside stations; information about their travel characteristics through the study area is also obtained. Some or all passing vehicles are stopped for the interviews. The vehicle drivers are requested to park at the side safely and the information is manually recorded at each road segment within the boundary of a study area. This process of data collection is termed the vehicle intercept survey technique.

A minimum of 25% of the sampling should be made during peak hours and the remainder may be suitably distributed to obtain representative information about the locality. Approximately 100 drivers can be interviewed per hour. The interview stations need not be operated at the same time.

Home interview survey

In this survey, the data pertaining to the OD study is collected by selecting sample households within a study area with varying levels of income, age of family members, different types of professions and vehicle types. Trained observers personally go to the individual houses in the selected area and record household information with their travel activity by conducting home interviews with them. The household surveys can also be conducted by telephone or mail. Such surveys are conducted for a certain time period, generally 24 or 48 hours. The information is aggregated with other similar household travel trips and summarised as average trip rates and trip lengths for that type of household. By combining this information with trip production and attraction analysis of travel demand modelling, an OD matrix for that study area is defined (Guy and Fricker, 2005).

This survey provides the most detailed data including the social and economic characteristics of transport users that can be used for deriving general travel patterns from a large sample of urban population for all the roadways in a study area. However, it does not provide detailed information on any particular roadway within a particular study area. Accuracy of the OD data depends on the sample interviews in the selected dwelling units.

Tag-on-car survey

This method has been found to be suitable for studying through trips. The vehicles tagged with specific colour ribbons at each entry station are tallied at the exit stations, and the resulting origin and destination data of the tagged vehicles is summarised for a study area. This method can be used to answer specific problems only. Matching the colour ribbons with the specific stations is very difficult, but this will dictate the accuracy of the data collected.

Carousel method

A team of trained observers with a driver travel in a survey vehicle on the roadway segment being studied more slowly than the general traffic stream. The observers enter the licence plate numbers of the passing vehicles in a palmtop. The survey vehicle may travel at speeds of 10–15 miles an hour below the prevailing traffic speed.

Licence plate survey

The licence plate numbers of vehicles at each entry and exit station on a road in a study area are recorded and tallied to summarise the OD data. The locations of such roadside stations are identified on border (i.e., cordon) line of a study area. The following methods are used to record the licence plate numbers of vehicles in this survey (Bradshaw, 1999):

1. An observer at a roadside station observes the vehicles crossing across the cordon line and calls out their licence plate numbers, as they pass. These numbers can be initially recorded on paper using a pencil, or may be directly entered into an Excel spreadsheet using a computer device, by a field helper.
2. An observer at a roadside station observes the vehicles crossing the cordon line and calls out their licence plate numbers; they are then recorded in a tape recorder or palmtop and finally, the data is entered into an Excel spreadsheet.
3. An observer at a roadside station observes the vehicles crossing the cordon line and dictates the licence plate numbers into a microphone; they are then recorded automatically in an Excel spreadsheet using computer-based voice recognition technology.
4. Roadside stations with mounted video recordable cameras are used to capture images of the moving vehicles with their licence plate numbers. Later, such video tapes are manually transcribed
5. **GPS tracing:** The vehicles equipped with GPS are traced and their trip data such as latitude and longitude with time are recorded for analysis of OD trips. This information is automatically recorded and analysed by a software package, without their knowledge. One of the major limitations of this technique is that only a small proportion of vehicles may be equipped with GPS. Hence, the sample of vehicles traced by this method is not likely to represent a sample of the total population.
6. **Wireless phone tracing:** The following two methods are used to obtain the location of vehicles (i.e., indirectly) and their moving paths. In both cases, the coordinates of the wireless phones with time in a study area are analysed to determine the OD trip data.
   - In the *handset method*, the coordinates of wireless phones equipped with a GPS receiver are obtained in a study time period and the data is analysed to determine the OD trip matrix.
   - In the *network-based method*, signals from the wireless phone in the study area with their time of occurrence are received from the nearest cell towers and the data is analysed based on triangulation principles to pinpoint the location of the wireless phone, either by measuring the angles of signal arrival or the difference in time when the signal reaches the respective towers. Finally, the paths traced by those vehicles are plotted to a GIS package to reveal the travel patterns of persons carrying cell phones.

Currently, these methods are being used in the USA. However, further technological improvements are required in the collection of travel information in a study area and their processing techniques to minimise errors.

Note: The OD data collected using the methods described above may be augmented with additional studies such as telephone survey, land-use study and classified population growth trend study for new routes or changes in existing routes by the local authorities. The data collected is geo-coded and presented using Geographic Information System (GIS) software. GIS is a standard approach used for storing, managing, classifying, manipulating and displaying all forms of travel survey data (Bradshaw, 1999).

Selecting a method

The objectives of conducting an OD survey determine the method selected. It must be noted that different methods of OD survey yield different types of information (number of trips between OD pairs, exact origins and destinations, socioeconomic data). Each method has its merits and limitations. Hence, the surveyor must select a suitable method so as to obtain the required information with minimum errors, time and cost. The traffic condition (i.e., vehicle type, speed and flow rate) on the road for which OD data is being collected is also an important factor.

Uses of OD data

The information collected from OD studies can be used for the following purposes:

- To locate expressways and plan for major routes on the desired lines
- To establish preferential routes and their associated needs for budgetary fund allocation
- To anticipate present and future trends of travel patterns, trip purposes, trip frequency and traveller demographics, based on existing traffic demand on the road network and subsequently, for development of transport models
- To find out the existing number of trips made within, through and across the study area or a specified space boundary, with reference to time of day, mode of travel and occupants per vehicle. Multimodal transport studies may also be considered as part of this study.
- To determine current travel patterns in terms of the areas that generate and/or attract more traffic. Such data, with traffic count, will be used to improve the efficiency of traffic flow per lane with respect to maximum allowable capacity, level of service and safety.
- To evaluate general road plans relating to current or future problems or needs based on analysis of travel characteristics such as travel time, delay and pollution
- To determine the need for revised traffic-flow patterns, alternative routes, bypass roads, ring roads and parking areas

Presentation of OD data

Several methods are available to represent the information obtained from OD studies. These include tabular form and graphical form (Fig. 4.27). The graphical form is widely used and comprises (i) Desire line diagrams and (ii) Route volume maps.

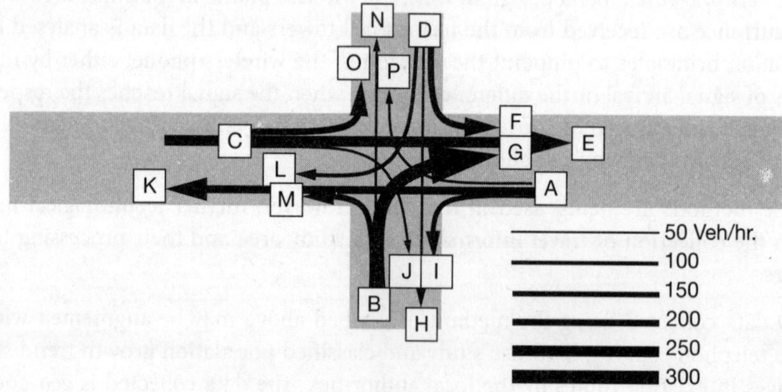

Figure 4.27 Typical presentation of traffic flow at intersections with traffic volume scaled lines.

In the case of desire line diagrams, various locations representing the origins and destinations in a study area are connected with straight lines. The thickness of these straight lines varies according to the thickness scale representing traffic volume in both the directions from one key area to the other. The desire line diagram represents the major traffic-flow patterns in a study area, and is used more often for transport planning purposes.

The route volume map is prepared by drawing lines of varying thickness on an actual topographic road map. The lines bend in the directions indicating the origin and destination of traffic-flow pattern between any two locations on the map.

EXAMPLE 4.11

A survey was conducted in a residential colony to establish a relationship by relating number of trips generated per household per day (N) with number of persons per household (P) and number of vehicles per household (V), in the colony. The relationship found by conducting statistical analysis as $N = 1.5P + 2.0V + 0.35$.

Based on the above condition, how many trips a neighbouring colony having 400 households is expected to generate if the input parameters are chosen as, average 3.8 persons and 2.3 vehicles per household?

Also determine, the number of car trips if the modal split of the V is assumed to be 40% car, 45% bus and local train, and 5% by other modes.

Answer

By substituting the given values as $P = 3.8$ persons/household and $V = 2.3$ vehicle/household,

The number of trips generated by 400 households per day $(N) = 400 \times (1.5P + 2.0V + 0.35)$

Number of trips generated per households per day $= N = 400 \times ((1.5 \times 3.8) + (2.0 \times 2.3) + 0.35)$

$$N = 400 \times 10.65 = 4260$$

Number of car trips per day $= 0.40 \times 4260 = 1704$

4.13 TRAVEL TIME AND DELAY STUDIES

The specific purpose of OD surveys is to identify the locations of delays (time spent in any form of stoppage and also due to slow speed) on various stretches of roads in urban and rural areas. In this process, the road facility is evaluated to determine the factors that cause delays and stoppages, and to make recommendations for improving the flow conditions of traffic.

4.13.1 Methods of Measurement of Travel Time and Delays on any Selected Length of a Route

Several methods/techniques are available to measure travel time and delays on a selected length of route (Fig. 4.28). Each method is based on a specific type of technique of field measurement and possesses its own advantages and limitations. Any one of the methods discussed below may be selected based on the nature and objectives of the study (Treadway, 1965).

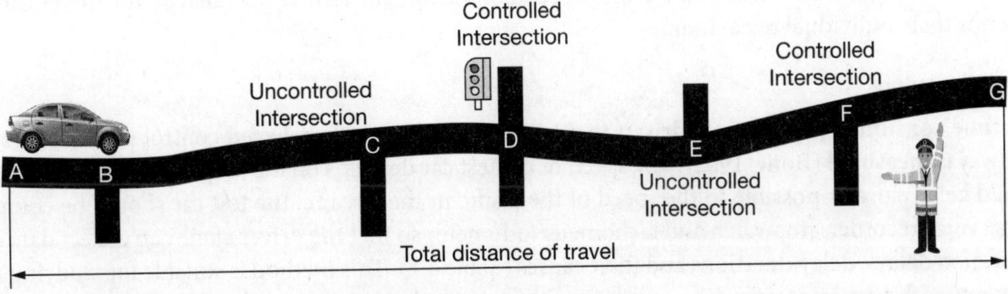

Figure 4.28 Schematic diagram of a route selected for conducting speed and delay survey.

4.13.2 Floating Car Method (Moving Observer Method)

In this method, the driver is instructed to maintain the speed of the test vehicle at the representative mean speed of all the traffic in the stream (i.e., floating in the traffic stream), and pass approximately the same number of vehicles that passed by the test vehicle. However, the driver does not try to overtake fast-moving vehicles (Bone, 1952). Therefore the test vehicle approximates the behaviour of an average vehicle in the traffic stream. This method is generally applied only on two-lane highways.

At least 12 trips need to be made in each direction to obtain reliable data about average speed and delays in that direction of traffic flow (for example, A→G and G→A, see Fig. **8.1**). As a part of data collection, three enumerators may be employed in the test vehicle to count the number of vehicles travelling in the opposite direction, the number of overtaking vehicles and overtaken vehicles and note the running speed, distance travelled, delays at each location and their causes. Stopwatches, survey sheets, notepads and voice recording devices may be used to record the speed and delay survey data.

The following equations may be used to analyse the collected data.

$$\text{Average travel time in a direction, } \mathrm{ATT}_{\mathrm{Direction}} = t_{\mathrm{a\_Direction}} - \left(\frac{N_{\mathrm{diff}}}{Q_{\mathrm{Direction}}}\right) \tag{4.32}$$

$$\text{Traffic volume in a direction, } Q_{\mathrm{Direction}} = (Veh_{\mathrm{Opp}} + N_{\mathrm{diff}})/(t_{\mathrm{a\_Direction}} + t_{\mathrm{a\_Opposite\ Direction}}) \tag{4.33}$$

$$\text{Space mean speed in a direction} = \text{Distance/Average time} = \text{Distance} / \mathrm{ATT}_{\mathrm{Direction}} \tag{4.34}$$

where $\mathrm{ATT}_{\mathrm{Direction}}$ = Average travel or journey time for *all* the vehicles in a direction of flow, in minutes

$t_{\mathrm{a\_Direction}}$ = Average travel time (in minutes) of the test vehicle travelling in the direction of $Q_{\mathrm{Direction}}$

$t_{\mathrm{a\_Opposite\ Direction}}$ = Average travel time (in minutes) of the test vehicle while travelling in the opposite direction

N_{diff} = Difference between the number of vehicles overtaking and the number of vehicles overtaken by the test vehicle, while travelling in the same direction

$Q_{\mathrm{Direction}}$ = Volume of vehicles per minute, in the direction of flow; and

Veh_{Opp} = Average number of vehicles travelling in the direction opposite to the test vehicle.

This procedure has been found to be reliable for two-lane two-way highways during low-to-moderate volumes and over long test runs. A major advantage of this method is that the locations of delays and their causes can be readily identified. Measurement of travel time for each individual section length over a long test distance is also possible (Berry, 1952; Cribbins et al, 1961). A major disadvantage of this method is that unreliable results may be obtained in the case of (a) high-volume traffic conditions and (b) multi lane highways because the average speed of the test car vehicle operated by the driver directly depends their individual behaviour.

Test car procedure

The time consumed by a test car driven in a traffic stream between selected control points of a study roadway is measured (Bone, 1952). The speed of the test car depends on the judgement of the driver and should be as close as possible to the speed of the traffic in the stream. The test car should be equipped with a voice recorder, stopwatch and tachometer/odometer so that the driver alone can collect data such as location delays, delay durations and their causes (Table 4.7). This method is suitable for studying road segments with very low traffic. The accuracy of the test results depends on the number of trips repeated over the study road segment; 8–12 such trips in each direction may be recommended to obtain reasonably good results.

Table 4.7 Sample speed and delay field sheet.

Site Name: Cotton Highway		Surveyors Names: Ahmed and Sony	Weather: Sunny		Route No: 133
Date: 06.08.15		Start Time: 10:00 am	Start Location: Milestone 2.0		Direction: East
			Per Section		
Check Point	Cumulative Distance along Route (km)	Cumulative Travel Time (minutes:seconds)	No. of Stops	Stopped Delay (Sec)	Reasons/Remarks
Library	1.22	1:45	0	0	–
Post-office	3.23	3:36	0	0	–
Tea-point	4.65	5:55	2	48	Stops due to parked vehicles
Sale Mall	6.78	9:14	1	15	Stop due to school van
Park Road	7.21	10:18	0	0	–
Pharma	8.21	11.34	1	11	Stop due to pedestrian
Bus Stand	9.54	13.45	3	58	Stops due to signals

Licence plate matching method

Licence plate number data can be collected in two ways: by the manual method or by the automatic number plate recognition (ANRP) technique.

In the case of the manual method, observers at the start and end points of a test length of a road record the licence numbers of vehicles along with their entry and exit times. Each observer located at such strategic points records the last three or four digits of the licence plates. The difference in the time values of a particular licence number of a vehicle is reported as their travel time. The travel time measured by this technique provides a *true measure* of the time that reflects the individual driver's characteristics (Cribbins et al.,1961). The limitations of this method are that only total travel time is measured, irrespective of the number of delays, with the locations and their causes. Also, it is time consuming to record the licence plate numbers of vehicles at each entry and exit point, and matching the same during calculation of travel time is not an easy task (Walker, 1957).

Elevated observation method

The travel information of any randomly selected vehicle movement along a specific short test length of road is obtained using elevated roadside video cameras or with a couple of observers stationed at elevated vantage points. The videographs are reviewed later in the office and the pertinent information along the control sections of the road is determined. This method is suitable for the study of delay at intersections and over short test sections only.

4.13.3 GIS and GPS Method

Global positioning system (GPS) is used to determine the exact location of a user on earth at any time and in any weather condition. There are 24 GPS satellites in orbit 11,000 nautical miles above the earth. They are continuously monitored by ground stations located worldwide. The radio signals transmitted by these satellites can be detected by any GPS receiver and can be used to determine the user's location with great precision. One limitation, however, is that traffic flow through tunnels and under dense trees and tall buildings may disturb signals and distort the user's current position (Gallagher, 1996).

Geographic information system (GIS) is a computer software package capable of storing, categorising, manipulating and displaying the collected data on a defined geographically referenced map. The geographic maps and data are stored as layers of information and the GIS algorithms enable the user to perform complex analysis for producing reports in a predefined format. In addition, GIS is used as a software platform that is capable of easily integrating/referencing travel time data with a spatial model database

GPS and GIS are the latest techniques by which travel time data collection is made easy. This method can eliminate the errors and inaccuracies that arise from the use of conventional study methods of travel time and delay. The GPS integrated signal data on a study area of a GIS map can depict the current location and movement of the test vehicle. The surveyor need not record the time elapsed between road junctions. The test vehicle location can be automatically traced through a GPS receiver mounted on the vehicle. The GPS signal data recorded on a digital map of GIS is used to analyse and summarise the travel time, distance travelled and delays with that location. As the total delay, total travel time and the total distance covered are known, the average travel and running speeds over the test segment of the road can be easily computed (Mauricio et al,. 2003).

EXAMPLE 4.12

A summary of the speed and delay study conducted by test car method over a road segment length of 2.14 km is given below. Seven runs were made in each direction.

East-bound trips				
Trip number	**Total journey time Minutes : Seconds**	**Number of vehicles met against the stream (i.e., towards West)**	**Vehicles overtaking the test car**	**Vehicles overtaken by the test car**
1	2:22	39	3	5
2	2:16	44	3	3
3	2:25	45	4	5
4	2:43	43	3	5
5	2:19	41	2	3
6	2:28	39	5	3
7	2:13	44	3	3
Total	16.77 Min.	295	23	27
Average	t_{a_East} = 2.40 Min.	Veh_{Opp} = (295/7) = 42.14	O_{diff} = (23-27)/7 = –0.57	

West-bound trips				
Trip number	**Total journey time Minutes : Seconds**	**Number of vehicles met against stream (i.e., towards East)**	**Vehicles overtaking the test car**	**Vehicles overtaken by the test car**
1	2:32	46	4	3
2	2:12	49	5	3
3	2:46	30	3	4

(Continued)

West-bound trips				
Trip number	Total journey time Minutes : Seconds	Number of vehicles met against stream (i.e., towards East)	Vehicles overtaking the test car	Vehicles overtaken by the test car
4	2:29	37	3	5
5	2:17	43	2	4
6	2:32	38	4	3
7	2:35	43	2	4
Total	17.38 Min.	286	23	26
Average	$t_{a\_West} = 2.48$ Min.	$Veh_{Opp} = (286/7) = 40.86$	$N_{diff} = (23\text{-}26)/7 \approx -0.43$	

Calculate (a) Traffic volume in the East-bound direction
(b) Travel (average journey) time in the East-bound direction
(c) Space mean speed in the East-bound direction
(d) Traffic volume in the West-bound direction
(e) Travel (average journey) time in the West-bound direction
(f) Space mean speed in the West-bound direction.

Answer

Speed and delay parameters of East-bound traffic:

(a) By Eq. 4.33

Traffic volume in the East-bound direction, $Q_{East} = (\text{Veh}_{Opp} + \text{N}_{diff})/(t_{a\_East} + t_{a\_West})$

On substitution, we get

$Q_{East} = (42.14 + (-0.57))/(2.40 + 2.48)$

$Q_{East} = (42.14 - 0.57))/(4.88)$

$Q_{East} = (41.57)/(4.88) \approx 8.52$ vehicles per minute

Or $Q_{East} = 8.52 \times 60 = 511.10$ vehicles per hour

(b) Average travel (average journey) time in the East-bound direction (ATT_{East}):

By 4.32

$$\text{ATT}_{East} = t_{a\_East} - \left(\frac{N_{diff}}{Q_{East}}\right)$$

$$\text{ATT}_{East} = 2.40 - \left(\frac{-0.57}{8.52}\right) = 2.46 \text{ minutes}$$

(c) By Eq. 4.34,

Space mean speed in the East-bound direction = Distance/Average travel time = 2.14/(2.46/60) = 52.19 km/hr

Speed and delay parameters of West-bound traffic:

(d) By Eq.4.33,

Traffic volume in the West-bound direction (Q_{West}) = $(Veh_{Opp} + N_{diff})/(t_{a\_East} + t_{a\_West})$

On substitution, we get

$Q_{West} = (40.86 + (-0.43))/(2.40 + 2.48)$

$Q_{West} = (40.86 - 0.43))/(4.88)$
$Q_{West} = (40.43)/(4.88) = 8.28$ vehicles per minute
Or $Q_{West} = 8.28 \times 60 = 496.8 \approx 497$ vehicles per hour

(e) Average travel (average journey) time in the West-bound direction (ATT_{West}):
By Eq. 4.32

$$\text{ATT}_{\text{West}} = \text{t}_{\text{a\_West}} - \left(\frac{\text{Ndiff}}{\text{QWest}}\right)$$

$$ATT_{West} = 2.48 - \left(\frac{-0.43}{8.28}\right) = 2.53 \text{ minutes}$$

(f) By Eq.4.34,
Space mean speed in the West-bound direction = Distance/Average travel time
= 2.14/(2.53/60) = 50.75 km/hr

EXAMPLE 4.13

A total of six test-car travel time and delay runs were conducted on a 3 Km route. Determine the average travel speed and average running speed for the test route based on the individual travel times and total delays of each run as given below.

Test run #	Travel time		Delay (sec)
	Minutes	Seconds	
1	6	55	55
2	7	20	95
3	8	15	115
4	6	10	5
5	8	35	165
6	6	50	25

The given data is arranged as:

Test run #	Travel time (sec)	Delay (sec)
1	$6 \times 60 + 55 = 415$	55
2	$7 \times 60 + 20 = 440$	95
3	$8 \times 60 + 15 = 495$	115
4	$6 \times 60 + 10 = 370$	5
5	$8 \times 60 + 35 = 515$	165
6	$6 \times 60 + 50 = 410$	25
Total = 2645		Total = 460

$$\text{Average travel speed} = \frac{\text{Distance travelled}}{\text{Average of total travel time}} = \frac{3000\text{ m}}{\left(\dfrac{2645\text{ sec}}{6}\right)}$$

$$= \frac{3000\text{ m}}{440.83\text{ sec}} = 6.80\frac{\text{m}}{\text{sec}} = 24.48\text{ km/hr}$$

$$\text{Running time} = \text{Total travel time} - \text{Total delay} = 2645 - 460 = 2185\text{ sec}$$

$$\text{Average running speed} = \frac{\text{Distance travelled}}{\text{Average of total running time}} = \frac{3000\text{ m}}{\left(\dfrac{2185\text{ sec}}{6}\right)}$$

4.14 INTERSECTION DELAY STUDIES

Delay studies are conducted at individual intersections and the data obtained is analysed to evaluate the efficiency or effectiveness of the existing traffic-control method. The outcome of the analysis is used to determine the congestion level and level of service of the intersection being studied.

Peak hour delay studies are commonly conducted at isolated intersections that have congestion problems. Average control delay studies at intersections are conducted to evaluate the delay conditions and to explore the option of signalisation.

Travel time delay

It is the difference between the actual time required to traverse a section of a street and the time corresponding to the average speed of traffic in uncongested/ideal conditions. This delay includes stopped delay and acceleration and deceleration delays.

$$\text{Travel time delay} = (\text{Actual travel time} - \text{Ideal travel time})\text{ through an intersection}$$

Ideal travel time through an intersection is determined as the average speed during uncongested traffic flow.

Methods of measuring delay through intersections

A major part of any delay occurs in urban localities at intersections, particularly at signalised intersections (considered as interrupted flow at controlled intersections). Two survey techniques are available to determine the delay or performance of intersections: viz., (*a*) Stopped vehicle count method and (*b*) Elevated observer method. A brief description of each of these techniques is given below.

4.14.1 Stopped Vehicle Method

The number of stationary queuing vehicles at the respective approaches of an intersection is counted at fixed time intervals (say, 15 or 30 seconds, for a duration of 5 or 10 minutes). The observation interval must not match the regular signal cycle time. The same fixed observation time is considered for all approach observations of an intersection. The selected duration (say, 10 or 5 minutes) of observation of delays at an instance may be used for different approaches surveyed in rotation. All the approaches at an intersection should be surveyed simultaneously for the same duration if the delays on these approaches are required to be correlated. Classified vehicle volume with turning movements is also counted at the same time. If the longest queues are observed during peak hours that extend back to the upstream intersection, then the floating car method or licence plate matching methods are adopted (TRL, 1993).

A minimum of two observers may be employed at each approach for data collection.

This method provides only the total delay (expressed as Vehicle-sec = Total number of queuing vehicles counted × Time interval in sec) and average delay in queue per approach vehicle (i.e., Total delay ÷ Number of approaching vehicles during that period) for each approach or for all approaches, for the period of time observed. The value obtained may be multiplied with an empirical adjustment factor (≈0.9) to account for errors due to sampling that tend to overestimate the delay in field studies.

Acceleration/Deceleration delay correction is applied to average delay in queue per approach vehicle with the TRB (2000) recommended values (Table 4.8).

Table 4.8 Partial table of acceleration–deceleration delay correction factors (*Source:* TRB, 2000).

Free flow speed (km/hr)	Number of vehicles stopping per lane, per cycle	
	≤ 7 vehicles	8–19 vehicles
≤ 60	+ 5	+ 2
> 60–71	+ 7	+ 4
> 71	+ 9	+ 7

A sample of such survey data is presented in Table 4.9. It does not provide details of the delays that occurred with turning movements. In addition, this method is not suitable for studying delays on roundabouts because a rolling queue of vehicles may be present. This method is recommended by the Highway Capacity Manual (TRB, 2000).

Table 4.9 A partial sample sheet of a completed delay by the stopped vehicle method.

Site name: Church Road			Surveyors' names: Sira and John			Weather: Sunny			Location sketch:		
Date: 06.06.15			Time: 9:00 to 9:30			Interval Time: 30 sec					
Approach Name: Park Lane West			Direction: East			Number of Lanes: 2 (One way)			Notes: A Road show held 1 km away		
Time [1]			Number of stopped vehicles, by class [2]			Time [1]			Number of stopped vehicles, by class [2]		
Hr	Min.	Sec	2W	3W	4W	Hr	Min.	Sec	2W	3W	4W
9	00	00	1	3	2	–	–	–	–	–	–
	30		9	0	4	–	–	–	–	–	–
	01	00	6	4	6	–	–	–	–	–	–
		30	4	5	0	–	–	–	–	–	–
	02	00	6	4	5		29	00	7	3	5
–	–	–	–	–	–			30	9	6	2
–	–	–	–	–	–	9	30	00	6	0	6

4.14.2 Elevated Observer Method (or Videographic Method)

In this method, an observer or a stationary elevated video camera is installed over a study road intersection and the number of vehicles that are stopped at each approach to the intersection is recorded during the study period. Such video cameras are capable of continuously recording traffic flow through an intersection. This data is later transcribed from the video tapes for each 15- or 30-second interval, over the period of the study. The classified vehicle volume count along with their turning movements is also recorded at the same time. A sample of such survey data is presented in Table 4.10.

Table 4.10 A sample form for conducting an intersection delay survey by the elevated observer method.

Site name: Beach Road Jn.		Surveyor's names: Sophia and Yen			Weather: Cloudy		Location sketch:	
Date:12-12-2015		Time: From 8:30 to 9:30			Carriageway widths: 7.5m each			
Approach name: School lane side		Direction: South			Number of Lanes: 2		Notes: Police help taken	
Vehicle type	Turning movement	Time of Passing			No. of Stops	Total stopped time		
		Hr.	Min.	Sec.		Min.	Sec.	Reason due to
Car	RT	9	00	10	3	00	6	Pedestrian
Truck	Thr	9	03	18	2	00	8	Bus
Van	LT	9	06	14	1	00	5	Police
2W	RT	9	08	32	1	00	4	Veh. Stopped
–	–	–	–	–	–	–	–	–
–	–	–	–	–	–	–	–	–

This survey method is suitable for determining the delays caused by turning movements at all types of controlled and uncontrolled intersections including roundabouts, where high-vantage video camera installation points are available. Care should be taken while mounting the video camera over an intersection so that three points are clearly visible from the shooting point, as described below:

- The *start timing point* on an approach, which is at least 10 m beyond the longest anticipated length of the vehicle queue, with an easily identifiable object on the ground
- The *sampling point,* which is about 20 m in advance of the start timing point
- The *finish timing point,* as the stop line representing the exit for the intersection

In addition, the cause of the delay can also be recorded by correlating with signal controls, vehicles parked on kerbs, stopped vehicles, cattle and pedestrians.

The total and average delays are derived for individual vehicles based on the difference in travel times observed from the video file and summarised for each approach or for all approaches to a junction.

Intersection delay is a measure of intersection performance, usually presented in the form of average delay per vehicle. The total stopped time of the vehicles (total stopped vehicle-time in units of total vehicle-seconds of delay) is computed by multiplying the total number of stopped vehicles by the time interval between the observations. It must be noted that the selected time interval should not be an exact multiple of the signal cycle length. This criterion enables sampling of data during different parts of the signal phase (Berry and Van Till, 1954).

EXAMPLE 4.14

The following details pertain to a vehicle delay study conducted on one approach to an intersection:

The stopped vehicle method was used for the sampling procedure. Vehicles occupying the intersection approach at successive intervals of 15 seconds for a study period of 5 minutes were counted (i.e., over 5 cycles). Take the signal cycle length as 116 seconds and the number of lanes as 2.

Vehicles occupying the intersection approach in each successive count of 15 seconds were counted by Observer-1. The number of vehicles passing through the intersection approach in the time period of 5 minutes was counted by Observer-2.

Observer	Observer-1 (Instantaneous density)				Observer-2 (Traffic volume)	
	Number of vehicles in the queue/ stationary in the approach, at time				Approach volume category	
Starting time	0 Sec	15 Sec	30 Sec	45 Sec	Number of vehicles that stopped in more than one observation	Number of non-stopped vehicles
17:30	0	6	8	7	11	7
17:31	4	7	4	6	7	13
17:32	6	9	10	7	12	7
17:33	5	11	6	11	14	6
17:34	7	3	1	5	6	12
Sub-totals	22	36	29	36	50	45
Total	123				95	

Compute (a) Total delay at the approach to the intersection; (b) Average delay per stopped vehicle; (c) Average delay in the queue per approach vehicle; (d) Percentage of vehicles stopped at the intersection; (d) Percentage of vehicles stopped at the approach; (e) Number of cycles surveyed; (f) Number of vehicles stopping per lane each cycle; and (g) Control delay per vehicle.

Answer

(a) Total delay at the approach to the intersection:

Total delay = Total number of vehicles in queue × Time interval

= 123 × 15 = 1845 vehicle seconds

(b) Average delay per stopped vehicle = Total delay ÷ Number of stopped vehicles

= 1845 ÷ 50 = 36.9 seconds

(c) Average delay in the queue per approach vehicle = Total delay ÷ Number of approaching vehicles

= 1845 ÷ 95 = 19.42 seconds

Note: This value may be multiplied by an empirical adjustment factor (≈ 0.9) to account for errors due to sampling that tend to overestimate the delay in field studies.

Average delay in queue per approach vehicle = 19.42 × 0.9 = 17.47 seconds

(d) Per cent of vehicles stopped at the approach = Stopped vehicles ÷ Approaching vehicles

$$= (50 \div 95) \times 100 = 52.63\%$$

Fraction of vehicles stopped (FVS) = 0.5263

(e) Number of cycles surveyed = Cycle length/15 = 116 /15 = 7.73

(f) Number of vehicles stopping per lane per cycle = 50/ (7.73 × 2) = 3.23 ≤ 7 vehicles

From Table 9.1 (TRB, 2000), the given value of free flow speed = 55 km/hr (i.e., ≤ 60 km/hr); so take acceleration/deceleration delay correction factor = + 5 seconds

Therefore, acceleration/deceleration correction delay = 0.52 × (5) = + 2.6 seconds

(g) Control delay per vehicle = (Average delay in queue per approach vehicle + Acceleration/ Deceleration correction delay)

Control delay per vehicle = (17.47 + 2.6) = 20.07 seconds

EXAMPLE 4.15

A vehicle wanted to cross an unsignalised intersection and make a right turn. In this process, the vehicle stops at the stop line of the intersection at 12:00:00 pm and starts entering the intersection at 12:00:06 pm. While taking right turn, it encountered 7 opposing vehicles (i.e., conflicting vehicles) and finally crossed the intersection by 12:00:13 pm. Calculate the instantaneous service delay and conflicting rate at the intersection.

Answer

Service delay at the stop line of the intersection = 12:00:06 – 12:00:00 = 6 seconds

$$\text{Conflicting flow rate} = \left(\frac{\text{Number of conflicting vehicles}}{\text{Total delay}}\right)$$

Number of conflicting vehicles = 7

$$\text{Conflicting flow rate} = \left(\frac{7}{12{:}00{:}13\text{-}12{:}00{:}00}\right) = \left(\frac{7\text{ veh}}{13\text{ sec}}\right) = 0.538\text{ veh/sec}$$

4.15 ROAD TRAFFIC ACCIDENT STUDIES

Traffic safety studies are conducted to evaluate, examine, identify and review the causes of road traffic accidents based on the information collected, and recommend necessary improvements to the road geometrics with their safety operations and appurtenances. The objective is to implement the necessary corrections to eliminate hazardous conditions in the highway right-of-way.

Traffic safety studies/surveys are conducted to collect information relating to traffic characteristics, vehicular characteristics, driver/pedestrian/user characteristics, road surface condition characteristics, road geometrics, deficient safety appurtenances, weather conditions and accident evidence.

There are two types of road safety studies (Fig. 4.29). The *traffic accident study* (TAS) uses police accident reports and investigates the causes of accidents based on the information and evidence collected. Reconstruction and investigation of the accidents recorded, with possible improvements in the roadway geometrics including traffic operations, are considered under this category of studies.

A *road safety audit* (RSA) is a formal study conducted by an independent audit team which evaluates the safety performance of a road or intersection. The safety audit can be carried out for an existing roadway or for one in the planning stage (FHWA, 2004).

The RSA is a formal examination and review of the safety performance of an existing or future road/intersection by an independent audit team. Based on the RSA report, which consists of identified

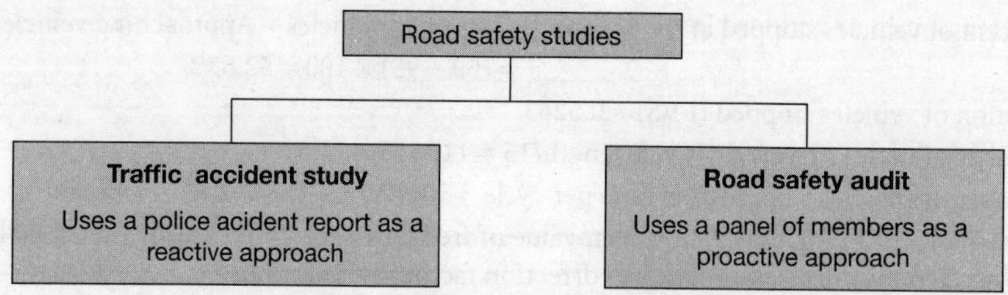

Figure 4.29 Types of road safety studies.

potential safety issues, assessment and recommendations, the jurisdiction owning the road responds to the problems identified. It is the sole responsibility of the owner of the road to determine the feasible action or to redesign/modify the road geometrics; otherwise, there is a need to document the reason for not acting on such RSA suggestions/recommendations. RSAs of existing roads are sometimes termed RSA reviews

4.15.1 Basic Report Information

The systematic recording and reporting of accident data is a prerequisite for any road accident analysis, as well as for the road safety management system. Relevant and reliable data is recorded to enable identification of the contributing factors for each accident and to offer the best way to explore the prevention of accidents, including ways to implement measures to reduce accident severity (PIARC, 2007).

Information and evidence must be documented as a report in a consistent and uniform practice. The following list includes the basic types of information that should be reported with adequate evidence and a description of the collision (Agent and Pigman, 1993).

- Location, time, weather and date of the accident
- Type of accident, along with the nature of the injury to the persons involved
- Information about the units involved, such as the driver, occupants, people, animals and witnesses
- Vehicles and their characteristics – interior, exterior, mechanical and electrical parts
- Vehicle/property damage
- Widened width of carriageway and super-elevation on curves
- Measurement of sight distance on the horizontal and vertical curves
- Geometrics, signs, markings, safety appurtenances and road surface characteristics of the location
- Vehicle occupant information
- Name of the hospital where the injured persons were admitted
- Contributing factors
- Accident description
- Physical location details of post-collision objects
- Collision/accident scene measurements sketched as a diagram
- Photographs/video clips
- Measurement of skid marks, yaw marks, scratches, gouges, drag factor, stop location of the vehicles/objects/pedestrians and dents on collided objects
- Distances measured relating to installation of advance signs, presence of delineators, height of signposts and dimensions of signs and lettering size
- Retro-reflectivity of pavement markings and road signs
- Statements from the vehicle occupants/witnesses and investigators

4.15.2 Assessment of Accident Data

Road safety deficiencies are determined from the available accident data. Locations where there is high frequency and concentration of occurrence of accidents are determined and such sections of the road or road locations are often termed *black-spots*.

Several methods are available to detect deficiencies based on the parameters assessed at a location, such as accident frequency, accident rate and accident severity. Accident data is assessed based on the following objectives:

- To determine the common characteristics of accidents and their counter measures
- To identify black-spots on road networks with reference to traffic volume data

4.15.3 Collision Diagrams

A collision diagram is a graphical representation of an accident that occurred on a road location. Each accident location is pictorially drawn, and present and past historical accident data with the date, time and type of collision, with the objects involved, are drawn to a convenient scale, say (1:200–1:500), or not drawn to scale.

Direction arrows with symbols are used to represent the nature of the collision, including the path traced by the vehicles and the pedestrians with reference to fixed objects on the roadway (Table 4.11). The fixed objects may be listed as electric poles, signal poles, fencing, kerb lines, compound walls and so on. The representing arrows may not be shown on the exact path and collision spot of the accident; this is because of overlap of accidents that may have occurred at the same spot. The arrows are shown with details of each accident, including the date and time of the day to the nearest hour. A typical collision diagram at an intersection of a road is shown in Fig. 4.30.

Table 4.11 Symbols used for the development of collision diagrams.

Symbol	Description
	Trees
—X—X—	Fence
	Guard rail
	Building
	Any fixed object on the roadway
	Signboard mounted on single post
	Signboard mounted on two posts

(Continued)

	Overhead sign
	Dotted line arrow indicating pedestrian movement in a direction
	Pedestrian push button
	Traffic signal
	Arrow indicating vehicle movement in a direction: The shape of the arrow shows the actual movement, which corresponds to the direction of travel just before the accident.
	Arrows with solid round marks indicate the approach speed of vehicles. Each solid round mark indicates the speed of the vehicle with a range of 10 km/hr. For example, an arrow with four solid round marks indicates that the approach vehicle is in the range of 40–50 km/h.
	Head-on collision
	Rear-end collision
	Side swipe
	Angle collision
	Retarding vehicle movement
	Reversing
	U-turn
	Uncontrolled vehicle movement/Vehicle out of control
	Overturned vehicle
	Non-fatal injury
	Fatal injury
	Parked vehicle
	Stationary vehicle

(Continued)

D	Dry road condition
W	Wet road condition
I	Ice/Snow road condition
O	Other road condition
R	Rain
F	Fog
C	Clear
→ B	Vehicle collision with a bicycle
→ P	Vehicle collision with a pedestrian
→ A	Vehicle collision with an animal
→ □	Vehicle collision with a fixed object
	Side swipe in the same direction
	Side swipe in the opposite direction

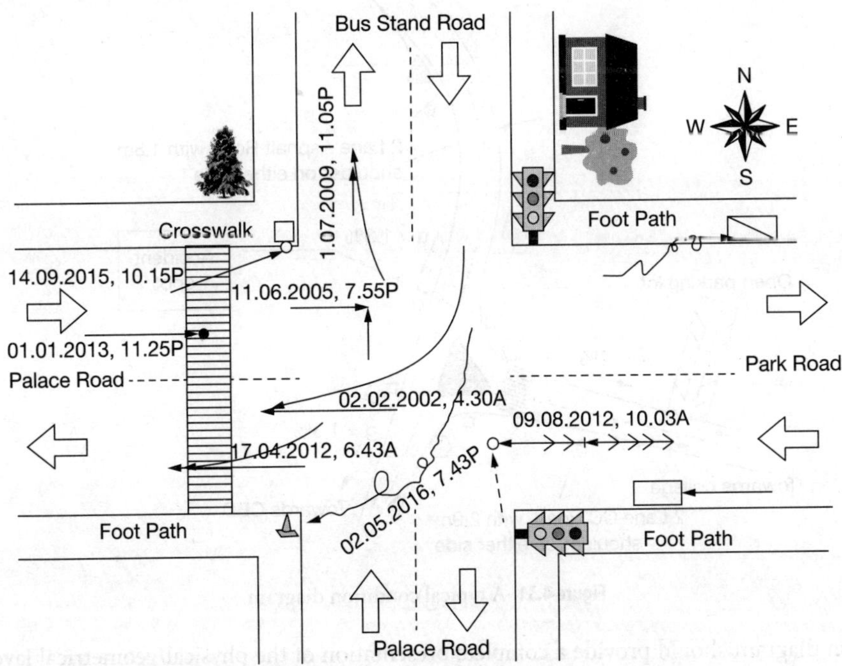

Figure 4.30 A typical collision diagram recorded for a location.

The main purpose of collision diagrams is to depict the accident type and pattern. They are used for evaluating the possible causes of an accident at a specific site. Such diagrams can be used for analysis within a single intersection (Nyerges and Cihon, 1989).

4.15.4 Condition Diagrams

To prepare a collision diagram, the layout of the intersection or road segment, along with its geometric features, must be known. Such a diagram is called a condition diagram. A typical condition diagram showing the pattern of accidents along with features of the site and its existing road geometric conditions, including details pertaining to the approach roads drawn to scale is shown in Fig. 4.31. Condition diagrams provide detailed information that may not be easily communicated in a tabular form.

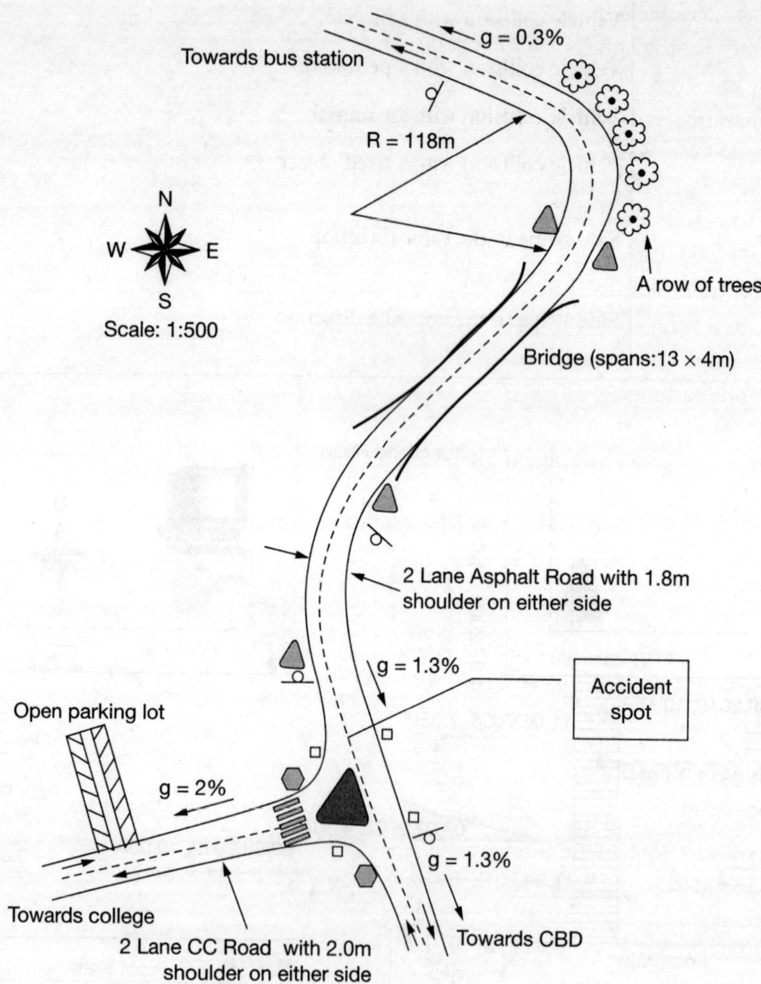

Figure 4.31 A typical condition diagram.

A condition diagram should provide a complete presentation of the physical/geometrical layout of the accident study location as it currently exists, including the appurtenances. Construction drawings done to scale can be used for this purpose. Condition diagrams may be developed to any convenient scale such as in the range 1:200–1:1000.

The following items should be included in the condition diagram to represent the on-site features (TxDOT, 2011):

1. A plan drawn to a scale with the North direction
2. Geometric components of the road site, including
    - Edges of pavement, kerb lines and shoulders with dimensions
    - Roadway limits
    - Widths and grades of approach roads
    - Surface characteristics such as skid resistance, pot holes, ruts and pavement edge drop-offs and other distresses
    - Channelisation details
    - Roadside parking and restriction details
    - Nearby bus stops and pedestrian crossings
    - Illumination details with locations of electric poles
3. Roadside developments
    - Obstructions to sight distance due to any object on the roadside such as trees, plants, signs, poles, fences, utility poles and guard rails
    - Adjacent land use
4. Position of traffic control devices
    - Traffic flow directions
    - Design speed and 85th percentile speed
    - Signs and markings
    - Traffic signals
    - Distance to adjacent traffic signals within 1.0 km
    - Any other traffic control devices
    - Level crossings with railways and type of control

Local law enforcement agencies use condition diagrams to investigate the current condition in a particular crash location. Condition diagrams are useful for correlating ground features on the road and detecting road deficiencies, for future road geometric design correction.

Collision diagrams and condition diagrams with 'before' and 'after' corrective measures are useful for conducting comparative studies and statistical analysis.

4.15.5 Road Safety Measures

Several safety measures are in used in different countries. Some are relatively simple and low-cost while others are less easy to implement and may require a larger budget. A short list of the road safety measures that are intended to mitigate the circumstances leading to accidents or otherwise is given below. These safety measures are categorised as engineering measures, enforcement measures and educational measures (DG TREN, 2007; Austroads, 2016).

Engineering measures

- *Land use and network planning*: The hierarchical integrity of the road network based on their functional objectives.
- Channelisation and directional control/regulation of the traffic streams, including morning and evening peak traffic regulations.

- Changes/Improvements in the geometric components of roadways, including providing safe stopping sight distance, overtaking sight distance, acceleration/deceleration tapers, and access control on high-speed lanes/highways/expressways.
- Provision of low-speed zones in residential areas.
- Provision of off-street parking facilities.
- *Improving/Modifying road intersections*: For example, replacing a junction with a roundabout. Roundabouts aim to lower junction speeds and thus eliminate right-angle and head-on collisions, with greater capacity than normal give-way or signalised junctions.
- Treatments to the centre line and medians for improved delineation/warning purposes; and treatments to other road infrastructure as part of speed management on curves, skid resistance and overtaking zones (Austroads, 2016).
- *Measures against tree collisions*: Decreased severity of road accidents by ensuring protection of rows of trees by providing guard rails wherever possible, or by felling the trees.
- *High-risk/Accident-prone site management*: Site visits should involve the relevant safety engineers, traffic police, emergency services and other stakeholders.
- *Signing and marking*: Maintaining the retro-reflectivity of signs and road markings, including installing them at standard recommended locations on the roadway.
- *Rumble strips*: These are milled into the asphalt surface of a road shoulder. Rumble strips vibrate and make a noise when a vehicle passes over them and alert drivers to the potential accident/hazard.
- *Variable message signs*: Speed limits and communication of warnings can be instantaneously provided on-site.
- *Road safety audits*: Audit is a formal proactive procedure by which independent assessments of the potential and likely safety performance of a specific design of a road or traffic scheme is carried out. The recommendations of the audit team are honoured by the client/municipality by incorporating those improvements in the designs drawings (IRC: SP: 88-2010).
- *Road safety inspections*: Safety inspection involves a periodic review or specific inspection of a road site by trained road safety experts. It involves visiting the road network to identify physical defects in the road infrastructure.
- *Vehicles and safety devices*: Safe car/vehicle designs including headlights, acceleration and braking capacity.
- *Daytime running lights* (DRL): Headlights equipped with DRL increases visibility and improves the distance and speed perception of motor vehicles so that the other road users can detect such motor vehicles during the day and adjust their driving behaviour.
- *Bicycle/Two-wheeler side reflection*: This implies that the front and rear wheels of a body frame are equipped with reflective material to increase visibility at night and at twilight.
- *Intelligent speed assistance* (ISA): The ISA systems establish the position of a vehicle and compare its current speed with the posted speed limit there. In case of excess speed, the system gives feedback to the driver and restricts the speed accordingly.
- *Alcohol ignition interlock* (Alcolock): It is an electronic device that prevents a vehicle from being started if the driver has drunk alcohol beyond the accepted level.
- *Event data recorder* (black box): It is an electronic system capable of monitoring a number of vehicle variables such as speed, acceleration and deceleration forces, use of lights, gears, seat belts and others. These data recorders can collect data just before and after an accident or/and for each trip. The data collected by these systems can be used for reconstructing a road accident, presenting evidence of accident contributing parameters and also for research purposes.

- *Promoting the implementation of eCall systems*: eCall is an automatic in-vehicle emergency call service developed in the European Union. An eCall-equipped vehicle has a terminal with a satellite-based wireless communication system by which accidents, rollovers and fire occurrences in the vehicle can be detected. When an accident occurs, the terminal sends information about the vehicle's position and type of accident to the nearest emergency response centre. This system is also capable of opening a voice connection between the vehicle occupants and the emergency response centre.

Enforcement measures

- Traffic law enforcement aims to prevent traffic offences by imposing appropriate rules and regulations which ensure safe and efficient movement of traffic.
- *Safety camera programme*: These cameras are mounted on the roadside at clearly marked locations so that road users see them well in advance and control their behaviour.
- *Automatic speed enforcement*: Speeding vehicles are detected by these automatic cameras and vehicles that violate the speed limit are photographed with their licence plate number. Such information is immediately passed on to a central processing office for identification of the vehicle owners.
- *Section controls*: Any particular length of road passing through a densely populated suburb is intended to operate as a low-speed zone. Such control sections may be identified to improve air quality along that road. For example, introduction of a reduced speed limit of 60 km/hr where it was previously 80 km/hr.
- Random breath testing (RBT) aims to identify drivers who exceed the legal alcohol limit. With the RBT device, drivers are stopped and tested for alcohol levels by the police.
- *Targeted seatbelt enforcement*: Police target drivers who do not wear their seat belts.
- Penalty point system (PPS) aims at prosecuting repeat offenders. When a traffic violation is detected, the violator receives one or more penalty points, and when the accumulated penalty points exceed the limit, the violator's licence will be suspended temporarily.

A list of **emergency services** is given below:

- *Efficient emergency responses*: The core of road safety measures is the agreement between insurance companies, hospitals and the police department that a rescue car/ambulance will be sent to the incident location as soon as an accident is reported.
- *Emergency lanes in congestion*: If there are more than two lanes, all the cars on the outer left lane must move to the left, while the others must move to the right. This free lane enables all emergency vehicles to provide fast and efficient help in congested traffic conditions.
- *Use of a mobile intensive care unit* (MICU): These provide pre-hospital treatment for road accident victims on the road site.
- *Transport of road crash victims by helicopter*: Severely injured road accident victims can be transported to a hospital in a trauma helicopter within a short duration.
- *Post-accident care*: This measure is important in rural areas where emergency services cannot arrive at the accident spot within 5–15 minutes (international standard). The first one hour after the accident is the most crucial and important for the accident victim's first aid, and hence, it is termed 'golden hour'.
- Psychological support for road crash victims should be provided at the first intervention stage, i.e., immediately after the crash and also during hospitalisation.

Educational measures

- Driver training is important for learner drivers as well as for regular drivers, to familiarise them with the latest traffic regulations and teach them to assess risks.
- Road safety halls containing educational aids designed to encourage active and accurate use of in-car safety equipment among novice drivers and school children.
- Rehabilitation seminars for novice drivers aim to avoid repeat offences by motivating them to behave more safely and responsibly.
- Newspapers, TV shows, mobile phone SMS alerts, internet-based awareness campaigns and school educational programs are widely used to create awareness among children and road users. Organising Road Safety Weeks is an example of educational measures in India.

4.16 EVALUATION OF EFFECTIVENESS OF ROAD SAFETY IMPLEMENTATION

To judge the merits of a road safety improvement and determine its effectiveness in reducing accidents, the following tests are conducted based on statistical significance. Accident data before and after improvements is obtained for this purpose.

4.16.1 Chi-square Test to Find Goodness of Fit

This test is conducted to determine whether the observed frequencies are close to the expected frequencies, or if they are really quite different from each other. The following formula is used for computation of Chi-square:

$$\chi^2 = \sum_{i=1}^{n}\left[\frac{(O_i - E_i)^2}{E_i}\right] \tag{4.35}$$

Chi-square (χ^2) is a single number summary that represents the discrepancy between obtained frequencies and expected frequencies. The bigger the calculated value of χ^2, the greater the difference between the observed and expected frequencies.

The calculated χ^2 value is compared to the critical χ^2 value (a value from Table 4.11) and conclusions are drawn (Graham, 2000):

1. If the calculated χ^2 > critical χ^2, any one of the following possibilities holds good:
   - **(a)** A high value of χ^2 implies a poor fit between the observed and expected frequencies.
   - **(b)** The fit is not acceptable at that level of significance.
   - **(c)** There is a difference between the observed and expected frequencies. The expected frequency (theoretical frequency) distribution is *not* a good description of the population.
   - **(d)** The discrepancy was not simply due to change.
   - **(e)** The discrepancy between the observed and theoretical frequency distribution is not a good description of the population. In other words, the observed sample poorly reflects the characteristics of the population from which it was taken.
2. If calculated χ^2 < critical χ^2, the following possibilities hold good:
   - **(a)** The fit is acceptable at that level of significance.
   - **(b)** There is no evidence to indicate that the true theoretical distribution differs from the observed distribution (Gerlough, 1955).
   - **(c)** Some relationship exists between the two variables in question. However, it does not indicate the type of relationship.
   - **(d)** The two variables are associated in some way, to be precisely determined by the investigator.

Table 4.11 A partial table of chi-square distribution for 95% confidence level ($\alpha = 0.05$).

Degree of Freedom (DoF)	1	2	3	4	5	6	7	8	9	10	11	12	13	14	15
χ^2	3.841	5.991	7.815	9.488	11.070	12.592	14.067	15.507	16.919	18.307	19.647	21.026	22.362	23.685	24.996

DoF = Number of data sets − 1 = $(n - 1)$

4.16.2 Chi-square Test to Determine Association between Two Variables

This test is used to determine the association between two independent and continuous variables that may be related to *before* and *after* road safety improvements. The observed frequencies are the recorded data obtained before road improvement. The formula used is the same as Eq. 4.35, but the expected frequencies are calculated as follows:

$$\text{Expected frequency} = \frac{\text{Row total} \times \text{Column value}}{\text{Grand total}}$$

$$\text{Degree of freedom} = (\text{No. of rows} - 1) \times (\text{No. of columns} - 1)$$

Take the null hypothesis as: "The differences between the two sample variables are only due to chance fluctuations in sampling".

The calculated χ^2 value is compared to the critical χ^2 value and conclusions are drawn:

1. If the calculated $\chi^2 >$ critical χ^2, reject the null hypothesis (there is a significance difference between the two variables due to improvement; that means that the road safety measures are effective)
2. If the calculated $\chi^2 <$ critical χ^2, accept the null hypothesis.

4.16.3 Chi-square Test to Determine Effectiveness of Before and After Improvements from One Group of Data

The following formula is used to determine the effectiveness of before and after improvements from one group of data:

$$\chi^2 = \left[\frac{\left(N_{\text{before}} \times t_{\text{after}} - N_{\text{after}} \times t_{\text{before}}\right)^2}{\left(t_{\text{before}} \times t_{\text{after}}\right) \times \left(N_{\text{before}} + N_{\text{after}}\right)}\right]$$

where t_{before} = period of time before the improvement for which accident data is available;
t_{after} = period of time after the improvement during for accident data is available;
N_{before} = number of accidents observed in the t_{before} time period; and
N_{after} = number of accidents observed in the t_{after} time period.

The calculated χ^2 value is compared to the critical χ^2 value.

EXAMPLE 4.16

A list of accidents observed in a city over a period of one year is given below. Find whether the observed accidents are uniformly distributed over the year using the chi-square test.

Month	Jan	Feb	Mar	Apr	May	Jun	Jul	Aug	Sep	Oct	Nov	Dec
Observed frequency	34	45	20	22	56	61	56	45	32	31	24	12

Answer

The computational details are as follows. Read from left to right.

Month	Frequency		$O_i - E_i$	$(O_i - E_i)^2$	$\frac{(O_i - E_i)^2}{E_i}$
	Observed, O_i	Expected, $E_i = 438/12$			
January	34	36.5	34 – 36.5 = –2.5	$(-2.5)^2 = 6.25$	6.25/36.5 = 0.1712
February	45	36.5	45 – 36.5 = 8.5	$(8.5)^2 = 72.25$	72.25/36.5 = 1.9795
March	20	36.5	20 – 36.5 = –16.5	$(-16.5)^2 = 272.25$	272.25/36.5 = 7.4589
April	22	36.5	22 – 36.5 = –14.5	$(-14.5)^2 = 210.25$	210.25/36.5 = 5.7603
May	56	36.5	56 – 36.5 = 19.5	$(19.5)^2 = 380.25$	380.25/36.5 = 10.4178
June	61	36.5	61 – 36.5 = 24.5	$(24.5)^2 = 600.25$	600.25/36.5 = 16.4452
July	56	36.5	56 – 36.5 = 19.5	$(19.5)^2 = 380.25$	380.25/36.5 = 10.4178
August	45	36.5	45 – 36.5 = 8.5	$(8.5)^2 = 722.25$	722.25/36.5 = 1.9795
September	32	36.5	32 – 36.5 = –4.5	$(-4.5)^2 = 20.25$	20.25/36.5 = 0.5548
October	31	36.5	31 – 36.5 = –5.5	$(-5.5)^2 = 30.25$	30.25/36.5 = 0.8288
November	24	36.5	24 – 36.5 = –12.5	$(-12.5)^2 = 156.25$	156.25/36.5= 4.2808
December	12	36.5	12 – 36.5 = –24.5	$(-24.5)^2 = 600.25$	600.25/36.5 = 16.4452
Total	438	438		Total =	**76.7397**

$$\chi^2 = \sum_{i=1}^{n}\left[\frac{(O_i - E_i)^2}{E_i}\right] = 76.7397$$

Degree of freedom = (12 – 1) = 11

Assume the significance level as 5% and the critical χ^2 value = 19.647 (from Table 4.11).

Null hypothesis H_o: Accidents are uniformly distributed over the year.

As the calculated χ^2 (76.7397) > critical χ^2 (19.674), the null hypothesis is *rejected*; and it is interpreted as the accidents observed in the city are *not* uniformly distributed over the year.

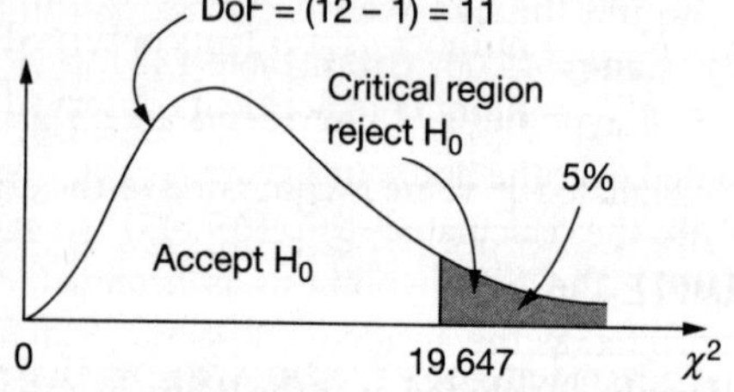

Figure 4.32 Example 4.16: Typical presentation of critical χ^2 value for 5% significance level.

EXAMPLE 4.17

A list of accidents observed in a city over a period of one year is given below. Find whether the observed accidents are uniformly distributed over the hours using the chi-square test.

Hours	00:00–03:00	03:00–06:00	06:00–09:00	09:00–12:00	12:00–15:00	15:00–18:00	18:00–21:00	21:00–24:00
Observed frequency	21	19	09	10	17	10	11	15

Answer

The computational details are as follows. Read from left to right.

Hours	Frequency		$O_i - E_i$	$(O_i - E_i)^2$	$\frac{(O_i - E_i)^2}{E_i}$
	Observed, O_i	Expected, $E_i = 112/8$			
00:00 – 03:00	21	14	21 – 14 = 7	$(7)^2 = 49$	49/14 = 3.5000
03:00 – 06:00	19	14	19 – 14 = 5	$(5)^2 = 25$	19/14 = 1.7857
06:00 – 09:00	09	14	09 – 14 = –5	$(-5)^2 = 25$	09/14 = 1.7857
09:00 – 12:00	10	14	10 – 14 = –4	$(-4)^2 = 16$	10/14 = 1.1429
12:00 – 15:00	17	14	17 – 14 = 3	$(3)^2 = 9$	17/14 = 0.6429
15:00 – 18:00	10	14	10 – 14 = –4	$(-4)^2 = 16$	10/14 = 1.1429
18:00 – 21.00	11	14	11 – 14 = –3	$(-3)^2 = 9$	11/14 = 0.6429
21:00 – 24:00	15	14	15 – 14 = 1	$(3)^2 = 1$	15/14 = 0.0714
Total =	112	112		Total =	**10.7143**

$$\chi^2 = \sum_{i=1}^{n}\left[\frac{(O_i - E_i)^2}{E_i}\right] = 10.7143$$

Degree of freedom = (8 – 1) = 7

Assume the significance level as 5% and the critical χ^2 value = 14.067 (from Table 4.11).

Null hypothesis H_0: Accidents are uniformly distributed over the hours in a year.

As the calculated χ^2 (10.7143) < critical χ^2 (14.067), the null hypothesis is *accepted;* and it is interpreted as the accidents observed in the city are uniformly distributed over the hours in a year.

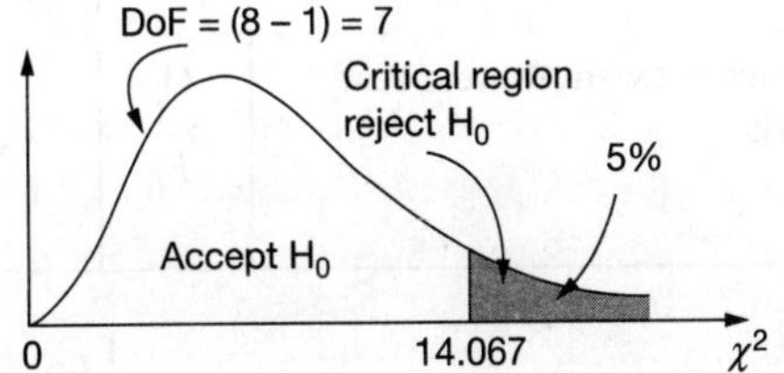

Figure 4.33 Example 4.17: Typical presentation of critical χ^2 value for 5% significance level.

EXAMPLE 4.18

The contingency table of two data sets obtained from a particular database belonging to the state revenue department is given below. Find whether the distributions of sample variables differ because of chance fluctuations, based on results obtained from a chi-square test.

Study	Before improvement	After improvement
Wants to wear seat belt	55	83
Does not want to wear seat belt	43	79

Answer

Let us calculate the subtotals of the row and column frequencies of the observed frequencies; the grand total is given below:

Study on	Before	After	Sum of row frequencies
Wants to wear seat belt	55	83	138
Does not want to wear seat belt	43	79	122
Sum of column frequencies	98	162	**260**

Now, calculate the expected frequencies using the following formula:

$$\text{Expected frequency} = \frac{\text{Row total} \times \text{Column value}}{\text{Grand total}}$$

$$\text{Expected frequency of those who want to wear seat belt, before improvement} = \frac{138 \times 98}{260} = 52.0$$

$$\text{Expected frequency of those who want to wear seat belt, after improvement} = \frac{138 \times 162}{260} = 86.0$$

The remaining expected frequencies of the observed frequencies are presented below:

Study		Before	After
Wants to wear seat belt	O_i	55	83
	E_i	$= \frac{138 \times 98}{260} = 52.0$	$= \frac{138 \times 162}{260} \approx 86.0$
Does not want to wear seat belt	O_i	43	79
	E_i	$= \frac{122 \times 98}{260} \approx 46.0$	$= \frac{122 \times 162}{260} = 76.0.$

Using the chi-square formula: $\chi^2 = \sum_{i=1}^{n} \left[\frac{(O_i - E_i)^2}{E_i} \right]$

Let us calculate the chi-square for each data set and sum:

$$\chi^2 = \left[\frac{(55-52.0)^2}{52.0}\right] + \left[\frac{(83-86.0)^2}{86.0}\right] + \left[\frac{(43-46.0)^2}{46.0}\right] + \left[\frac{(79-76.0)^2}{76.0}\right]$$

$$\chi^2 = \left[\frac{(3.0)^2}{52.0}\right] + \left[\frac{(-3.0)^2}{86.0}\right] + \left[\frac{(-3.0)^2}{46.0}\right] + \left[\frac{(3.0)^2}{76.0}\right]$$

$$\chi^2 = [0.1730] + [0.1046] + [0.1956] + [0.1184] = 0.5916$$

Calculated $\chi^2 = 0.5916$

Degree of freedom = (No. of rows – 1) × (No. of columns – 1) = (2 – 1) × (2 – 1) = 1

Assume the significance level as 5% at DoF = 1, and the critical χ^2 value = 3.841 (from Table 4.11).

As the calculated χ^2 (0.5916) < critical χ^2 (3.841), the null hypothesis is *accepted;* and it is interpreted as there is no significant difference between the accidents observed and the expected frequencies.

EXAMPLE 4.19

The frequency/contingency table (cross tabulation) for two independent variables is given below. The accident data observed in a year is obtained from a District State Highway Accident Records file. Find whether the road safety improvements are effective or not based on the results obtained from a chi-square test.

Predominance accident types	Annual average of the no. of accidents obtained	
	Before improvement	After improvement
Rear end	44.2	50.1
Angle collision at intersections	18.6	21.6
Right-turn head-on	10.9	9.9
Sideswipe	21.7	29.0
Other	51.1	20.9

Answer

Let us rewrite the given frequencies and sum of the column and row frequencies, as shown below:

No. of accidents	Observed frequencies	Rear end	Angle collision at intersections	Right-turn head-on	Sideswipe	Other	Sum of row frequencies
Before	O_i	44.2	18.6	10.9	21.7	51.1	146.5
After	O_i	50.1	21.6	9.9	29.0	20.9	131.5
Sum of column frequencies		94.3	40.2	20.8	50.7	72	278

Details of the calculation of expected frequencies (E_i) against observed frequencies are presented below:

No. of accidents	Freq	Rear end	Angle collision at intersections	Right-turn head-on	Sideswipe	Other
Before	O_i	44.2	18.6	10.9	21.7	51.1
	E_i	$\frac{146.5\times94.3}{278} = 49.69$	$\frac{146.5\times40.2}{278} = 21.18$	$\frac{146.5\times20.8}{278} = 10.96$	$\frac{146.5\times50.7}{278} = 26.71$	$\frac{146.5\times72}{278} = 37.94$
After	O_i	50.1	21.6	9.9	29.0	20.9
	E_i	$\frac{131.5\times94.3}{278} = 44.60$	$\frac{131.5\times40.2}{278} = 19.01$	$\frac{131.5\times20.8}{278} = 9.83$	$\frac{131.5\times50.7}{278} = 23.98$	$\frac{131.5\times72}{278} = 34.05$

Using the chi-square formula: $\chi^2 = \sum_{i=1}^{n}\left[\frac{(O_i - E_i)^2}{E_i}\right]$

No. of accidents	Rear end	Angle collision at intersections	Right-turn head-on	Sideswipe	Other
Before	$\left[\frac{(44.2-49.69)^2}{49.69}\right]$	$\left[\frac{(18.6-21.18)^2}{21.18}\right]$	$\left[\frac{(10.9-10.96)^2}{10.96}\right]$	$\left[\frac{(21.7-26.71)^2}{26.71}\right]$	$\left[\frac{(51.1-37.94)^2}{37.94}\right]$
After	$\left[\frac{(50.1-44.6)^2}{44.6}\right]$	$\left[\frac{(21.6-19.01)^2}{19.01}\right]$	$\left[\frac{(9.9-9.83)^2}{9.83}\right]$	$\left[\frac{(29-23.98)^2}{23.98}\right]$	$\left[\frac{(20.9-34.05)^2}{34.05}\right]$

No. of accidents	Rear end	Angle collision at intersections	Right-turn head-on	Sideswipe	Other
Before	0.6066	0.3142	0.0003	0.9397	4.5647
After	0.6783	0.3528	0.0004	1.0508	5.0784

Grand total of the above values = 13.5862

$$\chi^2 = \sum_{i=1}^{n}\left[\frac{(O_i - E_i)^2}{E_i}\right] = 13.5862$$

Degree of freedom = (No. of rows – 1) × (No. of columns – 1) = (2 – 1) × (5 – 1) = 5

Assume significance level as 5% at DoF = 5, and the critical χ^2 value = 11.070 (from Table 4.11).

As the calculated χ^2 (13.5862) > critical χ^2 (11.070), the null hypothesis is *rejected;* and it is interpreted as there is a significant difference between the accidents before and after improvement. It may be concluded that the improvements and road safety measures are effective.

EXAMPLE 4.20

Before realignment of a road intersection, 11 accidents occurred in 7 years. After the realignment, it was observed that only 5 accidents occurred in 1 year. Determine the effectiveness of realignment of the road intersection using the chi-square test.

Answer

The given values are substituted in the following formula:

$$t_{\text{before}} = 7, t_{\text{after}} = 1, N_{\text{before}} = 11, N_{\text{after}} = 5$$

$$\chi^2 = \left[\frac{\left(N_{\text{before}} \times t_{\text{after}} - N_{\text{after}} \times t_{\text{before}}\right)^2}{\left(t_{\text{before}} \times t_{\text{after}}\right) \times \left(N_{\text{before}} + N_{\text{after}}\right)}\right] = \left[\frac{\left(11 \times 1 - 5 \times 7\right)^2}{\left(7 \times 1\right) \times \left(11 + 5\right)}\right] = 5.14$$

The calculated χ^2 value (5.14) > critical χ^2 value (3.841 at 95% confidence level); hence, it is concluded that the realignment improvement was effective.

Review Questions

1. State the components of traffic engineering, along with the characteristics affecting the performance of any traffic flow system.
2. Classify and list the behaviour factors affecting driver performance.
3. Classify road users and list the factors affecting each type.
4. Write an overview of the physical characteristics of road users and explain their relevance to the geometric design of roads and traffic flow operations.
5. Explain the visual characteristics of road users relating to (a) visual acuity, (b) peripheral vision, (c) eye movement and visual concentration, (d) colour vision, (e) glare effect and (f) perception space.
6. Write an overview of pedestrian characteristics and explain their relevance to the geometric design of roads and traffic flow operations.
7. Explain the PIEV theory? What are the factors affecting PRT?
8. Define design driver and explain the design driver characteristics.
9. Write details about the maximum permissible gross weight of vehicles in India, with appropriate sketches of the vehicles for each case.
10. Classify and state the vehicle types.
11. State the concept of speed limit and draw a sketch.
12. Write the factors and explain how braking performance and the deceleration characteristics of vehicles are used for the safety evaluation of vehicles.
13. Explain the following vehicle characteristics that influence the design elements of a roadway:
    (a) Inertial force
    (b) Rolling resistance
    (c) Air resistance
    (d) Grade resistance
14. Define traffic engineering with reference to road transportation.
15. What are the five possibilities that may be explored to ensure the systematic development of traffic facilities and infrastructure needs?
16. State the reasons for collecting traffic data.
17. State the type of traffic studies conducted with their intended purpose.
18. What are the objectives of conducting traffic volume studies?
19. Write the broad classification of traffic studies and briefly explain each.
20. Distinguish between ADT and AADT?
21. Define DHV and illustrate the relations between percentage AADT and highest hourly volume of the year?
22. Define K-factor and state how it is useful.
23. Classify and enumerate the traffic counting methods with a sketch.
24. Define spot speed and write the distance considered with average operating speed of traffic.
25. Explain how spot speed surveys are conducted with a sketch.
26. Write the applications of spot speed data.

27. Define time mean speed, space mean speed, running speed and journey speed.
28. Define design speed, operating speed, median speed, modal speed and pace.
29. Define space headway and time headway. Explain how they are computed from the observations obtained by conducting field surveys.
30. Define lane occupancy and explain how it is computed from the observations obtained by conducting field surveys.
31. Define traffic density and state how it is represented by an equation.
32. Define lane capacity and explain the different types of traffic capacities.
33. Define (a) Basic capacity (b) Possible capacity, and (c) Design capacity.
34. State the factors affecting practical capacity.
35. Derive and describe an expression that represents the standard relationship between speed and density.
36. Derive and describe an expression that represents the standard relationship between speed and flow.
37. Derive and describe an expression that represents the standard relationship between traffic density and flow.
38. Draw the typical trend of curves showing the fundamental relationships between speed, flow and density of traffic flow characteristics
39. Explain the speed–density relationship.
40. Derive the parabolic flow–density model.
41. What is meant by level of service? Describe with operating speed and volume-to-capacity ratio.
42. Write the details of general operating conditions of the six LoS with volume-to-capacity ratio and the speed condition.
43. State the factors affecting the LoS of a roadway.
44. What are the factors affecting LoS on urban roads?
45. State the six factors that are used to determine LoS.
46. Write the HCM-recommended criteria for LoS of signalised and unsignalised intersections based on control delay and delay, respectively.
47. Define passenger car equivalent and explain any method of estimation of PCU values for a study area.
48. Explain the factors affecting PCU values.
49. Write the IRC: 106-recommended PCU factors applicable on urban roads in plain areas.
50. Write the IRC: 64-recommended PCU factors that are applicable on rural roads.
51. Traffic volume on a section of a rural road during morning peak hour is 800 veh/hr and during evening peak hour is 800 veh/hr. The details of traffic composition are given below:
For morning peak: 52% cars, 28% motor cycles and scooters, 9% buses and trucks, 6% cycle rickshaws, 3% agricultural tractor-trailers and 2% bicycles.
For evening peak: 49% cars, 27% motor cycles and scooters, 13% buses and trucks, 5% cycle rickshaws, 3% agricultural tractor-trailers and 3% bicycles.
Determine which peak hour traffic is the congested condition.
52. Calculate the capacity of traffic flow if the average velocity on the lane is 40 km/hr and the stopping sight distance (SSD) is 25 m. Assume length of rigid wheel base as 6.0 m.
53. A study was conducted on a road section and found that one-way traffic stream speed was 42 km/hr. Average length of the vehicle wheel base = 6.0 m.
Determine (i) theoretical capacity and (ii) practical capacity of the study road in terms of veh/hr/lane. Assume reaction time (t_p) of driver as 0.8 sec for determination of theoretical capacity; average length of vehicle (a_1) = 6 m, reaction time of the drivers (a_2) = 2.5 sec and coefficient of longitudinal friction (f) = 0.36.
54. Determine the mean values of time headway and space headway of vehicles if the traffic volume in a lane is 1200 veh/hr, assuming there are no mopeds and motor cycles in the traffic stream.
55. Vehicles in a traffic stream moved at an average time headway of 2.5 sec. The average speed of the vehicles was found as 60 km/h. Estimate the flow rate and density of the traffic stream.
56. In a 60-sec period, a loop-detector is occupied by vehicles for the following times: 0.30, 0.32, 0.43, 0.55 and 0.51 sec. Take the distance between the loop-detectors = 3 m and average length of vehicle = 6.0 m. Calculate the values of lane occupancy, density, speed and flow rate.
57. Free-flow speed of vehicles on a selected road-section is 75 km/hr and the jam density is 220 veh/km. Estimate the (a) traffic density in maximum traffic flow condition, (b) possible maximum flow rate and (c) average speed of

vehicles at maximum traffic volume. Determine the maximum flow for the above condition per lane if the free mean speed = 50 km/hr and average spacing between the vehicles under jam condition was 5.4 m.

58. The speed–density relationship for a study stretch of highway was found as V = 42 – 0.4 K, where V = speed (km/hr) and K = density (veh/km). Determine the value of maximum flow (i.e., capacity) of the highway, mean density and mean speed. Also sketch three plots with these traffic parameters and label the obtained values on these curves.

59. Determine the values of jam density, maximum flow rate, critical density and the corresponding speed at maximum flow, from the following traffic survey data:
Average length of vehicles in a stream of traffic = 6.6 m
Average gap between them = 2.0 m
Average time headway at maximum flow = 1.5 sec.

60. Calculate the volume data and the corresponding time headway and space headway values for the following traffic data obtained on a road segment.

Speed, V (km/hr)	60	55	47	42	35	30	22	15	11	5
Density, K (veh/km)	12	21	32	42	52	63	70	80	90	100

61. What are the points/factors to be considered before collecting OD data?
62. Define cordon line and explain its uses in conducting OD surveys in a study area.
63. Explain the various methods of OD data collection with their suitability and limitations.
64. Explain the registration questionnaire survey method of OD data collection with its advantages and limitations.
65. Explain the mail-back questionnaire survey method of OD data collection with its advantages and limitations.
66. Explain the roadside interview survey method of OD data collection with its advantages and limitations.
67. Explain the home interview survey method of OD data collection with its advantages and limitations.
68. Explain the tag-on-car survey method of OD data collection with its advantages and limitations.
69. Explain the carousel method of OD data collection with its advantages and limitations.
70. Explain the licence plate survey of OD data collection with its advantages and limitations.
71. Explain the methods used for recording of licence plate numbers of vehicles during OD surveys.
72. Write a short note on the applicability of ITS-related techniques in OD data collection.
73. Write the most commonly used equation for estimating the sample size for conducting traffic surveys by roadside interviews.
74. Write an overview of the equations used for estimating the sample size for conducting OD surveys.
75. What are the applications and uses of OD data?
76. Write an overview and explain the usability of various methods of presentation of OD data.
77. State the objectives of conducting travel time and delay studies.
78. What are the most commonly adopted improvements made for reducing delays?
79. Explain the methods of measurement of travel time and delays on any selected length of a route.
80. Explain the licence plate matching method of conducting travel time and delays on any selected length of a route.
81. Explain the floating car method of conducting travel time and delays on any selected length of a route.
82. Differentiate the floating car method and test car procedures of conducting travel time and delay studies on any selected length of a route.
83. Briefly explain the following methods of conducting travel time and delays on any selected length of a route:
    (a) Fixed time interval photographs method
    (b) Elevated observer method
    (c) Field interviews survey
    (d) GIS and GPS method
84. Write the details of presentation of speed and delay data collected along a route, with appropriate illustrations.

85. Describe the speed profiles method of presentation of speed and delay data collected along a route, with appropriate illustrations.
86. Write an overview of road safety studies.
87. State the types of road safety studies and explain their significance in reduction/prevention of road accidents.
88. State the objectives of conducting investigations and studies on road accidents.
89. List the basic types of information that should be reported with adequate evidence and give a description of collision as basic accident information.
90. Write the various methods adopted for locating road accidents. Also state their merits and limitations.
91. State the objectives based on which accident data is assessed.
92. What is a collision diagram? Explain how it is useful.
93. Draw any 20 widely used symbols for the development of collision diagrams.
94. Draw a typical collision diagram recorded at a location showing the collision of parked vehicle, swipe collision, uncontrolled vehicle hitting a traffic cone and a fatal accident on a zebra crossing. Indicate using different types of symbols. Provide sample date and time for each case.
95. What is a condition diagram? Explain how it is useful.
96. List the items that should be included in a condition diagram.
97. Draw a typical condition diagram showing a few salient features that are essential.
98. Explain the three widely used methods for measuring on-scene details. Write their merits and limitations.
99. Draw a typical diagram using the coordinate system of measurement and showing the measured distances and angles.
100. What are the measures that can be taken to reduce/avoid road-accidents?
101. Define road safety measures. Explain the (a) Engineering measures, (b) Enforcement measures, and (c) Education measures.
102. Explain how the effectiveness of road safety implementation is evaluated based on the chi-square test and validated by the student *t*-test.
103. What are the possible conclusions with reference a comparison between calculated χ^2 and critical χ^2 value?

Multiple Choice Questions

1. What is the percentile speed used as design of pedestrian crossing facilities?
   (a) 15th (b) 50th
   (c) 85th (d) 98th
2. What is the value of ADT if the measured value of traffic was 10,000 during 50 days of survey?
   (a) 50 (b) 100 (c) 200 (d) 300
3. The value of speed that is found to be occurring at the highest frequency is
   (a) Model speed
   (b) Median speed
   (c) Running speed
   (d) Spot speed
4. During a spot speed study, 100 vehicles were observed and it was found that 85 percentile vehicles and 15 percentile speeds were obtained as 50 km/hr and 16 km/hr, respectively. The number of vehicles travelling at speeds between 50 km/hr and 16 km/hr would be
   (a) 66 (b) 70 (c) 44 (d) 50
5. Speed regulation and control of traffic flow along a street is imposed based on the studies conducted by
   (a) Traffic volume studies
   (b) Spot speed studies
   (c) Speed and delay studies
   (d) Saturation flow studies
6. Match List-I (Traffic surveys) with List-II (Method) and select the correct answer using the codes given below the lists:

List-I	List-II
A. Spot speed	1. By video tape
B. Traffic volume	2. By carousel method
C. O-D survey	3. By Doppler radar
D. Parking survey	4. By pneumatic tube

	A	B	C	D
(a)	4	1	3	2
(b)	4	3	1	2
(c)	3	4	2	4
(d)	3	1	2	4

7. The following method of OD survey technique does not give important details such as delay causes, duration and the number of delays within a test section
   (a) Roadside windshield method
   (b) Photographic surveillance
   (c) License plate survey
   (d) Carousel method.
8. In a moving car study conducted over a road stretch of 10 km at a constant speed of 80 km/hr, it was observed that the test car is passed by 20 vehicles more than the number of vehicles it passes. The number of vehicles that met against stream was 300. The traffic flow rate is ______ veh/hr
   (a) 1280 (b) 2240
   (c) 2120 (d) 1160
9. Design capacity is often provided as
   (a) Basic capacity (b) Practical capacity
   (c) Ideal capacity (d) Possible capacity
10. As a common practice, the condition diagrams may be developed to any convenient scale such as in the range between.
   (a) 1:1000 and 1:5000
   (b) 1:1000 and 1:50000
   (c) 1:1000 and 1:10000
   (d) 1:200 and 1:1000

Answers to Multiple Choice Questions

1. (a)
2. (c) Traffic volume during 1 year = (365/50) × 10,000 = 73000
   ADT = 1,09,500/365 = 200
3. (a)
4. (b) Total number of vehicles observed = 100
   Number of vehicles corresponding to 85 percentile speed of 50 km/hr = 85
   Number of vehicles corresponding to 15 percentile speed 16 km/hr = 15
   The number of vehicles travelling between speeds 50 and 16 km/hr = (85 – 15) = 70
5. (b)
6. (c)
7. (c)
8. (a) Traffic volume in East direction (Q_{East}) = (Veh_{Opp} + N_{diff}) / (t_{a_East} + t_{a_West})
   On substitution, we get, Q_{East} = (300 + 20) / (10/80 + 10/80) = 1280 veh/hr
9. (b)
10. (d)

CHAPTER 5 Traffic Regulation, Control and Devices

5.0 INTRODUCTION

One of the principal challenges of traffic engineering is to control and regulate traffic flows in a safe and efficient order. Efficiency can be thought of as a measure of effective utilisation of available road-space and the related infrastructure in all conditions at an affordable cost. To achieve this objective, there will be a need to plan the activities of diversion, control and utilisation of conventional and non-conventional traffic in a phased manner.

The following are the details of classification based on specific methods of control/operation which are useful to regulate the flow of traffic.

Classification based on methods of traffic control at intersections

Intersections can also be classified based the control methods used (Fig. 5.1). Details of these are presented in subsequent chapters.

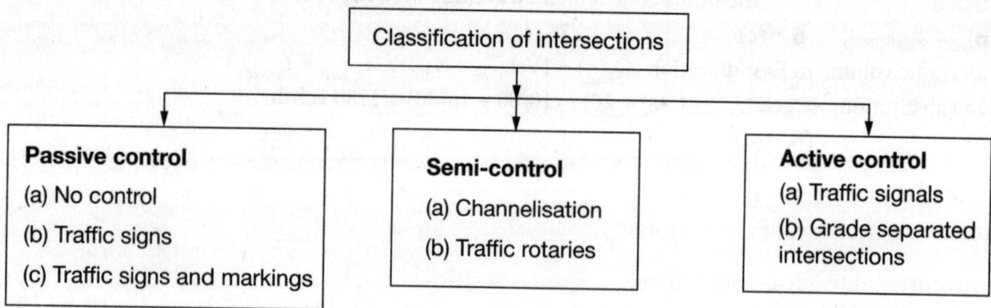

Figure 5.1 Operational classification of intersections (based on control type).

Classification of traffic control type

Based on the movement of traffic on a selected length of a route, the flow in at-grade intersections maybe classified as uninterrupted flow through uncontrolled intersections or interrupted flow through controlled intersections. In controlled intersections, traffic flow is controlled manually or using signals (Fig. 5.2).

Classification of traffic management techniques/methods

There are wide varieties of traffic management systems available for traffic control in urban locality (Fig. 5.3). Among them, traffic signals, pavement markings, signs and delineators are considered as conventional traffic control methods and the intelligent transportation system is considered as the advanced method. These methods will be elaborately discussed in subsequent paragraphs.

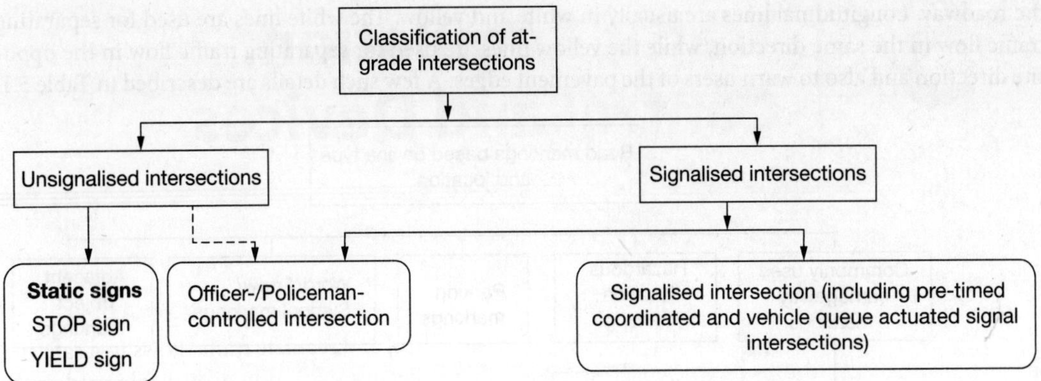

Figure 5.2 Classification of at-grade intersections based on traffic flow controls.

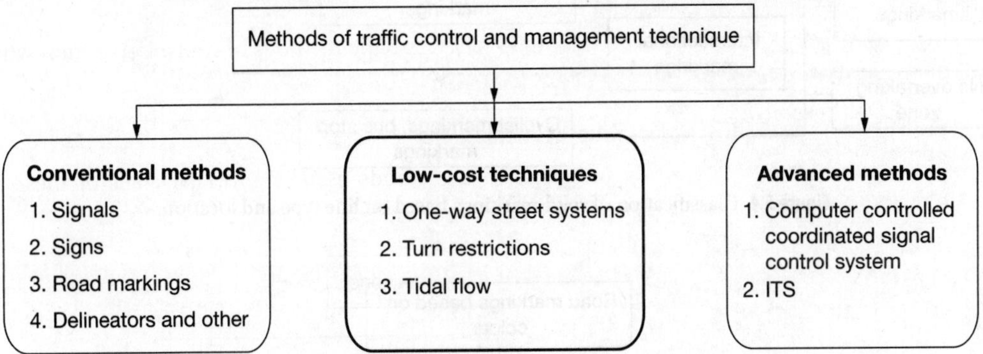

Figure 5.3 Methods of traffic control and management techniques.

The following sections of this chapter will discuss the various types of:

1. Conventional traffic control devices: Road markings, Signs and Signals and
2. Advanced traffic control devices: Intelligent transportation system.

5.1 ROAD MARKINGS

Road markings provide guidance and advance information, and convey regulations and warnings to road users. Road markings are the coloured paint lines on the road surface, symbolic objects and word messages, which are applied systematically with standard geometrics and intended control of traffic flow on different locations of roads.

5.1.1 Classification of Road Markings

Based on the location (Fig. 5.4), the intended function/control, colour (Fig. 5.5) and thickness (Fig. 5.6), road markings can be categorised into different types.

5.1.2 Longitudinal (Line) Markings

Line markings placed along the direction of traffic flow on a roadway belong to this group. These lines are used for separating traffic flow towards a specified lane; they also indicate the driver position on the roadway. Longitudinal lines are usually in white and yellow. The white lines are used for separating traffic flow in the same direction, while the yellow lines are used for separating traffic flow in the opposite direction and also to warn users of the pavement edges. A few such details are described in Table 5.1.

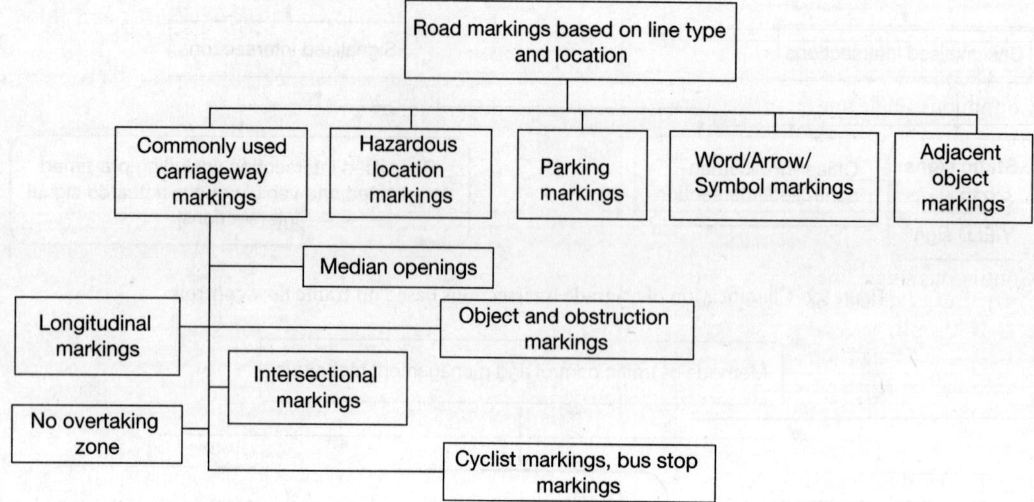

Figure 5.4 Classification of road markings based on line type and location.

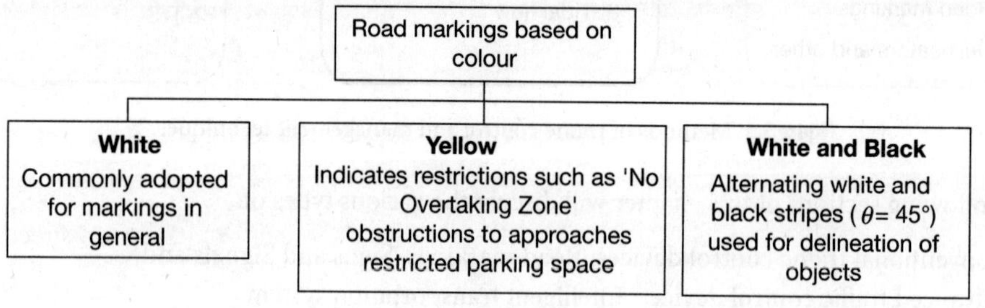

Figure 5.5 Classification of road markings based on colour.

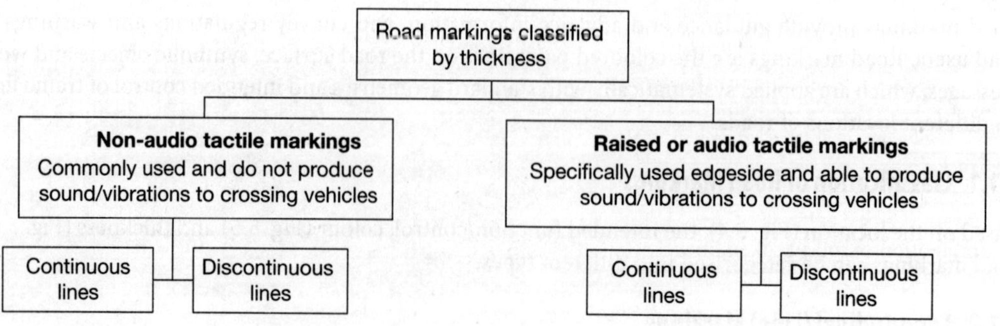

Figure 5.6 Classification of road markings based on thickness.

Table 5.1 Guiding principles of line markings (DDA, 2009).

Type with description	Line colour	Use type
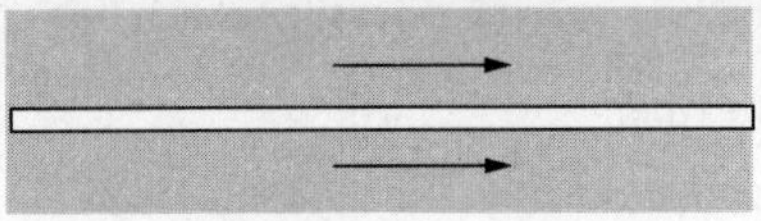 Continuous single line	White	• Restrictive in character • Does not allow crossing of vehicles, except for entry or exit to a side road • Used for separating traffic flow in the same direction.
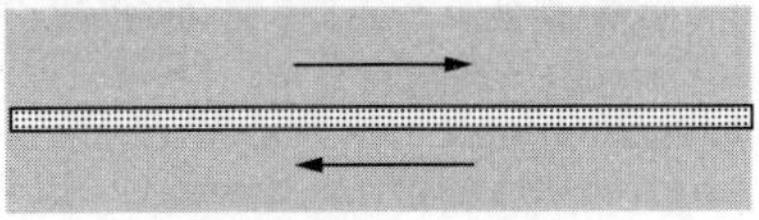 Continuous single line	Yellow	• Restrictive in character • Does not allow crossing of vehicles, except for entry or exit to a side road • Used for separating traffic flow in the opposite direction.
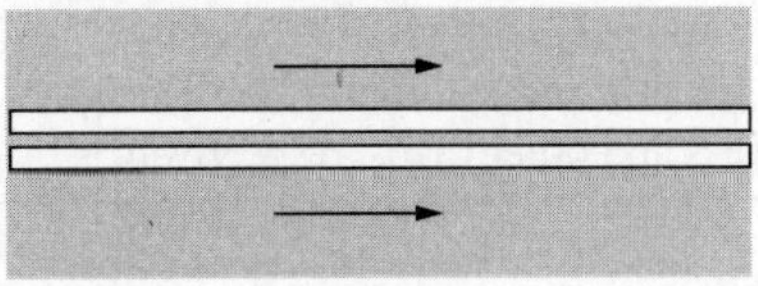 Continuous double line	White	• Seriously restrictive in character • Does not allow crossing of vehicles • Used for separating traffic flow in the same direction
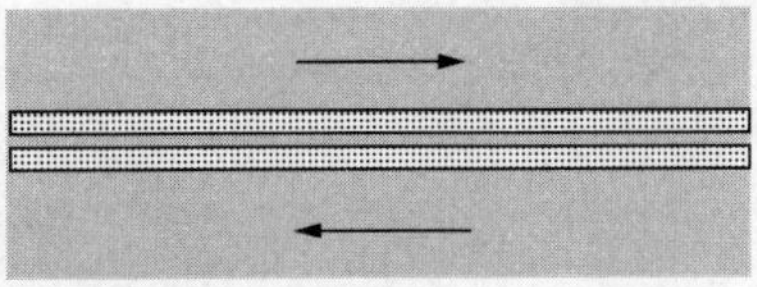 Continuous double line	Yellow	• Seriously restrictive in character • Does not allow crossing of vehicles, except in an emergency • Used for separating traffic flow in the opposite direction
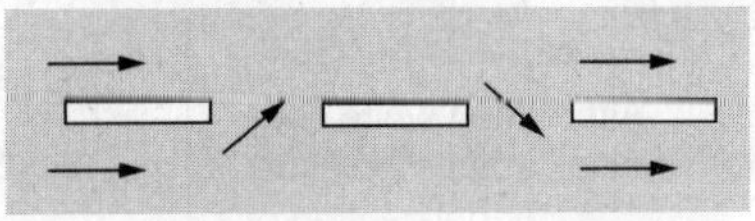 Broken single line 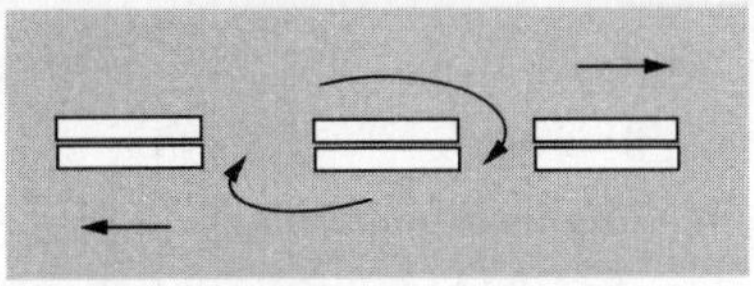 Broken double line 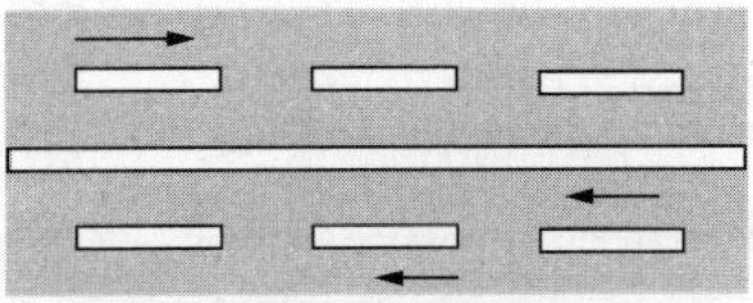 Combination of solid/broken lines	White	• Permissive in character • Allows crossing with user discretion based on the traffic situation • Used as centre line for one-way streets, ramps and slip roads

(*Continued*)

Type with description	Line colour	Use type
Continuous and broken line (Segment =1000; gap =1000 mm; width = 100 mm, separated by 100 mm)	Both white or White with yellow	• The solid line may be crossed with discretion, if the vehicle is crossing from the broken line lane • Vehicles from the opposite lane are not permitted to cross from the solid line lane • Used at continuity lines between auxiliary lanes and through lanes
Solid single edge line	White/ Yellow	• Pavement edge/border is indicated • Warns traffic to keep away from the edge • Desirable on narrow brides/culverts, pavement width transitions, fog/mist-covered pavement localities • Marked 200 mm and 150 mm from the edge of expressways and other roads, respectively.

5.1.3 Centre Line

Longitudinal markings such as centre line and edge line are provided along the length of traffic flow. The centre line demarcates the centre of the carriageway, along its length. It streamlines traffic flow and separates the same and opposite directions of traffic movement.

Continuous solid lines have the following variants:

The continuous single solid line in yellow is provided to strictly prohibit straddling by vehicles. This line is generally provided as an edge or verge line (Fig. 5.7).

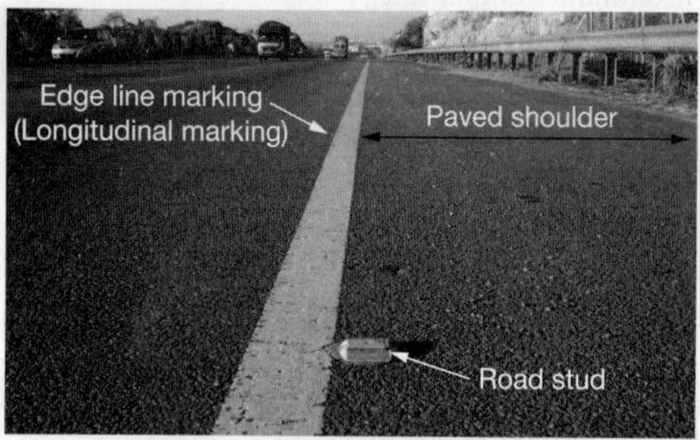

Figure 5.7 Edge line marking with a stud.

5.1.4 Transverse Markings

These markings are provided across the length of the carriageway. Continuous single/double lines and broken lines accompanied by appropriate signs are used to control traffic. The Stop line (Fig. 5.8) and Give Way markings (Fig. 5.9 and Fig. 5.10) are commonly used.

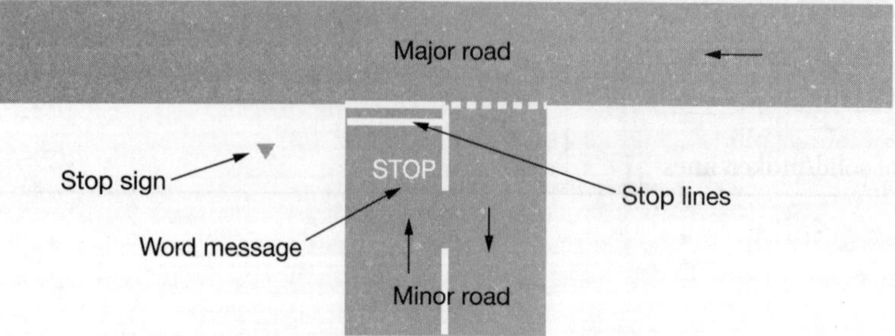

Figure 5.8 Stop line markings (schematic).

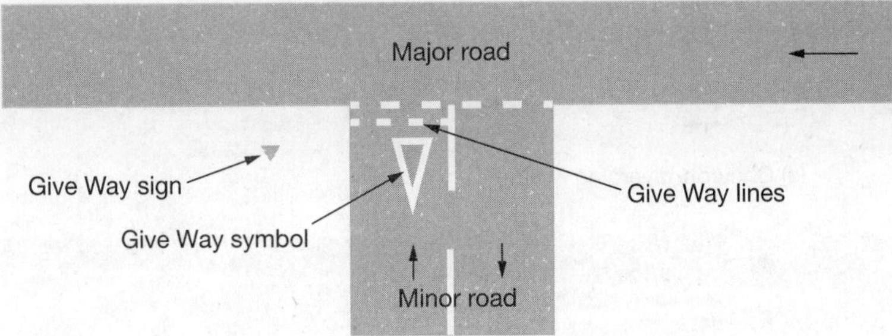

Figure 5.9 Give Way markings (schematic).

Figure 5.10 Two views of Give Way markings on a road.

- A set of bar markings across the full width of the carriageway (yellow): Six stripes of 300 mm each separated by 600 mm with a depth of 5 mm (Fig. 5.11). The 5-mm depth is useful for guiding visually challenged pedestrians when they cross the road, while the crossing vehicles will experience vibrations and slow down while passing.

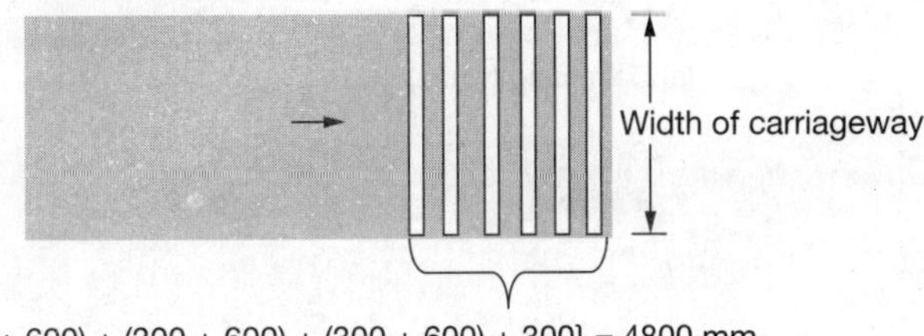

{(300 + 600) + (300 + 600) + (300 + 600) + (300 + 600) + (300 + 600) + 300} = 4800 mm

Figure 5.11 A set of bar markings across the width of a carriageway (Redrawn based on IRC: 35-2015).

5.1.5 Hazard Markings

Prohibited and hazardous locations at diverging and merging points and any object-obstructions on the carriageway are indicated and users are warned using chevron markings, diagonal markings, hatch markings and prohibitory markings. A few widely used hazardous markings are described below.

- Chevron-diverging (Fig. 5.12 (a)): The area covered by chevron markings at the nose of a channelising island is demarcated as a neutral area. It diverts the entering traffic at the proper angle so that it acts as a transition to the acceleration lane.
- Chevron-converging (Fig. 5.12 (b)): It smoothly converges the exiting traffic at a proper angle so that it acts as a transition to the deceleration lane.

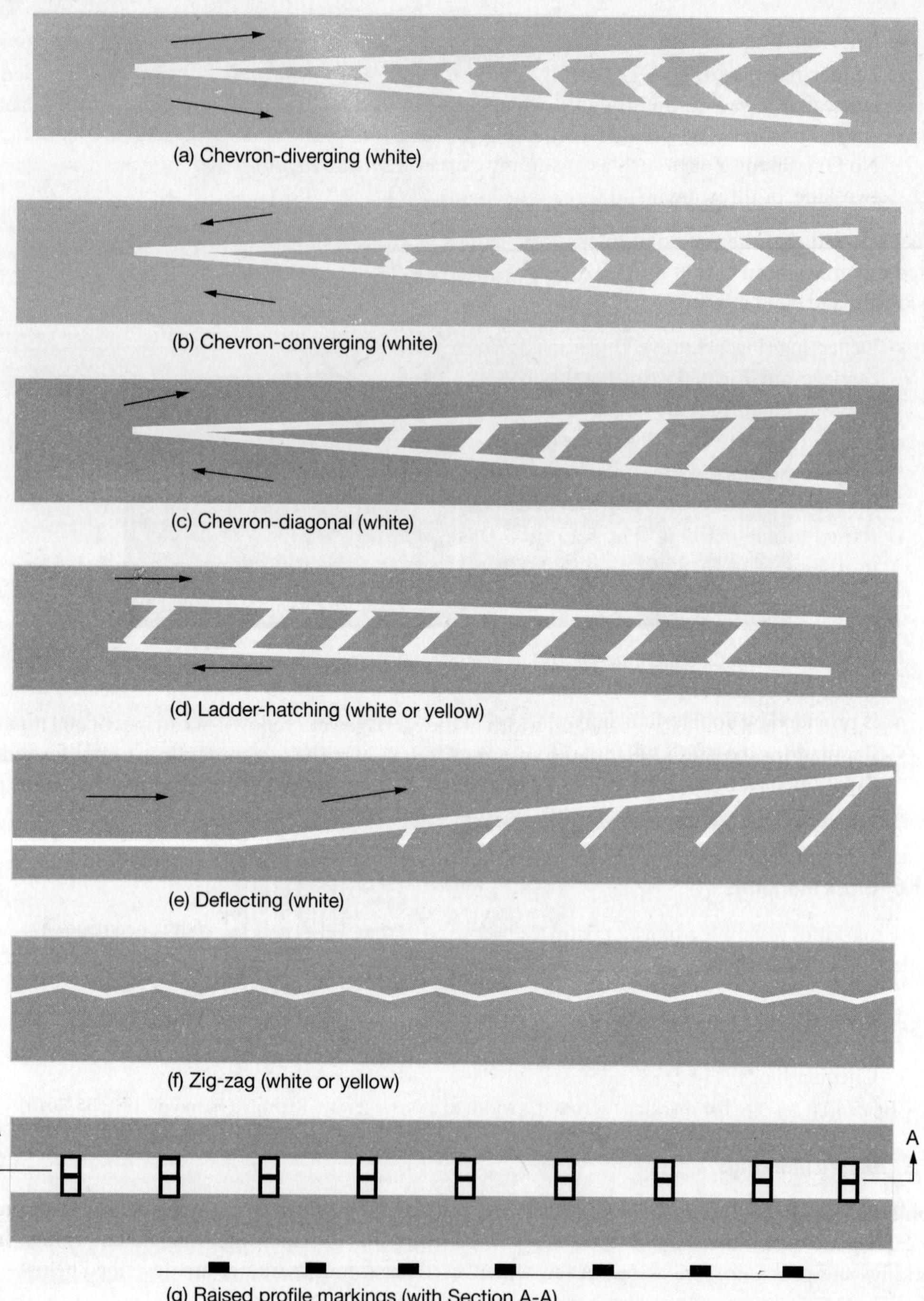

(a) Chevron-diverging (white)

(b) Chevron-converging (white)

(c) Chevron-diagonal (white)

(d) Ladder-hatching (white or yellow)

(e) Deflecting (white)

(f) Zig-zag (white or yellow)

(g) Raised profile markings (with Section A-A)

Figure 5.12 Types of hazardous markings (schematic).

- Chevron-diagonal (Fig. 5.12 (c)): It separates and diverts traffic flow in the opposite directions.
- Ladder-hatching (Fig. 5.12 (d)): It is used to separate opposing traffic flow on undivided two-lane carriageways in the straight segments and on curves. This marking is also effective on acute curved portions that provide adequate gap between opposite adjacent traffic flows and between No Overtaking Zones such as at summit curves, horizontal curves and elsewhere on undivided, two-lane, bi-directional and three-lane highways.

It is preferred that the two solid lines of the ladder be either 100 mm or 150 mm in length and separated by a minimum of 600 mm. A diagonal of length 4000 mm should be provided for speeds up to 60 km/hr, and 6000 mm for speeds exceeding 65 km/hr.

- Deflecting (Fig. 5.12 (e)): This is used to deflect traffic flow away from an edge or any portion of a carriageway. The deflecting hatchline marking is provided when the edge line marking is ineffective. Such markings are also used to create refuge islands and for achieving lane reduction.
- Zigzag (Fig. 5.12 (f)): These are used as centre line and edge line markings in vulnerable road sections, such as in areas near speed breakers where human activities and pedestrian presence necessitate speed reduction. These lines also help keep vehicles at a distance from extremities.
- Raised profile marking (Fig. 5.12 (g)): This is used as an alternative to edge markings. The vertical dimension (height raised to 4–8 mm) facilitates clearly visibility above water sheets in the wet season. In addition, it also provides audible and vibratory warnings when vehicles cross over them.
- Single continuous solid line of 100 or 150 mm width; yellow/Double continuous solid line of width 100 or 150 or 200 mm, separated by 100 mm and painted yellow: A No Overtaking stretch is provided by double-line demarcation.
- Bar markings of width 300 mm spread across the carriageway, leaving a gap of 150 m on either end to facilitate run-off; height = 4–5 mm. This is often provided in series at railway level crossings.

More details of line width and layout can be obtained from IRC: 35-2015.

5.1.6 Block Markings

These markings resemble a block of paint and are used for different locations depending on the intended control function. Details of different types of block markings are as follows (IRC: 35-2015):

- Rectangular block (Fig. 5.13 (a)): Used as zebra crossing
- Triangular block (Fig. 5.13 (b)): Used for speed breakers/road humps
- Chequer block (Fig. 5.13 (c)): Used for speed breakers/road humps

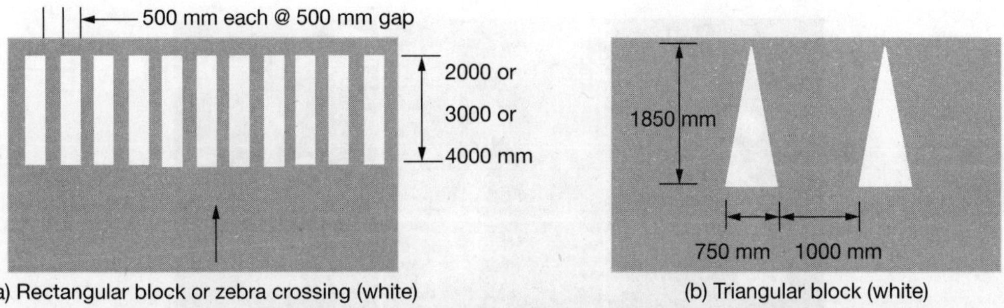

Figure 5.13 Different types of block markings (Redrawn based on IRC: 35-2015).

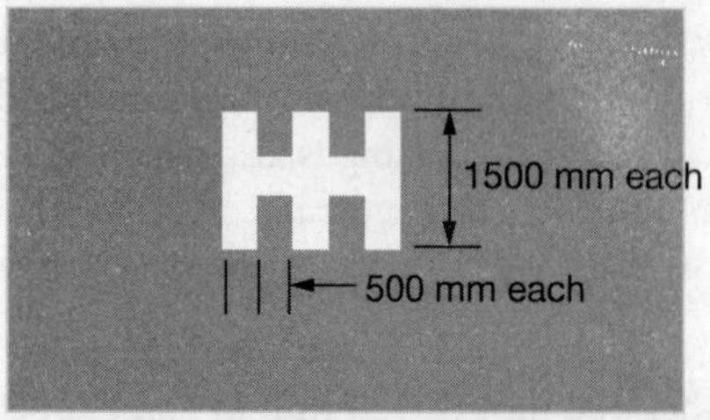

(c) Chequer block (white)

(d) Box marking adjacent to the kerb line (yellow)

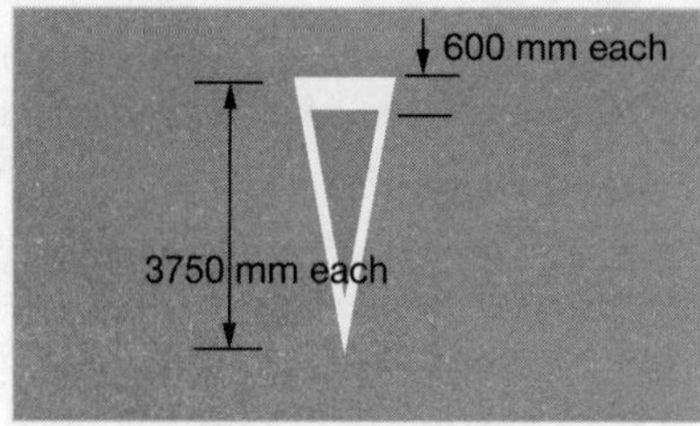

(e) Give Way symbol (white)

(f) Bicycle symbol (white)

Figure 5.13 Different types of block markings (Redrawn based on IRC: 35-2015).

- Yellow box marking (Fig. 5.13 (d)): Indicates prohibitory intersection area in which no vehicle is permitted to stop/halt
- Give Way symbol (Fig. 5.13 (e)): Provides priority of flow
- Cycle symbol (Fig. 5.13 (f)): Used to indicate dedicated cycle lanes

5.1.7 Arrow Markings

Arrow paintings (all in white) indicate the mandatory direction of travel and are categorised as follows (Fig. 5.14):

- Straight
- Left
- Straight and left
- Straight and right
- Right and left
- Straight, left and right
- Forward
- Deflection
- Bifurcation
- Roundabout rotation

Figure 5.14 Arrow road markings: left, straight, and straight and left.

Directional arrows are marked on the pavement surface in advance to guide drivers to the correct lane by avoiding confusion/ambiguities relating to entry restrictions at busy intersections and lane restrictions such as one-way flow.

5.1.8 Directional Markings

Word messages of a directional nature are categorised under this type of marking:

- STOP
- SLOW
- SPEED LIMIT
- KEEP CLEAR
- SCHOOL KEEP CLEAR
- NO PARKING
- NO ENTRY
- TURN LEFT
- TURN RIGHT
- KEEP LANE

Arrows supplemented with directional markings are termed mandatory turn arrows. For example, a left arrow followed by a TURN LEFT word message marking.

5.1.9 Facility Markings

The facilities provided for different purposes and their intended uses are indicated by these symbols and word messages. A few examples are listed below:

- Parking markings: Mid-block, Last Block
- Word message markings: BUS, BUS STOP, BUS LANE, CYCLIST, TAXIS and DISABLED

5.1.10 Object Markings

Any physical obstruction on or adjacent to a carriageway which is considered hazardous to traffic flow is delineated by being painted with dual-coloured hatched stripes. The main goal of painting alternating white-black or yellow-black stripes on these objects is to provide more visibility at night to avoid collision of vehicles with these objects.

The objects may be listed as any type, for example, piers, abutments, restrictive vertical clearance beams, electric poles, sign/signal posts, trees guard rails/stones, central islands/medians, culvert walls and subways/ramps supporting structural columns.

The objects within the carriageway are marked with yellow-black stripes sloping downward at an angle of 45° towards traffic (Fig. 5.15). The diagonal stripes should be uniform in width which shall not be less than 300 m. Appropriate chevron markings are painted on the carriageway to cover these objects.

Figure 5.15 Views of object markings at a toll booth.

The objects adjacent to the carriageway, such as abutments and side walls, are marked with white-black stripes sloping downwards at an angle of 45° towards traffic. Similarly, the width of the diagonal stripes shall not be less than 300 m and they should be supplemented with appropriate chevron markings.

Electric poles are marked with alternating horizontal black-white stripes, not less than 100 mm wide, up to a height of 1.25 above the road surface. Guard rails/stones/drums/trees are painted a solid white colour up to a height of 1.25 m above the road surface. A black band 300 mm wide is marked in the middle of the 1.25 m height of white paint.

Kerbs and concrete barriers on bridges/flyovers shall be painted with vertical black-white stripes of 500 mm width. All objects located within 2.4 m of the shoulders/kerbs shall be painted, in addition to the object markings. The object marker shall be at least 1.2 m above the nearest traffic lane.

5.1.11 Road Studs

Road studs are surface-mounted devices used to delineate road edges, centre lines and pedestrian crossing areas (Fig. 5.16). Based on their intended use, they are classified as follows:

- **Non-reflecting road studs:** These are used on pedestrian crossings as a tactile demarcation. Non-reflective road studs are cast using hard-grade aluminium alloys or stainless steel. They are available as a round or square (100 mm) top profile in aluminium or stainless steel and installed in small holes drilled on the pavement surface using hot-poured bitumen.
- **Reflective road studs:** These are used to supplement line markings, longitudinal and transverse. Based on the direction of reflectivity, they can be uni-directional or bi-directional. The uni-directional studs are used on one-directional carriageways of undivided roads, operated as one-way routes or slip roads. The bi-directional studs can be used for two-directional roads to delineate the centre line and edge line of pavements. Reflective road studs are called raised reflective pavement markers (RRPM).

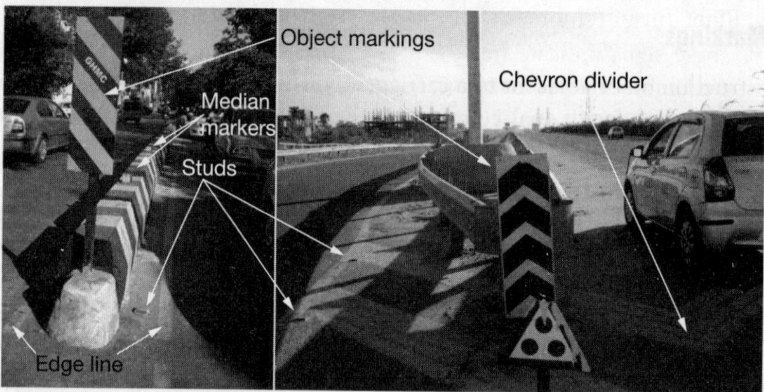

Figure 5.16 Road studs with other markings.

The most commonly used stud in India is the bi-directional road stud made of abrasive resistant plastic and fixed directly onto the pavement surface using hot melt bituminous/epoxy material. They provide guidance at night, including conditions of inclement/adverse weather conditions, when other traffic markings have been obliterated. In addition, the studs fixed across the carriageway ahead of junction/crossing/end of flyover ramp acts as a speed arrestor and as warning by creating a rumbling sensation to the driver through tyre interaction.

5.2 ROAD SIGNS

They are used for traffic operation and control, provide information and ensure safety to vehicular and pedestrian traffic. The signs used shall be easily recognisable, readable and understandable symbols, pictures, number and words.

5.2.1 Classification of Signs

Traffic signs can be classified based on their purpose (Fig. 5.17).

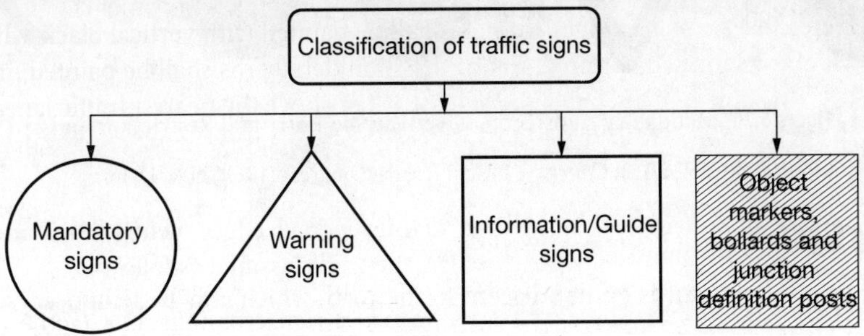

Figure 5.17 Classification of traffic signs.

5.2.2 Regulatory Signs

Traffic movements are regulated by assigning priority and indicating the regulations that are legally in force. Any failure in complying with these mandatory traffic regulations is treated as a constitutional offence. The regulatory signs may be categorised into three groups (MOTSAM, 2010):

- **Prohibitory:** These signs indicate that road users must not take a particular action. For example, No Parking, Right Turn Prohibited, Bullock Carts Prohibited.
- **Mandatory:** These signs indicate that road users must take a particular action. For example, Minimum Speed Limit, Compulsory Turn Right, Pedestrian Only.
- **Operational:** These signs indicate that road users must abide by a particular restriction. For example, Axle Load Limit, Length Limit, Maximum Speed Limit.

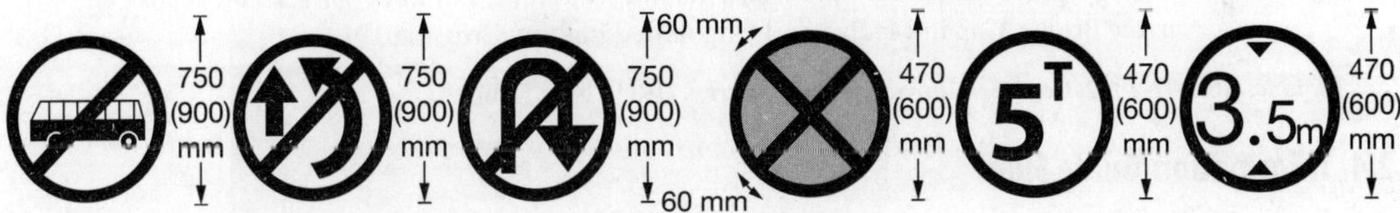

Figure 5.18 Sample regulatory signs: Bus Prohibited, Overtaking Prohibited, U-Turn Prohibited, No Parking and Standing, Weight Limit and Height Limit.

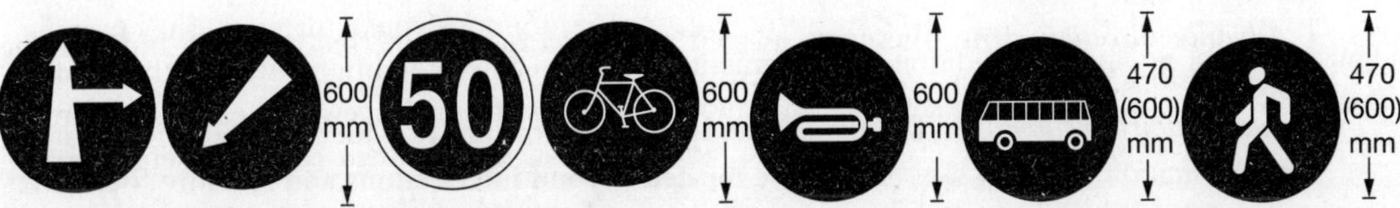

Figure 5.19 Compulsory signs: Ahead or Right Turn, Keep Left, Minimum Speed Limit, Cycle Track, Sound Horn, Buses Only and Pedestrians Only.

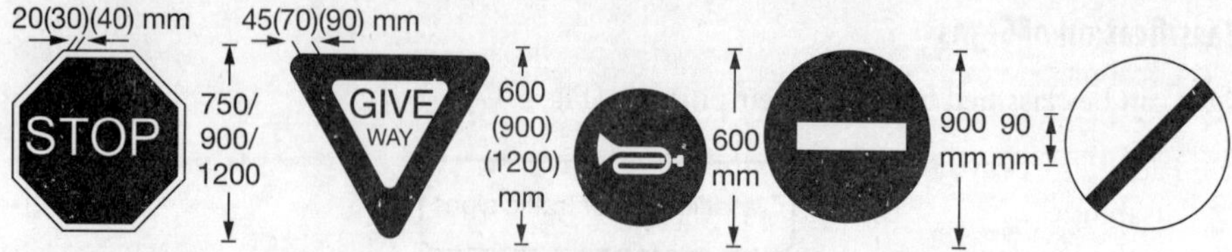

Figure 5.20 Mandatory signs: Stop, Give Way, No Entry and Restriction Ends.

(*Source:* IRC: 67-2012; *Courtesy*: Indian Roads Congress (IRC), New Delhi)

5.2.3 Warning Signs

Road users are warned of hazards on or adjacent to the road, which may be temporary, unexpected or unusual (Fig. 5.21).

Figure 5.21 Warning signs: Right-hand Curve, Side Road Left, Roundabout, Merging Traffic from Left, Narrow Bridge, Gap in Median and Unguarded Railway Crossing.

(*Source:* IRC: 67-2012; *Courtesy*: Indian Roads Congress (IRC), New Delhi)

5.2.4 Information/Guide Signs

Road users are guided along their journey with information about route directions, distances, destinations, location services and other important/related information along the roadways (Fig. 5.22). These signs may be categorised into three groups:

1. Direction signs
    - *Advance direction signs*: These are located at approaches to intersections and indicate the destinations ahead.
    - *Intersection direction signs*: These are located at intersections and provide directional information.
    - *Confirmation direction signs*: These are located beyond intersections and reassure the drivers about the route on which they are travelling.
2. Motor services and tourist signs
    - *Motorist services signs*: These indicate the services available ahead relating to motorist needs, such as repair services and fuel filling stations.
    - Tourist information signs

3. General information signs
   - Local place/feature names
   - Local authority boundaries
   - Facilities provided for persons with disabilities
   - Route marker signs
   - Other miscellaneous information

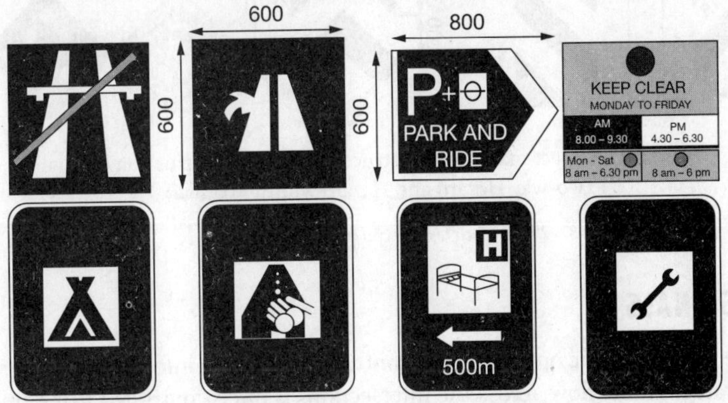

Figure 5.22 Sample facility signs: End of Expressway, Entry Ramp to Expressway, Park and Ride, Parking Restriction, Camp Site, Toll Road Ahead, Hospital and Repair Facility.

(*Source*: IRC: 67-2012; *Courtesy*: Indian Roads Congress (IRC), New Delhi)

5.2.5 Hazard Markers

Road users are informed about the presence of obstructions or change in directions, if any, ahead on the route. These marker signs may be categorised into three groups:

- Chevron markers (Fig. 5.23 and Fig. 5.24)
- Object hazard markers (Fig. 5.25)
- Two-way hazard markers
- Delineation posts
- Guide posts
- Miscellaneous devices

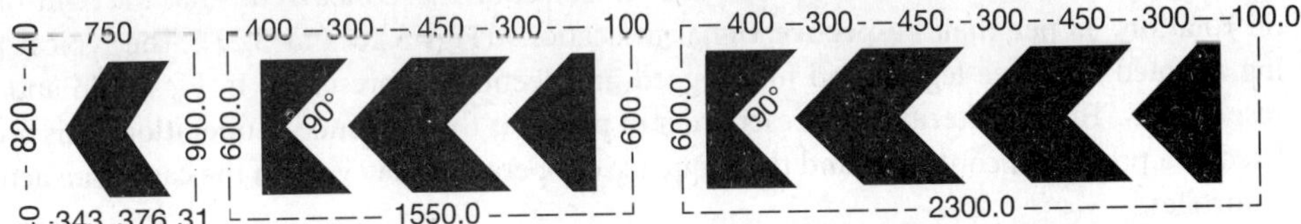

Figure 5.23 Chevron markings: single, double and multiple.

(*Source*: IRC: 67-2012; *Courtesy*: Indian Roads Congress (IRC), New Delhi)

Figure 5.24 Chevron markings along a left-hand curve.

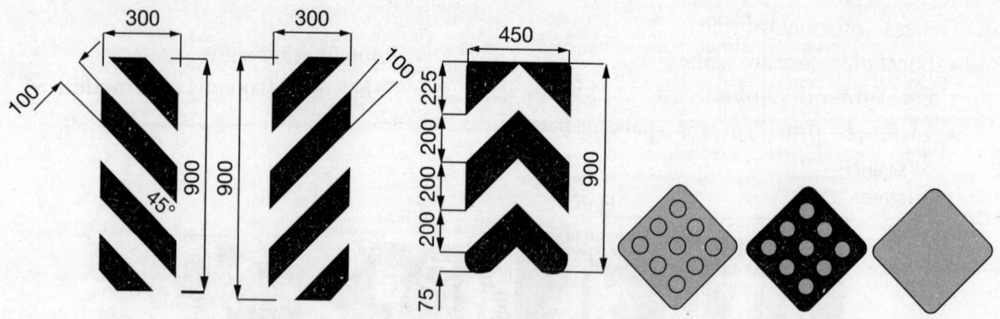

Figure 5.25 Object markings: Obstruction on Left, Obstruction on Right, Two-way Hazard and Hazard within Roadway.

(*Source:* IRC: 67-2012; *Courtesy:* http://mutcd.fhwa.dot.gov)

5.3 TRAFFIC SIGNALS

Traffic signals are used for traffic operation and control, and provide information and safety to vehicular and pedestrian traffic. Traffic flow across the intersections is often controlled using traffic lights or signals (Red, Green and Amber). Appropriate signal timing is to assign the right of way to alternating traffic movements in order to minimise the average delay to any group of vehicles or pedestrians; and to reduce the probability of accident-producing conflicts.

5.3.1 Commonly Used Terminology

- **Cycle:** It represents one complete sequence of phases or intervals. It is a period of time (in seconds) in which a complete rotation of signal indications takes place (Red→ Green→ Amber) → (Red→ Green→ Amber). The Amber indication that appears between Green and Red is termed clearance Amber. As the approach speed increases (from 50 km/hr to 80 km/hr), Amber time is also suitably increased (from 3.0 seconds to 5.0 seconds).
- **Cycle length:** The time taken to complete one sequence of signal phases (a cycle).
- **Phase:** A set of movements of traffic flow at an intersection that concurrently has the right-of-way on rotation, as per their respective signal indications (Figs 5.26 and 5.27). The typical phasing adopted for three-legged and four-legged intersections is presented in Figs. 5.26 and 5.27, respectively. The predetermined/preset order of phases is the sequence of operation. This order is fixed in a pre-timed controller and the sequence of operation may vary in the case of an actuated controller.

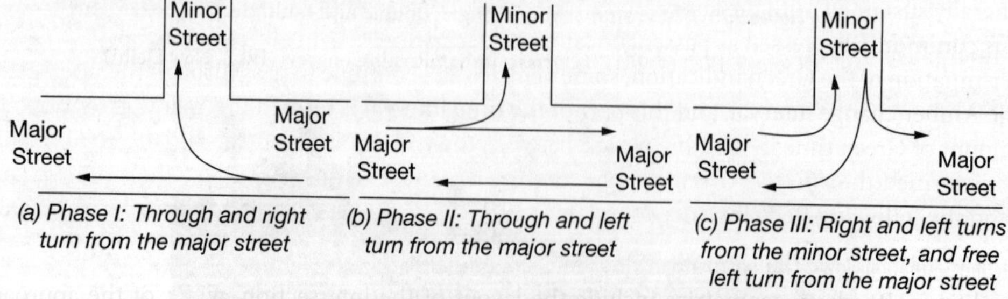

Figure 5.26 Typical phasing diagram for a three-legged intersection.

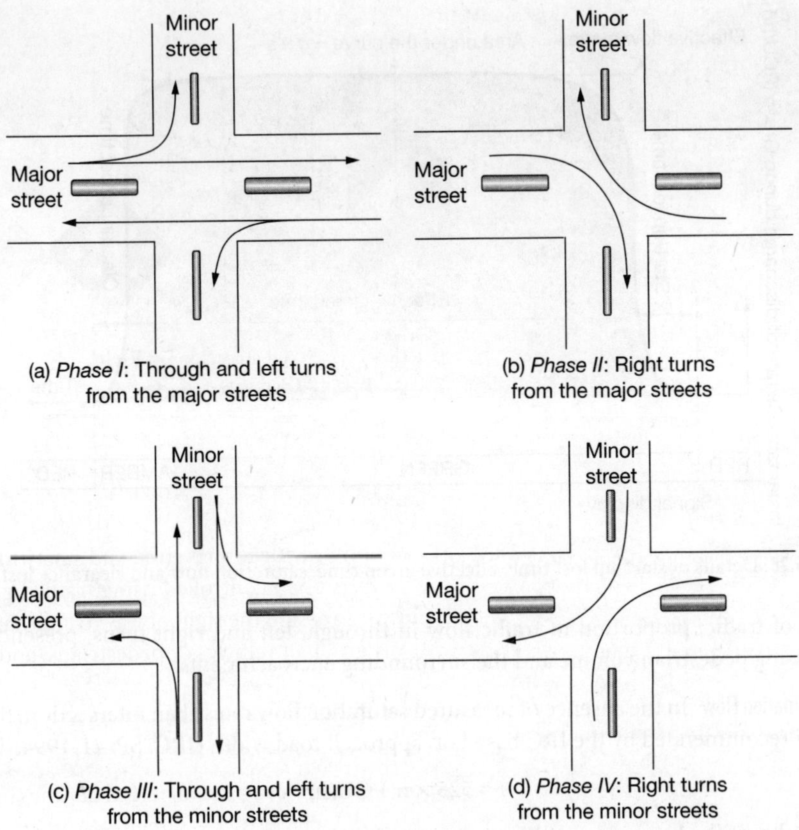

Figure 5.27 Typical phasing diagram for a four-legged intersection.

- **Saturation Flow Rate:** On receiving a Green indication, it takes a few seconds for the driver of the first vehicle to recognise the Green and get the vehicle in motion. The next few vehicles also take some time to accelerate. The time period during which no flow takes place and delay occurs during initial start-up of vehicles before attaining normal running speed is often accounted as *start-up lost time* (Fig. 5.28). Start-up lost time or start-up delay is commonly assumed to be approximately 2 seconds.

After approximately the fourth vehicle moves forward from the queue, the flow rate tends to stabilise at the *maximum discharge* rate as the conditions will allow, known as the saturation flow rate. This flow rate is generally sustained until the last vehicle in the queue departs from the intersection. The saturation flow rate is commonly expressed as passenger car units per hour (PCU/hr) or PCU/sec.

On termination of the Green indication, some vehicles will continue to pass through the intersection during the Amber change interval, and this part of the time interval is termed the Yellow extension. The usable amount of Green time in the time period between the end of start up delay and the end of Yellow extension is termed the *effective Green time* for the movement. The unused part of the Yellow change interval and the following Red clearance interval is called the clearance lost time (Koonce et al., 2008).

Factors Affecting Saturation Flow The saturation flow rate depends on the prevailing roadway geometrics and traffic conditions. Roadway geometrics include the layout of the intersection, width of the approach, number of lanes and their widths, approach gradient and site conditions. The traffic conditions include

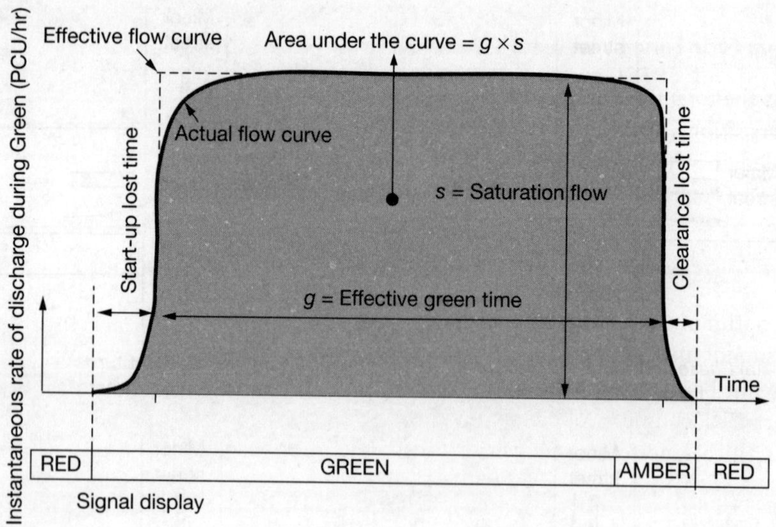

Figure 5.28 Details of start-up lost time, effective green time, saturation flow and clearance lost time.

composition of traffic, proportion of traffic flow in through, left and right turns, presence of parked vehicles, crossing pedestrian volume and the surrounding encroachments.

Estimation of Saturation Flow In the absence of measured saturation flow rates at an intersection, the following equations are recommended by the IRC based on approach road width (IRC: SP: 41-1994; TRL, 1996):

$$s = 525 \times w \text{ PCU/hr} \tag{5.1}$$

where w = width of the approach road (m), measured from the kerb to the nearest median or divider or centre line, whichever is closer. This expression is valid when the w value is in the range 5.5–18.0 m. If $w < 5.5$ m, the following values of saturation flow are estimated (TRL, 1996):

w (m):	2.70	3.0	3.5	4.0	4.5	5.0	5.5
s (PCU/hr):	1790	1850	1890	1950	2250	2550	2990

Estimation of saturation flow based on other geometric and traffic factors may be obtained from IRC: SP: 41.

- **Lost Time:** It is the sum of time periods in a cycle during which no traffic flow takes place in each phase when all the signals in the intersection show Red or Red-Amber. The delays occur during startup of Green and falling of discharge during Amber-Red. The following equation is used to estimate total lost time per cycle (L):

$$L = (n \times l + R) \tag{5.2}$$

where n = number of phases in a cycle
l = average time lost per phase, excluding the time during all Red or sequent Ambers; and
R = all Red period during each cycle when all the signals in an intersection show Red or Red with Amber.

Average lost time during startup delay in an Amber period may be assumed as 2 seconds per phase. The Highway Capacity Manual defines a default value of 4 seconds per phase as total lost time (sum of startup lost time and clearance lost time).

5.3.2 Optimum Cycle Time by Webster's Method

Cycle length is the total time period during which one complete sequence of signalisation takes place around an intersection. A complete cycle comprises all the phases with their respective signal indications.

The cycle length with minimum delay is often termed *optimum* cycle length, and it can be determined by Webster's equation:

$$C_o = \frac{1.5L_i + 5}{1 - Y_i} \tag{5.3}$$

where C_o = optimum cycle length (sec); and
L_i = sum of the lost time (sec) for all phases = $l_1 + l_2 + \ldots + l_i$, up to *i* number of phases,
Lost time per phase = Startup time (usually 1–2 sec) + Clearance lost time
Clearance lost time = A portion of the Amber interval (usually 1–2 sec) + All-Red interval
Y_i = Critical lane volume (in veh/hr) divided by saturation flow, summed over the *i* phases

$$Y_i = \sum_{i=1}^{n}(y_1 + y_2 + y_3 + y_4 + \ldots\ldots\ldots\ldots y_i)$$

$$y_i = \text{Flow ratio} = \frac{\text{Maximum flow through a lane during the } i\text{th phase}}{\text{Saturation flow for phase } i} = \frac{Q_i \max}{s_i} \tag{5.4}$$

Notes:

- Eq. (5.4) is used for isolated pre-timed signal locations only.
- Webster's formula is applicable if the volume/capacity ratio ≤ 1.0.
- Eq. (5.4) indicates that delay is not significantly increased if cycle length variations are in the range 0.75–1.5C_o.
- A detailed network analysis should be performed (using a suitable software program) to determine cycle length in a coordinated system. Software programs (with their inbuilt models) are capable of performing iterative procedures of varying cycle combinations across a series of signalised intersections, and ultimately finding the optimum signal timing parameters (MnDOT, 2013).

Minimum cycle length (C_m) allows just sufficient passing time to the traffic, and it is calculated using the following equation (TRL, 1996):

$$C_m = \frac{L_i}{1 - Y_i} \text{ sec} \tag{5.5}$$

C_m may lead to excessively long delays, but this cycle length may be selected to coordinate signals control when an adequate time margin needs to be maintained for key intersections. However, it is appropriate to put on load up to 90% of its capacity, and the resulting cycle length is termed practical capacity time (Cp). Cp is calculated as follows (TRL, 1996):

$$C_p = \frac{0.9L_i}{0.9 - Y_i} \text{ sec} \tag{5.6}$$

This value of C_p may be adopted to facilitate minimum delays to pedestrian crossings where the use of short cycle length is preferred with high pedestrian volume.

Effective green time

It the net time period of Green with Amber after excluding the total lost time per phase, calculated as follows:

$$g_i = G_i - L_i \tag{5.7}$$

where g_i = effective green time for the *i*th phase;
G_i = (Green period + Amber period) per phase; and
L_i = total lost time per phase.

Webster (1958) suggested dividing Green time (g) in the proportion of flow ratio (y) for each phase and recommended the following equation to establish the relationship between the ratios of effective Green time (g_i) and the corresponding *yi* values of each phase. The relationship can be extended to three or more phases as well.

$$\frac{g_1}{g_2} = \frac{y_1}{y_2} \tag{5.8}$$

where g_1 and g_2 are the effective Green times of Phases 1 and 2, respectively; y_1 and y_2 are the ratios of maximum flow to saturation flow of Phases 1 and 2, respectively, as explained in Webster's method of cycle time (Eq. 5.3). This is a simple rule useful for Green time allocation to obtain the least overall delay.

Green time

Effective Green time (g_i) for the respective *i*th phase can be calculated from the following expression:

$$g_i = \frac{y_i}{Y_i}(C_o - L_i) \tag{5.9}$$

where g_i = effective Green period of phase *i*;
y_i = flow factor of phase i;
C_o = cycle length;
L_i = total lost time;
Y_i = sum of y_i factors;
and $C_o - L_i$ = total effective Green time.

EXAMPLE 5.1

Determine the optimum cycle time based on Webster's formula. Take the sum of ratios of normal flow to saturation flow of two directional traffic flow = 0.6. The total lost time per cycle =10.0 sec.

Answer

The given input values are expressed as standard Webster notations.

$$L = 10.0 \text{ sec and } Y = 0.65,$$

Optimum cycle length as per Webster's formula $C_o = \frac{1.5L+5}{1-Y} = \frac{1.5\times10.0+5}{1-0.65} = 57.12$ sec.

Rounded to 60.0 sec., as the cycle time is commonly expressed in multiples of 5 sec.

Note: The delay is not so sensitive for a certain range of cycle length, and hence, the cycle length may be rounded to a multiple of 5 seconds.

EXAMPLE 5.2

A pre-timed four-phase signal has the following critical lane flow rates observed for the first three phases: 180, 192 and 198 veh/hr. Take saturation flow for all the phases as 1500 veh/hr. The time lost per phase is 4 sec.

Answer

If the optimum cycle length obtained by Webster's formula is 67 sec, the effective Green time (sec) of the fourth phase would be

$$Y_1 = \frac{\text{Normal flow Phase 1}}{\text{Saturation flow for Phase 1}} = \frac{180}{1500} = 0.12$$

$$Y_2 = \frac{\text{Normal flow Phase 2}}{\text{Saturation flow for Phase 2}} = \frac{192}{1500} = 0.128$$

$$Y_3 = \frac{\text{Normal flow Phase 3}}{\text{Saturation flow for Phase 3}} = \frac{198}{1500} = 0.132$$

$$Y_1 + Y_2 + Y_3 = 0.12 + 0.128 + 0.132 = 0.38$$

Total time lost during 4 phases = 4 × 4 sec = 16 sec

$$\text{Optimum cycle length as per Webster formula } C_o = \frac{1.5L + 5}{1 - Y}$$

$$67 = \frac{1.5 \times 16.0 + 5}{1 - Y}$$

$$Y = 0.567164$$

Therefore, $Y_4 = Y - (Y_1 + Y_2 + Y_3) = 0.567164 - (0.38) = 0.187164$

$$\text{Effective green time for 4th phase} = \frac{Y_4}{Y}(C_o - L) = \frac{0.187164}{0.567164}(67 - 16) = 16.83 \text{ sec}$$

EXAMPLE 5.3

The data given below is collected for designing a two-phase fixed signal at an intersection having North–South and East–West approach roads. Straight ahead traffic is permitted to operate in these approach roads.

Parameter	Approach road			
	North	South	East	West
Design hr. flow (PCU/hr)	845	750	850	490
Saturation flow (PCU/hr)	2500	2500	2100	2100

The time lost per cycle = 12.0 sec

Determine the optimum cycle length as per Webster's formula. Also calculate (a) effective Green time per phase North–South and (b) effective Green time per phase East–West.

Answer

Standard notations of Webster are used to represent the input values as follows

$$y_{\text{North-South}} = \frac{\text{Critical value of normal flow phase North – South}}{\text{Saturation flow for phase North – South}} = \frac{845}{2500} = 0.338$$

$$y_{\text{East-West}} = \frac{\text{Critical value of normal flow phase East – West}}{\text{Saturation flow for phase East – West}} = \frac{850}{2100} = 0.4048$$

$$Y = (y_{\text{North-South}} + y_{\text{East-West}}) = (0.4048 + 0.4048) = 0.7428$$

Optimum cycle length as per Webster's formula $C_o = \frac{1.5L+5}{1-Y} = \frac{1.5\times12.0+5}{1-0.7428} = 89.41$ sec.

Rounded to 90.0 sec., as the cycle time is commonly expressed in multiples of 5.

Effective green time per cycle = $C_o - L = 90.0 - 12.0 = 78.0$ sec

Let effective green time per phase North–South = g_1 sec

Let effective green time per phase East–West = g_2 sec

The cycle length is distributed as two Green times to the two phases on the basis of the following proportion:

$$\frac{g_1}{g_2} = \frac{Y_1}{Y_2} = \frac{0.338}{0.4048} = 0.835$$

We know, $g_1 + g_2 = C_o - L$

From the above proportion, we can write $g_1 = 0.835 \times g_2$

Substituting, we get

$$(0.835 \times g_2) + g_2 = 78.0 \text{ sec}$$

$$g_2 = (78.0/1.835) = 42.5 \text{ sec}$$

Therefore, $g_1 = (78.0 - g_2) = (78 - 42.5) = 35.5$ sec

$$\text{Cycle} = \text{Phase 1} + \text{Phase 2} = 35.5\ (45.5\%) + 42.5\ (54.5\%) = 78 \text{ sec}$$

5.4 LOW-COST TRAFFIC CONTROL AND MANAGEMENT TECHNIQUES

Apart from road markings, road signs and traffic signals, the following is a list of low-cost traffic control techniques which are commonly adopted in most of the urban roads. The low-cost traffic flow management techniques are appropriate for use in urban roads where there is high traffic volume, particularly when the funds available for operating the roads are insufficient. These techniques/ strategies are implemented with a broad objective of smoothing traffic flow with increased safety by reducing number of accidents and for optimal usage of existing infrastructure.

5.4.1 One-way Street Systems

The traffic flow is permitted in one direction only. There will be no opposite traffic operated in these streets. This technique is used when lane capacity is exceeded or to divert the traffic flow to a particular direction to avoid traffic delays.

5.4.2 Turn Restrictions

This strategy is applicable to traffic that turn across the opposing traffic stream by banning left or right or both turnings at a signalised intersection. As a common practice, right turn crossing at signalised intersection is banned so that crossing movements can be eliminated and also the number of signal phases can be reduced. To avoid diverging and merging conflict, restriction on U-turns at intersections or median opening is also another optional traffic control measure.

5.4.3 Dedicated Bus, Taxi, Cycle Lanes

Each lane is exclusively assigned to use for specific mode of vehicle. It is technique adopted so that the slow moving, fast moving or high occupancy vehicles are segregated from the mixed traffic flow condition.

5.4.4 Shared Lanes

At a few locations of urban roads, left lane of a multi-lane road is permitted to be used exclusively for loading and unloading of goods during a specified period of a day while the remaining lanes are be operated as usual for the through traffic flow. Variable message signs or changeable message signs may be used for this purpose.

5.4.5 Staggered Timings

To avoid excessive traffic density during peak periods, the school timings and the office timings in a locality are staggered. The same lanes can be used effectively with the staggered use of the lanes at the assigned timings for the specific trip types. Variable work-hours strategy is one of the variants of this technique.

5.4.6 Odd or Even Vehicle Registration Numbers

Restricting vehicle use based on odd or even number of registration of vehicles is one of the techniques of avoiding traffic volume and environment pollution. This technique encourages the commuters to use car-pooling or van-pooling facilities or forces them to use mass transit systems such as buses or local trains.

5.4.7 Contra Flow Lanes

On high-speed arterial roads, only buses (as high occupancy vehicles) are permitted to travel in reverse direction on specified lanes. This is a common practice in a few cities in India as a part of Bus Rapid Transit (BRT) system.

5.4.8 Tidal Flow Technique

In multi-lane urban arterial roads, the centre lanes are sometimes assigned for moving traffic in one direction in the morning peak-period and in the opposite direction during evening peak period. This technique is adopted on a daily basis during peak periods when there is recurrent traffic in excess of lane capacity in one direction and minimum in the opposite direction.

Gantry-mounted over-head signals are used for indicating and controlling the traffic flow in the desired directions. Variable-lane assignment-signs may also be used for this purpose as a system of dynamic reversal of lane directions. As an alternative, special equipment is used to shift the central safety barriers across the lane either to left- and right-side lanes. As a standard practice, the change of lane direction is generally implemented at a fixed-time basis, daily during all working days, for predetermined hours.

5.4.9 Congestion Charging or Congestion Pricing

It is a system of surcharging road users during peak periods so that traffic congestion is reduced by discouraging the traffic flow on certain busy lanes in urban roads, particularly during working days. The road pricing strategy regulates traffic demand and manages the traffic without enhancing capacity of lanes. This is a widely known practice in a few busy roads in London.

5.4.10 Traffic Calming Measures

These are intended to reduce speed of vehicles, thereby reducing number and severity of accidents. Such measures are generally taken on residential roads. These include restriction of traffic movement, often by erecting barriers across the road. Some of the traffic calming options commonly used are

(i) Installing speed limit signs
(ii) Installing and operating speed cameras which enable self-enforcement and are visible as evidence
(iii) Channelisation at intersection area including wider central island refuge for pedestrian crossing at mid-blocks. The wider central medians restrict the speed at narrow carriageways
(iv) Mini-roundabouts which can reduce the speed of the through traffic and
(v) Speed breaker/humps.

Traffic Control at Work Zone Traffic separation from the site of construction area which is adjacent to the roadway is a mandatory requirement of the work executors and the road administrators. Adequate separation in horizontal and vertical direction has to be provided to avoid conflict with the traffic flow and avoid accidents. Traffic cone, arrow boards, flag trees, barriers, barricades, traffic barrels, flashing chevrons and delineators are installed when there is work in progress on the roadside by using transition tapers, buffer space, work space and terminal area. Standard templates of these installation plans are available and suitably used from IRC: SP: 55 (2014)

5.5 INTELLIGENT TRANSPORTATION SYSTEM (ITS)

Intelligent transportation systems (ITS) are information, electronic and communication technologies applied to road infrastructure, vehicles and users in order to improve efficiency in traffic operation and safety, with reduced traffic congestion (Eq. 5.10), delays and impact on the environment. ITS is installed to regulate traffic based on real-time flow.

$$\text{Congestion criterion: } \frac{\text{Traffic demand on the road at a point in time}}{\text{Traffic capacity of the road}} > 1.0 \quad (5.10)$$

5.5.1 Benefits of ITS

The following benefits can be obtained by deploying various subsystems of ITS (Nexco, 2008).

Area-wise traffic control or traffic management and control

Traffic signal timings in urban areas are controlled based on the traffic flow demand observed at various intersections using real-time data collection and analysis. To this end, vehicle detectors, roadside transmitters and variable message sign (VMS) boards operations are integrated.

Such an integrated multifunctional enforcement system can automatically detect signal violations, lane violations and speed violations by a vehicle. The collected violation data (image, violation type, place, time and more) is sent to the control centre and violation tickets are automatically issued to the vehicle owner.

- **Advance driving information and route guidance:** Prior information about road conditions, rest areas with/without fuel/electric recharging facilities, hospitals and restrictions due to congestion and inclement weather makes drivers maintain better control. It also prevents fluctuations in traffic speed such as a sudden decrease in speed at sag curves and turnings, thereby reducing CO_2 emissions.
- **Variable speed control and safety information service:** On-board car navigation monitors connected using wireless communication from on-road mounted sensors/devices can provide spatial information about route alignment, traffic volume, merging vehicles, fallen objects and any incidents ahead.
- **Sophisticated graphical message signage:** Multi-colour LED variable message sign (VMS) boards are used to display alternative routes, congestion information, weather information and accurate travel times.

- **Traffic video cameras and incident monitoring:** Video cameras installed at suitable intervals along a road network provide location details and information about the severity of unusual circumstances or incidents/accidents. Such information can be quickly sent to drivers and rescue teams with appropriate medical/equipment.
- **Accurate travel time estimation:** Travel time estimates for each segment of a roadway are displayed based on current traffic data with up-to-the-minute precision. Traffic jams and delays can be accurately estimated with this approach.
- **ETC toll booths:** Stopping, queuing and paying of toll/fee manually at toll booths can be eliminated by using an electronic toll collection (ETC) system. The toll payment is made using prepaid cards or deducted from bank accounts. The accounts and vehicle database can automatically store this information for future use.
- **Automatic vehicle location:** On-board GPS enables owners to locate their vehicles travelling in a road network.
- **Logistics information:** In-vehicle mounted GPS or OBE-ETC devices can be used to locate the current position of trucks/cargos of distribution companies. Such information can foster business productivity and reduce delays in cargo delivery.
- **Truck traffic consolidation/separation:** Strategic travel planning at interchange ramp meters, rest areas and cargo-distribution centres close to expressways make trucks travel/halt efficiently in groups. Large truck traffic can be separated from the general traffic by establishing freight-only lanes, which provide safety and comfort.
- **Customer/Passenger/Driver inquiries/emergencies:** Dedicated telephone landlines are provided with default connection to the customer care centre.
- **Doctor-heli:** In case of road accidents, medical airlifts can be made by doctors, assisted by nurses.

5.5.2 ITS Design

Provision for and location of various components of road infrastructure and geometric elements with the proposed electronic sensor/display devices must be planned in advance. Planning of the following ITS subsystems, along with their deployment design, is essential for any typical road network in an area:

- Traffic flow and road user data collection systems
- Road condition data collection system
- Weather data collection system
- Road network system with local infrastructure facilities
- Traffic management/speed, height and weight control systems
- Toll collection and enforcement systems
- Incident management systems
- Heavy-vehicle management systems
- Advance travel information system
- Public transport priority/emergency vehicle pre-emption systems

These subsystems should be integrated properly to achieve the objectives of ITS. These systems use different technologies (wired and wireless) and sensors such as vehicle detectors (intrusive/non-intrusive), pedestrian detectors, various types of signboards/display monitors, closed-circuit televisions (CCTV) and wired/wireless systems. The data collected is passed on to a nearby traffic control centre. Then, information from the different subsystems is analysed in real time by software and translated into navigational information (volume, speed, congestion length, saturation traffic flows and estimated travel time). This data is finally passed back to the local roadside device – VMS, OBE, in-vehicle navigation unit or traffic signal.

The in-vehicle unit used to receive information may be any one or a combination of the following:

- **Character-based display:** Used to display the speed limit, over-height limit, and distance to specific facilities
- **Diagram-based display:** Used to display route alignment with distance and on-route stations using multi-coloured lines and symbols
- **Map-based display:** Used to display an area map of the road network showing the origin-destination and other related details

As part of the Japanese Universal Traffic Management System (UTMS21), on-road traffic is controlled and managed by the integrated traffic control system centre (ITCS). ITCS can collect traffic information in real time and control signals in real time by two-way communication with in-vehicle equipment.

Based on requirements, the traffic operation management of subsystems can be evaluated by traffic supervisors with real-time information through computer interfaces such as dynamic wall maps, GIS map display walls and CRT/LCD displays. Traffic supervisors are authorised to change the system control parameter settings and regulate the entire system so that it is operationally stable and consistent (IUTR, 1998).

5.6 DESIGN OF PARKING FACILITIES

The classification of parking facilities, based on the location and the associated facilities is shown in Fig. 5.29. The importance/attractiveness of these parking facilities is determined based on the delay, difficulty/ease and cost of parking.

5.6.1 On-street/ Kerb/Roadside Parking

Parking bays are located alongside the kerb of the road. It is usually controlled by the local municipality. On-street parking can be categorised based on the *angle* at which the vehicles are parked in relation to the kerb (Fig. 5.30).

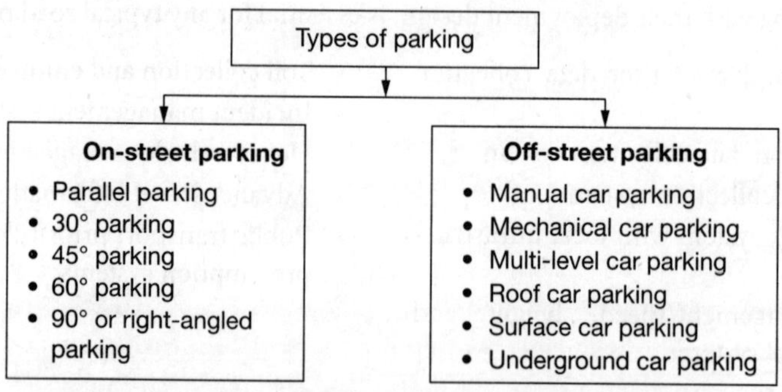

Figure 5.29 Classification of parking facilities.

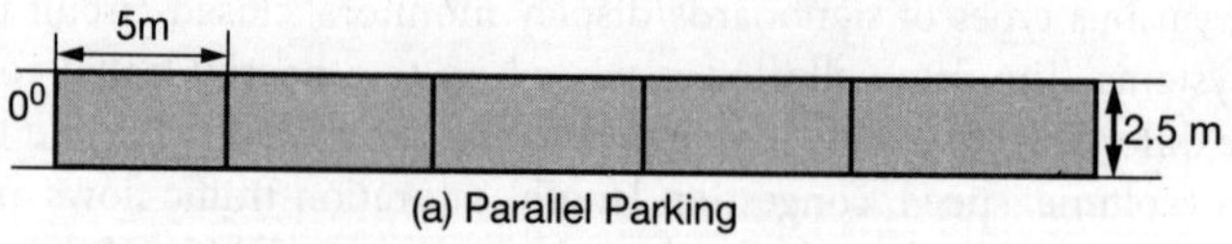

Figure 5.30 Types of on-street parking.

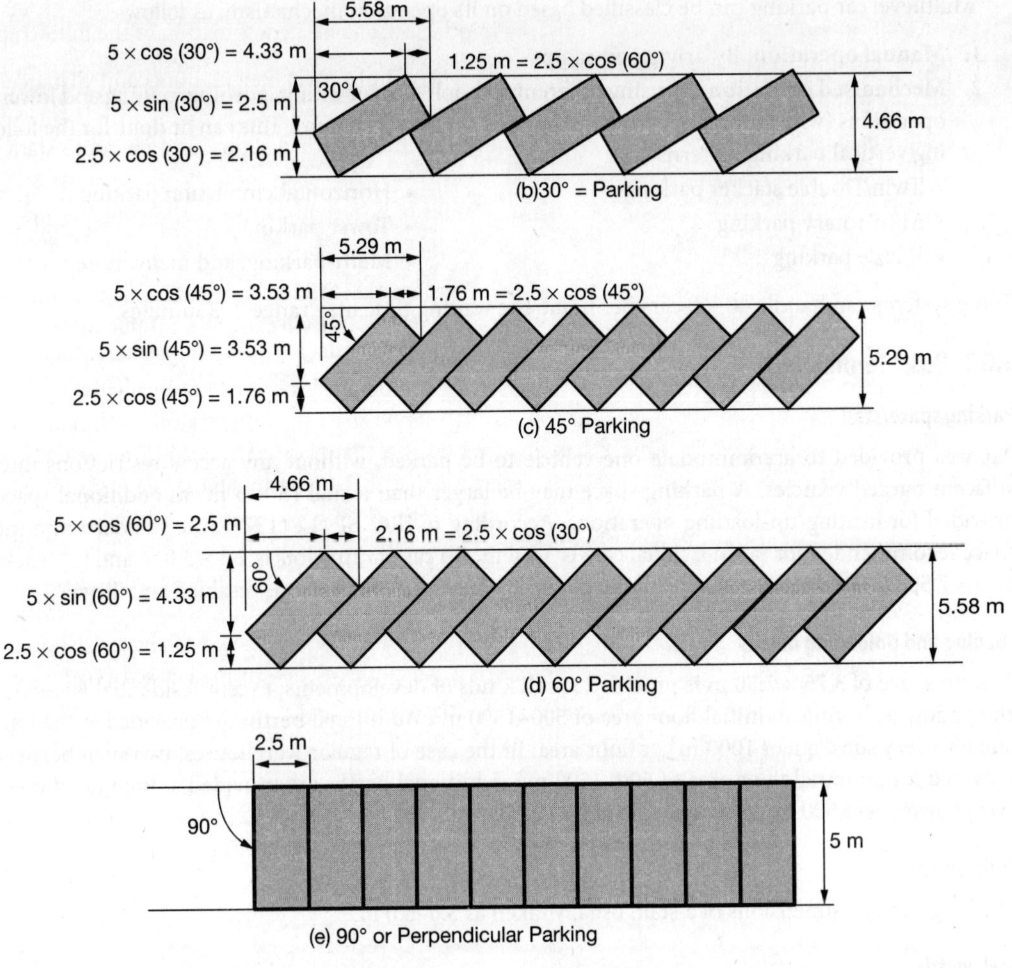

Figure 5.30 Types of on-street parking.

5.6.2 Off-street Parking

These parking spaces are provided *away* from the roadside.

1. **Lateral parking system**
   - *Open surface lots*: This is suitable for commercial trucks with/without loading/unloading bays.
   - *Closed garages or parking plazas*: This is also suitable for commercial trucks.
2. **Multi-storied car parking buildings or multiple level car parking**
   - Underground
   - Elevated

Multilevel car parking is an efficient means of parking vehicles, by which there is no disruption to the neighbouring roadway traffic flow. This type of parking is preferred particularly in urban areas where land is scarce. This system provides maximum car parking capacity by utilising vertical space.

Multilevel car parking can be classified based on its operation mechanism, as follows:

1. **Manual operation:** By driving on ramps
2. **Mechanised operation:** By using different technologies for lifting, pushing, pulling and lowering operations (with automated vehicle storage and retrieval system). This can be done for the following vertical parking patterns
   - Twin/Double stacker parking
   - Mini-rotary parking
   - Puzzle parking
   - Horizontal circulation parking
   - Tower parking
   - Multi-parking, and many more

These systems can handle 40–60 cars per hour. The waiting time may range 2–3 minutes.

5.6.3 Basic Terminology

Parking space/stall

An area provided to accommodate one vehicle to be parked, without any access restrictions due to adjacent parked vehicles. A parking space may be larger than a stall (3 × 6 m) as additional space is provided for loading/un-loading operations. According to IRC: SP: 12 (1973), the standard minimum space recommended for (a) individual cars is 3 × 6 m, (b) car parking lots is 2.5 × 5.0 m and (c) trucks is 3.75 × 7.50 m. Parking spaces may be categorised as free, paid, for persons disabled and illegal.

Loading and unloading berths

A berth space of 3.75 × 7.50 m is provided for all kinds of developments, except residential warehouses and godowns, having an initial floor area of 500–1500 m^2. Additional berths are provided at the rate of one for every subsequent 1000 m^2 of floor area. In the case of regular warehouses, two such berths are provided for an initial floor area of 500–1500 m^2. Additional berths are provided at the rate of one for every subsequent 500 m^2 floor area (IRC: SP: 12-1973).

Stall length

The longitudinal dimensions of a stall, usually taken as 5.6–6.0 m.

Stall width

The width of a stall measured crosswise to the vehicle parked; usually taken as 2.6–3.0 m.

Headroom

The vertical clearance provided at the entrance and inside a parking plaza. The minimum headroom is 2.4 m for private cars and taxis.

Other vehicle types

The space requirement details for other vehicles (length × width × headroom) are as follows:

- Light goods vehicles: 7.0 × 3.5 × 3.6 m
- Medium/Heavy vehicles: 11.0 × 3.5 × 4.7 m
- Big buses:12.0 × 3.5 × 3.8 m
- Mini-buses: 8.0 × 3.0 × 3.3 m
- Container vehicles: 16.0 × 3.5 × 4.7 m
- Motorcycles: 2.5 × 1.0 m

Turning radii

The largest turning radius of the design vehicle or any specified vehicle that is the largest is provided as a function of parking angle and end-island design. The turning radius should not be less than 5.5 m.

Study period

The time during which a study is conducted, usually during peak parking hours. The time period may be between 9:00 AM and 6:00 PM for offices and between 2:00 AM and 4:00 AM in residential areas. However, this time period may change based on the type of the location.

Aisle

The portion of the driveway provided in a parking lot which is intended to allow immediate access to the parking spaces (Fig. 5.31). Adequate aisle width is provided to accommodate turning movements of vehicles during parking on to the stall and reverse movement. The following aisle width (m) may be suitably provided (CoP, 2011):

	X1	X2	X3
30° parking	3.6	4.2	6.3
45° parking	4.2	4.8	6.3
60° parking	4.8	4.8	6.6

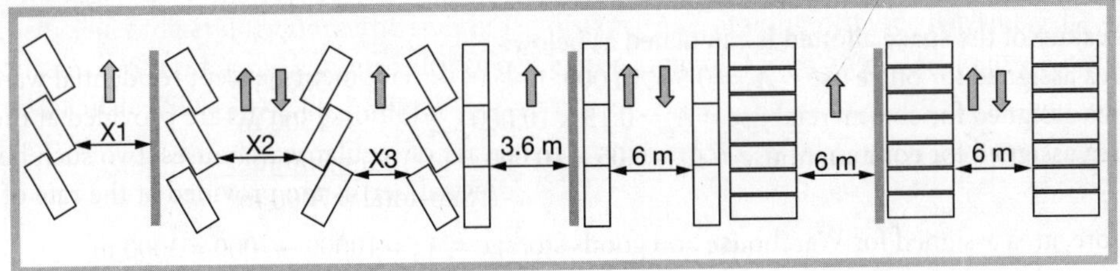

Figure 5.31 Typical parking aisle (Redrawn from CoP, 2011; *Courtesy*: www.lta.gov.sg).

EXAMPLE 5.4

Approximately how many car parking spaces can be accommodated in a total paved area of 2392 sq. yards with the below required configuration of the parking slots:

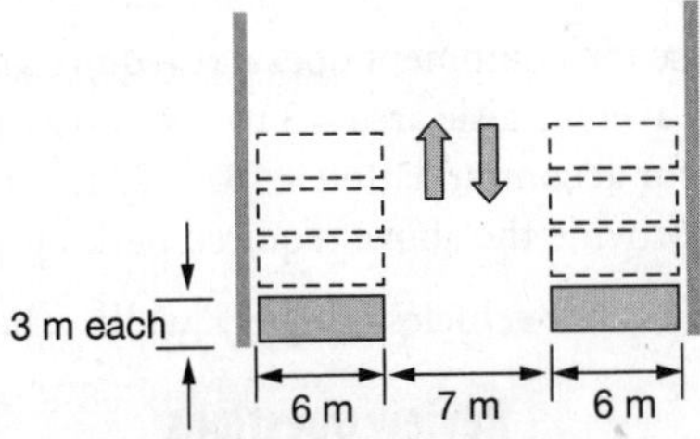

Note that, 5% of the paved area is required to be provided for drainage and other purposes. Take conversion factor as 1 sq. yard = 0.836127 sq.m.

Answer

Area required for one car parking space including the aisle area = $[(6 + 7 + 6) \times 3] = 57\ m^2$

Area required for one car parking space including the aisle area = $\frac{1}{2} \times 57 = 28.5\ m^2$

The paved area = 2392 sq. yards × 0.836127 sq. m = 2000 m^2
Net paved area available after deducting 5% for other purposes = (1 – 0.05) × 2000 = 1900 m^2
Number of car spaces = 1900/28.5 = 66.66
Therefore provide 66 number of car parking spaces in the given area.

Note: The above method of determination of number of parking spaces may be suitably used with the size of vehicle, angle of parking and aisle width with or without handicapped van spaces.

EXAMPLE 5.5

A commercial building is proposed to be constructed with a total carpet area of 10 000 m^2. This total area is allotted for different uses as 50% for office space, 15% for commercial shops space, 5% for installation of air-conditioners and other equipment space and the remaining will be used for warehouse and goods storage. Determine approximate number of parking spaces is required based on the minimum space requirement criteria given below.

Use type of building carpet area	Parking spaces requirement
Offices	1 space/ 70 m^2
Commercial	1 space/ 30 m^2
Warehouse and goods storage	1 space/ 100 m^2

Answer

The breakup of the space allotted is calculated as below:

Area assigned for office use = A_o = 0.5 × 10 000 = 5000 m^2
Area assigned for commercial use = A_c = 0.15 × 10 000 = 1500 m^2
Area assigned for equipment use = A_e = 0.05 × 10 000 = 500 m^2
Sub-total = 7000 m^2

Therefore, area assigned for Warehouse and goods storage = A_w = 10000 – 7000 = 3000 m^2

The total number of parking spaces and its breakup use details can be written as

$$N_T = N_o + N_c + N_w$$

Number of spaces required for office use = N_o = 5000/70 ≈ 72
Number of spaces required for commercial use = N_c = 1500/30 = 50
Number of spaces required for warehouse & goods storage = N_w = 3000/100 = 30
N_T = 152

Note: N_e value will be zero because the equipment does not require any parking space.

If 1 car parking space with an average aisle area = 3 m × 9 m = 27 m^2

Total parking space required for accommodation = 152 × 27 m^2 = 4104 m^2

Therefore, the builder has to provide the above required parking space and design either an underground or rooftop parking facility.

Review Questions

1. Write the various classifications based on specific methods of traffic control/operations which are useful to study and adopt appropriate techniques and devices.
2. State the objectives and functions of road markings.
3. Classify road markings based on function/location and the colours used.
4. State the intended use of road marking colours for specific control types. Write a few examples for each case.

5. Describe the guiding principles of line markings.
6. Describe the types of longitudinal markings, with their intended control type.
7. State the locations where centre line markings are provided on carriageways.
8. Describe the transverse line variants with broken and continuous solid lines as recommended by IRC: 35.
9. Describe and write the salient details of markings used as hazardous marking variants, as recommended by IRC: 35.
10. Draw a schematic view of a ladder-hatching marking on a horizontal curve, labelled with the main components.
11. List and give details of the different types of block markings used for different locations.
12. List and give details of the different types of arrow markings used for different locations.
13. List the different types of directional markings used for different locations.
14. Write the details and standard specifications of object markings.
15. What are the functions of road studs? Classify them and explain their details.
16. Write the limitations of road markings.
17. Classify road signs under the broad categories and subgroups.
18. Write about the subgroups of regulatory signs with three examples for each case.
19. Write about the four subgroups of information/guide signs with three examples for each case.
20. What are hazard markers? Write about any six types of hazard markers.
21. What are the seven standardised shapes of signs? Write about their intended use.
22. Write about the details of letters and numerals of different heights used on various highway signs.
23. Where are the regulatory/mandatory/compulsory signs used? Write about the four subgroups of regulatory/mandatory signs with examples for each case.
24. Where are warning /cautionary/precautionary signs used? Write about the three subgroups of warning /cautionary/precautionary signs with examples for each case.
25. What are the advantages and disadvantages of signal control of road intersections?
26. Define bandwidth of signal time space. State the factors affecting bandwidth.
27. Explain how the bandwidth can be increased with reference to the offset provided in a coordinated control system.
28. What are the factors affecting the design of a traffic signal system?
29. What data is collected for the design of signal timings at proposed intersections?
30. Define the terms relating to traffic signal systems: (a) Platoon (b) Protected turn (c) Cycle (d) Cycle failure.
31. Define the terms relating to traffic signal systems: (a) Interval (b) Phase (c) Change interval (d) Lost time (e) Clearance lost time (f) Start-up lost time.
32. Define the terms relating to traffic signal systems: (a) Clearance interval (b) Intergreen (c) Offset (d) Minimum Green.
33. Define the terms relating to traffic signal systems: (a) Saturation flow rate (b) Saturation headway (c) Effective Green time (d) Protected movement.
34. Draw typical sketches showing the possible phasing adopted for (a) three-legged intersections and (b) four-legged intersections.
35. Draw a typical sketch showing the signal time–space diagram indicating details of start-up lost time, effective Green time, saturation flow and clearance lost time.
36. Define and state the factors affecting saturation flow.
37. Write the details of estimation of saturation flow based on geometric and traffic factors related to (a) Effect of approach width (b) Effect of gradient (c) Effect of right-turning traffic for all four cases (d) Effect of mixed traffic conditions.
38. What is lost time? What is the formula used for estimation of total lost time per cycle?
39. Explain Webster's method of determining optimum cycle length. Also write the formulas used for estimation of minimum cycle length and practical capacity time as per TRL practice.
40. What is effective Green time? How is it determined?
41. What are the parameters considered for traffic signal performance measurement? Explain any two in detail.

42. What is degree of saturation and how is it determined? Also write remarks on flow conditions related to volume-to-capacity ratio.

43. Determine the Yellow time based on the fundamental principles for the given dimensions of an intersection and approach speed:

Average length of vehicle = 6.1 m; Longest width of the intersection area = 25 m; Approach speed at the intersection, S = 46 km/hr.

Assume perception time of driver = 2.5 sec.

44. Determine the optimum cycle time based on Webster's formula. Take the sum of ratios of normal flow to saturation flow of two directional traffic flows = 0.55. The total lost time per cycle =11.0 sec.

45. The following data is collected for designing a two-phase fixed signal at an intersection having North–South and East–West approach roads. Straight ahead traffic is permitted to operate in these approach roads.

Parameter	Approach road			
	North	South	East	West
Design hour flow (PCU/hr)	815	750	850	495
Saturation flow (PCU/hr)	2100	2100	2500	2500

Take the time lost per cycle = 10.0 sec. Determine the optimum cycle length as per Webster's formula. Also calculate effective Green time per phase North–South and East–West.

46. What are the low-cost traffic control and management techniques? Explain them briefly.

47. Define ITS and state its need.

48. Write about the historical development of ITS with the main developments in different countries.

49. What are the objectives of the deployment and design integration of ITS?

50. What are the benefits of ITS? Briefly explain and give an example for each case.

51. State and define the concept of ITS design.

52. Draw a typical sketch showing the components involved in a multifunctional enforcement system.

53. Explain the technologies used in ITS.

54. What are the objectives of conducting parking studies?

55. Define these terms and write the standard space used for (a) parking stall, (b) loading and unloading berths, (c) stall length and width, (d) headroom, and (e) turning radii.

56. What is the standard space (Length × Width × Headroom) requirement for

(a) Light goods vehicles

(b) Medium/Heavy vehicles

(c) Big buses

(d) Mini-buses

(e) Container vehicles

(f) Motorcycles

57. Define an aisle. Draw the typical layout of parking aisle and indicate the aisle space requirements for 30°, 45° and 60° parking.

58. Draw a sketch representing the off-street lateral car parking system layout and describe the space requirements of each parking unit with minimum manoeuvring space and minimum clear driveway width for one-way and two-way traffic.

59. Classify and describe multilevel car parking systems based on operation mechanism.

60. Draw a typical sketch layout showing the clearway dimensions recommended for off-street parking.

61. Write details of the minimum parking space requirements based on the purpose of use, as per IRC: SP: 12.

62. Explain how parking studies are carried out in four steps.

Multiple Choice Questions

1. The width of longitudinal markings in rural sections for design speeds greater than 100 km/hr is
   (a) 50 mm (b) 100 mm
   (c) 150 mm (d) 200 mm
2. As per IRC: 35, what would be the minimum preview distance required for a design speed of 80 km/hr?
   (a) 44 m (b) 50 m
   (c) 56 m (d) 61 m
3. The road stud marker height shall not be less than _______ mm
   (a) 5 (b) 10
   (c) 15 (d) 20
4. As per IRC: 67, for better visibility, the signs relating to parking should be fixed at an angle of _______ to the carriageway.
   (a) 90^0 (b) 95^0
   (c) 15^0 (d) 0^0
5. The light reflecting devices used to guide the driver along proper alignment are called
   (a) Traffic signs (b) Attenuators
   (c) Delineators (d) Road markings.
6. As per IRC:67, a traffic sign indicating 'merging traffic ahead' should be of
   (a) circular shape with white background and red border
   (b) triangular shape with white background and red border
   (c) triangular shape with blue background and red border
   (d) circular shape with blue background .
7. Compulsory signs are exhibited in what shape?
   (a) Circular shape
   (b) Triangular shape
   (c) Rectangular shape
   (d) Equilateral triangle (with one point downward).
8. The triangular sign means _______
   (a) a warning sign
   (b) a prohibitive sign
   (c) an indicative sign
   (d) a compulsory sign.
9. The total lost time on a traffic signal is noted to be 6 sec, the actual green time is 32 sec and the amber time is 3 sec. How much is the effective green time.
   (a) 35 (b) 21
   (c) 33 (d) 29
10. The commonly adopted turning radius provided in parking plazas should not be less than
   (a) 5.5 m (b) 5.4 m
   (c) 5.0 m (d) 4.8 m

Answers to Multiple Choice Questions

1. (c) 3. (b) 5. (c) 7. (a)
2. (a) 4. (c) 6. (a) 8. (a)
9. (d) Effective green time = Actual green time + Amber time – Total lost time
   = 32 + 3 – 6 = 29 sec
10. (a)

CHAPTER 6

Pavement Materials: Soils and Stone Aggregates

6.0 INTRODUCTION

Various materials with appropriate engineering properties are used for road construction and they play a vital role in design, construction and maintenance operations. Suitability of different materials according to their availability and their specific application should be thoroughly examined before undertaking a design methodology. Different pavement design methods need different input parameters. Hence, it is important to determine the engineering properties of materials that are used in pavement construction. Figure 6.1 shows the natural and artificial materials generally used for construction of road pavements.

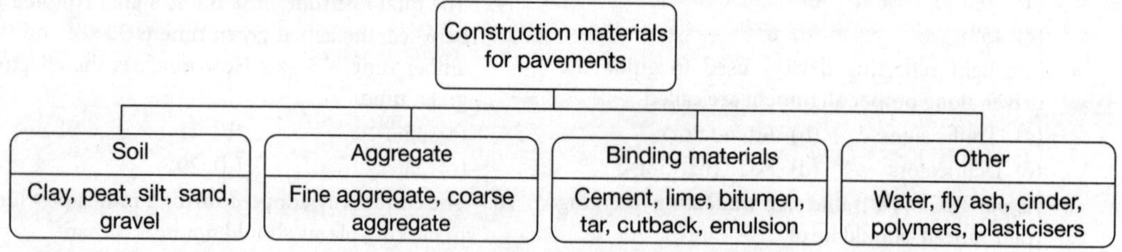

Figure 6.1 Material used for highway construction.

In this chapter, a few important methods that are generally used in India for testing highway materials are described.

Soil consists of various inorganic–mineral residues of rocks from the earth's crust produced as a result of geological processes like weathering. It is one of the major components of sub-grade, base course foundations, embankment and soil aggregate mix and is often used as mixtures with other materials, viz., cement, bitumen, cutback, emulsion, fly ash, tar etc., in the construction of roads. To understand the behaviour of soil alone and with other admixtures, it is essential to determine its characteristics under similar climatic and environmental conditions as that prevailing at the proposed construction site. Engineering properties of a particular type of soil are found to have correlation with its mineral content and particle size distribution.

6.1 PARTICLE SIZE ANALYSIS OF SOILS

The given sub-grade soil particles are separated by sieving through a set of sieves of standard sizes. The soil content retained on and passed through respective sizes of sieve are weighed and recorded for further analysis to interpret the results in terms of soil character (Burmister, 1939; IS: 2386 (Part 1)-1963; 2004). Particle size analysis of soil is classified as

(i) **Dry-mechanical analysis:** The limitation of the test is up to 75 microns size of soil particles.

(ii) **Wet-mechanical analysis:** The limitation of the test is up to 2 microns size of soil particles.

For classification of soil grain size less than 75 microns, sedimentation analysis is performed using a hydrometer (Fig. 6.2) or the pipette method based on Stoke's law. The above test methods are described in detail in IS: 2720 (Part 4)-1985.

Soil classification charts were developed by various agencies based on the percentage of soil grains passing through different sizes of sieves. Table 6.1 enlists the general methods used to classify soil based on the diameter of the particle. The terminology used to designate the different types of soil particles based on their size slightly differs in different countries (Fig. 6.3).

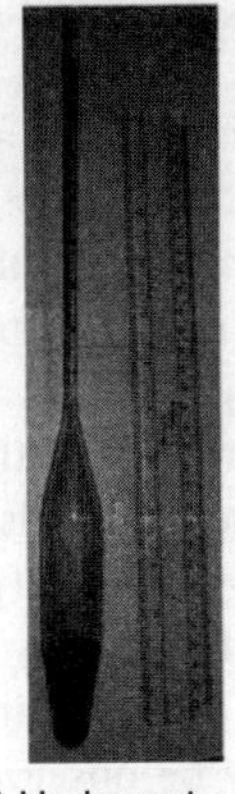

(i) Hydrometer

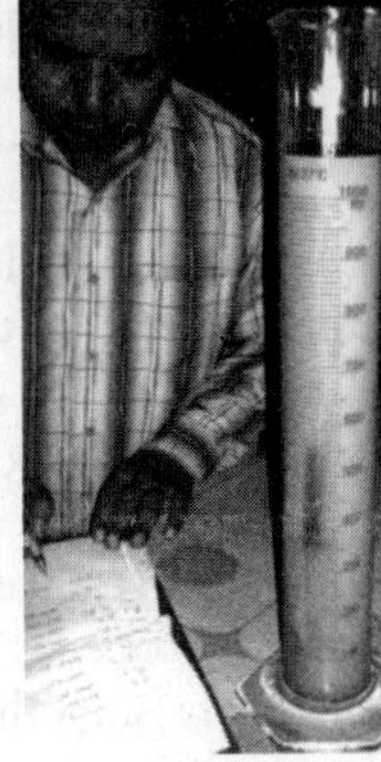

(ii) Test in progress

Figure 6.2 Grain size distribution of fine grained soil using hydrometer.

Table 6.1 Soil texture based on particle size (according to IS: 1498-1970).

Sl. No.	Soil texture		Particulars	
			Diameter of particle (mm)	**IS sieve size (mm)**
1	Gravel (G)		75–4.75	Retained on 4.75
2	Sand (S)		4.75–0.075	Passes 4.75 and retained on 0.075
3	Fines	Silt (M)	0.075 to 0.002	Passes 0.075; For detailed investigation, a hydrometer analysis may be carried out.
4		Clay (C)	Less than 0.002	

Notes:

1. One micron = 1×10^{-3} mm
2. Sand and silt are further classified as coarse, medium and fines based on particle size
3. Clay is further classified as colloids if their particles size is less than 0.001 mm.

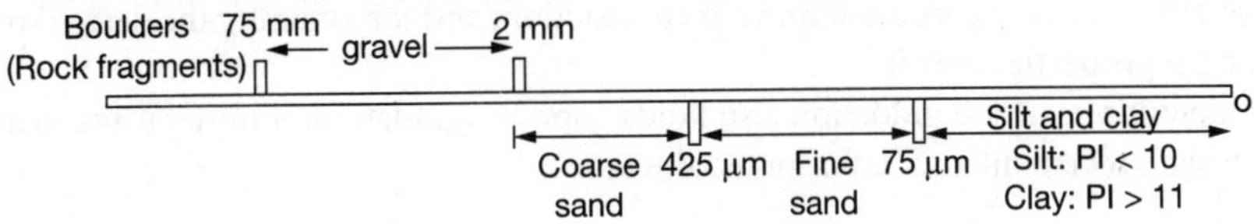

Figure 6.3 Terminology used by ASTM designation: D3282-93.

Fine-grained soil consists of grains of equal size. Their behaviour under service conditions depends not only on their grain size composition but also on the following factors (Fadum, 1948; Feld, 1948).

- Different mineralogical composition
- Variability in shape of soil particles
- Cohesion effect due to wetness caused by electrostatic charge with different intensities on grain surfaces

For all these reasons, fine-grained soil is classified not only according to the size of the grains (or gradation) but also on based on its degree of plasticity and dry strength.

6.2 SOIL GRADATION

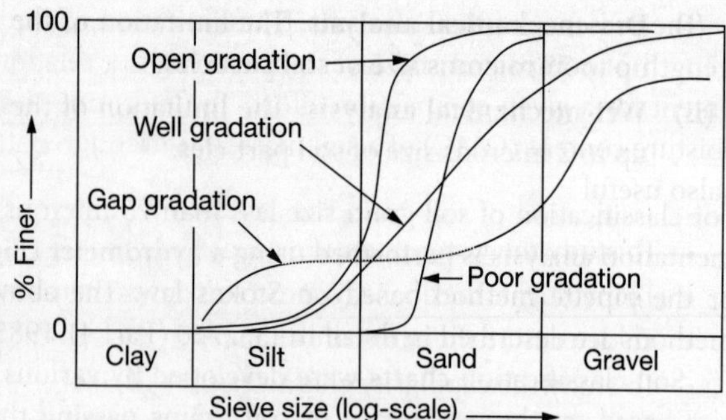

Figure 6.4 A typical gradation chart.

The percentage of material passing through and retained on the respective sieves (or a set of sieves) are noted down in a tabular form and the cumulative percentages finer than each sieve size are calculated. These values are plotted on a semi-log graph, with the grain sizes on logarithmic scale as abscissas and cumulative percentages by weight (passing) on ordinary scale as ordinates. By joining all the points smoothly, a curve similar to an elongated S-shape may be observed. This curve is called a grading curve or a grain size distribution curve and is a method of international expression, or a universal language presenting grain size data graphically (Campbell, 1939). Based on geometric characteristics such as the shape, slope and position of the grading curve, the properties of soil are categorised as poorly graded, open graded, well graded and gap graded (Fig. 6.4).

Well or Dense Graded Soil under this gradation consists of all sizes of soil particles ranging from fine to coarse and has the maximum density. The S-curve spreads almost evenly along the *x*-axis covering all sizes. This type of soil is generally preferred for constructing sub-grades or embankments and other pavement layers where heavy traffic is anticipated. The value of C_u will be > 6.

Gap or Skip Graded Certain sizes of particles in the middle-size range are available only in small quantity (%) compared with the other sizes. Hence, the shape of the S-curve appears to be flat in the middle-size range. To make this soil useful, it has to be blended with other soils for supplementing the missing sizes.

Open Graded The finer size of particles are present only in a small quantity. Hence, there are more air voids between coarser particles which would otherwise have been occupied by fines of appropriate size. The S-curve appears to be flat at the finer particle size zone.

Poorly or Uniformly Graded In this type, the soil mass consists of similar-sized particles resulting in more voids due to less percentage of finer particles. The S-curve appears to be steep in a limited size range.

Gradation analysis of the particle sizes of soil mass helps the soil engineer to blend soils to achieve satisfactory gradation such as dense soil mass in compaction, and for attaining the desired permeability and shear strength properties of soil.

The above definitions of soil gradation also holds good for gradation of mineral aggregate, which is used for mix design of cement and bituminous concrete.

6.3 MOISTURE CONTENT

The influence of soil characteristics on the sub-grade layer in the pavement structure is an important consideration in highway engineering. The presence of moisture in different soils of sub-grade continuously changes their characteristics and the degree of change depends on the nature of the soil. Therefore, determination of moisture content in sub-grade soils is essential. Certain soils are capable of retaining high percentages of moisture. The amount of moisture may freeze or swell or shrink in different types of soil (Goldbeck, 1925).

The following test is conducted to determine the amount of moisture present in a soil mass. Soil specimen weight is taken before and after drying the sample in an oven at about 110°C for 24 hours. It is expressed as the percentage by weight of the moisture content present in dry weight of the soil (IS: 2720 (Part 2)-1973; 2004).

The presence of moisture and its quantity in a soil mass plays a very important role in studying the strength of compacted soil layers. There exists a relationship between density of soil mass and water content for a given compaction effort. To determine the maximum dry density (MDD) and optimum moisture content (OMC) of a soil mass, it is useful to obtain the moisture content relationship. This test is also useful

- To study the water retention capacity of the soil.
- To investigate the quantity of moisture present in a compacted layer during quality control operations.
- To determine the moisture changes in the sub-grade soil in different seasons of a year.
- To study the capillarity characteristics of soils.
- To evaluate the density of compacted pavement layers and embankments.

The oven drying method is a standard approach to determine moisture content in a soil sample. To carry out this test, about 50 to 100 grams of soil specimen have to be collected from a soil layer. If the soil sample is collected from the field, it is packed in an air-tight cup and carried to the laboratory without any loss of moisture. The weight of the empty cup with its cap is taken and recorded as W_1 g. The collected soil is filled in the cup and weighed along with the cap as W_2 g. The cup filled with soil along with its opened cap is kept in an oven at 105 to 110°C for about 24 hours. After drying, the soil cup is placed in a desicator to allow the contents to cool down to room temperature without any absorption of atmospheric humidity. The dried sample along with its cup and cap are weighed and recorded as W_3 g. The moisture content in the soil is calculated as (IS: 2720 (Part 2)-1973; 2004):

$$\text{Moisture content in the soil } (w) = \frac{\text{Weight of moisture present in soil } (W_2 - W_3)}{\text{Dry weight of the soil } (W_3 - W_1)} \tag{6.1}$$

At least 3 samples are tested under identical conditions to get an acceptable average value.

The following alternative methods are generally employed for rapid determination of the moisture present in a given soil sample.

1. Sand bath method
2. Alcohol method
3. Infrared lamp drying method
4. Calcium carbide method
5. Nuclear radiation method
6. Vibratorymeter

6.4 CONSISTENCY TESTS OF SOIL

6.4.1 Liquid Limit Test

The liquid limit (LL) is the minimum water content (%) at which the soil will flow on application of a very small shearing force, i.e., at this limit, the soil possesses very low shear strength and begins to exhibit liquid characteristics. The stress history and other characteristics of a soil can be well predicted by its liquid limit value. Further, the compression index and settlement analyses can also be estimated based on liquid limit values. This test is also used to determine the flow index value and for soil classification. There are two methods of conducting the LL test. They are (i) the mechanical method and (ii) the cone penetration method. The test procedures are explained below.

Mechanical Method (as per IS: 2720 (Part 5)-1985; 2004)

To carry out this test, about 120 grams of dry pulverised soil sample passing through a 0.425 mm IS sieve size is to be taken. The soil should be thoroughly mixed with distilled water in a porcelain evaporating dish to make a soil paste of uniform consistency. The apparatus used for this test is given in Fig. 6.5(a). The brass cup is set to have a free fall of 10 mm above its base.

A portion of the prepared soil paste is placed in the brass cup so as to almost half-fill it. A spatula is used to smooth the paste to have a horizontal surface. The maximum depth at the centre of the brass cup should be 10 mm. A 'V' shaped groove is made along the symmetrical axis of the cup (Fig. 6.5(b)) using a grooving tool. To lift and drop the cup on to the base, the handle of the cup is rotated at a rate of 2 revolutions per second. The number of blows to the cup needed to bring the two parts of the soil paste into contact with each other (because of the impact) at the bottom for a length of above 10 to 12 mm by flow, is recorded. To determine the moisture content in the soil, a portion of soil paste may be taken from the brass cup.

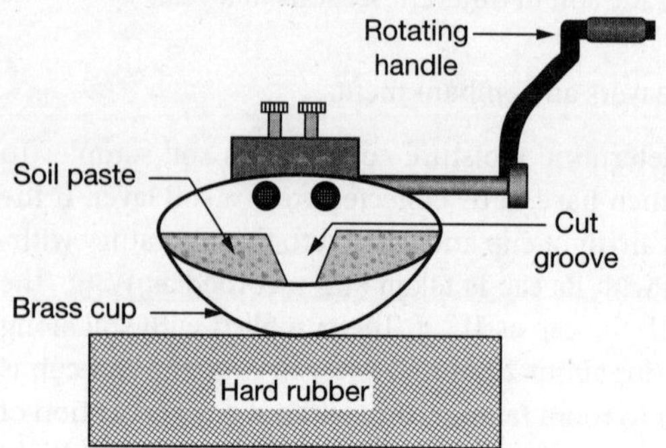

Figure 6.5(a) Liquid limit test device (with manual operation).

Figure 6.5(b) Cutting a groove in the soil paste (electrically operated device).

The above procedure may be repeated for at least three trials, preferably with different moisture contents of soil paste resulting in the number of blows ranging from 15 to 35. The soil paste prepared for the first trial should be with lower water content. In later trials, the water content can be increased in suitable increments. At lower water content, the soil specimen records more blows and vice versa.

The test data is plotted on a semi-log graph by taking the number of blows on the log scale (x-axis) and water content in percentage on the ordinate scale (y-axis). A straight line is obtained by joining all the points. The line obtained is called a flow curve. The water content corresponding to 25 blows is taken as the liquid limit of the soil (Fig. 6.6).

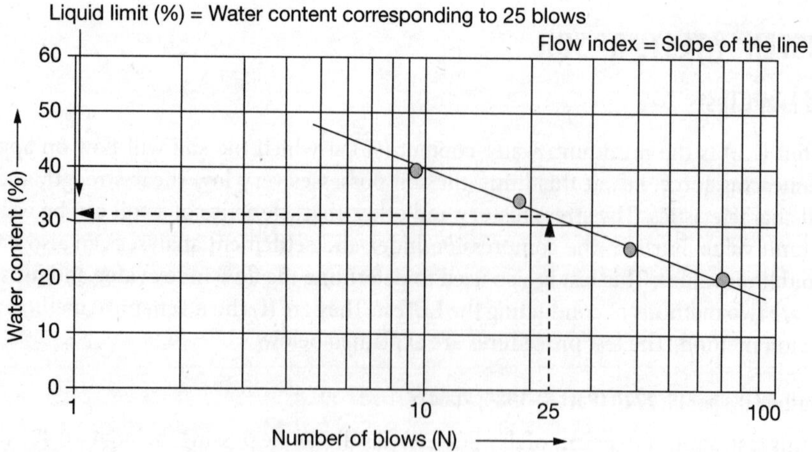

Figure 6.6 Flow curve.

From the liquid limit test data, soil flow index property can be calculated using the following equation:

$$\text{Flow index} = \text{Slope of flow curve} = FI = \frac{W_2 - W_1}{\log_{10} {}^{N_1}\!/_{N_2}} \tag{6.2}$$

where, N_1 and N_2 are the number of blow counts corresponding to the water contents of W_1 and W_2.

Cone penetration method

In this test, about 150 g of soil sample passing a 0.425 mm IS sieve is taken. Distilled water is mixed to obtain a uniformly consistent soil paste. A standard cone of 30° angle with 35 mm depth, having an 80 g plunger is allowed to penetrate into the soil paste kept in a standard size cup for a duration of 5 seconds (IS: 2720 (Part 5)-1985 and IS: 11196-1985; 1985). The principle of the cone penetration method is presented in Fig. 6.7. At least four tests are conducted on the soil sample by varying the water content in the soil. The measured penetration values of the cone are plotted on the x-axis and the corresponding water contents (%) are plotted on the y-axis. The points are connected by a straight line and the liquid limit value in percentage is determined at 20 mm penetration value (Fig. 6.8).

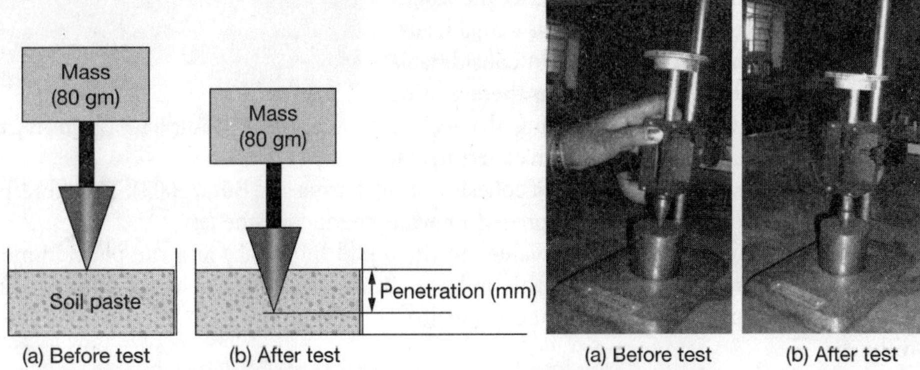

Figure 6.7 Principle of the cone penetration method.

The characteristics and suitability of soils for pavement construction can be interpreted based upon LL (Fig. 6.9).

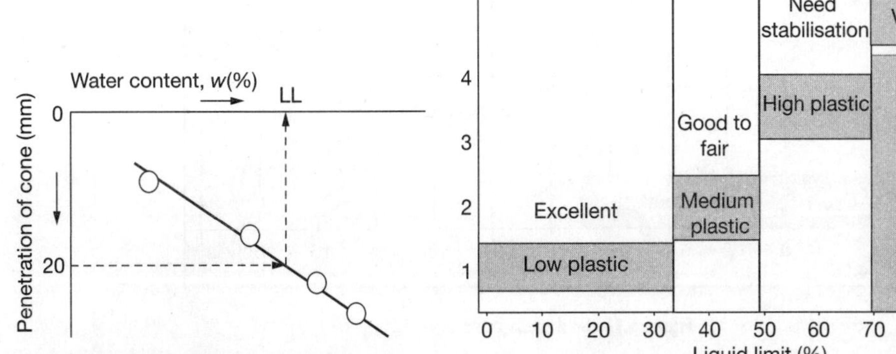

Figure 6.8 Determination of LL value from testresultsofconepenetrationmethod.

Figure 6.9 Suitability of soil based on LL.

6.4.2 Plastic Limit Test

The plastic limit (PL) of a soil is the water content at which the soil begins to crumble when rolled into a thread of approximately 3 mm diameter (IS: 2720 (part 5)-1985). It is the water content which indicates the boundary between the plastic and semi-solid states of consistency of the soil. This test is carried out to determine the semi-solid state of soil. The PL value is widely used to classify soil.

About 20 g of dry pulverised soil sample passing through a 0.425 mm size IS sieve is taken and mixed with distilled water in a porcelain evaporating dish till it becomes plastic enough to be easily rolled with the fingers. In case of clayey soils, the soil paste should be kept for 24 hours in order to ensure uniform distribution of moisture throughout the soil mass.

About 8 g of the prepared soil paste is taken to make a small ball with the fingers. The soil ball is rolled on a plane glass plate to form a thread of slightly more than 3 mm diameter. The soil is pressed together and again rolled until the soil thread of 3 mm diameter just starts crumbling (Fig. 6.10). The crumbled pieces of soil are taken into a moisture container to determine the water content. The test is repeated at least three times and the average of the results is rounded off to the nearest whole number as a final result.

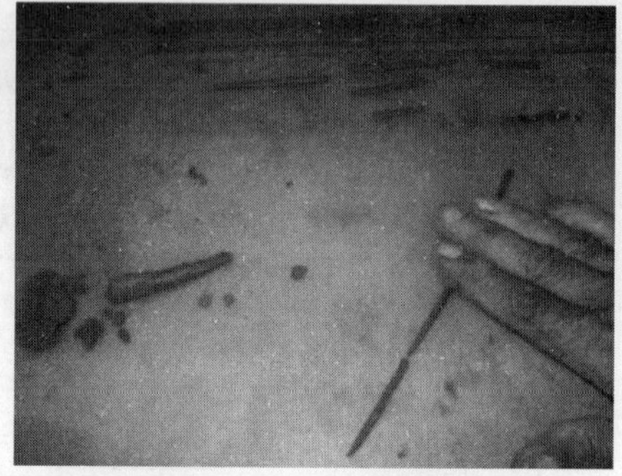

Figure 6.10 Plastic limit test in progress.

This thread rolling method may result in considerable error in the measured value of plastic limit as there is more involvement of human activity while rolling the soil to 3 mm diameter of thread. Occurrence of such experimental error is a serious concern, especially in case of cohesive soils (Sharma and Bora, 2003). Therefore, proper care should be taken to follow the standard procedure while conducting the test.

The numerical difference between the values of the liquid limit (LL) and the plastic limit (PL) of soil is known as plasticity index (PI). This index indicates the range of water content through which the soil remains plastic, the upper and lower limits being in the liquid and semi-solid states respectively.

$$\text{Plasticity index (PI)} = \text{LL} - \text{PL} \tag{6.3}$$

Soils can also be classified as follows based on plasticity and liquidity indices (Fig. 6.11).

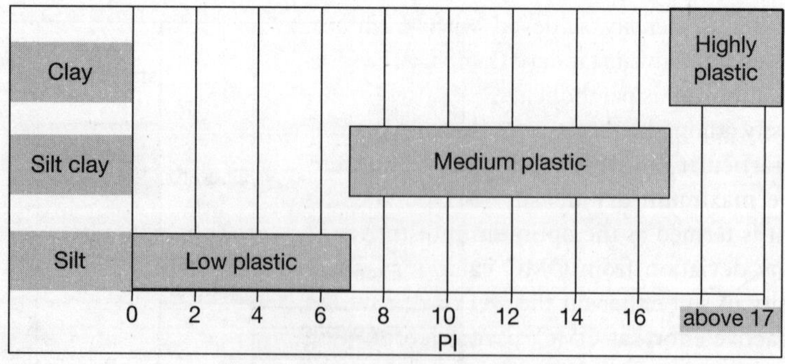

Figure 6.11 Soil class based on PI.

Determination of specific gravity of soil by pycknometer (Fig. 6.12):

Figure 6.12 Determination of specific gravity of soil using pycnometer.

EXAMPLE 6.1

A soil sample was tested in a pycnometer to determine its specific gravity. Calculate the specific gravity of the soil using the following data:

Weight of empty pycnometer, $W_1 = 515$ g
Weight of pycnometer with soil, $W_2 = 1025$ g
Weight of pycnometer with soil and water, $W_3 = 1866$ gm
Weight of pycnometer completely filled with water, $W_4 = 1548$ gm

Answer

$$G_s = \frac{\text{Weight of soil}}{\text{Weight of water equal to the volume of soil displaced}} = \frac{W_2 - W_1}{W_4 - W_3 + (W_2 - W_1)}$$

$$= \frac{1036 - 515}{1548 - 1866 + (1025 - 515)} = 2.71$$

6.5 DETERMINATION OF SOIL COMPACTION

In a soil compaction test, the relationship between the quantity of moisture or water content present in a soil sample and the maximum density achieved with minimum voids is determined for a given standard compactive effort. Generally, the relationship is parabolic.

The dry density attained by soil on compaction reaches maximum at a particular moisture content only. This density is called the maximum dry density (MDD) and the moisture content is termed as the optimum moisture content (OMC). Any deviation from OMC value may result in low dry density of soil either on the wet or dry side of OMC. By compactive effort, at OMC, particles of soil are forced to orient themselves closely in such a fashion that maximum density is obtained with minimum voids. As the compactive effort increases, the value of maximum dry density increases even at lower values of moisture content (Fig. 6.13) (Lambe, 1960).

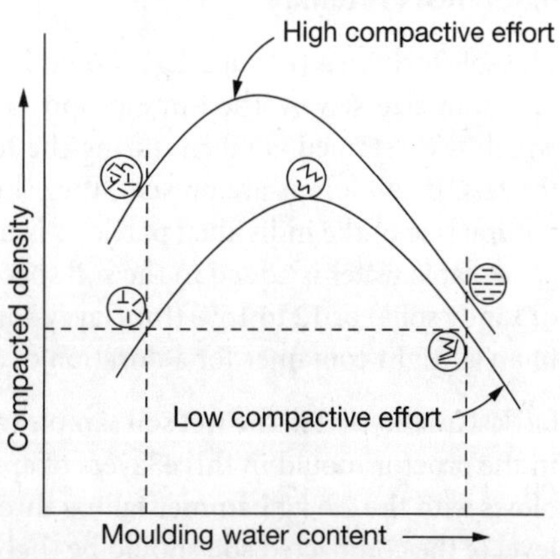

Figure 6.13 Effects of compaction on soil particles (after Lambe, 1960).

Due to compaction, the following useful changes in soil properties are expected to happen (Mullis, 1939; Lambe, 1960):

- Increase in density
- Increase in shear strength characteristics
- Decrease in permeability characteristics
- Decrease in swell and shrinkage characteristics
- Decrease in compressibility characteristics

6.5.1 Principles of Compaction Test Methods

1. The knowledge of MDD helps to improve in situ ground conditions which may otherwise be unsuitable for construction purposes.
2. The compaction energy usually employed as
   (i) Light weight compaction = 592.77 kJ/m^3
   (ii) Heavy weight compaction = 2697.26 kJ/m^3
   (Ratio = 4.55)

   Magnitude of compaction energy is recommended based on the importance of roads, the intensity of traffic loads anticipated, the number of pavement layers and type of soil. Table 6.2 shows the details of dynamic compaction recommended by the Bureau of Indian Standards.

Table 6.2 Details of dynamic compaction (IS: 2720 (Part 7)-1980 and IS: 2720 (Part 8)-1983).

Compaction type	Number of layers	Hammer weight (kg)	Hammer free-fall height (mm)	Number of blows on each layer	Imparted energy (kJ/m^3)
Light compaction	3	2.60	310	25	592.77
Heavy compaction	5	4.9	450	25	2697.26

3. The laboratory compaction and field compaction methods usually conform to similar energy conditions.

6.5.2 Test Procedure

The soil specimen (about 3 kg) should be air dried, thoroughly mixed and passed through 20 mm and 4.75 mm size sieves. The same proportion of soil which passed through the 4.75 mm sieve and that which was retained is taken during the test. The fraction retained on the 20 mm sieve is not used for the test. If soil lumps are present, they should be thoroughly broken, but care should be taken that the natural size of the individual particles is not reduced.

Potable water is added to the soil specimen such that its water content is about 3 to 5% (for gravelly or sandy soils) or 12 to 16% (for clayey soils) less than the plastic limit of the soil. This soil is then placed in an air tight container for a duration of about 16 hours to achieve homogeneity.

For Light Compaction Fill the wet soil sample (about 2.5–3.0 kg which passes through the 20 mm size IS sieve) in the proctor mould in three layers of approximately equal thickness. Compact each layer by giving 25 blows with the 2.6 kg hammer falling through a height of 31 cm of free fall by gravity (Fig. 6.14(a)). The level of the compacted soil should be slightly above the top of the mould and project into the collar after the compaction of the third layer (IS: 2720 (Part 7)-1980; 2004).

For Heavy Compaction Fill the wet soil sample (with size as specified above) in the proctor mould in five layers of approximately equal thickness. Compact each layer by giving 25 blows with the 4.9 kg hammer falling through a height of 45 cm under gravity. The level of the compacted soil should be slightly above the top of the mould and project into the collar after the compaction of the fifth layer (IS: 2720, Part 8-1983; 2004).

Water is added to the soil sample to increase the moisture content by about 2 to 3 per cent and the above procedure is repeated 5 to 6 times. The values of dry densities obtained for each trial test are calculated using following equations.

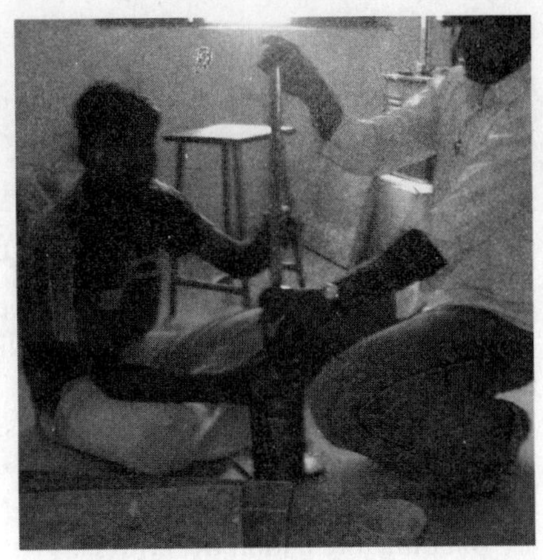

Figure 6.14(a) Soil compaction test in progress.

$$\text{Bulk density, } \rho = \frac{M}{V} \text{ (g/cc)} \tag{6.4}$$

$$\text{Dry density, } \rho_d = \frac{\rho}{1+w} \text{ (g/cc)} \tag{6.5}$$

where,

M = weight of compacted soil in g
w = weight of moisture present in g
V = Volume of mould in cm^3

A curve is plotted with the values of the moisture content *(w, %)* on the *x*-axis and dry density on the *y*-axis. A smooth curve is drawn by joining these points. On this curve, the MDD on the *y*-axis and the corresponding value of the moisture content (*w*) on the *x*-axis, which gives the OMC of the soil, are noted. Figure 6.14 (b) shows a typical example of MDD and OMC determination using experimental data.

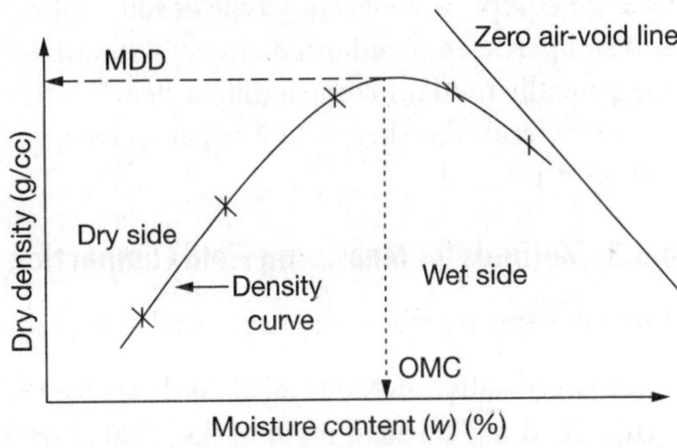

Figure 6.14(b) Soil density curve.

A zero air-void line is also drawn on the same chart. This line suggests the degree of saturation value obtained at the OMC condition. Therefore, it is customary to plot the zero air-void line or 100 per cent saturation curve using the following equations.

$$\text{Dry density, } \rho_d = (G\,\rho_w)/(1+e) \tag{6.6}$$

where,

e = Void ratio = $w\,G/S_r$
G = Specific gravity of water
ρ_w = Unit weight of water (9.81 kN/m^3 or 1 g/cc)
S_r = Degree of saturation (for 100 per cent degree of saturation, $S_r = 1$)

The method of determining MDD and OMC of a given soil by the standard proctor density test is illustrated below (Fig. 6.14).

The lower and higher values of MDD indicate low and high strength of soils respectively. Therefore, it is always preferred to have soils that have higher MDD values at optimum conditions of moisture content.

6.5.3 Degree of Compaction

Degree of compaction is the percentage of the maximum field dry density (FDD) obtained to that of laboratory test conditions. The degree of compaction (DoC) may be expressed as follows.

$$\text{DoC} = \frac{\text{MDD obtained at field}}{\text{MDD obtained at laboratory}} \times 100 \tag{6.7}$$

The DoC is specified based on the type of construction, anticipated loads, nature of loads, climate and other field conditions. In general, the following are the standard range of DoC specified for different applications.

Base courses of highways and airfields:	98 to 100% modified proctor
Road embankments:	95 to 100% standard proctor
Earth dam constructions:	95 to 100% standard proctor
Building foundation work:	Above 90% modified proctor

6.5.4 Field Compaction Methods

A smooth wheel roller is generally used for compacting sub-base and base layers of gravelly soils. They are available in different capacities ranging from 50 kN to 100 kN. Pneumatic and sheep's foot rollers are used for compaction of clayey type of soils. Vibratory rollers produce multiple pressure waves in the soil and soil particles get adjusted closely without any internal friction between the particles. These rollers are generally used for compacting or densification of cohesion-less soils.

In general, the sheep's foot roller is recommended for cohesive soils and the vibratory roller for cohesion-less soil.

6.5.5 Methods for Measuring Field Compaction of Soil Layers

The following are the methods generally adopted to measure field densities.

(i) Core cutter method (IS: 2720 (Part 29)-1975; 2006)
(ii) Sand replacement method (IS: 2720 (Part 28)-1974)
(iii) Ring and water replacement method (IS: 2720 (Part 33)-1971)
(iv) Rubber balloon method (IS: 2720 (Part 34)-1972) (Fig. 3.26)
(v) Penetration meter (puncture needle)
(vi) Nuclear density meter (NDM).
(vii) Vibratory meter

The first two methods are generally employed during field compaction control. A brief explanation of these two methods is given below.

1. The *core cutter method* is suitable for fine-grained soil, of which not less than 90 per cent passes through a 4.74 mm size IS sieve (IS: 2720 (Part 29)-1975; 2006). In this test, a sharp-edged steel cylindrical hollow mould having internal diameter 100 mm and depth 130 mm with suitable thickness is used. A steel dolly of suitable size is placed over the mould during the testing. The mould with the dolly is driven into the soil by hammering on its top. The mould is then dug out from the soil and excess soil from the top and bottom of the mould is trimmed off with a knife. By establishing the weight of the soil in the mould and volume of the empty mould, the bulk density of the soil is determined. A certain amount of soil is taken from the mould and weighed before and after drying in the oven. The moisture content in the soil is expressed as the ratio of the weight of water present in the soil to the initial weight of the soil sample kept in oven.

2. *Sand replacement method* is suitable for coarse-grained soils (Fig. 6.15). A sand-pouring cylinder of suitable capacity is used for this test. A hole of 100 mm diameter having depth less than 150 mm is dug into the soil and the material is carefully collected and weighed. Later, the hole is filled with sand of known density using a sand-pouring cylinder. By ascertaining the volume of the sand and weight of the soil collected, the bulk density of the *in situ* soil is determined. A suitable amount of soil sample is taken for moisture content determination.

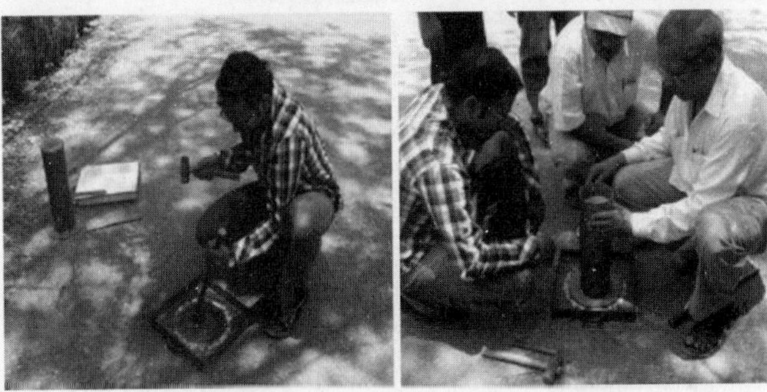

Figure 6.15 Sand replacement test in progress.

EXAMPLE 6.2

A sand replacement test was conducted on a sub-grade soil. Calculate the bulk density and dry density of the soil from the following data:

Mass of soil removed from hole = 2311 g
Mass of the soil after oven fried = 2095 g
Mass of sand pouring cylinder completely filled with sand = 5875 g
Mass of sand pouring cylinder after test = 3182 m
Volume of cone in sand pouring cylinder = 250 cc; pouring sand density = 1.97 g/cm^3

Answer

Mass of sand filled in the hole = (5875 – 3182) = 2693 g

Mass of sand in the cylinder cone= 250 cm^3 × 1.97 g/cm^3 = 492.5
Mass of sand used to fill the hole = 2693 – 492.5 = 2200.5 g
Volume of the hole = 2200.5/1.97 = 1117 cm^3
Bulk density of in situ soil = 2311 g/1117 cm^3 = 2.03 g/cm^3

$$\text{Moisture content in the soil} = = \frac{(2311-2095)}{2095} \times 100 = 10.3\% \ \ (2311\text{-}2280)$$

$$\text{Dry density of the soil} = \frac{2.03}{1+\frac{10.3}{100}} = \frac{2.03}{1.103} = 1.84 \text{ g/cm}^3$$

6.6 STRENGTH DETERMINATION OF SOIL

6.6.1 California Bearing Ratio (CBR) Test

O J Porter (1938) of the California Division of Highway developed the CBR test for strength evaluation of sub-grade soil, base course materials and strength properties of flexible pavements. During Second World War, the US Army Corps of Engineers (USACE) first adopted the CBR test for designing the base course of airfield pavements and found it suitable for testing flexible pavement designs.

CBR tests were performed on a large number of typical crusher-run materials which were deemed representative of high-quality base course materials. The average load values corresponding to different penetration values are recommended as standard load values (Porter, 1950). The standard load values for the CBR test are given in Table 6.3.

Table 6.3 Standard load values for CBR test for different penetration values.

S. No.	Standard load		Standard pressure		Penetration of plunger (mm)
	(kg)	(N)	kg/cm^2	MPa or N/mm^2	
1	1370	13439.70	70	6.84	2.5
2	2055	20159.55	105	10.27	5.0
3	2630	25800.30	134	13.14	7.5
4	3180	31195.80	162	15.89	10.0
5	3600	35316.00	183	17.99	12.5

About 20 kg of air-dried soil specimen which has passed through the 20 mm size IS sieve is taken and compacted at OMC with an imparted energy of compaction as per specifications to obtain maximum dry density (MDD). Compaction methods have already been discussed in detail in Section 6.5. After the soil sample is prepared, the sample-filled mould is kept in a soaking bath for four days. Before testing, the specimen is drained for 15 minutes (Jervis and Eustis, 1950).

Figure 6.16 CBR test in progress.

A hydraulically loaded cylindrical plunger of 50 mm diameter is allowed to penetrate vertically into the test soil at the rate of 1.25 mm per minute (Fig. 6.16). The values of the load on the proving ring are noted at plunger penetration values of 0.0, 0.5, 1.0, 1.5, 2.0, 2.5, 4.0, 5.0, 7.5, 10.0, and 12.5 mm. The values of load applied are recorded corresponding to the plunger penetration values at 2.5 mm and 5.0 mm. The penetration values are indicative of the combined influence of cohesion and internal friction and the CBR is an index of the shearing strength of soil (Turnbull, 1950). These recorded test loads are used to calculate the percent ratios with respect to standard loads at 2.5 mm and 5 mm as 1370 kg and 2055 kg respectively (IS: 2720 (part 16)-1987; 2004).

To calculate the CBR value, the recorded load is expressed as a percentage of the standard load value for the respective penetration value (Fig. 6.17). The CBR values are calculated using the following formulae.

$$CBR_{2.5\,mm} = \frac{\text{Test load at 2.5 mm penetration(kg)}}{\text{Standard load at 2.5 mm penetration(1370 kg)}} \times 100 \quad (6.8)$$

$$CBR_{5.0\,mm} = \frac{\text{Test load at 5.0 mm penetration(kg)}}{\text{Standard load at 5.0 mm penetration(2055 kg)}} \times 100 \quad (6.9)$$

The following interpretations can be made regarding CBR values.

- The CBR value calculated at 2.5 mm penetration is generally found to be more than the CBR value calculated at 5.0 mm. For flexible pavement design purposes, the CBR value calculated at 2.5 mm penetration is adopted.

- If the CBR calculated at 5.0 mm penetration is constantly more than the CBR value at 2.5 mm penetration, the CBR at 5.0 mm should be taken as the design CBR value (IS: 2720 (part 16)-1987; 2004).

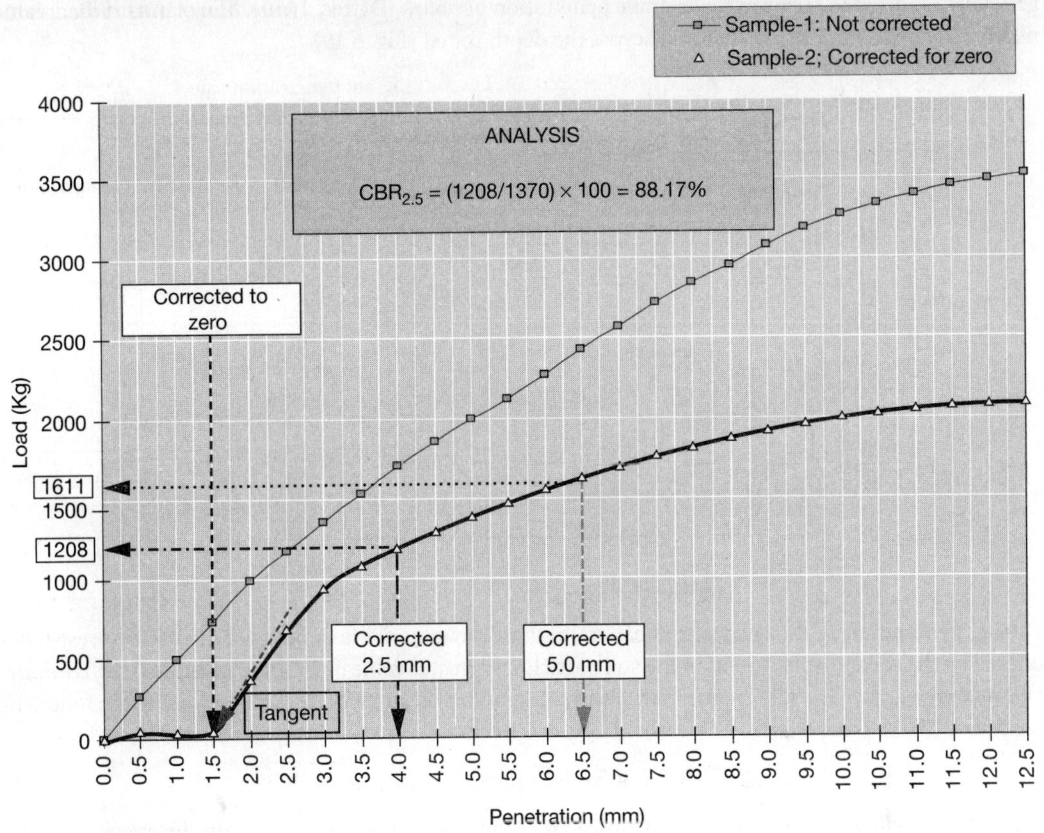

Sample-1: Not corrected
Sample-2; Corrected for zero

ANALYSIS

$CBR_{2.5}$ = (1208/1370) × 100 = 88.17%

Figure 6.17 A typical CBR Load-Penetration Curve.

According to IRC (*IRC Pocket Book* 2006), the sub-grade soil should conform to the minimum CBR requirement. In general, the CBR value of sub-grade soil should be more than 2%. If not, the soil should be suitability stabilised to improve the CBR value. In case of embankment, after compaction, the soil used as sub-grade should preferably have a CBR value more than 7%. Soil having less than 5% CBR is not suitable for sub-grade.

6.6.2 Dynamic Cone Penetrometer (DCP) Test

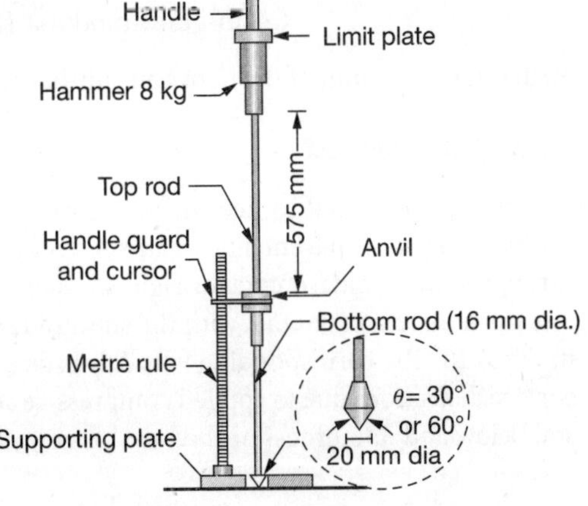

Figure 6.18 The dynamic cone penetrometer.

The DCP is a portable equipment that can be used to evaluate unbound granular base, sub-base, sub-grade and also pipeline congested/ narrow trenches (Fig. 6.18). This device can be used to structurally evaluate unbound soil layers and determine *in-situ* moduli of the tested layers. The basic principle involved in the operation of this apparatus is the measuring of resistance offered by the pavement layers to the penetration of a standard cone with a diameter of 20 mm (with 60° apex angle)[‡] driven by an 8 kg hammer freely falling through a height of 575 mm. The amount of penetration (in mm) of the cone is generally reported in terms of the average penetration per blow, $DCPI_{60°}$ ***(mm/blow). This indicates*** the relative shear strength of the material across the depth tested (Fig. 6.19).

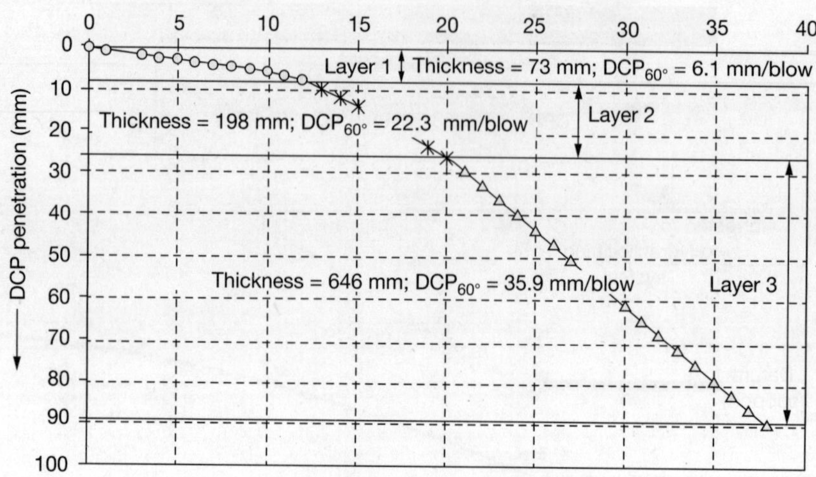

Figure 6.19 Typical analysis of DCP test data.

During penetration of the cone, the shear strength of the material is mainly accrued due to resistance offered by the shear displacement of the soil particles. A greater value of $DCPI_{60°}$ ***indicates a softer material and vice versa.*** The CBR value (%) of the sub-grade soil can be determined by using the following formula (ASTM-D6951-09):

$$\text{Log}_{10}\ \text{CBR} = 2.465 - 1.12\ \text{Log}_{10}\ (\text{DCPI}_{60°}\ \text{mm/blow}) \tag{6.10}$$

Sub-grade modulus value can be determined by using the following relationship developed between the DCP test values with the sub-grade modulus as back-calculated from falling-weight deflectometer (FWD) deflection blows (Srinivasa Kumar et al., 2003).

$$\text{Sub-grade modulus (MPa)} = 357.2 \times (\text{DCPI}_{60°})^{-0.645} \tag{6.11}$$

Structural evaluation of the base and sub-base can also be carried out by using the DCP.

6.6.3 Plate Load Test

This test is conducted to determine the modulus of sub-grade reaction (*k* value) of in situ sub-grade soil for analysis of pavements. In this test, sub-grade soil is subjected to different compressive stresses through some standard circular rigid stacked plates (75cm at the bottom, 60, 45 and 30 cm at the top). The main plate is in contact with the sub-grade soil and the remaining plates are placed over it as shown in Fig. 6.20. The corresponding vertical surface deflections of the plate are measured. The magnitude of surface deflections due to applied compressive stresses depends mainly on the supporting capacity of the soil below and its stiffness or the load bearing capacity of the soil.

[‡] The initial DCP cone apex angle was taken as 30° (Scala, 1956; van Vuuren, 1969). Later, the cone apex angle was revised to 60° (Kleyn et al., 1982; Livneh and Ishai, 1985; Webster et al., 1992). The revised cone angle of 60° was adopted by many standard institutions (FHWA, TRL-UK, ARRB, ASTM, IRC etc.) and formalized as a national standard in different countries.

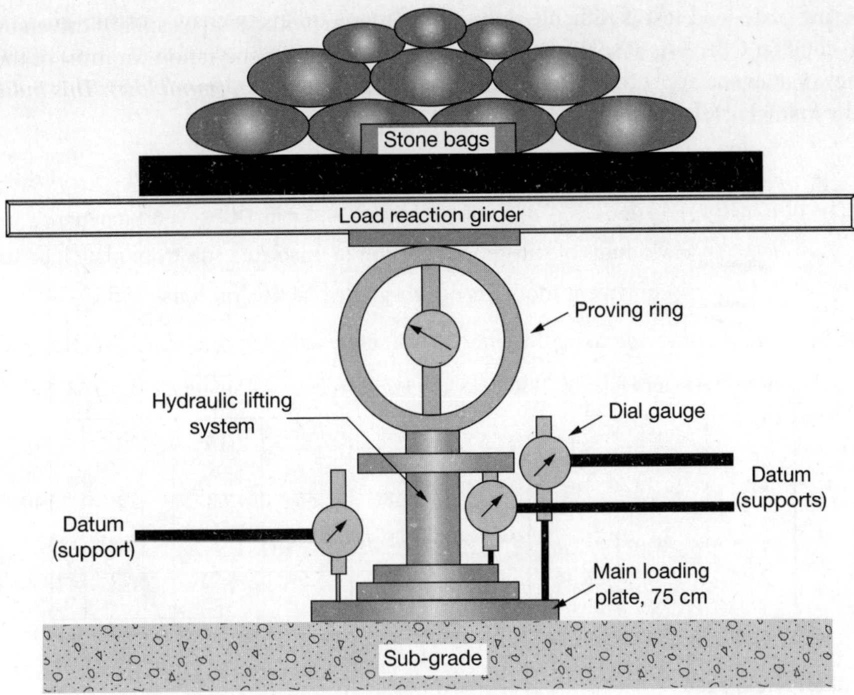

Figure 6.20 A typical plate load test set-up.

The values of sub-grade reaction are dependent on the type of sub-grade soil, density, frost action, saturation and intensity of stress applied. To analyse pavements, we need to find a representative value of the modulus of sub-grade reaction. From different studies and based on past experience, it has been found that the most representative value of sub-grade reaction (k-value) can be obtained at a pressure of 0.7 kg/cm^2 (T 222-81, AASHTO; Yoder and Witczak; IS: 9214-1979, 1980) or by taking the corresponding pressure found at 1.25 mm of average deflection value (IRC:58, 2015).

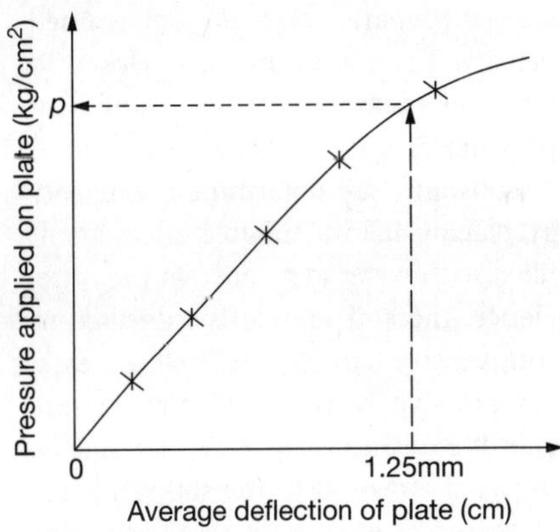

Figure 6.21 Pressure–deflection curve of plate load test.

Plot a graph with mean deflection of plate (in cm) on the x-axis and applied pressure (in kg/cm^2) on the y-axis. Draw a smooth curve joining the points. On the graph (Fig. 6.21), draw a dotted horizontal straight line from the x-axis at 1.25mm. From that point, draw a horizontal straight line to the x-axis.

$$\text{Modulus of sub-grade reaction } (k \text{ value}) = \frac{\text{Pressure on 75 cm dia. plate found at 1.25 mm deflection}}{\text{1.25 mm of average plate deflection}} \qquad (6.12)$$

As conducting plate-load test is difficult along the roadway on every change of sub-grade soil, it is a practice to conduct CBR tests at soaked and unsoaked conditions and correlate the results with the corresponding k-values and apply corrections. The correction factor for k of a soaked specimen is calculated by using the following relationship.

$$k_{\text{soaked}} = \text{Correction factor} \times k_{\text{unsoaked}} \tag{6.13}$$

where Correction factor $= d_u/d_s$

k_{unsoaked} = Modulus of sub-grade reaction of unsoaked soil from plate load test

k_{soaked} = Equivalent modulus of sub-grade reaction on soaked soil.

The CBR values can be converted to equivalent of k-values by using any standard practice (Table 6.4).

Table 6.4 Relationship between values of CBR and k of a homogeneous soil sub-grade (IRC: 58-2002 & 2015).

CBR (%)	2	3	4	5	7	10	15	20	50	100
k-value (MPa/m)	21	28	35	42	48	55	62	29	140	220
k-value (kg/cm^3)	2.08	2.77	3.46	4.16	4.84	5.54	6.21	6.92	13.85	22.16

6.6.4 Tri-axial Shear Test

In this test, a cylindrical soil sample with its height equal to twice its diameter is prepared at a pre-determined density. The soil sample is enclosed in a rubber membrane. The confinement pressure around the sample is kept at a constant pre-determined intensity by placing it in a suitable encasement filled with water at a constant pressure. Hence, the soil sample is subjected to compressive stresses as well as axial compressive stress until the sample fails due to shear (Fig. 6.22). Gradually, the axial stress on to the soil sample is increased at a constant rate (pre-determined). The values of vertical deformation (change in original depth) of the sample corresponding to the applied axial loads are recorded throughout the length of the test. A suitable arrangement is made to measure the pressure due to pore water of the soil sample. The shear failure of the sample may be observed as slip surface at an angle of (45° + ϕ/2) to the vertical axis or a bulge or a combination of these deformations. The detailed test procedure may be found elsewhere (Bishop and Henkel 1962; Head 1982; Vickers 1983; BS: 1377-1990; IS: 2720 (Part 12)-1981).

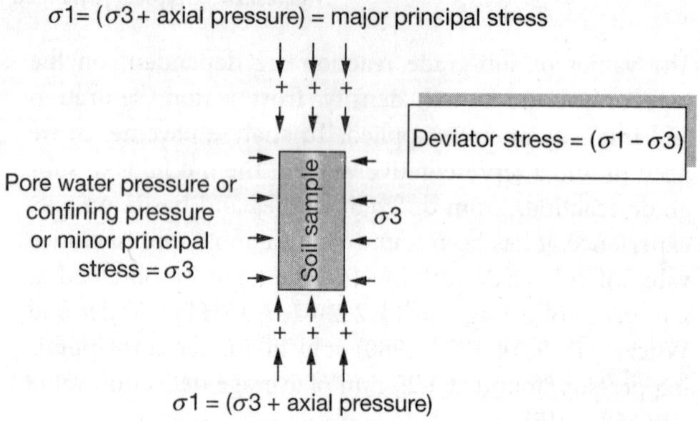

Figure 6.22 Principle of tri-axial shear test.

Typical tri-axial test set-ups are presented in Figs. 6.23 and 6.24. Drainage conditions prevailing in the field are best simulated in a tri-axial test. The only limitation is that shear failure of the soil sample takes place under equal compressive stress conditions but in the field, the soil may be subjected to anisotropic stress conditions.

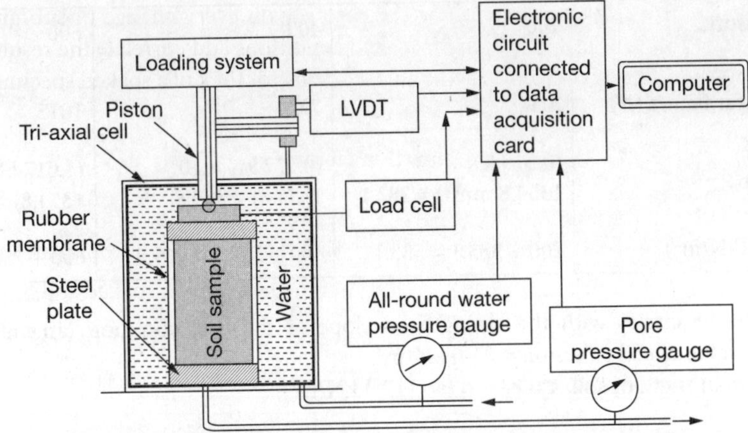

Figure 6.23 A typical computer-based tri-axial shear test set-up (schematic).

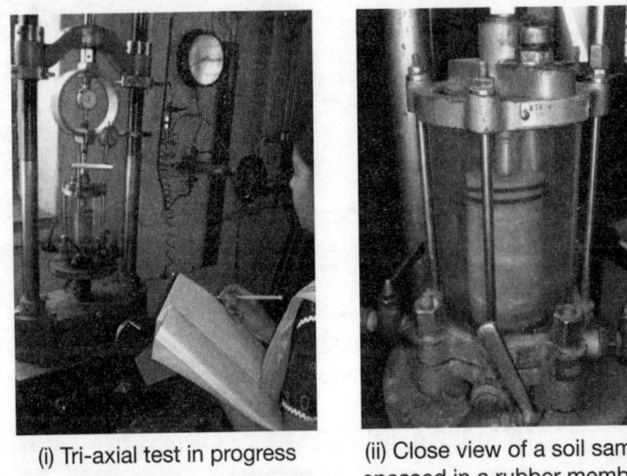

(i) Tri-axial test in progress

(ii) Close view of a soil sample encased in a rubber membrane

Figure 6.24 Tri-axial test set-up (conventional type).

EXAMPLE 6.3

Determine the parameters of Coulomb's equation by drawing Mohr's rupture of envelope of a soil sample. A sample 40 mm in diameter and 80 mm depth was cut and an un-drained tri-axial test conducted. Shear failure was observed at 5.8 mm of vertical deformation. Following are the details of stresses applied on the sample during testing.

Pore water pressure (kN/m^2)	220	440	660
Additional axial load at shear failure of the soil sample (N)	519	772	1017

Answer

Volume of the soil sample = $(\pi/4) \times (40\text{ mm})^2 \times 80\text{ mm} = 100530.9\text{ mm}^3$

The cross-sectional area at shear failure = $(100530.9\text{ mm}^3) \div (80 - 5.8) = 1354.8\text{ mm}^2$

Pore water pressure, σ_3 (kN/m^2)	200	400	600
Axial load at shear failure (N)	519	772	1017
Deviator stress, $(\sigma_1 - \sigma_3)$ (kN/m^2)	(0.519 kN $\times$ 10^6 $\div$ 1354.8 mm^2) = 383.1	(0.772 kN $\times$ 10^6 $\div$ 1354.8) = 569.8	(1.017 kN $\times$ 10^6 $\div$ 1354.8) = 750.6
Axial stress, σ_1 (kN/m^2)	200 + 383.1 = 583.1	400 + 569.8 = 969.8	600 + 750.6 = 1350.6

The details of Mohr circles with the strength envelope of the best common tangent are given in Figure 6.25. The value of soil cohesion = 81.0 kN/m^2.

$\phi = 17°$; Shear strength of soil, $\tau = 81 + \sigma \tan(17°)$ kN/m^2

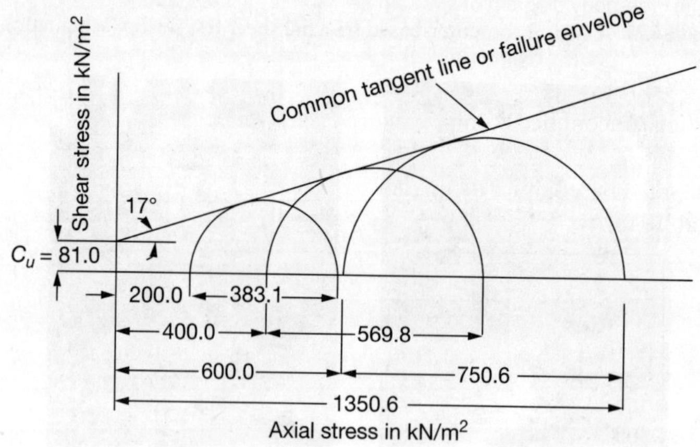

Figure 6.25 Diagram for Example 6.3.

6.6.5 Unconfined Compression Test

In this test, the soil sample is subjected to axial stress only (Fig. 6.26), without pore water pressure and the soil sample is not enclosed in a rubber membrane. It is not suitable for granular soils as no confinement is provided. This test is more suitable for cohesive or saturated clays. It is also simple and easy to perform in the field. The soil sample prepared for this test may be re-moulded or undisturbed. The axial stress applied on the soil sample is increased gradually at a constant rate (pre-determined). The values of axial deformation (change in original depth) of the sample corresponding to the applied axial loads are recorded throughout the length of the test. The shear failure of the soil sample may be observed at the point where the value of axial stress drops suddenly. But in case of highly plastic saturated clays, there will be no sudden drop of the axial stresses. In such cases, an appropriate value of failure stress may be taken based on the soil type, field conditions and purpose of use, or until an axial strain of 20% is reached (IS: 2720 (Part 10)-1991). The cohesion value of the soil is calculated as one-half of the axial stress recorded at sample failure. The angle of shearing resistance is equal to zero ($\phi = 0$), since the line of failure envelope is assumed to be horizontal.

The unconfined un-drained compression tests conducted in the laboratory on samples of *in situ* saturated clay show quite different shear strength properties based on direction of axial stress experienced by the soil under *in situ* conditions and unequal pressure prevailing around the soil (Duncan and Seed, 1966).

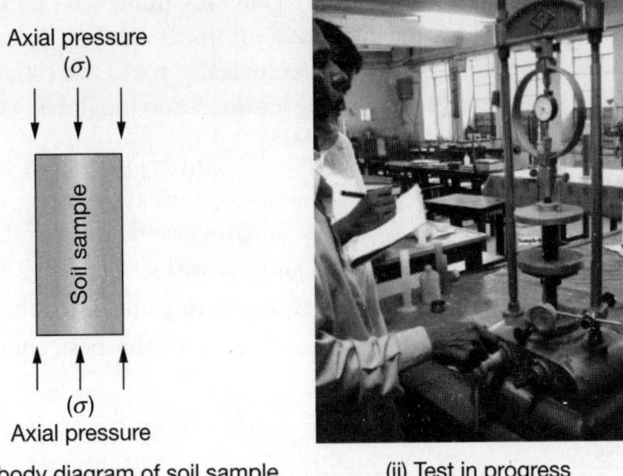

(i) Free-body diagram of soil sample (ii) Test in progress

Figure 6.26 Unconfined compression test.

The details regarding unconfined compressive strength of cohesive soils are given in IS: 2720 (Part 10)-1991.

Soil may be classified based upon its unconfined compressive strength. The normal range is given below (Cooling et al, 1948).

Class	Unconfined compressive strength	
	(kPa)	PSI
Very soft	< 34.5	< 5
Soft	34.5 to 68.9	5 to 10
Firm	68.9 to 137.9	10 to 20
Stiff	137.9 to 275.8	20 to 40
Hard	> 275.8	> 40

EXAMPLE 6.4

An unconfined undrained compression test was conducted on a saturated clay sample. The failure stress of the sample was found to be 204 kN/m^2. Calculate its shear strength properties.

Answer

The value of the angle of shearing resistance (ϕ) may be assumed to be zero, since the soil sample is saturated clay.

The undrained cohesion value (C_u) of the soil = failure stress (σ)/2 = 204/2 = 102 kN/m^2.

The above answer may be also obtained using Mohr's circle.

6.7 STRENGTH PROPERTIES OF MINERAL AGGREGATES

Mineral aggregates consist of any type of rock (igneous, sedimentary and metamorphic) possessing strong, hard and tough properties. Generally they are crushed stone, gravel, sand, slag, rock dust or rock powder. Aggregates of different sizes are obtained from different sources. The characteristic strength and durable properties of aggregate materials are generally influenced by their origin of occurrence. Knowledge about the origin of the rock is important for understanding the mineral content of the rock and deciding its suitability as road aggregates. Geologically, rocks are classified into three types, viz., igneous, sedimentary and metamorphic rocks. The hardness and toughness of these rocks depends upon the following fundamental properties (Hunting 1945).

- Composition of minerals
- Chemical composition (relates to hardness, toughness and cleavage of individual minerals)
- Texture (relates to size and shape of grains and porosity)
- Freshness of minerals in terms of degree of weathering and hydrothermal alteration of original minerals
- Fracturing and jointing

The following tests are conducted on the stone aggregates to determine its suitability for construction of pavements.

6.7.1 Water Absorption and Specific Gravity Tests

Strength properties of road aggregates are indirectly related to its specific gravity and water absorption characteristics. In general, it is observed that an aggregate sample having high specific gravity and low water absorption is dense, less porous, durable and possesses high strength. This concept is used while judging suitability of aggregates for construction of pavements. Further, specific gravity is an index value which indicates the aggregate origin and constituent minerals. Its value is used for computation of densities and amounts of voids while designing the mix of cement and bituminous concrete.

Specific gravity

It is the ratio between the dry weight of aggregates in air to the weight of water equivalent to volume of the aggregates tested at a standard temperature.

Water absorption

It is the percentage by weight of water absorbed by the aggregates to the oven dried weight of the aggregates at standard temperature.

$$w\,(\%) = \frac{\text{Weight of water absobed by aggregates}}{\text{Dry weight of the aggregates}} \times 100 \tag{6.14}$$

The aggregates which possess higher values of density and specific gravity with less porosity are found to be strong, tough and more resistant against the action of abrasive forces. Hence, it is desirable to select aggregates having higher specific gravity values.

$$\text{Specifi gravity} = \frac{\text{Dry weight of aggregates}}{\text{Weight of water equivalent to the volume of aggregates}} \tag{6.15}$$

Determination of specific gravity and water absorption of aggregates can be performed based on the size of the aggregates.

Method-1 by using wire basket (IS: 2386 (Part 3)-1963)

This method is useful when the aggregate size > 10 mm.

About 2 kg of washed aggregates placed in a wire basket is immersed in water and the entrapped air between them is removed by jolting the basket in water up and down 25 times. After immersing it in the water for a duration of 24 hours, its submerged weight is recorded as M_1 g.

Then, the basket with aggregates is removed from water and the loose aggregates are drained for few minutes on a dry cloth. The weight is recorded as M_2 g. The surface of the wet aggregates is wiped with a cloth and completely dried by blowing air, if required. During the drying process, the aggregates are not exposed to direct sun light or any kind of heat. After air-drying, the weight of the aggregates is taken and recorded as N g.

The aggregates are then kept in an oven at a temperature between 100 to 110°C for 24 hours. The weight of the aggregates is recorded as P g. Specific gravity of the aggregate sample is determined by the following equations (IS: 2386 (Part 3)-1963).

$$\text{Specific gravity} = \frac{P\text{ g}}{(N-M)\text{ g}} \tag{6.16}$$

$$\text{Apparent specific gravity} = \frac{P\text{ g}}{(P-M)\text{ g}} \tag{6.17}$$

$$\text{Water absorption by per cent weight of aggregates(\%)} = \frac{(N-P)\text{ g}}{P\text{ g}} \times 100 \tag{6.18}$$

Where,

M_1 = Total weight of the basket filled with saturated aggregates in water.
M_2 = Empty weight of the wire basket in water.
$M = (M_1 - M_2)$ = Net weight of saturated aggregates in water.
N = Weight of saturated surface-dry aggregates in air.
P = Weight of oven dried aggregates in air.

Method-2 by using Pycnometer (IS: 2386 (Part 3)-1963)

This method is useful when the aggregates size < 10 mm.

About 1000 g of aggregate of size in between 10 mm and 4.75 mm or 500 g of aggregates of size less than 4.75 mm is taken in a tray and immersed in distilled water for 24 hours. Later, they are dried and the weight of the saturated surface-dry aggregates is recorded as Q_1 g.

The saturated aggregates are placed in a pycnometer and the remaining portion of the pycnometer is filled with distilled water without any air bubbles in it. The outer surface of the pycnometer is allowed to dry and its weight recorded as Q_2 g.

The aggregates in the pycnometer are removed and drained on to a tray. During this process there should not be any loss of aggregate particles. The drained aggregates are kept in an oven and dried for 24 hours at 100 to 110°C. The weight of the oven-dried aggregates is recorded as R g. Finally, the pycnometer is completely filled with distilled water and the weight recorded as S g. Fig. 6.27 depicts the details of the weight measurements to be taken to conduct the test.

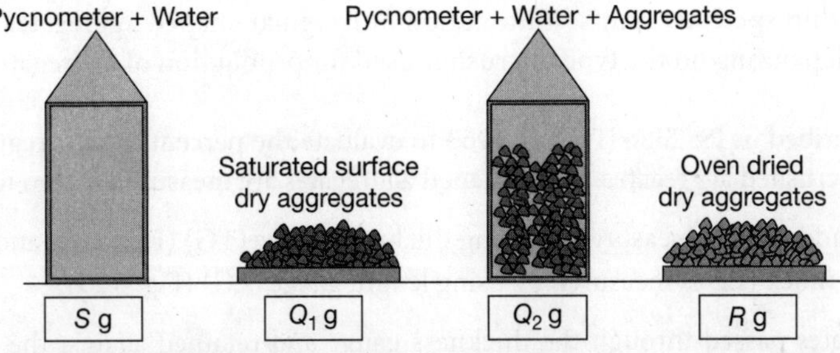

Figure 6.27 Details of weight measurements.

The specific gravity of the agggregates sample is calculated as given below:

$$\text{Specific gravity} = \frac{R\,\text{g}}{Q_1 - (Q_2 - S)\,\text{g}} \tag{6.18}$$

$$\text{Apparent specific gravity} = \frac{R\,\text{g}}{R - (Q_2 - S)\,\text{g}} \tag{6.19}$$

$$\text{Water absorption by per cent weight of aggregates} = \frac{(Q_1 - R)\,\text{g}}{R\,\text{g}} \times 100 \tag{6.20}$$

Where,

Q_1 = Weight of saturated surface-dry aggragates.
Q_2 = Total weight of pvcnometer filled with saturated aggregates and water.
S = Weigth of surface –dry [ycnometer filled with water.
R = Weight of oven dried aggregates.

According to the IRC (*IRC Pocket Book 2006*), the maximum allowable limit of water absorption for any aggregate used for highway pavement construction should not be more than 2% the weight of the aggregate. However, the maximum allowable limit of water absorption is relaxed to 5% for aggregates used in CC slab pavements of rural roads (IRC: SP: 62-2004).

6.7.2 Shape Tests

Generally, coarse aggregates may be in various shapes such as granular, spherical, cubical, flaky (thin chips or platy) and elongated (long). The interlocking of aggregates is achieved by using various sizes and graded cubical, flaky or elongated aggregates having rough fractured surfaces. Use of cubical shaped aggregates is preferred in order to achieve high-compacted dry density and strength. Flaky and elongated aggregates are not useful because of their tendency to break and split easily on application of loads leading to the creation of more new voids during compaction.

Pavement performance is also affected by shape characteristics of aggregates during pre- and post-construction stages. Irregular-shaped aggregates do not accommodate and adjust themselves into the standard gradation limits. Thin and lengthy aggregates tend to break down easily even under small external loads. Consequently, broken aggregates tend to form more voids and alter gradation. In cement–concrete mix, flat and elongated particles (*i*) tend to decrease workability, (*ii*) require more sand, cement and water, (*iii*) reduce bulk weight and (*iv*) decrease compressive strength (Blanks, 1950). They are also prone to formation of micro-cracks in asphalt concrete and cement concrete layers. Because of these reasons, the shape of aggregates is important in the construction of pavements.

Further, during quality control of road construction works, proper care should be taken to keep the aggregates within specified limits. Elimination of irregular shaped aggregates may not be completely possible, depending on the type of crusher used for production of aggregates, type and origin of rocks.

Tests are prescribed in IS: 2386 (Part 1)-1963 to evaluate the percentage of irregular-shaped aggregates in a mass of crushed aggregates. Flaky shaped aggregates are measured in two folds. Viz.,

1. Flakiness index (FI) is measured by using thickness gauge (TG) (Fig. 6.28) and
2. Elongation index (EI) is measured by using length gauge (LG) (Fig. 6.29)

Note: The aggregates passed through the thickness-gauge and retained against the gaps of respective length-gauge are used for calculation of combined index (CI).

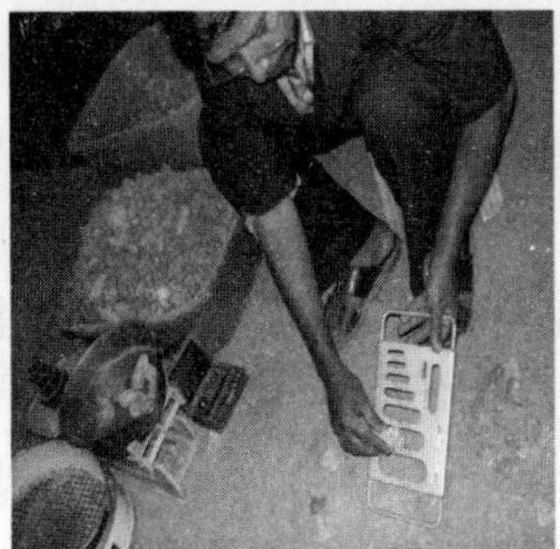

Figure 6.28 Flakiness test using a thickness gauge.

Figure 6.29 Elongation test using length gauge.

Flakiness test

Flakiness index (FI) of a given size of aggregate sample is expressed in terms of percentage by weight of aggregates, whose smallest dimension or thickness is less than 3/5 or 0.6 times their average size (IS: 2386 (Part 1)-1963; 2004). In other words, FI is the percentage by weight of flaky aggregate particles whose least thickness is less than 0.6 times the average size of the aggregates.

The thickness gauge consists of a steel plate having 9 slots or openings of suitable dimensions by which an aggregate can be measured (Fig. 6.28).

Following is a typical example for calculating the least dimension of the slot of TG for the size of aggregates used.

For an aggregate size in between 63 mm to 50 mm, least thickness = $0.6 \times \dfrac{(63+50)}{2} = 33.9$ mm

The given aggregate sample is sieved through standard sieves of nine sets. It is then checked whether each aggregate particle from all the sets passes through the slot designated. The aggregate particles which pass through the respective slots are weighed separately. FI may be calculated using the formula given below.

$$\text{FI}\ (\%) = \frac{\sum_{i}^{n} w_i}{\sum_{i}^{n} G_i} \times 100 \qquad (6.20)$$

where,

w_i = Weight of various size of aggregate particle passed throught the respective slots of TG, i.e., 63-50 mm, 50-40 mm, 40-25 mm, 31.5-25 mm, 25-20 mm, 20-16 mm, 16-12.5 mm, 12.5-10 mm, 10-6.3 mm.

G_i = Total weight of aggregates considered for the test.

Elongation test

Length evaluation of coarse aggregates is conducted using a length gauge (LG) and expressed in terms of the elongation index (EI) of a given aggregate sample. EI of a given size of aggregate sample is expressed in terms of percentage by weight of aggregates whose longest dimension is more than 1.8 times their average size (IS: 2386 (Part 1)-1963, 2004). In other words, EI is the percentage by weight of long aggregate particles whose length is more than 1.8 times average size of the aggregates.

The LG is a handy and portable tool (Fig. 6.29) which consists of a rectangular steel plate attached to a suitable size of wooden plate. Eight steel rods of suitable diameter are fixed in vertical direction in a row on the steel plate. The method of calculating clear distances (CD) between successive rods according to the size of aggregates used is shown below.

For aggregate sizes in between 50 mm to 40 mm,

$$\text{Longest dimension} = 1.8 \times \frac{(50+40)}{2} = 81.00 \text{ mm}.$$

The given aggregate sample is sieved through the standard sieves of 7 sets. For all the sets, it is checked whether each aggregate particle passes in between the spaces of respective rods of designated sizes. The aggregate particles that *do not pass* in between the respective rods are weighed separately. The EI is expressed in terms of percentage by weight of aggregate particles not passing between the rods of the sets of sieves as compared to the total weight of the aggregates used for the test. EI may be calculated as mentioned below.

$$\text{EI } (\%) = \frac{\sum_{i}^{n} m_i}{\sum_{i}^{n} G_i} \times 100 \tag{6.21}$$

where,

m_i = Weight of various sizes of aggregate particles which are not passed between the respective slots of LG, i.e., 50-40 mm, 40-25 mm, 25-20 mm, 20-16 mm, 16-12.5 mm, 12.5-10 mm, 10-6.3 mm.

G_i = Total weight of aggregates taken for the test.

Flakiness and elongation tests are not applicable when the aggregate size is smaller than 6.3 mm (IS: 2386 (Part 1)-1963, 2004).

The specified maximum allowable limit of combined flakiness and elongation indices (i.e., combines index) of all aggregates used for road construction, according to IRC, is 35 %.

6.7.3 Aggregate Crushing Test

The performance of paving aggregates having hardness or resistance to withstand external pressures is evaluated in the aggregate crushing test. It is an important test conducted on road aggregates. Under a standard load of 40 tonnes, the given aggregates are allowed to be crushed.

Dry aggregates of 3.5 kg, passing through 12.5 mm size sieve and retained on 10 mm size sieve, are taken and filled in a standard measuring cylinder in three equal thick layers. Each layer is tamped 25 times by a tamping rod with a rounded end. Excess aggregates on the top of the measuring cylinder are levelled off and weighed as (W_1 g) up to 1 g accuracy. The aggregates are transferred to the test mould in three layers of equal thickness, tamped 25 times for each layer (Fig. 6.30 (a)). The piston assembly with the mould (Fig. 6.30 (b)) is put in position for loading and a vertical compressive load is applied on it at the rate of 4 tonnes per minute, gradually up to a maximum of 40 tonnes.

After releasing the load, tested aggregates are sieved through 2.36 mm sieve and the powder is weighed and recorded as W_2 g, to an accuracy of 0.1 gm. The above test procedure is repeated with fresh aggregates. The aggregate crushing value is calculated.

$$\text{Aggregate crushing (AC) value } (\%) = (W_2/W_1) \times 100 \tag{6.22}$$

where,

W_1 = Weight of the compacted aggregate filled in the test mould in g.

W_2 = Weight of aggregate powder passing through a 2.36 mm size sieve in g.

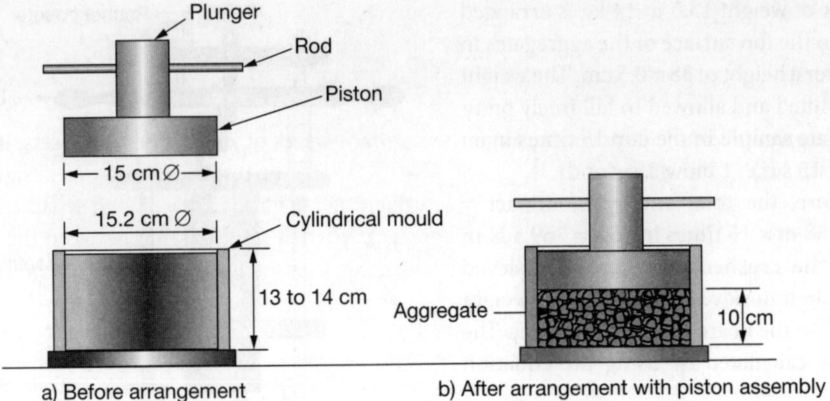

Figure 6.30 Details of mould and plunger.

If the tested aggregate sample is hard in nature, the numerical AC value will be low and vice versa. Generally, aggregates with lower crushing values are selected for the surface course of pavement layers where direct heavy compressive forces are anticipated. The limits of aggregate crushing values according to the IRC for various types of pavement layers are presented in Figure 6.31.

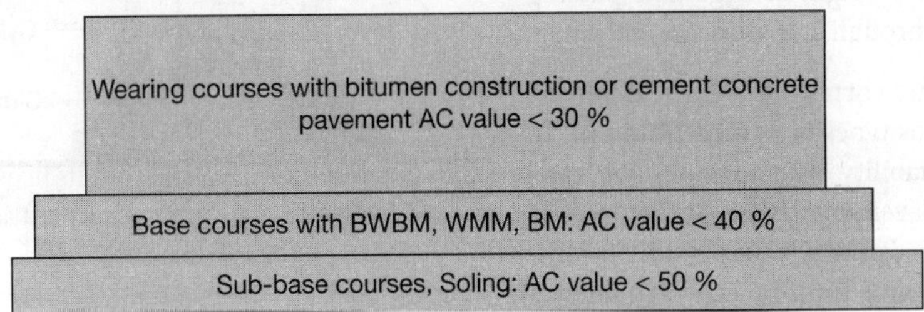

Figure 6.31 Aggregate crushing values for different pavement layers.

The test procedure followed for non-standard sizes of aggregates (i.e., not in between 12.5 mm to 10 mm) is same as for the standard-size aggregates. Details of the apparatus used for the test, like the cylindrical mould, tamping rod and sieve sizes may be obtained from IS: 2386 (Part 4)-1963.

6.7.4 Aggregate Impact Test

Repetitive traffic wheel loads cause damage to the pavement layers by compression, impact and abrasion. Traffic loads are mainly supported by aggregates and transferred safely by the aggregate interlocking. Therefore, the aggregates should be tested for their resistance against impact or shock loads caused due to the movement of vehicles at high speed. The property of toughness of aggregates and the pavement performance to sustain these loads is evaluated by the aggregate impact (AI) test.

An oven-dried aggregate sample which passes12.5 mm sieve and retained on 10 mm sieve is used for the test. The aggregate is filled in a measuring cylinder in three equal layers and each layer is tamped 25 times with the round end of a tamping rod. Excess aggregates in the measuring cylinder are struck off by a straight edge. The net weight of the aggregates is recorded (W_1 g) to 1 g accuracy. The measured aggregates are transferred to a cylindrical cup and tamped 25 times. The cup with the aggregate sample is placed in the centre of the base plate of the impact test set-up (Fig. 6.32) and screwed firmly to it.

A hammer of weight 13.5 to 14 kg is arranged to fall on to the top surface of the aggregates in the cup, over a height of 38 ± 0.5 cm. The weight should be lifted and allowed to fall freely on to the aggregate sample in the cup 15 times in an interval of 15 s (i.e. 1 blow/1 second).

Therefore, the total energy of impact = 135 N × 0.38 m × 15 times in 15 s = 769.5 N m

Then, the crushed aggregates are sieved through 2.36 mm sieve and the powder weight is recorded to the nearest 0.1 gm accuracy. The AI value is calculated by using the equation below.

$$\text{Aggregate impact value} = 100 \times W_2/W_1 \quad (6.23)$$

where,

W_1 = Weight of aggregate filled in measuring cylinder, in g.

W_2 = Net weight of aggregate passing through 2.36 mm size sieve in g.

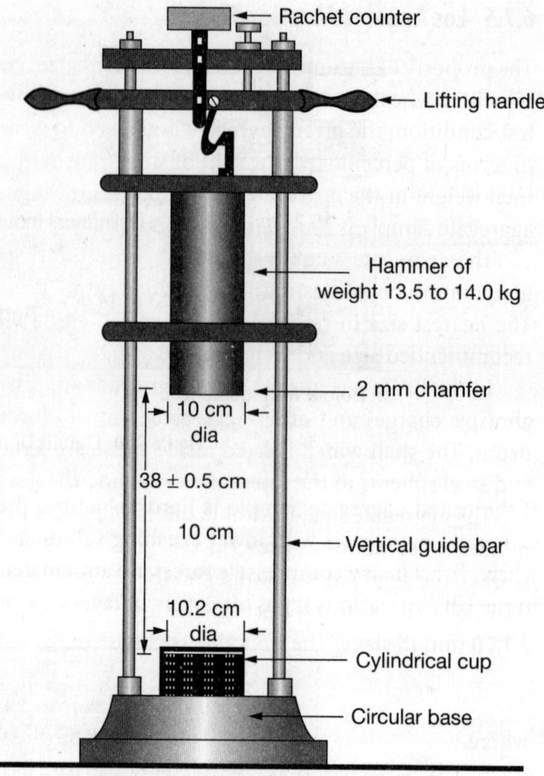

Figure 6.32 Aggregate impact test apparatus.

The AI values are correlated with field performance of various types of pavement layers. To decide the suitability of aggregates for various types of pavement layers, Indian Roads Congress (*IRC Pocket Book*, 2006) recommends the following limiting values (Table 6.5).

Table 6.5 Maximum allowable aggregate impact value (*Pocket Book*, 2006).

Sl no.	Description	Maximum allowable aggregate impact value (%)
1	Wet mix macadam for sub-base or base	30
2	Crushed run macadam base	30
3	Bituminous base	30
4	Dense graded bituminous macadam	27
5	Semi-dense graded bituminous macadam	27
6	Bituminous concrete	24
7	Aggregate used for concreting according to IS:383- 1993:	
	Other than wearing courses	45
	Wearing courses	30

6.7.5 Los Angeles Abrasion Test

The property of hardness or resistance to withstand the combined effect of abrasion and impact forces by repetitive wheel load on aggregates are evaluated by the Los Angeles abrasion test. Under the standard test condition, the given aggregate is allowed to wear. The Los Angeles abrasion (LA) value is expressed in terms of percentage by weight of stone powder passing through a 1.70 mm sieve as compared to the total weight of the aggregate sample taken for the test. If the LA value is low, it indicates that the tested aggregate sample is hard and vice versa.

The aggregate sample should be clean and oven dried at 105 to 110°C (allowed to cool to room temperature before testing). It is divided into 7 types based on the size, namely A, B, C, D, E, F and G. The nearest size in the table matching the grade requirement of a pavement layer is selected and the recommended size and weight of the aggregate sample is taken for testing.

In the LA test (Fig. 6.33), the aggregates are tested for abrasion by interaction with steel spheres (as abrasive charge) and other aggregates. Impact forces are applied on the aggregates by rotation of the drum. The shelf which is fixed inside the drum along its length facilitates the transfer of the aggregates and steel spheres to the top of the drum and their sudden drop to the bottom of the rotating drum.

The aggregate loaded steel drum is allowed to rotate at the rate of 30 to 33 rpm to the desired total number of rotations as prescribed (Fig. 6.33). After a prescribed number of rotations (say 500 for road works), the aggregates, powder and the steel spheres are collected. The dust on the steel balls is also carefully collected by using a brush. The fine dust particles from the aggregate sample are separated through a 1.70 mm IS sieve. The LA value is obtained by using the equation below:

$$\text{Aggregate crushing value} = 100 \times (W_1 - W_2)/W_1 \tag{6.24}$$

where,

W_1 = Weight of the selected grade of the aggregate used for the test in g.
W_2 = Weight of the aggregate retained on a 1.70 mm size IS sieve in g.

Generally, aggregates with lower LA values (%) are selected for surface course or for pavement layers where direct heavy abrasive forces are anticipated. According to IRC (*Pocket Book*, 2006), allowable limits of aggregate LA values for various types of pavement layers are presented in Fig. 6.34.

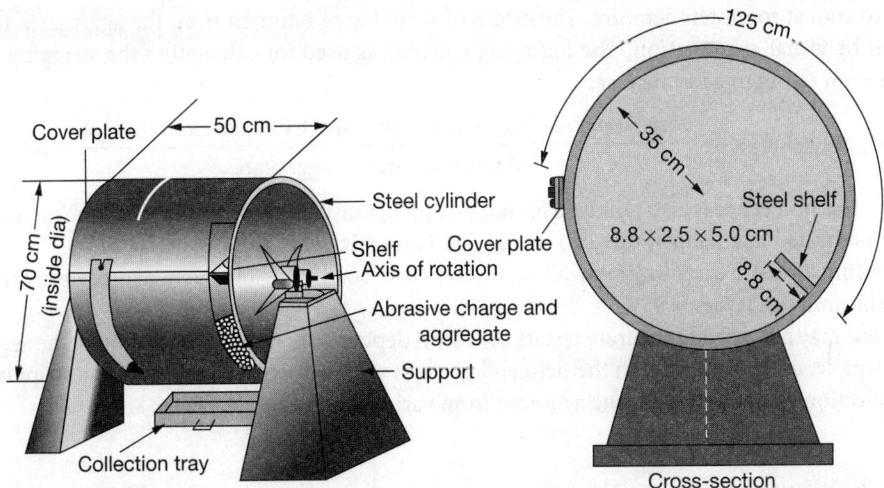

Figure 6.33 The Los Angeles test set-up.

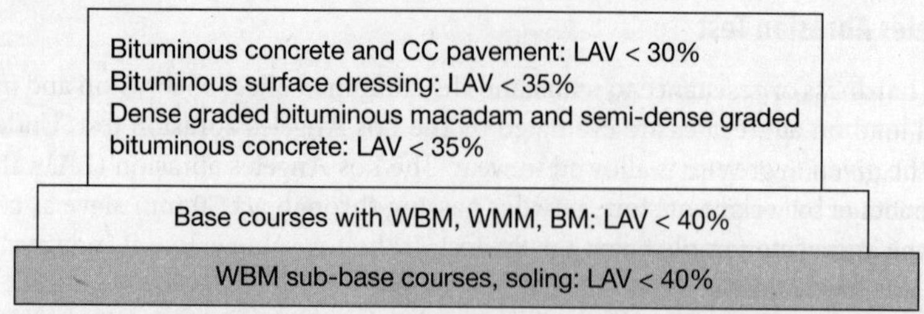

Figure 6.34 Los Angeles abrasion test limiting values for various pavement layers.

6.7.6 Stripping Value of Aggregates

The adhesion property of bitumen to the surface of freshly crushed aggregates is determined using the stripping value of aggregates. During the construction of bituminous layers, the aggregates are mixed with a suitable quantity of bitumen at a standard temperature. But the surfaces of aggregates may not be coated fully with bitumen. Following are the reasons for partial bitumen coating of aggregates.

- The temperature (or viscosity) of bitumen may not be maintained at a standard value during mixing, or the aggregates may not be fully coated with bitumen.
- The aggregates may be wet/damp.
- Aggregate surface may be contaminated by dust or mud or clay.

In the above cases, the compacted pavement surface may not have good adhesion or interlocking bond between bitumen-coated aggregate particles.

In the static immersion test, about 200 g of aggregates passing a 20 mm sieve and retained on a 12.5 mm size IS sieve and about 5 per cent by weight (10 g) of bitumen is taken. The bitumen and aggregates are preheated separately at 160 and 150°C respectively (IS: 6241-1971, 2006). Then they are mixed thoroughly in order to coat the aggregate surfaces completely with bitumen. This mixture is transferred to a beaker and allowed to cool at room temperature for 2 hours. The beaker is filled with distilled water just sufficient to immerse the aggregate–bitumen mix. It is then covered and kept in a water bath at a temperature of 40°C for about 24 hours. Afterwards, the beaker is taken out of the water bath and allowed to cool at room temperature. The extent of stripping of bitumen from the aggregates surface is estimated by visual examination. The following equation is used for calculating the stripping value of bitumen from the aggregates surface.

$$\text{Stripping value (\%)} = \frac{\text{Uncovered surface area of aggregates by visual examination}}{\text{Total surface area of aggregates}} \times 100 \qquad (6.25)$$

The above test procedure is also followed for tar, except that instead of 160 and 150°C, the tar and aggregates are preheated to a temperature of 110 and 100°C respectively (IS: 6241-1971; 2006).

According to IRC, the minimum allowable retained coating of bitumen–aggregate mixtures for different bituminous layers is 95%.

This test may not provide accurate results because it depends on visual examination of the aggregates. But it is simple, easy to perform in the field and gives an approximate estimation of the stripping value during selection of suitable aggregate acquired from various sources.

Review Questions

1. What are the materials used for highway construction?
2. Define gradation of soil. What are the different types of grades of soil or mineral aggregate?
3. Draw a typical gradation chart showing different types of soils and their grades.
4. Briefly describe how moisture in a soil mass is determined.
5. State the uses of moisture content present in a given soil.
6. What are the different methods used for determining the moisture present in the soil? State the limitations of each method.
7. Distinguish between field moisture content (FMC) and optimum moisture content (OMC).
8. Define (i) liquid limit, (ii) plastic limit, (iii) liquidity index, (iv) consistency index, (v) toughness index and (vi) activity of soil.
9. How is the degree of plasticity of soil measured?
10. Briefly explain the test standards and procedure adopted for the determination of
    (i) liquid limit of soil by (a) mechanical method and (b) cone penetrometer method
    (ii) plastic limit of a given soil.
11. What are the limitations and suitability of mechanical and cone penetrometer methods for determination of liquid limit of soil?
12. What are the generally accepted interpretations with reference to liquid limit, plasticity and liquidity index?
13. State the significance of LL and PI values in characterisation of the properties of soil used for road construction.
14. Cone penetrometer tests conducted on a soil sample gave the following test data:

Moisture content (%)	32.0	33.4	34.4	36.1	37.3
Cone penetration (mm)	16.5	18.0	19.5	20.9	22.8

Draw the plot and determine the liquid limit of the soil.

15. Liquid limit tests were conducted on a clay sample and the following observations were obtained.

Test No.	1	2	3	4
Water content (%)	69	63	48	43
No. of blows (N)	5	8	9	44

Draw the flow curve on a semi-log paper and determine the LL value.

16. The following observations were recorded from an LL test.

Test No.	1	2	3	4
Wt. of container (g)	13.3	13.1	13.4	12.9
Wt. of container + wet soil (g)	26.3	21.8	21.4	22.7
Wt. of container + dry soil (g)	22.5	18.1	18.3	19.7
No. of blows (*N*)	12	18	21	29

Determine the LL value by drawing the density curve. What is the value of PI if the plastic limit value is 22%.

17. Define specific gravity of soil and state where this value is useful.
18. Determine the values of dry density and void ratio if a sand sample has a porosity of 34.5% and the specific gravity is given as 2.70.
19. What happens when soil is compacted (i) at different moisture levels and (ii) at different compactive efforts? Explain with a neat sketch.
20. What are the useful changes that occur in a compacted soil mass?
21. What is the test principle of (i) a standard soil compaction method and (ii) the modified soil compaction method?
22. Define degree of compaction of soil. What is its relevance to field compaction levels?

23. State the different methods of compaction of soil and explain core cutter and sand replacement methods.
24. State the test principle of CBR test. Write the standard specifications followed for preparation of soil sample and for testing.
25. Why is the CBR value determined at 2.5 mm penetration generally greater than the CBR determined at 5.0 mm penetration? Why is surcharge weight used for testing of soil sample by the CBR method?
26. Comment on CBR as an empirical value of soil strength.
27. Draw sketches showing standard details of CBR penetration test and expansion test.
28. State the allowable values of CBR of soil that can be used for road construction and embankment
29. With the help of a figure, explain the plate load test procedure.
30. State the uses of the plate load test. What are the standard specifications of the test?
31. What is the drawback of PLT? State how the k value does not represent the stress distribution due to load on flexible pavements, base and sub-grade.
32. Why are corrections made to the k value determined by the plate load test? Describe the different corrections made?
33. State the general classification of soil based on the k value.
34. State how shear strength parameters are useful for pavement and road embankment design in general.
35. Explain how the tri-axial test best simulates field loading conditions. Describe with a neat sketch the test principle, set-up and procedure.
36. What are the factors affecting unconfined compressive strength of a given soil?
37. Compare the different shear strength tests carried out on sub-grade soil.
38. State the sources of mineral aggregate used for road construction.
39. Explain how specific gravity and water absorption values of aggregate are used by the pavement engineer.
40. Briefly describe the test procedure adopted for determining the specific gravity of aggregates when (i) the aggregate size is greater than 10 mm and (ii) the aggregate size is less than 10 mm.
41. Enumerate the purpose of DCP testing. State how the test results can be interpreted for (i) structural evaluation, (ii) design, (iii) quality control for construction of earth work, embankments and granular layers and (iv) pavement performance.
42. Describe the salient parts of the DCP and explain its working principle.
43. What are the factors affecting test results of DCP?
44. State the DCP–CBR and DCP–Sub-grade modulus equations which are applicable for Indian soil.
45. Explain the significance of shape test of aggregates used for (i) cement concrete and (ii) bituminous concrete.
46. Write a short note on (i) flakiness test and (ii) elongation test of road aggregates.
47. What is the significance of the angularity test of road aggregates? Describe the test principle.
48. What is the significance of the (i) aggregate crushing test, (ii) Los Angeles abrasion test, (iii) aggregate crushing test, (iv) ten per-cent fines test and (v) aggregate impact test? Describe the principle behind the tests.
49. State the reasons for partial coating of bitumen on aggregate surfaces.
50. What is the significance of the stripping value test? Describe the principle behind the test.

Multiple Choice Questions

1. The wet-mechanical analysis is applicable if the size of soil particles is:
   (a) between 75 and 2 microns
   (b) between 2.36 mm and 75 microns
   (c) greater than 2 microns
   (d) greater than 75microns
2. The sieve size of soil sample considered for determination of LL by cone penetration method is
   (a) passing 425 microns
   (b) passing 75 microns
   (c) between 425 and 75 microns
   (d) passing 1.7 mm

3. In the cone penetration method, as per IS: 2720 (Part 5), a standard cone of 30° angle with 35 mm depth, having an 80 g plunger is allowed to penetrate into the soil paste kept in a standard size cup for a duration of ________ seconds.
(a) 5 (b) 10 (c) 15 (d) 20

4. As per IS: 2720 (Part 7), in light compaction of soil, what is the free-fall height of the rammer considered?
(a) 300 mm (b) 310 mm
(c) 450 mm (d) 260 mm

5. As per IS: 2720 (Part 8), in heavy compaction of soil, what is the weight of the rammer considered?
(a) 2.6 kg (b) 10 kg
(c) 4.9 kg (d) 5.0 kg

6. As per IS: 2720 (Part 29), the core cutter method is suitable for fine-grained soil, of which not less than 90 per cent passes through a ________ mm size IS sieve.
(a) 2.36 (b) 4.75
(c) 1.7 (d) 0.075

7. The standard load values considered for CBR test at 2.5 mm penetration value is.
(a) 70 MPa (b) 6.84 MPa
(c) 10.27 MPa (d) 2.5 MPa

8. By dynamic cone penetrometer (DCP) test, a standard cone having a diameter of 20 mm (with 60° apex angle) is driven by an 8 kg hammer freely falling through a height of ________ mm.
(a) 975 (b) 475
(c) 575 (d) 675

9. The most representative value of sub-grade reaction (*k*-value) by plate load test can be obtained at a pressure of ________ kg/cm^2.
(a) 0.070 (b) 0.700
(c) 0.007 (d) 70.000

10. The details regarding unconfined compressive strength of cohesive soils are given in ________.
(a) IS: 2720 (Part 7)
(b) IS: 2720 (Part 8)
(c) IS: 2720 (Part 9)
(d) IS: 2720 (Part 10)

Answers to Multiple Choice Questions

1. (a)	3. (a)	5. (c)	7. (b)	9. (b)
2. (a)	4. (b)	6. (b)	8. (c)	10. (d)

CHAPTER 7 Pavement Materials: Bituminous Binders and Bituminous Paving Mixes

7.0 INTRODUCTION

Bituminous materials are used for constructing wearing courses. They are also used as stabilising agents for other layers of flexible pavements, such as emulsions and foamed bitumen. The properties of bituminous materials, viz., adhesion, cementation, ductility, strength, imperviousness and resistance to oxidation are useful for paving purposes. Basically, bituminous materials are thermoplastic hydrocarbons which possess high molecular weight. They become soft on heating and harden on cooling. They are obtained from natural sources or prepared artificially. Based on the origin, type of mineral composition and method of processing in refineries, bituminous materials are classified into different types and grades. A few of the different types used for highway construction are given below:

Bitumen Bitumen is obtained from natural sources or vacuum distillation of crude petroleum products at a temperature of about +400°C. It is a black or brown-to-black, non-crystalline material in solid or semi-solid state having good adhesive property. Bitumen is completely soluble in *carbon disulphide* but largely non-volatile and resistant to most acids, alkalis and salts (Dickinson, 1984).

Asphalt Based on the origin of occurrence (by nature or artificial preparation), the bitumen may contain a certain amount of mineral matter. The bitumen associated with inert mineral matter is known as asphalt.

Note: In some parts of Europe and North America, bitumen is referred to as asphalt.

Straight-run Bitumen The bitumen extracted as a residue from crude oil refinery without any diluents is called straight-run bitumen. It is generally found to have higher viscosity. Bitumen cutbacks are prepared by adding soluble volatiles to straight-run bitumen to obtain bitumen which has lower viscosity and is more workable.

Mastic Asphalt Coarser filler material (stone grit) is mixed with bitumen to remove any internal voids. This mixture, called mastic asphalt, can be used for constructing impervious layers for surfacing, bridge decks and roofs.

Bitumen blends of different types (Crumb Rubber Modified Bitumen–CRMB, Polymer Modified Bitumen–PMB, Emulsions and Cutbacks) and grades (Viscosity Grades–VG) will be discussed later in this chapter.

7.1 DESIRABLE PROPERTIES OF BITUMEN FOR PAVING PURPOSES

Following are the important desirable properties of bitumen used for constructing pavements.

- Low cost
- Workability
- Strength
- Durability
- Imperviousness
- Minimal sensitivity to aging
- Cohesion
- Adhesion

Bitumen is the most cost-effective binding material which can effectively bind aggregate particles of different sizes together into a strong, stable and impervious compact layer. However, the bitumen may be subjected to an ageing process in two situations, viz., due to longer duration of storage and by oxidation. Bitumen gets hard due to long-term storage, although it can be made soft by heating. However, when bitumen oxidises, i.e., the hydrogen present in the bitumen reacts with the oxygen in atmosphere, it becomes permanently brittle due to change in chemical properties (Shen J et al., 2006) and cracks. Bitumen oxidises due to sunlight raising the temperature of the pavement. The hydrogen ions are lost via water molecules. It is noted that the phenomenon of oxidation of bitumen only takes place in the presence of oxygen (Dickinson 1982). The oxidised bitumen may be physically identified by its change in colour from black to grey.

The following criteria should be considered for selection of a suitable binder of bituminous material for pavement construction.

(a) The bitumen binder should satisfy the criteria of strength and durability for the prescribed loading and temperature variations.

(b) The workability of the bitumen should be studied carefully at temperatures similar to climatic conditions of the in-service pavement. The viscosity of bitumen is largely dependent on its working temperatures during mixing, laying and compaction. Hence, the equipment used for mixing of bitumen with aggregate, the method of storage, transportation, laying and compaction of a bituminous layer should be decided in advance.

(c) Susceptibility and durability of bitumen under prevailing environmental conditions should be evaluated throughout its period of serviceable life. To prevent the bitumen from cracking and undergoing undue deformations, it should be neither brittle nor too soft during cold and hot climates respectively. The bitumen should allow elastic deformations in the range of temperature changes that may occur in different seasons of a year.

(d) Partial coating and stripping off of bitumen film from the surfaces of aggregates due to poor adhesion property should also be studied under prevailing climatic and drainage conditions.

The above requirements were set and standardised by various institutions in different countries based on the intensity of traffic loads, temperature, climatic conditions, materials availability, design criteria and methods of construction. The suitability of bitumen for road construction as per design criteria should be evaluated by suitable tests.

7.2 TESTS ON BITUMINOUS MATERIALS

Materials which are to be used as binders in constructing pavements should have desirable properties such as high strength, stability and durability to withstand repeated wheel loads even under adverse environment and climatic conditions. To characterise bitumen before using for construction, it is necessary to evaluate its strength under different test conditions in the laboratory. Temperatures maintained for conducting the different tests in the laboratory and the relevant working range of temperature of pavements are presented in Fig. 7.1.

7.2.1 Rolling Thin Film Oven (RTFO) Test

Simulation of hardening of bitumen that occurs during mixing operations was studied by the California State Highways Department in 1963. The test developed by them was called the rolling thin film oven (RTFO) test (*Handbook of Highway Engineering*, Baker R F, Byrd L G and Mickle D G, 1975; Dickinson, 1984). Later, with minor modifications to suit their specific needs, this test was used by many agencies around the globe and also adopted by IS: 73, since 2006.

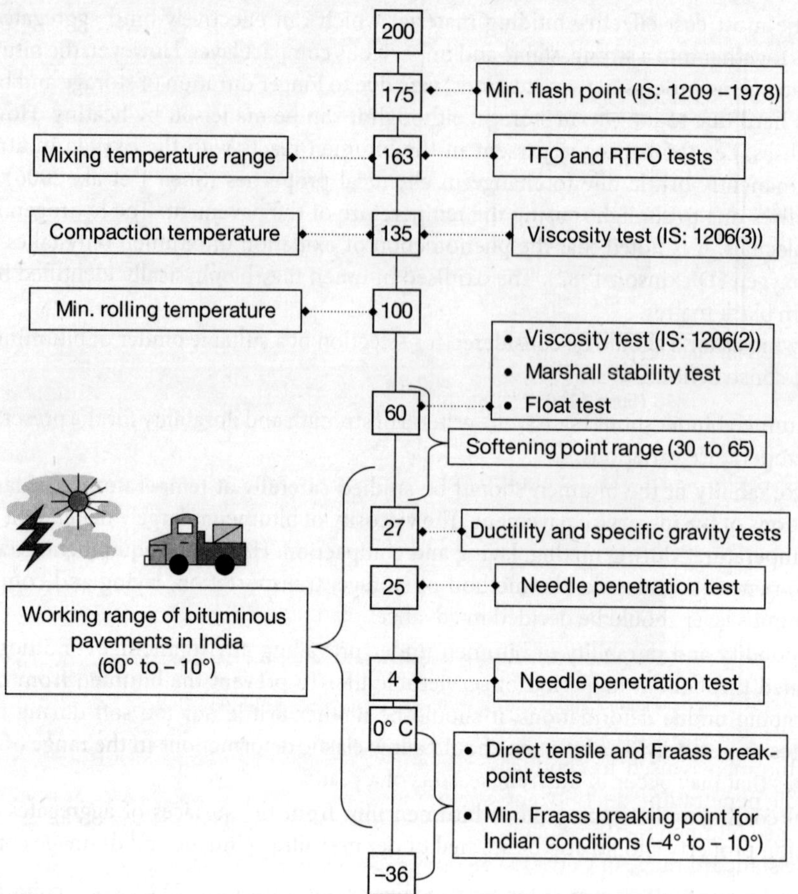

Figure 7.1 Working temperatures (°C) with bitumen and bituminous pavement in India.

The paving binder is subjected to high temperature and exposed to blasts of hot air thus simulating accelerated aging process.

In this test, a bitumen sample of 35g is filled in a standard glass bottle and suitably fixed onto a rotating rack (Fig. 7.2(a) and (b)). The rack can accommodate up to 8 glass bottles and it rotates about its centre in a vertical plane at a predetermined speed of 15 rpm. The bottle filled with bitumen rotates about its horizontal axis while the rack rotates in vertical plane. The rotating rack system is assembled in an oven which is maintained at a constant temperature of 163°C for a duration of 85 minutes (TAI, 1983). While the bottle is rotating, the liquid bitumen also rolls inside the bottle and is exposed to heat. To accelerate the hardening, hot air is blown into the bottle by means of an air jet (@4000 ml/minute) after each rotation of the rack. After the test, the residue bitumen from the glass bottle is collected by scratching the inside surface (about 90%) and is measured to the nearest value of 0.001 g.

After the test, the percentage loss in weight of the bitumen is determined, as below:

$$\text{Mass loss (\%)} = \frac{\text{Original mass} - \text{Aged mass obtained after RTFO test}}{\text{Original mass}} \times 100 \quad (7.1)$$

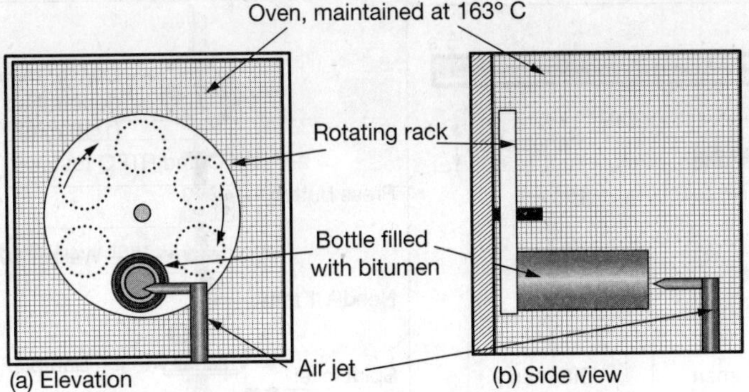

Figure 7.2(a) Rolling thin film oven test (schematic).

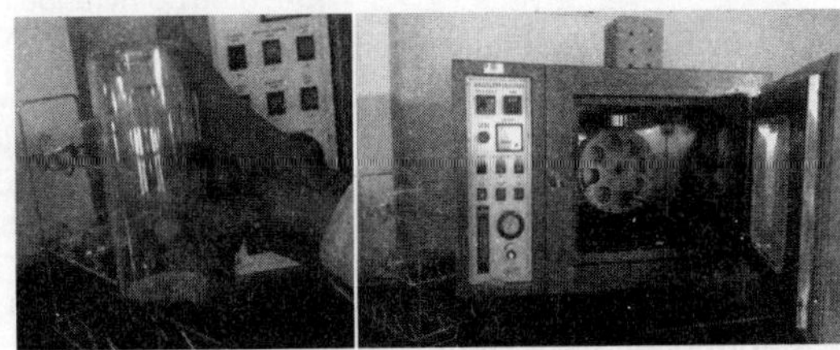

Figure 7.2(b) Rolling thin film oven and the glass bottle.

The residual bitumen is used to conduct further tests for determination of change in the values of viscosity, needle penetration, ductility and softening point. The original bitumen-tested values and the test values of aged bitumen (i.e., residual bitumen obtained after conducting RTFO test) are finally compared with the standard range of values, as recommended in IS: 73.

7.2.2 Needle Penetration Test

Bowen (1889) designed a bitumen test for determining the consistency range or grade of bitumen based on the penetration of a standard needle under standard test conditions. In this test, a standard needle of 1 mm diameter, bearing 100 g of weight is allowed to penetrate into a bitumen sample at 25°C for 5 seconds (Figure 7.3(a)). The apparatus used for the test is known as the needle penetrometer (Figure 7.3(b)). The needle penetration is measured using a dial gauge having a least count of 0.1 mm (IS: 1203-1978 and AASHTO T 49).

The tested bitumen is graded according to the needle penetration value. For example, if a bitumen sample is graded as 30/40, it means that the penetration of the needle is in the range of 30 to 40 times of one-tenth of a millimeter (i.e. 3 to 4 mm). If bitumen registers a higher penetration value, it is classified as soft-grade bitumen and vice versa. In India, according to IS:73-1992, the bitumen used for paving was classified into six grades viz. S-35, S-45, S-55, S-65, S-90 and S-200 having penetration grade values 30/40, 40/50, 50/60, 60/70, 80/100 and 175/225 respectively (MORTH 2001). These grades are known as penetration grades.

For further characterisation of bitumen under cold conditions, the penetration test can be conducted on a bitumen sample maintained at 4°C. A needle of standard size with a 200 g weight is applied on the sample for 60 seconds. To study the behaviour of bitumen with respect to temperature, an index parameter termed as the penetration ratio (PR) is determined (IS: 1203-1978; 2002). The equation used for calculation of PR is given below.

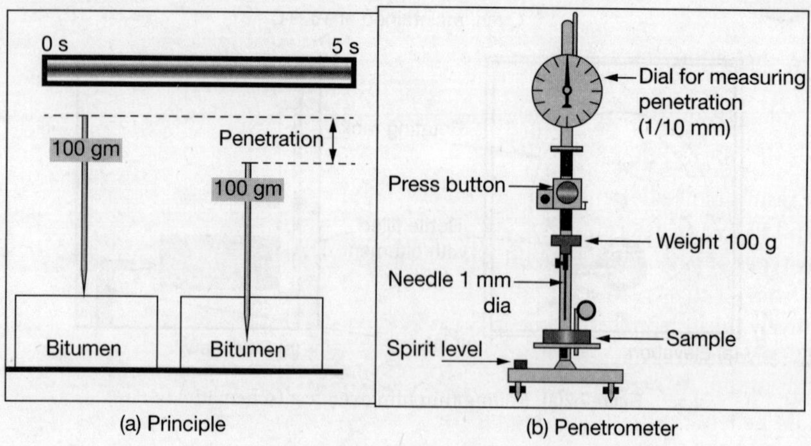

Figure 7.3 Bitumen penetration test.

$$\text{PR} = \frac{\text{Penetration value found at 4°C, under 200 g of weight, loaded for 1 minute}}{\text{Penetration value found at 25°C, under 100 g of weight, loaded for 5 seconds}} \times 100 \quad (7.2)$$

According to IS: 73-1992, the minimum value of PR required for most grades of paving bitumen is 35.

A linear relationship was found between temperature and logarithm of the penetration value (Fig. 7.4) (Pfeiffer and Doormaal, 1936). This was used to determine the temperature susceptibility of bitumen. The relationship may be written in the following form.

$$\text{Log}\,(Pen) = A\,T + I \quad (7.3)$$

where,

Log (Pen) = Logarithmic value of penetration of bitumen
T = Temperature
A = Slope of the line or temperature susceptibility of bitumen and
I = Intercept

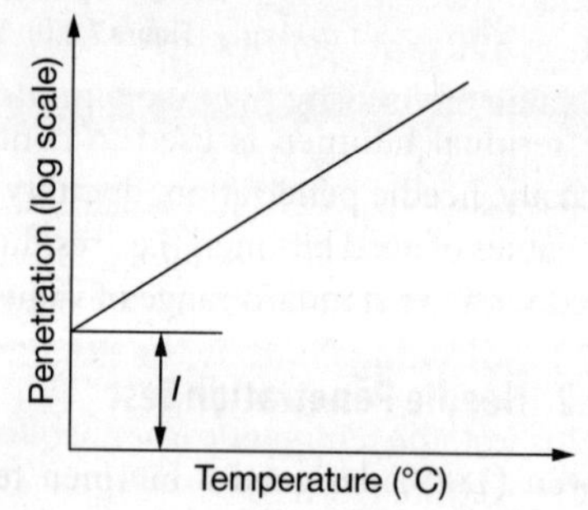

Figure 7.4 Relationship between temperature and log (pen.) of bitumen.

Pfeiffer and Doormaal (1936) initiated a concept called the penetration index (PI) based on the following relationship between the logarithmic value of penetration and the softening point value of bitumen.

$$\text{PI} = \frac{500\log(\text{Pen}_{25}) + 20\,\text{T}_{\text{R\&B}} - 1951.55}{\text{T}_{\text{R\&B}} - 50\log(\text{Pen}_{25}) + 120.15} \quad (7.4)$$

where,

Pen_{25} = Bitumen penetration in 1/10 mm at 25°C and
$T_{\text{R\&B}}$ = Bitumen softening point (°C) by ring-and-ball test

This concept was further developed by Heukelom (1973). The relationship used for determining temperature susceptibility of bitumen is as follows:

$$A = \frac{1}{50}\left[\frac{20 - PI}{10 + PI}\right] \quad (7.5a)$$

where,

PI = penetration index.

$$A = \frac{\log_{10} (Pen) \text{ at } T_1 - \log_{10} (Pen) \text{ at } T_2}{T_1 - T_2} \quad (7.5b)$$

Based on the results of needle penetration tests conducted at two different temperatures (T_1 and T_2), the values of A and PI can be determined by using Eqs. (7.4), (7.5(a) and (b)).

For many types of paving bitumen, the slope (A) value ranges between 0.015 and 0.06 and the value of PI ranges from –1 to +1. However, it has been observed that bitumen having PI value greater than zero are relatively unsusceptible to temperature changes and those having PI values less than zero are susceptible to temperature changes. The PI values for the most commonly used bitumen range from –0.5 to +0.5 (Ullidtz and Peattie, 1980).

The needle penetration test is an empirical test and suffers from many drawbacks (The Asphalt Institute, 1983; MORTH, 2001). Bitumen obtained from different sources and manufacturing processes and having the same penetration grade at 25°C may have a wide range of viscosity from 500 to 3000 poises at a standard temperature of 60°C (Discussions, IRC, 1994). Bitumen obtained from different sources may also contain significant quantity of wax. Thus, the test results may vary. Only 2 to 3% wax can be tolerated in paving bitumen (Nair, 1977). Therefore, to determine the true consistency of bitumen, in addition to the needle penetration test, viscosity and other relevant tests should be conducted.

The change in characterisation of bitumen is inevitable as field temperatures seldom remain constant at 25°C. Also, any slight variation in test/field temperature will greatly influence the bitumen behaviour. Hence, different agencies such as HRB, FHWA, Asphalt Institute, AASHTO and ASTM insist on using the value of viscosity to grade bituminous binders rather than grading them by the needle penetration value. In the third revision of IS:73-2006, the paving bitumen is classified into four types of grades based on viscosity, viz., VG 10, VG 20, VG 30 and VG 40 having standard needle penetration test values of 80/100, 60/70, 50/60, and 30/40, respectively.

7.2.3 Softening Point by the Rind-and-Ball Method

In this test, the temperature of the bitumen is measured at the point when it becomes soft. As a standard test procedure (IS: 1205-1978; 2002), bitumen sample is heated until it become completely fluid. This fluid is filled into two brass rings of 10mm diameter. Two steel spears of weight 3.5 g and diameter 9.5 mm are placed over the bitumen filled brass rings. The brass rings, each with a spear, are placed on the top plate of a frame (Fig. 7.5). The frame is immersed in a distilled water bath which is maintained initially at 5°C for 15 minutes. The height of water above the steel spears should be 50 mm and the vertical distance between the two plates of the frame should be 25 mm. The water is heated at a uniform rate of 5°C per minute throughout the length of the test. When the steel spears sink through the rings along certain portions of melted bitumen and just touch the bottom plate of the frame, the water temperatures are recorded. The softening point of the bitumen tested is reported as the average temperature of both spears (i.e., spears 1 and 2). If the expected softening point of bitumen is above 80°C, *glycerin* is used instead of *water* at an initial bath temperature of 35°C.

If the softening point (temperature) of bitumen is high, it indicates a harder grade of bitumen and if it is low, it indicates softer grades of bitumen. The value of the softening point of bitumen is very useful for determining the penetration index, temperature susceptibility, temperature shear resistance and other visco-elastic properties of bitumen.

According to IS:73-2006, the minimum temperature of the softening point for paving bitumen of grades VG 10, VG 20, VG 30 and VG 40 should be 40, 45, 47 and 50°C respectively.

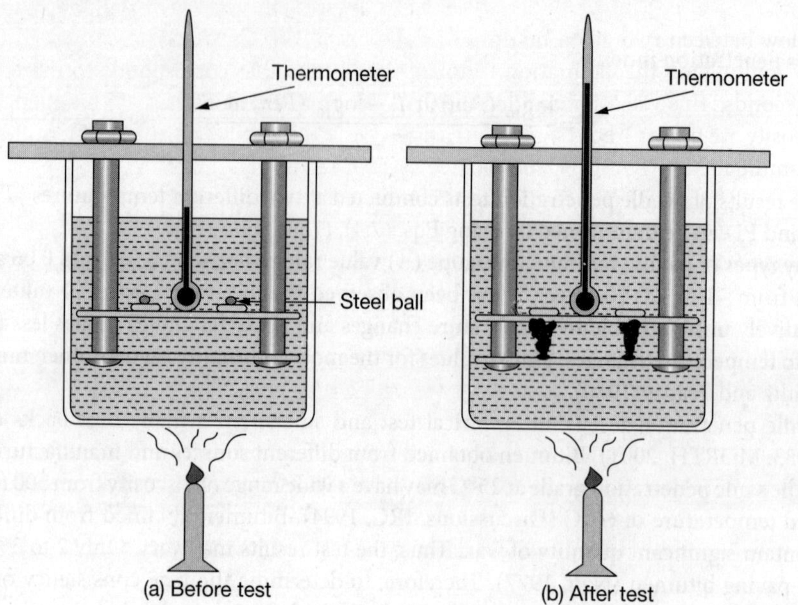

Figure 7.5 Bitumen softening point test.

The softening point of different types of bitumen may vary from 25 to 120°C. From literature, it is found that most bitumen types have a needle penetration value of 800 at the softening point.

7.2.4 Viscosity Test

Viscosity is defined as the internal resistance offered by the molecules of a substance as it flows with time.

To understand the behaviour of bitumen under different temperatures and rates of loading, it is essential, as mentioned before, to determine its viscosity by a more reliable method than the needle penetration test with its penetration values.

As we know, resistance to flow is the measure of viscosity of any liquid. By warming the bitumen, it gets soft. Further heating turns it to liquid. Based on this fundamental principle, the viscous property of bitumen is determined at various working temperatures (Fig. 7.6).

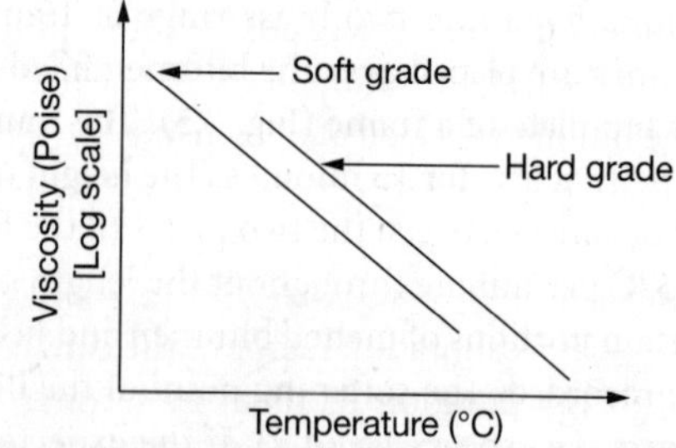

Figure 7.6 Typical relationships between temperature and viscosity of bitumen.

There are many methods available for determining the viscosity of bituminous materials used for construction of pavements. Depending on the principle used to determine viscosity, different apparatus are used, namely, (i) the glass tube capillary viscometer, (ii) the orifice or efflux viscometer, (iii) the rotational viscometer and (iv) the sliding plate viscometer.

Glass tube capillary viscometer

Liquid bitumen is filled in one limb of a pre-calibrated U-shaped glass tube to a pre-determined level and allowed to flow further into its other limb (Fig. 7.7). The time (in seconds) required for the liquid bitumen to flow between two given markings on the limb of the U-tube is recorded. The viscosity is calculated by multiplying the calibration factor of the U-tube that corresponds to the recorded time duration in seconds. This viscosity is called the dynamic or absolute viscosity of bitumen. The units of dynamic viscosity are poise (1dyne.s/cm^2) or Pascal second (N s/m^2). The unit is determined by a standard test recommended by AASHTO, designated as T202 (AASHTO, 2001).

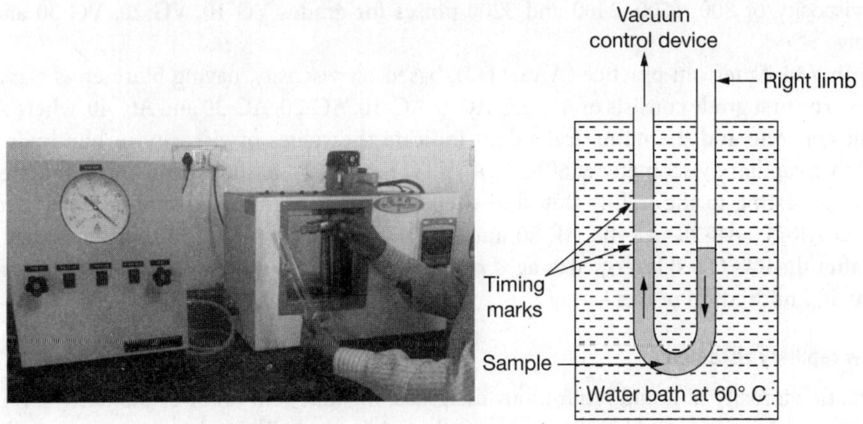

Figure 7.7 The Asphalt Institute vacuum viscometer and its U-Tube manometer.

These tests evaluate the consistency of the bitumen at 135°C and 60°C, which represent the temperature of bitumen during construction and service life of pavement respectively (TAI 1983).

Based upon geometrical differences, there are different types of capillary glass tube viscometers. They are described below.

Vacuum tube capillary viscometer

Pre-heated bitumen (in fluid state) is allowed to flow through a limb of a calibrated capillary glass U-tube which is kept in a water bath at 60°C. Timing marks are placed on one of the limbs of the U-tube. A certain amount of vacuum is created at the end of the U-tube limb which carries the timing mark and the time duration (in seconds) for the bitumen to flow between the timing marks is recorded (Fig. 7.7). During the test, the bitumen should be maintained at a constant temperature of 60°C. Absolute viscosity of the bitumen is calculated by using the following equation.

$$\text{Absolute viscosity (in poise)} = \text{CF} \times t \tag{7.6}$$

where,

CF = calibration factor of the capillary glass tube

t = duration of time, in seconds, required for the bitumen to flow in between the timing markings of the capillary glass tube.

From the following relationship, the viscosity in poises can be converted to Pascal seconds.

$$1 \text{ Pascal second} = 10 \text{ poises} \tag{7.7}$$

As a matter of interest, at 20°C, the viscosity of water is 10^{-2} (in poise) whereas for different types of bituminous materials it ranges between 10 and 10^{12}.

To determine the absolute viscosity of tar and bituminous materials, the Bureau of Indian Standards has recommended the following types of capillary viscometers. These viscometers can test fluids having viscosity in the range of 42 to 200 000 poises (IS: 1206 (Part 2)-1978; 2002).

- Cannon–Manning vacuums capillary viscometer
- Asphalt Institute vacuum capillary viscometer
- Modified Koppers vacuum capillary viscometer

Details of the above capillary-type viscometers are given in IS: 1206 (Part 2)-1978; 2002.

According to Indian practice (IS:73-2006), at 60°C, the paving bitumen should have a minimum absolute viscosity of 800, 1600, 2400 and 3200 poises for grades VG 10, VG 20, VG 30 and VG 40, respectively.

According to American practice (AASHTO), based on viscosity, paving bitumen is classified into two grades. The first grade consists of AC-2.5, AC-5, AC-10, AC-20, AC-30 and AC-40, where AC stands for asphalt concrete and the numerical values indicate the values of viscosity in hundreds of poises measured by a capillary viscometer at 60°C (AASHTO test T202). In the second grade, bitumen residue obtained after testing in the rolling thin film oven (RTFO) is used to measure viscosity at 60°C and classified as AR-10, AR-20, AR-40, AR-80 and AR-160 (AASHTO test T240). The residual bitumen obtained after the RTFO test is termed as aged residue (AR) and the numerical values indicate the values of viscosity in hundreds of poises.

Gravity flow capillary viscometer

The kinematic viscosity of liquid bituminous materials, in stoke (cm^2/s) or centistoke (or 1/100th of a stoke = $0.01 cm^2/s$), is measured by using a gravity flow viscometer. The unit stoke is named after George Gabriel Stokes.

There is a standard relationship to calculate the kinematic viscosity of any liquid. It uses the values of absolute viscosity and density of the material, measured at the same temperature. The relationship is given below.

$$\text{Kinematic viscosity} = \frac{\text{Absolute viscosity}}{\left(\begin{array}{l}\text{Density of the material, at the same temperature}\\ \text{as measured for absolute viscosity}\end{array}\right)} \tag{7.8}$$

To determine the kinematic viscosity of liquid bitumen, there is no need to apply vacuum. This is because at 135°C, the bitumen is less viscous and free to flow by gravity. Hence, there is no need to apply vacuum at the other end of the capillary tube.

According to the AASHTO test procedure (T201), liquid bitumen at a pre-determined height is allowed to flow by gravity through a calibrated capillary tube, of standard size and shape, kept in an oil bath maintained at 135°C. The time taken for the liquid to flow in between the two timing marks of the capillary tube is recorded. The kinematic viscosity of the bitumen in stokes is calculated by multiplying the calibration factor of the capillary tube with the duration of time (in seconds) taken for the liquid bitumen to flow in between the pre-determined timing marks.

To determine the kinematic viscosity of paving grade cutbacks of bituminous materials and distillation residues of cutbacks, the Bureau of Indian Standards recommends the following capillary type viscometers. These viscometers can be used to test materials having viscosity in the range of 30 to 100 000 centistokes (IS: 1206 (Part 3)-1978; 2002).

- Cannon–Fenske viscometer for opaque liquids
- BS U-tube modified reverse flow viscometer

Details of the above capillary type viscometers are given in IS: 1206 (Part 3)-1978, 2002.

According to Indian practice (IS: 73-2006; 2006), at 135°C, paving bituminous materials should have a minimum kinematic viscosity of 250, 300, 350 and 400 centistokes for grades VG 10, VG 20, VG 30 and VG 40 respectively.

Orifice or efflux viscometer

A specified volume of liquid bitumen maintained at standard temperature is allowed to flow by gravity through an orifice of standard diameter. The time in seconds is recorded. Based on this principle, different types of equipment are available for determination of viscosity of bitumen. Following are some of the apparatus widely used to measure the viscosity of bitumen, tar and bitumen cutback.

- Tar viscometer [10 mm dia. of orifice; 50 ml of volume; at 35, 40, 45 or 50°C (IS:9026-1985) and for cutbacks: 4 mm dia. of orifice; 50 ml of volume (IS:1206 (Part I)-1978)]
- Saybolt Furol viscometer [3 mm dia. of orifice; 60 ml of volume; at 25 or 50°C (ASTM D88 & E102)] (Fig. 7.8).

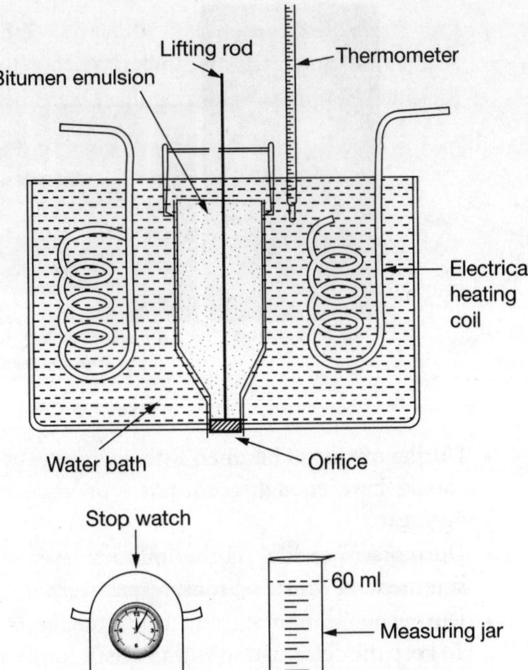

Figure 7.8(a) Schematic set-up diagram of a Saybolt Furol viscometer.

Figure 7.8(b) Saybolt Furol viscometer set-up.

Using orifice viscometers, the duration of time measured does not provide any scientific relationship to understand the visco-elastic behaviour of bituminous materials. However, the tests are very simple and the values (in seconds) obtained by the orifice or efflux viscometers can be correlated to test values obtained using other standard viscometers.

Rotational viscometer

A spindle is vertically inserted into a chamber filled with bitumen liquid and allowed to rotate at a given rate. The torque required to maintain the specified rate of rotations of the spindle per minute at a standard test temperature gives a measure of the viscosity. This kind of viscometer is gaining importance because it is simple to use and the test can also be carried out on modified binders at a wide range of temperature. The Brookfield viscometer (Fig. 7.9) and the portable Haake rotational viscometer (Kandhal, 2006) are good examples of this kind of viscometer.

7.2.5 Ductility Test (IS: 1208-1978)

A thin film of bitumen coated over and in between aggregates undergoes elastic deformation under repetitive axle loads. The elastic homogeneity of the bituminous mix must be maintained at all working temperatures and stresses. Ductility is an essential property of bitumen in view of the following requirements.

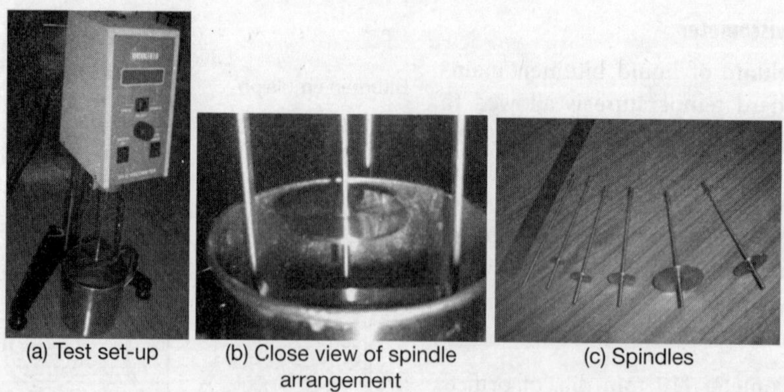

(a) Test set-up (b) Close view of spindle arrangement (c) Spindles

Figure 7.9 Rotational viscometer.

- During mixing of bitumen with aggregates at a given standard temperature, the molten bitumen should have enough consistency or viscosity to evenly spread over and coat the surface of aggregates.
- During service life, the bituminous layer should become neither too soft nor brittle during summer and winter seasons, respectively.
- During application of traffic loads, the bituminous pavement undergoes repeated deformations. To keep the deformation within elastic limits or allowable limits in its design life, the bitumen film coated around the surface of the aggregates should have enough ductility.

If the bitumen is obtained from different sources or refineries, it may satisfy the minimum criteria of penetration value but may fail to have the required ductile property. Hence, the property of stretchability (or ductility) of bituminous materials must be ascertained during selection of suitable binder.

In the ductility test, bitumen is heated to become completely fluid and poured into a briquette mould of standard size (Fig. 7.10). It is allowed to cool to room temperature. The minimum cross-sectional area of the bitumen filled in the briquette mould is 1 sq. cm. The mould with bitumen is suitably fixed to two jaws in a water bath which is maintained at 27°C (Fig. 7.11(a)). One side of the briquette mould is attached to a fixed jaw and the other end is attached to a movable jaw which is allowed to stretch away in the horizontal direction at a constant rate of 5 cm per minute using a suitable mechanical pulling system. During the process, the bitumen stretches to become like a thin thread and at a certain point breaks on further pulling. The length of the bitumen thread just before breaking is measured on a scale (Fig. 7.11(b)).

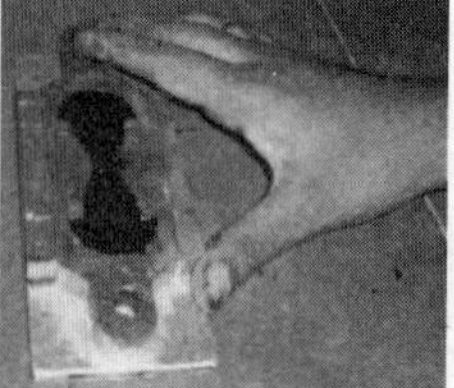

Briquette mould filled with bitumen

Figure 7.10 Ductility test in progress.

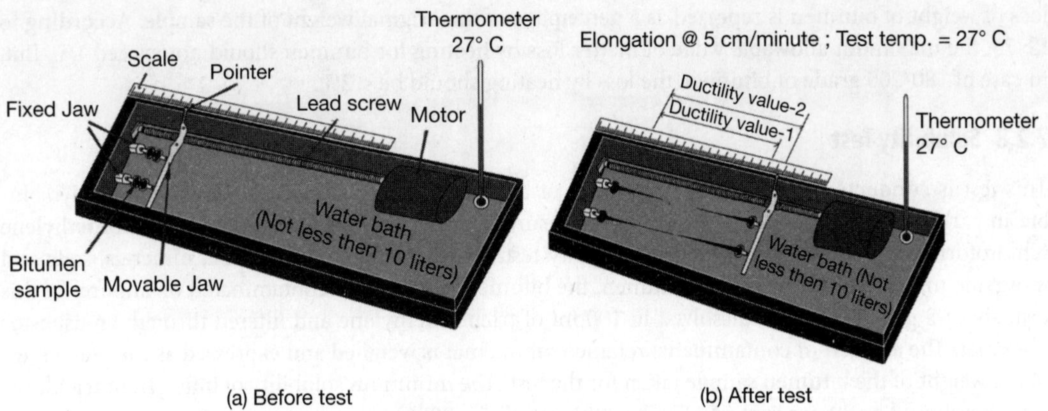

Figure 7.11 Bitumen ductility test.

According to Indian practice (IS: 73-2006), the ductility test should be conducted at 25°C on a residual bitumen obtained after an RTFO test. The minimum ductility value for the bitumen grades of VG 10, VG 20, VG 30 and VG 40 are 75, 50, 40 and 25 cm length respectively. A pavement constructed with bitumen having a higher ductility value will serve longer. However, the ductility property of bitumen may reduce with age and under different service conditions.

The bitumen ductility test is also conducted at a lower temperature. A temperature of 4°C should be maintained (this is done using a water bath) and the rate of pulling is 10 mm/minute (IS: 1208-1978; 2002).

7.2.6 Flash and Fire Point Tests (IS:1209-1978; 2002)

Flash and fire point tests are conducted on bitumen to determine at what temperatures the surface of heated bitumen can flash and be ignited. The test results are useful to prevent fire hazards during heating, mixing and compaction of bituminous layers.

The flash point test may be conducted using Pensky Marten's apparatus (IS: 1209-1978; 2002). To determine the flash point (temperature) of bitumen, a suitable quantity of bitumen is taken in a cup and heated at a rate of between 5 to 6°C per minute. The molten bitumen is stirred at a rate of 60 rpm. A thermometer is kept inserted in the molten bitumen to take constant temperature measurements. During the process of heating, a small test flame is frequently brought over the surface of the molten bitumen. The temperature is recorded at the earliest spark or flash. Stirring is temporarily stopped while applying the test flame. This minimum temperature of the earliest flash is reported as the flash point of the bitumen.

For the fire point test, the above test is continued by further heating of the bitumen until the bitumen surface ignites by itself at least for 5 seconds. At this exact point of ignition, the temperature on the thermometer is reported as the fire point of the bitumen. According to the Indian Standard Bureau (IS: 73-2006), the minimum requirement of flash point for all types of paving grades of bitumen is 220°C.

7.2.7 Loss on Heating Test

Loss of volatile matter of bitumen (in percentage by weight) is determined in this test. About 50 g of bitumen is weighed in to a container and the container is placed in an oven for 5 hours with the temperature maintained at 163°C (IS: 1212-1978). After the test, the weight of bitumen is recorded and the loss of weight of bitumen is reported as a percentage of the original weight of the sample. According to IS: 73, the maximum allowable value of matter loss by heating for bitumen should not exceed 1%. But, in case of 180/200 grade of bitumen, the loss by heating should be < 2%.

7.2.8 Solubility Test

This test is conducted to determine the purity of bitumen. As we know, bitumen is completely soluble in carbon disulphide (CS_2). As a safety measure, instead of carbon disulphide, trichloroethylene (chloroform) is used as solvent for the solubility test. If any insoluble matter, viz., minerals, salts and inorganic matter are present in the bitumen, the bitumen is said to be contaminated or impure. In this test, about 2 g of bitumen is dissolved in 100 ml of trichloroethylene and filtered through an asbestos fibre mat. The amount of contaminants retained on the mat is weighed and expressed as the percentage of the weight of the bitumen sample taken for the test. The minimum solubility of bitumen in trichloroethylene should be 99 per cent (AASHTO 2001 and IS: 73-2006).

7.2.9 Specific Gravity Test

The specific gravity of bituminous material may be determined by using a pycnometer. This is otherwise known as the balance method (IS: 1202; 1978). The knowledge of the specific gravity of bitumen is useful for calculating the volume of bitumen required in the density–void relationship of the bitumen–aggregate mixture.

$$\left(\text{Volume} = \frac{\text{Weight}}{\text{Specific gravity}} \right)$$

Normally, the specific gravity of paving bitumen and tar are in the range of 0.97 to 1.02 and 1.10 to 1.25 respectively.

7.3 OTHER BINDERS

Bitumen is available as a solid or semi-solid material. But to make it workable, it is necessary to make it less viscous by heating, adding cutbacks and blending with petroleum solvents or making emulsions by mixing with water. Ordinary bituminous binders fail to perform satisfactorily under increased over-loaded truck traffic, in extreme temperature conditions and intensive abrasive conditions (viz. roundabouts, junctions and at signal intersections). To improve the performance of these traditional bituminous materials, some compounds are admixed to improve the strength and durability of the mix constituents. Various types of bituminous materials and their modifications obtained by adding different type of additives are presented in the following sections. Here, the word 'bituminous binder' is used for both modified and unmodified bituminous materials.

7.3.1 Bitumen Cutbacks

The viscosity of bitumen is reduced suitably by mixing or blending it with a volatile diluent or petroleum distillate. Based upon the rate of volatility, the bitumen cutbacks are divided into three types, viz.,

(i) Rapid curing (RC70, RC250, RC 800 and RC3000),

(ii) Medium curing (MC30, MC70, MC250, MC800 and MC3000) and

(iii) Slow curing (SC70, SC250, SC800 and SC3000) cutbacks.

The above designated numerical values indicate *initial kinematic viscosity* (IS: 217-1988; 1993). The diluent used to prepare RC (highly volatile, low viscosity), is gasoline; for MC (medium viscosity), it is kerosene and for SC (lower volatility; highly viscous), gas oil is used. Cutbacks are found to be costlier than ordinary bitumen because extra cost is involved in the use of petroleum solvents even though they do not play a vital role in making the bitumen stronger. Moreover, use of cutbacks with a little extra amount of petroleum solvents may pose problems related to safety and environmental pollution.

For further details regarding specification for cutback bitumen please refer to IS: 217-1988; 1993.

7.3.2 Bitumen Emulsions

A bitumen emulsion is a liquid blend of water and emulsifier (an additive) containing a suspension of dispersed minute particles of bitumen (i.e. oil in water). A colloidal mill may be used to prepare the bitumen emulsion. Based on why and where the emulsion is used, the amount of bitumen to be mixed with water may range between 40 to 70%. A suitable emulsifier (depending on the ionic nature of the mix) may be added to the bitumen–water mix to keep the minute particles of bitumen in the water away from each other and in a dispersed form. Emulsified bitumen is used in cold mix preparations for paving bituminous layers in difficult and inaccessible conditions.

Based upon the type of surface charge, bitumen emulsions are classified into two types. They are (i) anionic bitumen emulsion (particles of bitumen are electro-negatively charged) and (ii) cationic bituminous emulsion (particles of bitumen are electro-positively charged) (Fig. 7.12(a)). Cationic emulsions are widely used in today's construction works.

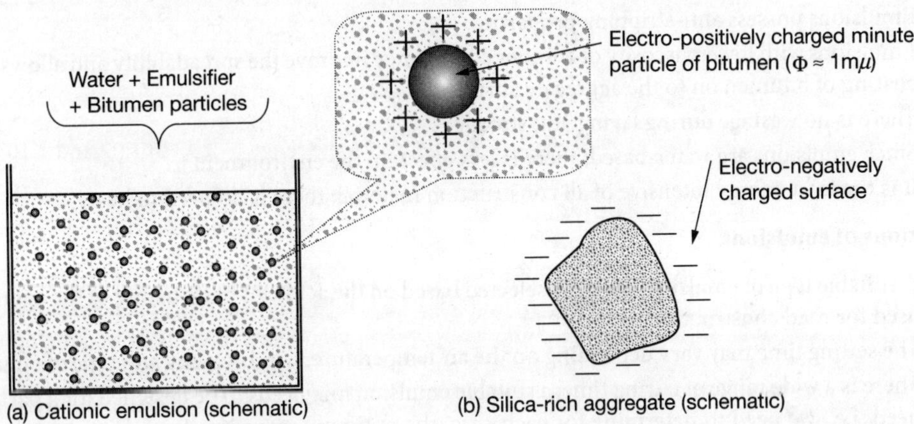

Figure 7.12 Surface charges on particles of bitumen in cationic emulsion and silica-rich aggregate.

The type of emulsion should be selected based on the mineral composition of aggregates used for construction. For example, if the aggregate used for construction is rich in silica (SiO_2), then the surface of the aggregate possesses strong electro-negative charge (Fig. 7.12(b)). In this case, cationic emulsions are found to be more suitable since the minute particles of bitumen in the cationic emulsion possess electro-positive charge. To become electro-statically stable, the positively charged particles of bitumen are strongly attracted to the negatively charged aggregates surface. Hence, the compacted layer becomes strong and stable.

Similar to bitumen cutbacks, bitumen emulsions are also classified into three types based on setting time and purpose of use. They are (i) rapid setting (RS), (ii) medium setting (MS) and (iii) slow setting (SS) emulsions. The word 'setting', here, means separation of the emulsion. In other words, when the water present in the emulsion evaporates, the minute bitumen present in the emulsion coats the surface of aggregates. Later, curing of the emulsion takes place by which the compacted layer of the emulsion–aggregate mixture hardens and attains strength. In this context, rapid setting emulsion for example, means that the emulsion sets and cures rapidly. Following are the factors affecting setting and curing of emulsion mixes.

- Percentage of water absorption of aggregates (depends on the surface texture)
- Dampness of the aggregates
- Amount of dust on aggregates
- Gradation of aggregates
- Mineral composition and intensity of charge on the surface of aggregates
- Ingredients of emulsion and quantity used
- Meteorological conditions, viz., weather, climate, temperature, humidity, wind velocity and altitude.
- Drainage conditions prevailing around the construction site.

The uses and advantages of emulsions for the construction of pavements (Discussions, IRC 1994) are enumerated below.

- The strength properties of bitumen are preserved in emulsion mixes since emulsions do not need hot mixing. Generally, cold mixing or just slightly warming the ingredients of the emulsion mix are done for construction of pavements.
- In cold and damp conditions, emulsions are also used as slurry seal, pre-mix carpet, or seal coat for repair, soil stabilisation and other works.
- Emulsions possess anti-stripping properties.
- Emulsions with lower viscosity or thinner consistency improve the spreadability and allows better coating of bitumen on to the aggregate surfaces.
- There is no wastage during laying and storage.
- Since emulsions are water-based, there is less effect on the environment
- It is the least energy intensive of all construction material; there is no need to heat.

Limitations of emulsions

- A suitable type of emulsion has to be selected based on the acidic or basic nature of the aggregates used for road construction.
- The setting time may vary depending on the air temperature, wind velocity and type of emulsion.
- There is a wide range of curing time; a suitable emulsion mix needs to be designed for a particular need, i.e., we need to determine for each case, the optimum quantity of emulsion based on the zone, type of emulsion, mix grade and desired mechanical properties.
- It has a restricted storage time (ranging from a few days to six months).

For fine-grained aggregates, the surface area is large and the time required to mix an emulsion for them is more. Hence, slow-setting emulsions are preferred. It gives ample time for blending the mix uniformly before it sets.

In general, bituminous pavements laid using emulsions are considered inferior compared to hot-mix bituminous pavements that cater to heavy traffic loads.

7.3.3 Tar

Tar is a black or brown to black, non-crystalline material in the viscous state having binding properties. It is obtained from the destructive distillation of organic materials (viz., coal, petroleum, oil, wood, shale, peat etc.) in the absence of air at about 1000°C. Tar is completely soluble in carbon tetrachloride. It is more susceptible to change in temperature than bitumen because it has of more volatile oils than bitumen (McGovern and Alderton 1972). Therefore, tar quickly hardens during the summer and become brittle during winter. Generally, tar is used for surface dressing on the wearing course since it possesses a comparatively good adhesive property in damp conditions.

Paving tar is classified into two types based on the ratio of the pitch/ anthracene (the materials obtained as a result of the distillation process of tar), viz., Types A and B. Again these two types of tars are further classified into five grades (viz. RT1, RT2, RT3, RT4 and RT5) which are intended to be used for specific paving purposes. The details of the requirements of these paving tars are given in IS: 215-1995; 2000.

7.3.4 Tar–bitumen Mixture

Blending tar with a suitable quantity of bitumen may result in a mixture which has good binding properties and lower susceptibility to temperature changes. However, the blend of tar–bitumen may pose problems of incompatibility since tar and bitumen has different chemical compositions. Thus, the blend of tar–bitumen may be used for pavement construction but with proper care and suitable specifications.

7.3.5 Sulphur-modified Bitumen

About 15 to 18 per cent of sulphur is added as a chemical modifier to bitumen (MORTH 2001). This modified bitumen provides better workability and higher elastic modulus of bituminous mixes and reduces the demand for bitumen (Deme, 1974; Fromm and Kennepohl, 1979). A few properties of sulphur are given below (Love 1979).

- The specific gravity of solid sulphur is about 2.0
- Three forms of sulphur are generally used for paving works. They are orthorhombic, polymeric and monoclinic sulphur. However, polymeric sulphur mixed with suitable plasticisers, termed as Sulfex, has the most desirable characteristics for use as pavement binders.
- For best workability and strength, the mixing temperature of sulphur should be in the range of 120 to 150°C. In this range, the sulphur reacts slowly and forms polysulphide resins that dissolve in un-reacted sulphur and impart highly desirable mechanical properties to the final mix.
- Sulphur becomes viscous at higher temperatures and maintains lowest viscosity at 150°C.
- When bitumen–sulphur mix is heated to higher temperatures, a dehydrogenation reaction occurs and sulphur forms hydrogen sulphide (H_2S) and sulphur dioxide (SO_2) which pose health hazards.
- However, emission of hydrogen sulphide can be avoided if the mixing temperature is controlled.

7.3.6 Lime-modified Bitumen

Hydrated lime is also used as a modifier in hot mix bituminous concrete. The advantages of using lime in hot mix bituminous concrete are enumerated below (Witczak and Bari, 2004).

- It is an excellent anti-stripping agent
- It reduces bitumen's susceptibility to change in temperature and moisture
- It enhances binder viscosity and adhesion
- It increases stiffness of the mix, thus making it resistant to shear and the formation of ruts
- It retards the oxidation process
- It improves durability of the compacted bituminous layer.

7.3.7 Polymer-modified Bitumen

Polymers are mixed with bitumen to improve the strength of bitumen so that the service life of bituminous pavements can be extended (IRC: SP: 53-2002; Taylor, 1991; Kumar et al., 2006; www.bpbitumen.com 2006).

Generally, there are two types of synthetic thermoplastic polymers which are widely used for pavement construction, plastomers and elastomers. The word 'thermoplastic' means that the polymers melt when heated and set to harden when they are cooled. Plastomers induce plasticity or viscosity or stiffness to bitumen. They consist of polyethylene, ethylene vinyl acetate (EVA), ethylene butyl acrylate, linear low-density polyethylene (LLDPE) (Kumar et al., 2006) etc. Elastomers induce elasticity as well as stiffness properties to bitumen. They consist of styrene–isoprene–styrene, styrene–butadiene–styrene (SBS), ethylene/propylene, styrene–butadiene, poly-butadiene, some types of rubber, etc. In India, three types of polymers are commonly available, viz., polyethylene, vinyl acetate, and styrene–butadiene–styrene. These three types of polymers are used to prepare modified bitumen as well as emulsions (Discussions, IRC 1994).

Different grades of bitumen may be obtained by changing their composition. The Central Road Research Institute (CRRI) has developed high-performance polymer-modified binders by using suitable additives for use in high-temperature regions and construction of heavy duty pavements.

Following are a few advantages of using polymer-modified binders for construction of pavements (IRC: SP: 53-2002; Vasudevan et al., 2003; Gopal Raju and Rengaraju, 2003).

- Reduces permanent deformation
- Increases viscosity and shear resistance
- Increases softening point
- Retards oxidation process
- Improves adhesive and cohesive properties
- Bitumen content can be reduced by around 10% (Vasudevan et al., 2006)
- The ill-effects of waxy bitumen such as susceptibility to temperature changes, bleeding in hot climates and brittleness during cold climates can be reduced by using bitumen blended with EVA, aromatic resin and SBS (Shukla and Jain, 1984).
- The Marshall Stability value, indirect tensile strength and durability of asphalt mixes can be considerably increased (Bose and Jain, 1984).
- The resistance of bituminous concrete against fatigue cracking, temperature cracking and stripping increases (Brule et al., 1998).
- Elastomer binders improve resistance against rutting but affect fatigue life while plastomer binders improve resistance against rutting without compromising the fatigue properties (Maccarrone et al., 1995).
- In surfacings, use of modified bitumen extends the service period by 50 per cent as compared to unmodified bitumen (IRC: SP: 53-2002).

However, polymers are costly; hence cost-effective economic conditions have to be evaluated while considering their use.

7.3.8 Rubber-modified Bitumen

Compounds of rubber materials are obtained either from natural sources or manufactured by using different types of polymers which undergo different processes. The inherent predominant properties of polymers are imparted to rubber materials. The rubbers are also called elastomers since their strain capabilities are very high—even more than hundred per cent in some cases. Following are the uses of rubber-modified bitumen

- Rubber introduces more elastic property to the bitumen.
- Improves ductility under low temperature and high rate of loading.
- Rubberised roads offer adequate skid resistance; therefore, it provides more safety.

- Useful for sealing cracks on roads.
- Improves the fatigue property of bitumen.
- Reduces permanent deformations to a greater extent.

A certain quantity of rubber compounds is mixed with bitumen to impart elastic property to bitumen. About 2 to 5 per cent of these compounds of rubber is suitably mixed with bitumen at a given temperature and laid using usual construction machinery. While mixing, the rubber swells initially and later spreads in to the bitumen (Discussion, IRC 1994).

Based upon the source of rubber, it may be classified as synthetic or belonging to other types (IRC: SP: 53-2002). Synthetic rubber include styrene butadiene rubber latex while latex, unvulcanised rubber powder, vulcanised rubber, ground rubber mix and ground tyre rubber or powder of rubber crumbs obtained from scrap tyres are a few of the other types of rubber used. Bitumen mixed with these types of rubber is termed as *Crumb Rubber Modified Bitumen* (CRMB). Use of natural rubber is preferred to synthetic rubber for pavement construction, as increased quantity of natural rubber in bitumen provides proportional improvement of elastic strain recovery (Oliver, 1981).

For preparation of any rubber-modified bitumen, the following factors have to be considered.

- Type and grade of rubber
- Quantity of rubber to be added to the bitumen
- Method of blending rubber with bitumen at hot or cold temperatures.

For detailed specifications of binder properties and mix design of rubber and polymer modified bitumen, please refer to IRC: SP: 53-2002.

7.4 ENGINEERING PROPERTIES OF BITUMINOUS MATERIALS

How bitumen behaves in pavement should be determined with reference to anticipated climatic conditions and intensity of traffic loads. These input parameters play a vital role during selection of suitable binders for construction of pavements. Thus, engineering properties of bitumen should be determined considering the following factors.

- Temperature (maximum, minimum and average values)
- Loading time
- Intensity of load

Bitumen behaves differently in different combinations of the above values. For example, bitumen behaves elastic under short duration of loading at normal temperature. But, in case of prolonged loading and as the temperature increases, it loses elasticity and flows like a fluid. This is the reason why bituminous pavements at higher temperature tend to deform without cracking.

7.4.1 Stiffness Modulus of Bitumen

In 1954, van der Poel developed the concept of a stiffness modulus for thermoplastic polymers which was useful for determination of the engineering properties of bitumen. The concept was modified by Heukelom and Klomp (1964) and Heukelom (1973). In this method, the bitumen is characterised by its stiffness modulus, (E_{bit}), the ratio between applied stress to the observed strain at a given time of loading (t in s) and temperature (T in °C). The value of the stiffness modulus of bitumen may be expression as follows.

$$E_{bit}\ (t, T) = \text{stress/strain at } (t, T) \tag{7.9}$$

The above equation holds good and is suitable for in-service pavements, if the assumed strain is not more than 1 per cent. If the loading time changes, preferably under variable temperatures similar to seasonal variations, the value of stiffness of bitumen also changes. To account for this variability of the stiffness modulus of bitumen, a stiffness nomograph was developed by van der Poel and modified by Heukelom based on three parameters, viz., time of loading (t), temperature difference (T-T800 pen) and penetration index (PI).

The equation developed by Heukelom and Klomp (1964) or a method prescribed by Bonnaure et al. (1977) may be used to determine the stiffness values of bituminous mixes. Measured stiffness values greatly agree with the values predicted by the Bonnaure's method and it is adopted by the Shell method (1977) of the asphalt pavement design (Ullidtz and Peattie 1980).

To understand the implication of the true engineering properties of bitumen at different ranges of temperatures from cold to hot under cyclic loading similar to field conditions, it is necessary to know its visco-elastic properties.

7.4.2 Visco-elastic Properties of Bituminous Materials

Bitumen becomes hard and soft by cooling and heating respectively. These states of bitumen may be reversed as and when required and this property of bitumen is known as *thermo-plasticity*. Based on temperature and rate of loading, the bitumen may behave like a dual characteristic material.

Elastic Material At a given temperature, the material undergoes immediate deformation or strain under loading and immediately recovers upon unloading. In other words, if the material is linear elastic, the stress is proportional to the strain according to Hooke's law, i.e., Stress = Young's modulus value × Strain (Wilhelm Flugge, 1975).

Viscous Material At a given temperature, the material undergoes delayed deformation or strain under loading but it may or may not recover to the initial condition. A Newtonian material will recover completely. But in Bingham, power law fluids and pseudo-plastic materials, complete recovery of strain will not take place due to plastic straining during loading. In other words, if the material is viscous, the stress is proportional to the strain rate (Wilhelm Flugge, 1975).

Bitumen exhibits intermediate characteristics ranging from elastic to viscous and vice versa due to change in temperature and rate of application of load. Because of this, bituminous mixes are known as visco-elastic materials. The behaviour of visco-elastic material depends on temperature, time and frequency of loading.

When a visco-elastic material is subjected to sinusoidal loading (Fig. 7.13), the occurrence of strain (ε_o) is delayed due to phase difference of the applied stress (σ_o) at a given time (t). This difference of response is known as phase angle (ϕ). As the material transforms from being elastic to visco-elastic during change of temperature, the value of the phase angle (ϕ) changes from 0° to 90°. The energy dissipated by the material is determined by using the phase angle value. The values of complex modulus and dynamic complex modulus of bitumen are determined from the following popularly known relationships (Huang, 2004; Di Benedetto et al., 2001 and Kallas, 1970).

Complex modulus (E^*) = Real part of modulus value + Imaginary part of modulus value

$$E^*(f, T) = (\sigma_o/\varepsilon_o)\sin(\omega t) + i\cdot(\sigma_o/\varepsilon_o)\sin(\omega t - \phi) \tag{7.10}$$

where,

f = frequency
T = temperature
σ_o = maximum stress
ε_o = maximum strain
ω = angular velocity or pulsation = $2\pi f$
ϕ = phase angle
t = a point of time with reference to stress application

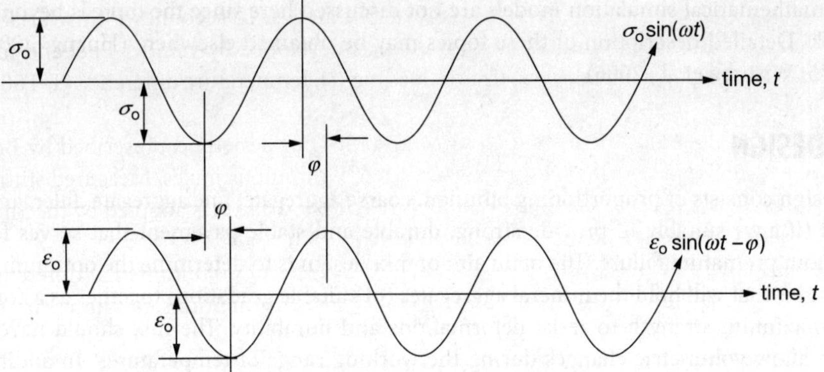

Figure 7.13 Phase difference between stress and strain (Yang, 2004; Relem, 2001).

$$\text{Dynamic modulus} = \text{Absolute value of complex modulus} = |E^*| \tag{7.11}$$

If the viscous effect is ignored, the ratio between the maximum values of stress to strain is termed as the elastic modulus.

The values of complex bulk modulus (K^*) and shear modulus (G^*) may be determined from the following popularly known relationships.

$$K^* = E^*/(3 - 6\mu) \tag{7.12}$$

$$G^* = E^*/[2(1 + \mu)] \tag{7.13}$$

where,

μ = Poisson's ratio

To characterise visco-elastic properties of binders at lower temperatures, another parameter called the creep compliance is used. The creep compliance may be calculated using following relationship.

$$D(t, T) = 1/E_{bit}(t, T) \tag{7.14}$$

Shear compliance is also used to characterise the visco-elasticity of bituminous materials. The following relationship is used to calculate shear compliance ($J(t, T)$).

$$J(t, T) = 3/E_{bit}(t, T) \tag{7.15}$$

The above equations holds good if the material is assumed to be linear visco-elastic and isotropic. The values of dynamic modulus, bulk modulus, shear modulus, creep compliance and shear compliance provide important information to explain the visco-elastic behaviour of the bituminous mix over a range of temperature and frequency of loading.

The visco-elastic properties of bituminous materials are characterised using the following mathematical models.

- Elastic model
- Viscous model
- Maxwell model
- Kelvin model
- Power law model
- Huet model
- Huet–Sayegh model
- Burger model
- 2S2P1D model
- Combined model consisting of all the above models.

The above mathematical simulation models are not discussed here since the topic is beyond the scope of this book. Detailed description of these topics may be obtained elsewhere (Huang, 2004; Wilhelm Flugge, 1975; Wistuba et al., 2006).

7.5 MIX DESIGN

The mix design consists of proportioning bitumen, coarse aggregate, fine aggregate, filler, air–voids and admixtures (if any) suitably to provide strong, durable and stable pavement that serves for a design period without premature failure. The main aim of mix design is to determine the optimum quantity of bitumen binder that will hold the mineral aggregates (of suitable gradation) together as a compact layer providing maximum strength to resist deformations and durability. The mix should have minimum air-voids to allow volumetric changes during the working range of temperatures. In addition to this, site-specific requirements should also be satisfied by the designed mix.

7.5.1 Factors Affecting Mix Design

The following factors should be considered while designing bituminous mixes for use in pavement construction.

- Stability
- Durability
- Workability
- Stress–strain relationship
- Permanent deformations
- Fatigue and low-temperature racking
- Imperviousness
- Susceptibility to moisture
- Cost-effectiveness

In addition to the above factors, in-service conditions of pavement, viz., seasonal temperature changes, traffic intensity, humidity, drainage, rainfall and soil class are also to be taken into account while designing bituminous mixes.

7.5.2 Desirable Properties of Bituminous Mixes

Physical requirements

The binding material and mineral aggregate (i.e., coarse aggregate, sand and filler) should be qualified with reference to their engineering properties (discussed in Chapter 6) according to the given standard specifications. The standard specifications should be that in the IS and IRC (in India) or BIS or AASHTO or any other standard specification in force.

Maximum size of aggregate

The maximum size of the aggregate should depend on the design thickness of the layer. As a rule of thumb, the maximum size of the aggregate should not exceed one-half of the compacted layer thickness.

Gradation

A mixture of different sizes, ranging from coarse to very fine mineral particles, should attain the desired density after compaction. Different types of aggregate gradations and the typical shape of gradation curves were discussed in Section 6.2.

Maximum density

Maximum density of the mix is determined based on the particle size distribution of the mineral aggregate. To examine the geometry of the curve with reference to a maximum mix density curve, a gradation curve is drawn for the proposed mixture of mineral aggregate. Based on theoretical considerations, the maximum density may be determined by using Fuller's equation.

$$P = \left(\frac{d}{D}\right)^n \times 100 \qquad (7.16)$$

where,

P = Per cent passing through the sieve (%).
d = Size of the sieve considered in mm.
D = Maximum sieve size used in a given sieve set in mm.
n = Adjustment parameter of the density curve (for maximum density, n is taken as ≈ 0.5 (Fuller and Thompson, 1907).

Proportioning of mineral aggregate

To get the desired density of mix, optimum utilisation of available materials by blending in suitable proportions is necessary. Once the specifications or target gradation is finalised, the process of proportioning the different sizes of aggregates is evaluated. There are many methods available for suitable proportioning of mineral aggregates (i.e., coarse aggregates, fine aggregates and mineral dust) obtained from different sources. The final proportion of the materials should satisfy and approach the mean values of the standard grading specification in force, viz., IRC, AASHTO, BS, Superpave, etc. In no case should the percentage of mineral aggregates passing fall off the given standard gradation limits (it must be within the upper and lower limits). Different methods of proportioning mineral aggregates are presented in Fig. 7.14.

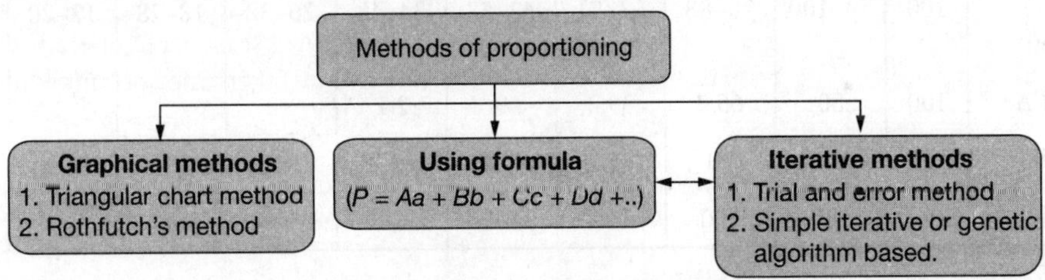

Figure 7.14 Proportioning methods of mineral aggregates.

Graphical Methods By triangular chart method, exactly three material types should be blended. This method is not useful if two or four types of materials are to be blended. In Rothfutch's method, an approximation is required in order to draw straight lines representing enclosures of equal area on either side of different given curves of particle size distribution. Graphical methods are not used very often as they need more time and become complicated as the number of blending materials increases (TAI 1983).

Using Formulae and Trial-and-Error Methods According to the Asphalt Institute, the following formula is used to determine the appropriate proportions of the given types of materials (TAI 1983).

$$P = Aa + Bb + Cc + Dd + \dots \qquad (7.17)$$

where,

P = Percentage passing of combined different type of aggregates ($A + B + C + D + \dots$
$A, B, C, D, \dots$ = Percentage of different type of aggregates passing through a given set of sieves
$a, b, c, d, \dots$ = contribution or proportion of the different type of aggregates in decimals, i.e., $a + b + c + d + \dots = 1.0$

For example, if two types of aggregates are to be blended, Eq. (7.17) can be re-written as

$$P = Aa + Bb \tag{7.18}$$

$b = (1.0 - a)$, since $a + b = 1.0$. By substituting the value of b in Eq. (7.18), we get,

$$a = (P - B) / (A - B) \tag{7.19}$$

In this method of proportioning, an aggregate type is examined from the given types of aggregates (*A, B, C, D*...) that will contribute a major portion of the proposed mix and a size of sieve in the given types of aggregate is identified such that a distinguishable value of per cent passing is observed. This means that a value of sieve size identified from the above types of aggregate may be labelled as a *separation sieve* which reasonably separates the given set of values of per cent passing of aggregates. By using Eq. (7.19), values of *a, b, c, d* ... are calculated by suitably substituting them in Eq. (7.17). With these *seed values* of *a, b, c, d* ... further pruning is made by trial-and-error method to get a closer match to the values of desired mean gradation. The following example presents the procedure adopted for proportioning different types of aggregates.

EXAMPLE 7.1

Determine proportion values and final gradation of the following per cent passing of three types of mineral aggregates for a proposed bituminous concrete mix.

Size of sieve (mm)	19	13.2	9.5	4.75	2.36	1.18	0.6	0.3	0.15	0.075
Desired gradation	100	79–100	70–88	53–71	42–58	34–48	26–38	18–28	12–20	4–8
Material A	100	80	65.2	39.2	9	2.1				
Material B	100	100	90	81.2	71.6	52.3	38.2	25	17.8	1.1
Material C	100	100	100	100	100	100	99	98	90	84

Answer

From examination of the per cent passing values, the separation sieve may be taken as 2.36 mm. Now let us calculate seed values of *a, b, c*.

$$P = (42 + 58)/2 = 50;\ A = 9;\ B = 71.6.$$

We know that $a = (P - B)/(A - B)$ from Eq. (7.19). On substitution, we get

$$a = (50 - 71.6)/(9 - 71.6)$$

$$a = 0.34$$

With reference to 0.075 mm sieve size,

$$P = Aa + Bb + Cc$$

$$((4 + 8)/2) = 9 \times 0.34 + 1.1b + 84c,$$

$$6 = 3.06 + 1.1b + 84c,\ (a + b + c = 1,\ b + c = 1 - 0.34,\ b + c = 0.66, \text{ or } c = 0.66 - b)$$

$$6 = 3.06 + 1.1b + 84\ (0.66 - b), \text{ therefore, we get, } b = 0.63 \text{ and } c = 0.03$$

Excel spread sheet or a simple calculator may be used to carry out the trials.

Trial 1 Take seed values of $a = 0.34$, $b = 0.63$ and $c = 0.03$.

Size of sieve (mm)	19	13.2	9.5	4.75	2.36	1.18	0.6	0.3	0.15	0.075
Desired gradation	100	79–100	70–88	53–71	42–58	34–48	26–38	18–28	12–20	4–8
Mid values	100	89.5	79	62	50	41	32	23	16	6
0.34 × Material A	34	27.2	22.168	13.33	3.06	0.714				
0.63 × Material B	63	63	56.7	51.15	45.10	32.95	24.06	15.75	11.21	0.69
0.03 × Material C	3	3	3	3	3	3	2.97	2.94	2.7	2.52
Total	100	93.2	81.87	67.48	51.17	36.66	27.03	18.69	13.91	3.21

Trial 2 Take $a = 0.32$, $b = 0.62$ and $c = 0.06$.

Size of sieve (mm)	19	13.2	9.5	4.75	2.36	1.18	0.6	0.3	0.15	0.075
Desired gradation	100	79–100	70–88	53–71	42–58	34–48	26–38	18–28	12–20	4–8
Mid values	100	89.5	79	62	50	41	32	23	16	6
0.32×Material A	32	25.6	20.864	12.544	2.88	0.672	0	0	0	0
0.62×Material B	62	62	55.8	50.344	44.392	32.426	23.684	15.5	11.036	0.682
0.06×Material C	6	6	6	6	5.7	6	5.94	5.88	5.4	5.04
Total	100	93.6	82.664	68.888	53.272	39.098	29.624	21.38	16.436	5.722

Similar trials may be made for more accurate results. The comparison between calculated gradation and desired gradation (mid-values) is presented in Fig. 7.15.

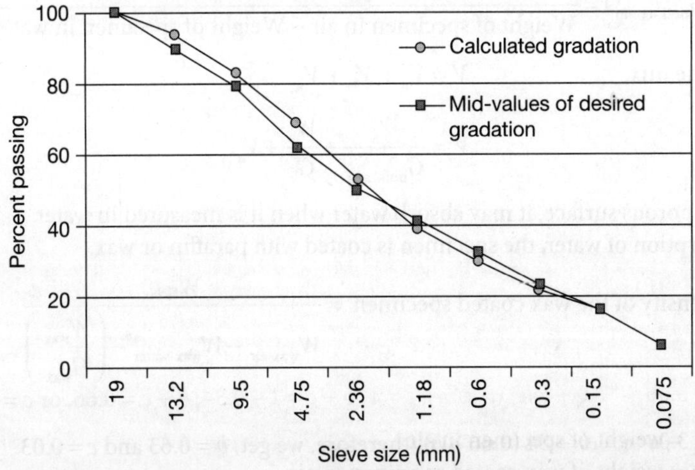

Figure 7.15 Comparison Between Calculate Gradation and Mid-values of Desired Gradation.

7.5.3 Volumetric Analysis of Bituminous Mix

Density and voids present in the mineral aggregate and compacted bituminous mix are important volumetric parameters that decide the performance of bituminous pavements. Bituminous mixes of high density are undesirable because there will be no room for expansion or internal adjustment of constituents during temperature changes. Therefore, the presence of a sufficient amount of air voids in a compacted bituminous mix is important for better performance of in-service pavements. In this connection, volumetric analysis of the material used is desirable before preparing a bituminous mix (TAI 1983; MORTH 2001). For volumetric analysis of bituminous mixes, Fig. 7.16 and the formulae given below are used.

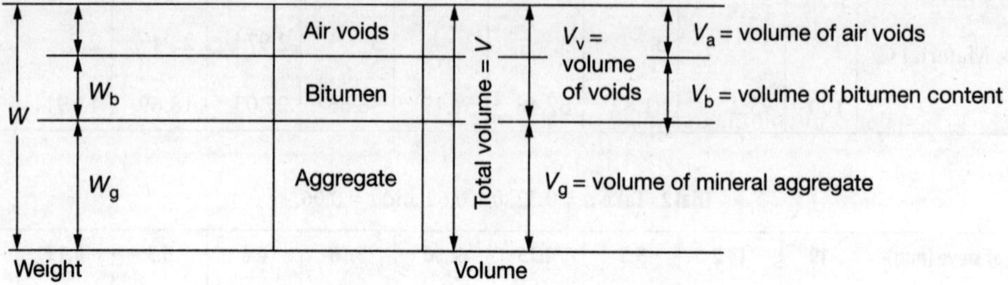

Figure 7.16 Elements in a bituminous concrete mix.

$$\text{Weight} = \text{Volume} \times \text{Specific gravity}$$

$$\text{Bulk specific gravity of combined aggregates } (G_{\text{bulk-agg}}) = \frac{p_1 + p_2 + p_3}{\dfrac{p_1}{G_1} + \dfrac{p_2}{G_2} + \dfrac{p_3}{G_3}} \tag{7.20}$$

where,

p_1, p_2, p_3 = per cent by weight of coarse, fine aggregate and filler respectively
G_1, G_2, G_3 = bulk specific gravity of coarse, fine aggregate and filler respectively

Bulk specific gravity of compacted bituminous mix:

$$(G_{\text{bulk-comp}}) = \frac{\text{Weight of specimen in air}}{\text{Weight of specimen in air} - \text{Weight of specimen in water}} \tag{7.21}$$

Total volume of the mix,
$$V = V_g + V_b + V_a \tag{7.22}$$

or
$$V = \frac{W_g}{G_{\text{bulk-agg}}} + \frac{W_b}{G_b} + V_a \tag{7.23}$$

If a specimen has porous surface, it may absorb water when it is measured in water.

To avoid absorption of water, the specimen is coated with paraffin or wax.

Then, the bulk density of the wax coated specimen
$$= \frac{W_{\text{spec.}}}{W_{\text{wax-air}} - W_{\text{wax-water}} - \left(\dfrac{W_{\text{wax}}}{G_{\text{wax}}}\right)} \tag{7.24}$$

where,

$W_{\text{spec.}}$ = weight of specimen in air
$W_{\text{wax-air}}$ = weight of wax coated specimen in air

$W_{wax\text{-}water}$ = weight of wax coated specimen in water
W_{wax} = weight of coated wax in air = ($W_{wax\text{-}air}$ – W)
G_{wax} = specific gravity of the wax

$$\text{Voids present in the mix in percentage} = 100 \times \frac{V_a}{V} \quad (7.25)$$

where,

$V_a = V - V_g - V_b$
V = Total volume of the specimen
V_g = Volumes of three mineral aggregates
V_b = Volume of bitumen

$$\text{Voids in mineral aggregate, VMA (\%)} = 100\% - V_g\% \quad (7.26)$$

where,

$V_g\%$ = per cent volume occupied by total mineral aggregates

$$V_g\% = (p_1 + p_2 + p_3) \times \frac{G_{bulk\text{-}comp}}{G_{bulk\text{-}agg}}$$

$$\text{Voids filled with bitumen, VFB(\%)} = 100 \times \frac{V_b\%}{VMA} \quad (7.27)$$

$$V_b\% = \text{volume of bitumen in percentage} = (\text{per cent of bitumen}) \times \frac{G_{bulk\text{-}comp}}{G_b} \quad (7.28)$$

Volume of bitumen absorbed by aggregates due to the presence of pores
On aggregates surface (ASTMD 2041)

$$V_{ab} = 100 \times \left(\frac{1}{G_{bulk\text{-}agg}} - \frac{1}{G_{se}} \right) \times G_b \quad \textbf{(7.29)}$$

G_{se} = effective specific gravity of mineral aggregates (AsTMD 2041)

$$G_{se} = 100 \times \left(\frac{100 - \text{per cent of bitumen}}{\frac{100}{G_{mm}} - \frac{\text{per cent of bitumen}}{G_b}} \right) \quad (7.30)$$

where,

G_{mm} = Max. specific gravity of mix without air voids according to ASTMD 2041

EXAMPLE 7.2

Calculate the bulk specific gravity of the combined aggregates from the following data.

Aggregate type	Bulk specific gravity	Proportion in mix (%)
Coarse aggregate	2.61	54.5
Fine aggregate	2.71	38.0
Mineral filler	2.69	7.5

Answer

From Eq. (7.20)

$$\text{Bulk specific gravity of combined aggregates } (G_{\text{bulk-agg}}) = \frac{p_1 + p_2 + p_3}{\dfrac{p_1}{G_1} + \dfrac{p_2}{G_2} + \dfrac{p_3}{G_3}}$$

On substituting,

$$G_{\text{bulk-agg}} = \frac{54.5 + 38.0 + 7.5}{\dfrac{54.5}{2.61} + \dfrac{38.0}{2.71} + \dfrac{7.5}{2.69}} = \frac{100}{20.88 + 14.02 + 2.78} = 2.65$$

EXAMPLE 7.3

Calculate the bulk specific gravity of a compacted bituminous concrete mix from the following data.

Specimen weight in air = 3055.1 g

Specimen weight in water = 1725.7 g

Answer

From Eq. (7.21),

Bulk specific gravity of compared bituminous mix:

$$(G_{\text{bulk-comp}}) = \frac{\text{Weight of specimen in air}}{\text{Weight of specimen in air} - \text{Weight of specimen in water}}$$

$$G_{\text{bulk-comp}} = \frac{3055.1}{3055.1 - 1725.7} = 2.298$$

$\therefore$ Unit weight of compared bituminous mix = 2.298 × 1 g/cc = 2.298 g/cc

EXAMPLE 7.4

A few laboratory tests were conducted to design a bituminous concrete mix. Using the following data, calculate the (i) bulk density of the compacted mix, (ii) bulk density of the compacted mix specimen coated with paraffin, (iii) combined specific gravity of aggregates, (iv) percentage of voids present in the compacted mix, (v) voids in mineral aggregates and (vi) voids filled with bitumen.

Weight of compacted uncoated specimen in air = 1195.4 g

Weight of compacted uncoated specimen in water = 684.4 g

Weight of compacted specimen (coated with paraffin) in air = 1240.9 g

Weight of compacted specimen (coated with paraffin) in water = 678.3 g

Take specific gravity of paraffin as 0.91 and maximum specific gravity of mix as 2.539 according to ASTM2041.

Material	Specific gravity	Composition of mineral aggregate (% by weight)	Proportion in mix (% by weight)
Coarse aggregate	2.61	54.5	51.5
Fine aggregate	2.71	38.0	35.9
Mineral filler	2.69	7.5	7.1
Bitumen	1.02		5.5

Answer

(i) Bulk density of compacted mix (without paraffin coating):

$$G_{\text{bulk-comp}} = \frac{1195.4}{1195.4 - 684.4} = 2.339$$

∴ Unit weight of compared bituminous mix = 2.298 × 1 g/cc = 2.298 g/cc

(ii) Bulk density of compacted mix specimen coated with paraffin:

$$\text{The bulk density of paraffin or wax coated specimen} = \frac{W_{\text{spec.}}}{W_{\text{wax-air}} - W_{\text{wax-water}} - \left(\dfrac{W_{\text{wax}}}{G_{\text{wax}}}\right)}$$

$$= \frac{1195.4}{1240.9 - 678.3 - \left(\dfrac{45.5}{0.9}\right)} = 2.33 \qquad (\because W_{\text{wax}} = 1240.9 - 1195.4 = 45.5\text{g})$$

(iii) Combined specific gravity of aggregates:

$$\text{Bulk specific gravity of combined aggregates } (G_{\text{bulk-agg}}) = \frac{p_1 + p_2 + p_3}{\dfrac{p_1}{G_1} + \dfrac{p_2}{G_2} + \dfrac{p_3}{G_3}}$$

$$\text{On substituting, } G_{\text{bulk-agg}} = \frac{54.5 + 38.0 + 7.5}{\dfrac{54.5}{2.61} + \dfrac{38.0}{2.71} + \dfrac{7.5}{2.69}} = \frac{100}{20.88 + 14.02 + 2.78} = 2.653$$

$$\text{Based on individual specific gravity} = \frac{51.5 + 35.9 + 7.1}{\dfrac{51.5}{2.61} + \dfrac{35.9}{2.71} + \dfrac{7.1}{2.69}} = \frac{94.5}{19.73 + 13.24 + 2.63} = 2.654$$

(vi) Percentage of voids present in the compacted mix:

$$= 100 \times \frac{V_a}{V}$$

$$V_a = V - V_g - V_b$$

$$V = \text{Total volume of the specimen} = 1195.4 \times \frac{1.0}{2.339} = 511.07 \text{ cc}$$

V_g = Volumes of three mineral aggregates

$$V_g = 1195.4 \times \left[\left(\frac{0.515}{2.61}\right) + \left(\frac{0.359}{2.71}\right) + \left(\frac{0.071}{2.69}\right)\right] = 425.78\text{cc}$$

$$V_b = \text{Volume of bitumen} = 1195.4 \times \frac{0.055}{1.02} = 64.45\text{cc}$$

$$V_a = 511.07 - 425.78 - 64.45 = 20.84 \text{ cc}$$

$$\%\text{ voids} = 100 \times \frac{20.84}{511.07} = 4.07\%$$

(vii) Voids in mineral aggregates:

Voids in mineral aggregate, VMA (%) = 100% – V_g%

$V_g\%$ = per cent volume occupied by total mineral aggregates

$$V_g\% = (p_1 + p_2 + p_3) \times \frac{G_{\text{bulk-comp}}}{G_{\text{bulk-agg}}} = (94.5) \times \frac{2.3339}{2.65} = 83.31\ \%$$

$$\text{VMA}\ (\%) = 100 - 81.31 = 16.69\%$$

(viii) Voids filled with bitumen:

$$\text{Voids filled with bitumen, VFB}(\%) = 100 \times \frac{V_g\ \%}{\text{VMA}}$$

$$V_b\% = (\text{Per cent of bitumen}) \times \frac{G_{\text{bulk-comp}}}{G_b} = 5.5 \times \frac{2.339}{1.02} = 12.61\%$$

$$\text{VFB}\ (\%) = 100 \times \frac{12.61}{16.69} = 75.56\%$$

(ix) Volume of bitumen absorbed by aggregates:

$$G_{se} = \left(\frac{100 - \text{Per cent of bitumen}}{\dfrac{100}{G_{mm}} - \dfrac{\text{Per cent of bitumen}}{G_g}} \right) = \left(\frac{100 - 5.5}{\dfrac{100}{2.539} - \dfrac{5.5}{1.02}} \right) = 2.779$$

$$V_{ab} = 100 \times \left(\frac{1}{G_{\text{bulk-agg}}} - \frac{1}{G_{se}} \right) \times G_b = 100 \times \left(\frac{1}{2.654} - \frac{1}{2.779} \right) \times 1.02 = 1.728\%$$

7.5.4 Marshall Mix Method

The Marshall Mix method is widely used in India as well as in many countries to design hot bituminous paving mixtures. This method was developed by Bruce Marshall of the Mississippi State Highway Department in 1940. Later, it was modified and improved by the U.S. Army Corps of Engineers and formulated as the mix design criteria ASTM D 1559 and AASHTO T 245 (TAI 2001).

The Marshall method is used to determine the optimum bitumen content which satisfies a given design criteria of stability, deformation or flow, density, minimum air void, VMA and VFB of hot mixed mineral aggregates and bitumen under standard test conditions.

Preparation of Specimens A number of specimens are prepared by changing the mix proportions of bitumen (at increments of 0.5% by weight) with a selected grade of aggregate samples whose maximum size is restricted to 25 mm.

About 1200 g of mineral aggregates consisting of coarse aggregates, fine aggregates and mineral fillers is taken as per a given standard grade. The mineral aggregates and bitumen are separately heated to a temperature of up to 154–160°C and 175–190°C respectively. The quantity of bitumen is measured by the weight of the mix and poured into the heated mineral aggregate. The mix is blended until all the aggregate surfaces are evenly coated with bitumen. During mixing, the bitumen viscosity should be maintained at 170 ± 20 centistokes. The hot mixture is placed in a standard cylindrical mould (of diameter size 102 mm and depth 64 mm) which is heated to 93–149°C. It is compacted by applying 75 blows/each side with 4.5 kg weight of hammer on each of its face over a free fall height of 457 mm. During compaction, the bitumen viscosity should be maintained at 280 ± 30 centistokes. The prepared specimen is allowed to cool at room temperature for 12 hours and later kept in a water bath at 60°C for about 30 minutes before being tested.

Tests on the Prepared Specimens The sample is taken out from the water bath and diametrically placed in position as shown in Fig. 7.17. Load is vertically applied on to the specimen at a constant rate of 5 cm per minute. The maximum load at failure (on the proving ring) is taken as the stability value (kN) and the corresponding deformation value obtained from the dial gauge is taken as the flow value (mm). The flow value is expressed in units of 0.25 mm.

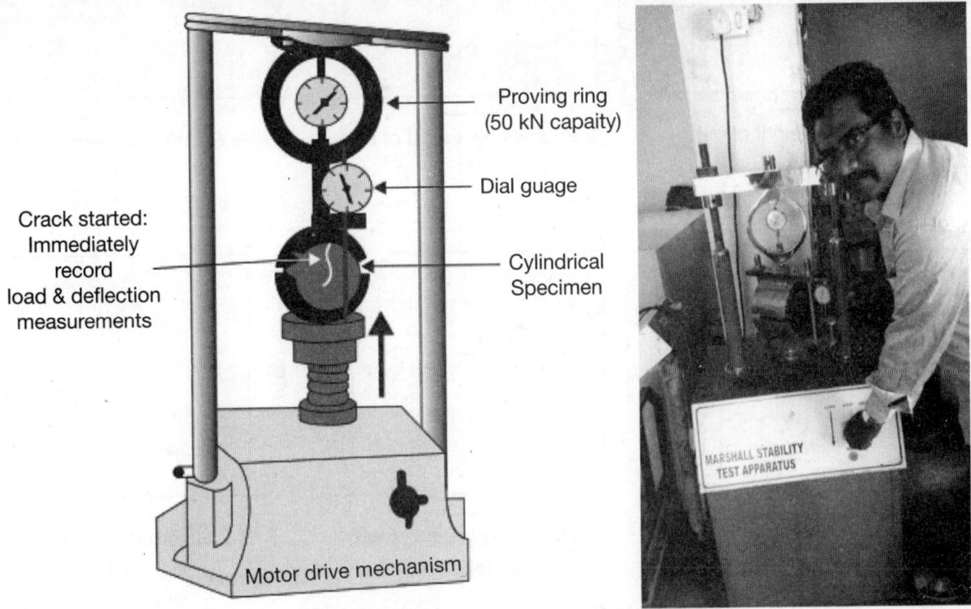

Figure 7.17 Marshall stability test set-up.

Presentation of the Test Data The data obtained from the Marshall tests should be analysed with reference to the voids–density relationship of the bituminous mixes as discussed in Section 7.5.3. From the results of the volumetric analysis, six curves are plotted. The details of the curves are shown in Fig. 7.18.

Interpretation of the Curves

Curve (a): Unit weight vs. bitumen content and Curve (e): Marshall Stability vs. bitumen content.

The above curves show similar trends. Initially, at lower content of bitumen, the values of unit weight and stability of mix is less but as the bitumen content increases, these values attain a peak. Afterwards, they show a decreasing trend. At lower content of bitumen, the specimen fails at smaller loads since the quantity of bitumen mixed with aggregates is not enough to develop sufficient bond between aggregates to withstand the applied load. Specimens with excess quantity of bitumen (above a certain level) also fail at smaller loads due to the thickness of the bituminous film at the interface of aggregates, which do not effectively transmit the applied load on to the aggregates. Therefore, it is only at a particular quantity of bitumen that maximum values of density and Marshall Stability are observed.

Curve (b): Air voids vs. bitumen content

As the quantity of bitumen increases, the air voids present in the mix decreases to a low value. Thereafter, it increases as the quantity of bitumen further increases. At a given temperature, though, stiffness of the bituminous concrete decreases as air void content increases (Mackenzie and Fletcher 1980).

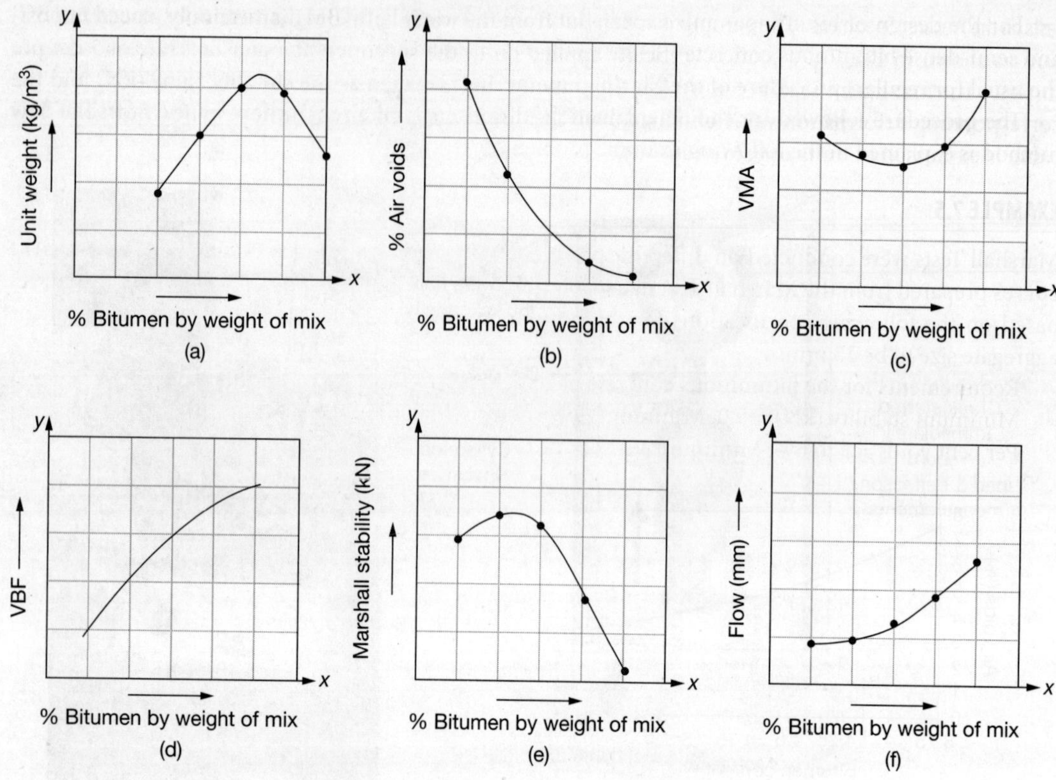

Figure 7.18 Typical curves of Marshall test data.

Curve (c): Voids in Mineral Aggregate (VMA) vs. bitumen content
As the bitumen content increases, slight variation may be observed in the shape of the curve. This may be due to adjustment of mineral aggregates with increased bitumen content.

Curve (d): Voids Filled with Bitumen (VFB) vs. bitumen content
VFB increases with increase of bitumen content since the bitumen fills all the available voids in the mix.

Curve (f): Flow vs. bitumen content
Flow (in mm) is a measure of the deformation due to applied load. The value of flow increases (the compacted specimen fails even at lower load) with increase of bitumen content because the presence of thick bitumen film around mineral aggregates prevents strong aggregate interlocking.

After obtaining the optimum binder content, another six mix-samples are prepared by adding the optimum binder content at the specified void level. The samples' moisture sensitivity and potential for stripping are evaluated by conducting Tensile Strength Ratio (TSR) test (ASTM D 4867). Later, three samples are saturated in water under vacuum, exposed to hot water (at 60°C for 24 hrs), soaked in water at 25°C for 2 hours and finally tested using the Marshall Stability testing machine (Fig. 7.17). The TSR can be calculated as below:

$$\text{TSR}(\%) = \frac{\text{Average wet tensile strength of the sample admixed OBC}}{\text{Average dry tensile strength of the sample admixed OBC}} \times 100 \qquad (7.31)$$

The minimum value of the TSR should not be less than 80%.

For the design of bituminous mixes, viz.,. bituminous macadam (BM), bituminous concrete (BC) and semi-dense bituminous concrete (SDBC), polymer modified binders can also be used according to the usual (or regular) procedure of the Marshall method of mix design (ASTM D 1559 1979).

The procedure followed for the determination of optimum bitumen content by the Marshall Mix method is explained in the following example problem.

EXAMPLE 7.5

Marshall Tests were conducted on different specimens with varying bitumen content. Following are the curves prepared from the Marshall Test data. Determine the optimum bitumen content from the curves based on the following specifications for the bituminous concrete layer. Take the normal size of the aggregate size to be 25 mm.

Requirements for the bituminous concrete pavement layer (Table 500-19. MSRT&H. 2006):
Minimum stability (kN) = 9.0; Minimum flow = 2 mm; Maximum flow = 4 mm;
Per cent voids = 3 to 6%; Minimum VMA = 12 to 14%; VFB = 65 to 75%.

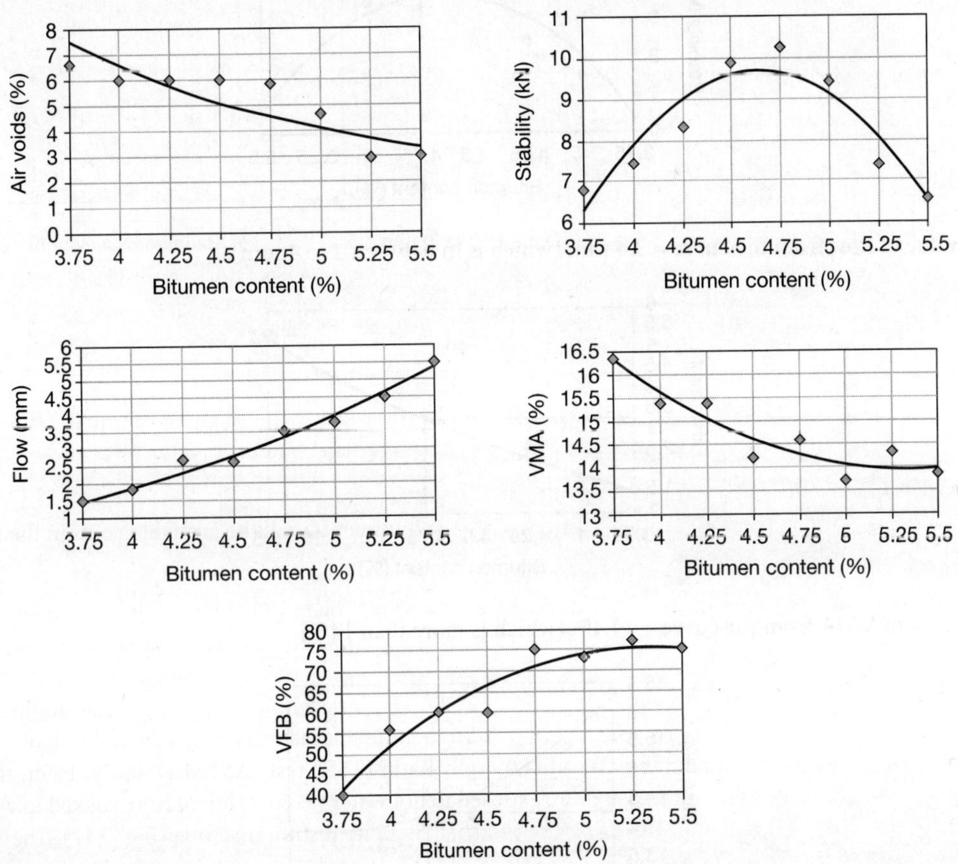

Answer

From the above curves and according to the given specifications, let us select a bitumen content at which the value of air voids = 4.5% (which is in between 3 to 6%). Then, the corresponding value of bitumen from the given curve is 4.75%. Now, check the other requirements of the specifications with respect to a bitumen content of 4.75 %.

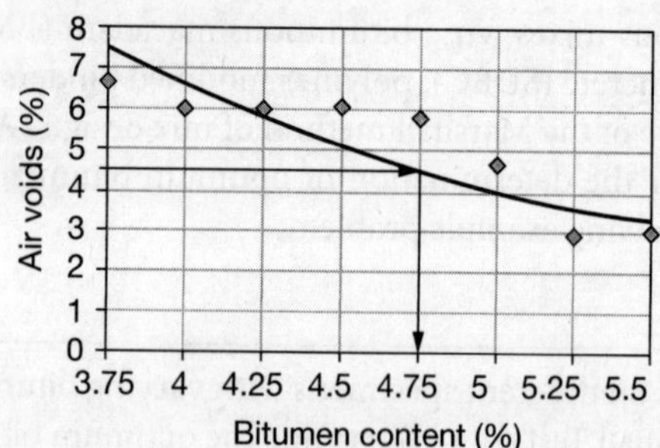

The value of stability from the curve (given below again for easy reference) = 9.6 kN (which is more than 9.0 kN).

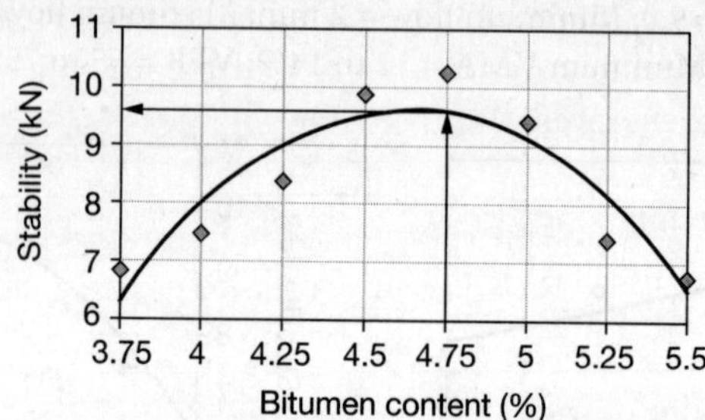

The value of flow from the curve = 3.4 mm (which is in between 2 to 4 mm).

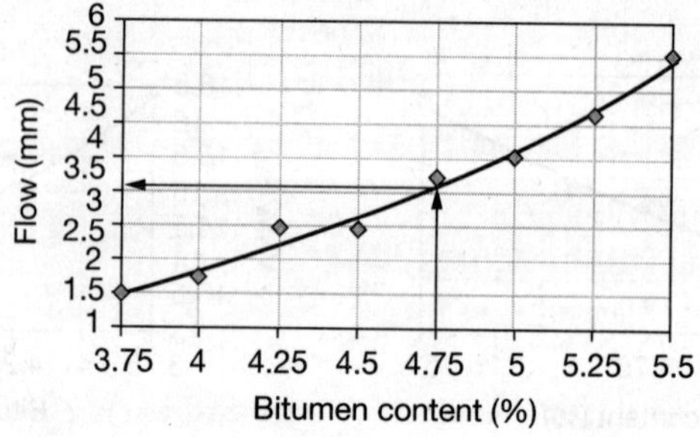

The value of VMA from the curve = 14.4% (which is more than 14%).

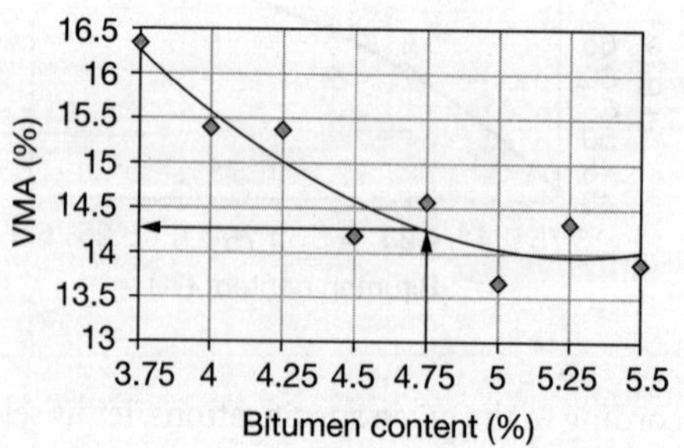

The value of VFB from the curve = 71% (which is in between 65 to 75%).

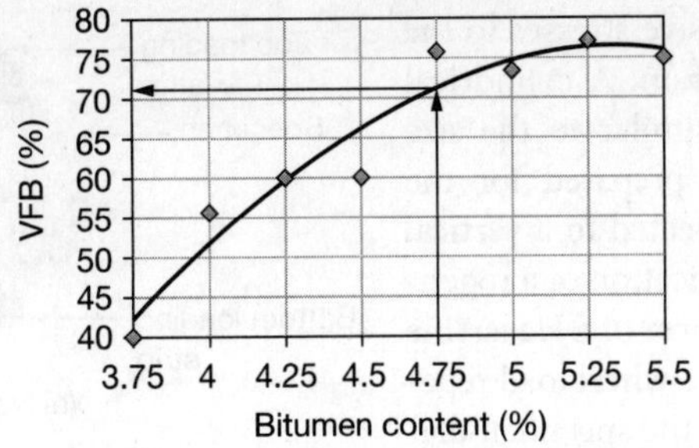

Hence, the design criteria are met. Therefore, optimum bitumen content (OBC) may be taken as 4.75%. However, further trials may be made to determine a more appropriate content of bitumen that satisfies the design criteria according to the given conditions of a project requirement better.

Alternatively, the optimum binder can also be calculated as follows.

Binder contents corresponding to the (i) maximum stability value, (ii) maximum unit weight and (iii) permissible air void content are determined from the curves prepared with the help of the Marshall Test results and their average value is reported as the optimum binder content. The optimum binder content should meet the required design criteria; otherwise, an appropriate value may be determined from the curve which satisfies the design criteria and mix economics. A computer programme to determine the optimum binder content by using this procedure is available in literature (Kumar et al., 2003).

Note: If the desired requirements are not met by the determined optimum binder content, gradation of the aggregates should be changed and the Marshall Mix design method should be followed again.

Limitations of the Marshall Mix Design Method (TAI 1997; TAI 2001; Pandey, 2003)

- The shear strength parameters of bituminous mixes are not considered.
- This is an empirical mix design method. There is no scientific analysis behind it to explain pavement failures due to fatigue, rutting, durability, etc.
- Degree of compaction achieved in the laboratory may not be comparable to real-life conditions as construction practices differ.
- This method is not found suitable for the precise prediction of various permanent deformations (Brown et al., 1981).

Despite these limitations, the Marshall Mix method is widely used in India as well as many other parts of the world due to its simple procedure of testing, analysis of results, past experience and its cost effectiveness. In USA, to overcome the limitations of the Marshall Mix method and other available mix design methods, a new system for the design of bituminous mix has been developed called the *Superpave mix design method.* The details of the Superpave method are out of the scope of this book.

7.5.5 Resilient Modulus of Bituminous Mix

The behaviour of bituminous material is complex. Its properties are visco-elastic and susceptible to temperature changes. The elastic modulus of a bituminous mix is normally determined in the laboratory by an appropriate test at representative temperatures similar to field conditions. The test is usually conducted under repeated constant load conditions or constant strain mode. The modulus of the bituminous mix is influenced greatly by test conditions, temperature and parameters related to the mix.

Resilient modulus (MR) or the elastic modulus value of a bituminous mixture is the ratio of repetitive vertical compressive stresses to the corresponding recoverable strain. A cylindrical bituminous mix specimen (similar to the size and shape of the specimen prepared for the Marshall Stability test) is subjected to a vertical compressive stress by the application of a repetitive compressive load in the form of a Haversine wave, which is similar to traffic wheel load repetitions. The load is applied on the specimen diametrically in opposite direction by loading strips (Fig. 7.19). To determine the resilient modulus of the bituminous mix, elastic deformation of the specimen ($D_d - D_0$) is measured. The values of repetitive load, duration of loading, rest period and test temperature should be taken in accordance with field conditions.

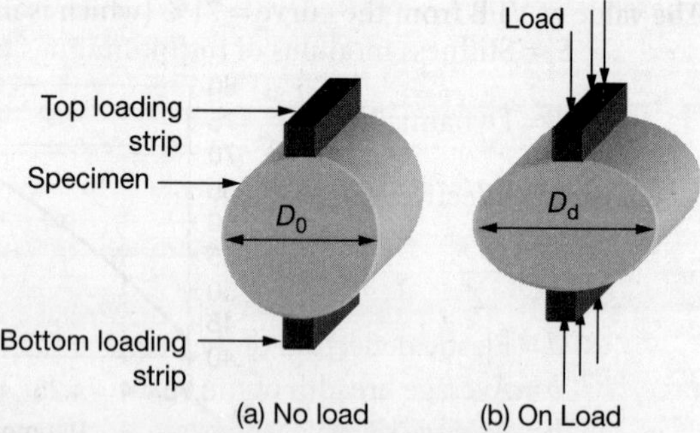

Figure 7.19 Principle of repeated load indirect tensile test (schematic).

The above test method is given in detail in ASTM D4123-82, titled 'Standard test method for indirect tensile test for resilient modulus of bituminous mixtures'.

7.5.6 Stiffness Modulus and Fatigue Life of Bituminous Mix

The stiffness or elastic modulus of a bituminous mix can be determined by fatigue testing of a beam specimen. In this test, a beam of standard size made up of bituminous mix is subjected to repetitive dynamic loading in the form of a Haversine wave (1/10 s loading and 4/10 s rest period), at 1/3rd distance from both of its ends (Fig. 7.20) (Kallas and Puzinauskas, 1972; Yoder and Witczak, 1975; Monismith and Deacon, 1969).

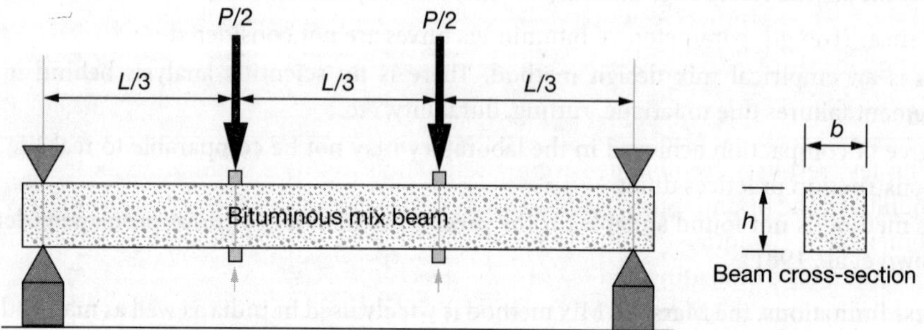

Figure 7.20 Fatig.20ue test on a bituminous mix beam (schematic).

About ten per cent of the load used is applied in the upward direction from the bottom side of the beam at the exact points where the beam is subjected to the downward load. This is done to reposition the beam horizontally after each load application. Then the beam is allowed to rest for a given duration of time. Such repetitive loading is performed in the laboratory to simulate the traffic wheel loading on in-service pavements. After initial straining or at a given number of repetitions, the elastic deflection (δ) of the beam at its centre (at a distance of $L/2$) is measured and the stiffness modulus value is determined by using the following equation which is derived from simple elastic principles.

$$E_s = \frac{23\,P\,L^3}{1296\,\delta I} \tag{7.32}$$

where,

E_s = Stiffness modulus of the bituminous beam specimen

P = Dynamic load $\left(\frac{P}{2}+\frac{P}{2}\right)$

L = Effective span of the beam

$I = \frac{bh3}{12}$

d = Elastic deflection of the beam at its mid-span

b = Average breadth of the beam

h = Average depth of the beam

The above test set-up (Figure 7.20) is also used for fatigue testing of bituminous mixes. In this test, the number of dynamic load repetitions is measured until the beam fails at a known range of stress. It is conducted at different temperatures similar to the temperatures experienced by in-service pavements. The fatigue test is conducted under controlled stress and strain conditions, which are suitable to evaluate thick (>150 mm) and thin (<50mm) bituminous pavement layers respectively (Yoder and Witczak, 1975; Monismith, Secor and Blackmer, 1961). Fatigue performance of a bituminous mix is generally expressed using the following relationship between fatigue life, the initial tensile strain at the bottom fibre of the beam and its elastic modulus (IRC:37-2001).

$$N_{fail} = C_1\left(\frac{1}{\varepsilon_t}\right)^{C_2}\left(\frac{1}{E_s}\right)^{C_3} \tag{7.33}$$

where,

N_{fail} = No. of dynamic load repititions to fail the beam

ε_t = Initial tensile strain at the bottom fibre of the beam

E_s = Stiffness modulus of bituminous mix

C_1, C_2 and C_3 = Constant values obtained from regression analysis of the fatigue test results.

EXAMPLE 7.6

The following test results were obtained from a fatigue testing of a bituminous beam. Calculate the stiffness modulus of the bituminous mix.

Beam dimensions: Breadth = 75 mm; Depth = 75 mm; Span length = 380 mm

Loading details: 0.1 s of loading and 0.4 s of rest period

Magnitude of loads at 1/3rd distance from both ends of the beam = 1.5 kN

Deflection at centre of the beam = 0.02 mm.

Answer

On substituting in equation (7.31), $E_s = \frac{23\,PL_3}{1296\,I\delta}$

$$P = \text{total load} = 1.5 + 1.5 = 3.0 \text{ kN}$$

$$L = 380 \text{ mm}$$

$$d = 0.02 \text{ mm}$$

$$I = \frac{bh^3}{12} = \frac{75\times(75)^3}{12} = 2636718.75 \text{ mm}^4$$

$$E_s = \frac{23\times 3.0\times(380)^3_3}{1296\times 2636718.75\times 0.02} = 55.4 \text{ kN/mm}^2$$

7.6 BITUMEN CONTENT DETERMINATION BY CENTRIFUGE EXTRACTION METHOD

This test is conducted as a post-construction quality control check of a constructed bituminous pavement. In this test, 500 gm (say W_1 g) of bituminous mix is cut from the constructed pavement layer or loose mix material may be obtained from the truck/ mixing-plant. This sample of the mix is added with benzene in a tray and allowed to melt for half an hour (Fig. 7.21). Later, this mix + benzene liquid

(a) Addition of benzene to the sample mix (b) Transferring the mix into the cup

(c) Placing of the cup in position (d) Rotation of the handle to spin mould

(e) Collection of the residual mix (f) Sieving and weighing of the residual mix

Figure 7.21 Different stages of bitumen extraction test and the mix-aggregates gradation check.

is carefully placed in the centrifuge extraction mould, sealed with filter paper and the metal cap of the mould is closed tightly. The extraction mould is then spun by rotating the handle about its vertical axis while benzene is continued to be poured until the draining benzene colour changes to clear and white.

The residue mix is then carefully taken out of the extraction mould, allowed to dry and its final weight is recorded as W_2 g. The bitumen content in the mix can be determined using the relation:

$$\text{Bitumen content in the mix (\%)} = \frac{W_1 - W_2}{W_1} \times 100 \qquad (7.34)$$

Later, the residual mix obtained from the extraction mould can be used for sieve analysis and verified against the standard table values for specified mix gradations.

Review Questions

1. Define the following
   a. Bitumen, (ii) asphalt, (iii) straight-run bitumen, and (iv) mastic asphalt.
2. State the desirable properties of bitumen and explain their significance on pavement design.
3. What is ageing process of bitumen? Define oxidation of bitumen.
4. What criteria should be considered for selection of a suitable binding material for construction of pavements?
5. Enumerate the various types of tests conducted on bitumen and explain the principle of each test.
6. What are the standard test temperatures at which bitumen is tested? What temperatures relevant to the service condition and construction of pavements?
7. State the significance of the RTFO test for characterising bitumen.
8. How does an RTFO test simulate adverse temperature conditions?
9. Explain the relationship between temperature and needle penetration of bitumen with standard equations.
10. State the limitations of the needle penetration test. Why is the viscosity test mandatory?
11. What is the difference between AASHTO and IRC methods of classification of bitumen? List the different types of bitumen with their viscosity values?
12. Discuss different types of viscometers and their working principles.
13. State the limitations of the orifice or efflux type viscometers.
14. With a neat sketch, explain the working principle of the Saybolt Furol viscometer?
15. State the difference between dynamic and kinematic viscosity.
16. Explain the working principle of a rotational viscometer with a neat sketch?
17. For different grades of bitumen, write the limiting values of (i) softening point, (ii) ductility values, (iii) loss on heating value, (iv) solubility value in CS_2 and (v) normal range of specific gravity according to IRC specifications
18. What is the significance of the Fraass breaking point in characterisation of paving bitumen? Write the limiting values of the Fraass breaking point for bitumen used in India.
19. What is meant by bitumen emulsions? State their merits and demerits in pavement construction.
20. Distinguish between cutback and emulsion. Explain the basic properties and uses of each.
21. What do you understand by bitumen emulsion? What are the different types of bitumen emulsions used for road construction?
22. What are the factors affecting the selection of suitable emulsion for pavement construction?
23. State the factors affecting setting and curing of emulsion mixes.
24. Explain with a figure the bonding phenomenon of emulsion with aggregates.
25. Differentiate between tar and bitumen. How are their predominant properties useful for pavement construction?
26. Write a note on the different types of road tars according to IS: 215.
27. How is sulphur used as a bitumen modifier? What are the useful properties of sulphur-modified bitumen for road construction?
28. State the advantages of lime-modified bitumen for road construction.

29. Write a short note on polymer modified binders for road construction.
30. What are thermosetting polymers? How they are useful in pavement construction?
31. What are the main differences between elastomers and plastomers?
32. Enumerate the different type of plastomers and elastomers.
33. What are the different types of thermoplastics used for pavement construction?
34. What are the uses of rubber-modified bitumen?
35. What is meant by performance-grade binders? Explain how it is superior to that of ordinary grade bitumen?
36. What are the salient features of performance-grade binders?
37. Compare ordinary grade and performance-grade binders.
38. Distinguish between ordinary grade and performance-grade binders used for road construction.
39. What tests are conducted for performance-graded binders and how are they relevant to conditions of in-service pavements?
40. How are the design temperatures determined in performance-graded binders? What other factors are considered before selecting a suitable grade of binder?
41. Write a short note on stiffness modulus of bitumen.
42. Define dynamic modulus, creep compliance and shear compliance of a bituminous material.
43. What are the factors affecting bituminous mix design?
44. State the desirable properties of bituminous mixes.
45. Explain the different methods of proportioning mineral aggregates.
46. Write the salient features of the Marshall Method for designing bituminous mix.
47. Using charts, interpret the results obtained from a Marshall test.
48. Write the limitations of Marshall Mix design method.
49. Explain the test method of resilient modulus of bituminous mix.
50. Explain how the stiffness modulus and fatigue life of bituminous mix are determined.
51. Explain how the bitumen content is determined by centrifuge extraction method.

Multiple Choice Questions

1. The needle penetration test is conducted on bitumen at what standard temperature?
(a) 4° C (b) 25° C
(c) 4° and 25° C (d) 27° C
2. In rolling thin-film oven (RTFO) test, the rack can accommodate up to 8 glass bottles and it rotates about its centre in a vertical plane at a predetermined speed of ________ rpm.
(a) 15 (b) 20
(c) 33 (d) 10
3. To determine the softening point of bitumen by the rind-and-ball method, the weight of steel spear used should be
(a) 2.5 g (b) 3.5 g
(c) 9.5 g (d) 10.5 g
4. By rolling thin-film oven (RTFO) test, to accelerate the hardening, hot air is blown into the bottle by means of an air jet at the rate of ________, after each rotation of the rack.
(a) 40 ml/minute (b) 400 ml/minute
(c) 4000 ml/minute (d) 4 ml/minute
5. According to IS:73, the minimum temperature of the softening point for paving bitumen of grades VG 10, VG 20, VG 30 and VG 40 should be ________ °C, respectively.
(a) 10, 20, 30 and 50 (b) 40, 45, 47 and 50
(c) 30, 45, 47 and 50 (d) 42, 45, 47 and 50
6. As per IS: 1206 (Part 2), vacuum tube capillary viscometers are recommended to test fluids having absolute viscosity in the range of ________ poises.
(a) 10^{-2} to 1000 (b) 10^{-3} to 100
(c) 10^{-2} to 10000 (d) 42 to 200000
7. By conducting ductility test on bitumen, the minimum cross-sectional area of the bitumen filled in the briquette mould is ________ sq. cm.
(a) 0.5 (b) 1.0
(c) 1.5 (d) 2.0

8. In the flash point test conducted using Pensky Marten's apparatus, the molten bitumen is stirred at a rate of ________ rpm.

(a) 30 (b) 33
(c) 60 (d) 163

9. For preparation of bituminous mix sample for conduct of Marshall Test in the laboratory, during mixing, the bitumen viscosity should be maintained at ________ centistokes.

(a) 170 ± 20 (b) 170 ± 10
(c) 280 ± 30 (d) 280 ± 10

10. According to IS: 1206 (Part 3), capillary type viscometers are recommended to test paving grade cutbacks of bituminous materials and distillation residues of cutbacks having kinematic viscosity in the range of ________ centistokes.

(a) 10^{-2} to 1000 (b) 10^{-3} to 100
(c) 30 to 100 000 (d) 42 to 200 000

Answers to Multiple Choice Questions

1. (c) **3.** (b) **5.** (b) **7.** (b) **9.** (a)
2. (a) **4.** (c) **6.** (b) **8.** (c) **10.** (c)

CHAPTER 8

Pavement Materials: Portland Cement and Cement Concrete

8.0 INTRODUCTION

Cement concrete is a versatile and commonly used product for civil engineering constructional works of the 20th and 21st century (A cement factory is shown in Fig. 8.1). A fresh cement–concrete mix consists of ingredients such as cement, minor aggregates, water and admixtures, if required. These ingredients are designed to provide good workability during placing and compaction when mixed in optimal proportions. After hardening, the concrete provides a strong, stable and durable compacted pavement layer which can resist repetitive impact wheel loads as well as withstand effects of environmental variations.

Therefore, it is imperative that the ingredients of a mix is properly selected based on their geological origin, physical, chemical and engineering properties. In the following sections of this chapter, various properties of these ingredients and methods of proportioning the mixes for cement concrete pavement construction will be briefly discussed.

Figure 8.1 Different views of a cement manufacturing factory.

8.1 CEMENT

Cement is the most important constituent of the mix. In the presence of water, it binds together all other ingredients, viz., mineral aggregates, in a compacted form. Cement mixed with water becomes a plastic paste and spreads all over the surface area of aggregates forming a thin film. It subsequently sets with time due to a chemical reaction in which the compounds of cement form aluminum silicates. The process which results in the hardening of cement concrete with time, in the presence of water, is known as the *hydration process*.

The first artificial cement was patented by Joseph Aspdin, in England, in the year 1824. This early cement had similar chemical composition of what is now known as *Ordinary Portland Cement* (OPC). Details about proportions of different chemical compounds in cement and the associated manufacturing processes with different mineral ingredients can be found in literature (Nawy, 1997; Neville, 2005).

Calcarious and silicate mineral compounds are blended and heated to high temperatures of the order of 1500°C to form clinkers of cement. During this process, new chemical compounds are formed. These clinkers are ground to fine particles. The resulting fine powder is called cement. The new compounds in the form of oxides are generally determined by Bogue's method (Taylor, 1991). Different types of chemical compounds in OPC with their approximate proportion are given below.

- Tricalcium silicate ($3CaO\ SiO_2$)..............................≈ 50%
- Dicalcium silicate ($2CaO\ SiO_2$)..............................≈ 22%
- Tricalcium aluminate ($3CaO\ Al_2O_3$)..........................≈ 9%
- Tetracalcium aluminoferrite ($4CaO\ Al_2O_3\ Fe_2O_3$).......≈ 9%
- Other compounds..≈ 10%

Tricalcium and dicalcium silicates contribute to the gain (rapid or slow) of strength in cement. The other compounds do not contribute to immediate strength gain but they do have an effect on long-term strength gain. Any slight change in proportioning of raw materials during the process of manufacturing the clinker form of cement may result in large variations in formation of the above-mentioned compounds. Thus, the chemical composition has a significant effect on cement properties and also on performance of the concrete.

Cements are classified into different types based on their chemical composition and their utility. The properties of cement can be modified by blending in different types of admixtures at different stages of the manufacturing process. Following are some of the different types of cements widely used in India.

- Ordinary Portland cement (OPC)
- Rapid-hardening Portland cement, RHPC (IS: 8041)
- Sulphate-resisting cement (IS: 12330)
- Hydrophobic cement (IS: 8043)
- High-alumina cement (IS: 6452)
- Low-heat cement (IS: 12600)
- White cement (IS: 8042)
- Portland Pozzolana cement (IS: 1489) (It reduces the heat generated in concrete by hydration and helps to gain early strength)
- Portland blast furnace slag cement (IS: 455) (It reduces the heat due to the hydration process and helps to gain strength. Setting time can also be altered)

According to the Bureau of Indian Standards (BIS), Ordinary Portland Cement (OPC) is classified based on its grade.

- OPC grade 33 (IS: 269)
- OPC grade 43 (IS: 8112)
- OPC grade 53 (IS: 12269)

The grade number indicates the compressive strength of cement mortar in MPa at 28 days. Generally, OPC grade 43 is widely used for construction of highway pavements since its rate of heat of hydration and shrinkage cracks are found to be considerably less compared to OPC grade 53. It also meets the other basic requirements (NAC 2001) elaborated in the following section.

Rapid-hardening Portland cement (RHPC) (IS: 8041) is preferred for remedial and rehabilitation works of concrete pavements where gain of strength in a short duration is desired to restore traffic corridors. The rapid strength gain of RHPC is due to the high specific surface area of the cement which is more than 325 m^2 per kg. In terms of chemical composition, the proportion of dicalcium silicate compound is kept lower than that of tricalcium aluminate.

8.2 DESIRABLE PROPERTIES OF CEMENT

As mentioned earlier, properties of cement are very sensitive to even a slight change in their chemical composition. The *desirable properties of cement* should be first ascertained in a laboratory before manufacturing it.

During construction, if a particular brand of cement is changed, its properties should be checked with reference to design specifications (Hall, 1932). The various tests conducted to determine the properties of cement and their limitations are briefly discussed below.

8.2.1 Fineness

The cement clinkers formed in kilns are pounded and ground to fine sizes of particles of the order of 90 micron and less. Fineness of cement particles is related to its grain size and specific area. The smaller the size of particles, the higher will be the specific surface area of cement. These are the indicative parameters of the fineness of cement.

Fineness is a desirable feature in cement—it helps cement concrete or cement paste to attain high strength within a required time interval. It is, thus, related directly to the cement's intensity of hydration and gain of strength. Excessive fineness of cement also poses problems related to cost of grinding and more water requirement for hydration, which may result in reduced strength and durability. Hence, an optimum value of fineness is a desirable feature to be considered at the time of manufacturing.

The following tests are performed to determine the fineness of the cement.

- Sieve analysis
- Wagner turbidimeter test
- Blain air permeability test

In the sieve analysis method, dry cement is sieved through a sieve of 90 micron size and the weight of cement retained on the sieve is reported in percentage. According to IS: 269-1967, the weight of OPC and RHPC retained should not be more than 10% and 5%, respectively. This test is not generally preferred since the percentage of cement retained on a standard size of sieve does not provide information about the proportions of other sizes of finer particles in the cement.

In the Wagner turbidimeter (ASTM C 115-93) and Blain air permeability tests (ASTM C 204-92), the surface area of cement particles per unit weight is determined. In the Blain air permeability test, the surface area of a sample of OPC and RHPC should not be less than 225 m^2/kg and 325 m^2/kg, respectively (IS: 269-1993 and IS: 8041-1998).

8.2.2 Setting Time

The term *setting time* is related to the preparation of a workable cement mortar paste with water using a desired cement–water ratio and the time taken to harden the paste (calculated from the point of mixing water to the cement). Thus, it is indirectly related to the initial chemical reaction of cement with water to form aluminium-silicate compounds.

The setting time of cement depends upon environmental factors such as temperature, climatic conditions and humidity. There should also be no physical disturbances such as vibrations. As a standard, setting times are determined at a temperature of 20 ± 2°C and at a relative humidity of 65%.

The degree of consistency of the cement–water paste is determined based upon the percentage of water needed to be mixed to obtain a standard consistency (IS: 4031, Part 4). The setting time of cement is basically classified into two types, viz. initial and final setting times. Vicat test apparatus is used to determine both the setting times.

Initial setting time

The duration of time required to transform a wet cement paste to an initial solid state is termed as *initial setting time* of the cement. The reckoning of setting time starts immediately after adding water to the cement. This time duration is essentially required to delay the process of hydration or hardening process. Determination of initial setting time is important for transportation, placing and compaction of cement concrete. The minimum initial setting time for OPC and RHPC should not be less than 30 minutes (IS: 269-1993 and IS: 8041-1998).

Final setting time

The final setting time is the time taken for the cement mortar paste or cement concrete to harden sufficiently and attain the shape of the mould in which it is cast. During this period of time, the primary chemical reaction of cement with water is almost completed and the secondary chemical reaction commences to permanently form aluminum silicate bonds. The determination of setting time period also facilitates safe removal of scaffolding or forms. The maximum duration of time for final setting time is generally about 10 hours for OPC and RHPC (IS: 269-1993 and IS: 8041).

8.2.3 Soundness

The term *soundness* of cement is used to describe the cement's ability to resist volume expansion as a result of physical and chemical changes when water is added.

The change of volume in cement is caused by unburnt free lime (CaO), dead burnt MgO and also $CaSO_4$, which forms calcium sulphonates. The presence of these materials is due to defective proportioning of raw materials and malfunctioning of the heating process in kilns during manufacturing of cements.

There are two types of tests to evaluate the expansion characteristics of cement, viz., the Le Chatelier test and the autoclave test. In the Le Chatelier test, volume expansion of cement due to free lime is evaluated under increased temperature of steam for 3 hours and then with 24 hours of immersion in water at 27°C and after aeration for 7 days (BS: 4550 (Part 3)- 1978; IS: 269-1993; IS:4031 (Part 3)). Under standard test conditions, the expansion after the first and second stages of testing should not be more than 10 mm and 5 mm respectively (IS: 269-1993 and IS: 8041). If the tested cement does not satisfy any one of the above specified limits, it is not recommended for use.

In the autoclave test (IS: 269 and ASTM C 151-89), under standard test conditions of humidity, at a room temperature of 27°C, cement cube samples are tested for 60 to 75 minutes at a steam pressure of 2.1 MPa. Longitudinal expansion due to the presence of free lime and magnesia are evaluated. The linear expansion in the autoclave test during the first and second stages should not be more than 0.8 and 0.6% respectively (IS: 269 and IS: 8041).

To determine the soundness of cement due to the presence of sulphates and other compounds, it may be necessary to resort to chemical analysis.

8.2.4 Strength

To determine the strength of cement, generally, the cement–sand mortar or cement concrete cubes are tested for compression, tension and flexural strength. In some cases, torsion tests are also prescribed.

The structural adequacy of a cement concrete pavement to withstand the pounding action of traffic wheel loads and resist environmental conditions will primarily depend upon the strength of the cement paste which binds the aggregates. Therefore, conducting strength tests on cement mortar or cement concrete is an important parameter to decide suitability of the cement.

Conducting strength tests on pure cement paste is not recommended since it will undergo considerable shrinkage on drying and micro-cracks will develop in its mass. Therefore, the strength tests are carried out either on hardened cement mortar cubes made of cement and standard sand or on cement concrete cubes. In case of cement mortar cubes, the standard sand comprises three equal proportions (33.33%) of sizes 2 to 1 mm, 1 mm to 500 microns and 500 to 90 microns (IS: 650-1991). In south India, Ennore sand is recommended for all testing laboratories.

The compression test is conducted on a cube of cement–sand mortar prepared by mixing cement with standard sand (IS: 650-1991) in a 1:3 proportion by weight and mixed with a quantity of water (($P/4 + 3$) per cent of the combined weight of cement and sand, where P is the water content at standard consistency limit for OPC and RHPC (IS: 269-1993; IS: 4031 (Part 6)-1988)). Generally, water–cement ratios of 0.4, 0.5 and 0.6 are also considered to ascertain the strength of cement mortar cubes. The compressive strength of OPC should be tested after curing under water for 3, 7 and 28 days. The strength should not be less than 16, 22 and 33 MPa respectively (IS: 269-1993). In case of RHPC, the compressive strength of cubes which are tested after curing for 1 and 3 days should not be not less than 16 and 27 MPa respectively (IS:8041).

If cement concrete cubes are used, the relationship of strengths of either cement mortar or cement concrete with different proportions is evaluated and correlated to 3, 7 and 28 day compressive strengths.

According to the American practice, the strength of cement can also be determined by conducting flexural tests on cement–sand mortar beams. The test details are given in ASTM C 348-93. The other important tests required by the IS standards to be conducted on cement are given in Table 8.1.

Table 8.1 Important tests on cement.

S. no.	Test name	Relevant IS code
1 a b c	***Fineness tests*** Dry sieve analysis Blain air permeability test Wet sieve analysis	IS: 4031 (Part 1) IS: 4031 (Part 2) IS: 4031 (Part 15)
2	***Setting time tests*** Initial and final setting times by the Vicat apparatus test	IS: 4031 (Part 5)
3	Soundness test by the Le Chatelier apparatus or autoclave test	IS: 4031, Part 3, 1988
4	Strength test by compressive strength test	IS: 650 and IS: 4031 (Part 6)
5	Specific gravity test	
6	Heat of hydration by calorimeter	IS: 4031 (Part 9)
7	Chemical composition by chemical analysis	ASTM C 348-93
8	Density using the Le Chatelier flask	IS: 4031 (Part 11)
9	False set using the Vicat apparatus	IS: 4031 (Part 14)

Nowadays, standard deviations of the strength of the cement mortar for various proportions and for each month of storage are also demanded by the users.

8.2.5 Hydration of Cement

When water is mixed with cement, heat is liberated due to hydration—the reaction is an exothermic chemical reaction. The heat of hydration may be of the order of 60 calories per gram in low-heat cements to 120 calories per gram in normal cements. The liberation of heat is greatest in the first 72 hours. A temperature rise of the order of 40–50°C has been reported in technical literature while building massive foundations and concrete dam works. This heat causes cracking and shrinkage of cement concrete on curing. The rate of liberation of heat due to the hydration process increases if the following quantities are increased.

- Fineness of cement
- Tricalcium silicate
- Tricalcium aluminate

The heat of hydration can be determined by the vacuum flask method or by using an adiabatic calorimeter. In case of low-heat cement, the quantity of heat generated is reported to be of the order of 65 calories per gram at 7 days and 75 calories per gram at 28 days.

In hot weather concreting, the following measures may be taken to control the excess liberation of heat due to hydration.

- The mixing water should be cooled — if required, crushed ice pieces may used.
- Minimum cement quantity should be used.
- Use of mineral admixtures such as fly ash, pozzolana or slag; or chemical admixtures such as retarders or water reducers will considerably reduce the heat of hydration.

8.3 MINERAL AGGREGATES

Mineral aggregates consist of coarse aggregates (above 4.75 mm), fine aggregates (between 4.75 to 0.075 mm) and stone dust (below 0.075 mm). The major portion by volume (about 75%) of the cement concrete mix comprises of mineral aggregate only. Most of the impact and abrasive forces generated by the movement of traffic are borne and resisted by coarse aggregates along with cement paste as a binder. The fine aggregates and stone dust, as a mass, assist to attain a dense and durable mix. The strength properties of the hardened cement concrete, viz., toughness, strength, stability, resistance to abrasion and durability against external harmful agents, are influenced largely by the properties of mineral aggregates. Hence, mineral aggregates for construction of cement concrete pavements should be carefully selected based on tests (Table 8.2). The tests have been described in detail in Chapter 6.

Table 8.2 Important tests to be carried out on mineral aggregate for cement concrete pavements.

S. no.	Test name	Property to be determined	Relevant IS code
1	Sieve analysis	Size distribution and gradation	2386 (Part 1) and 383
2	Aggregate crushing test	Aggregate crushing value	2386 (Part 4)
3	Aggregate impact test	Aggregate impact value	2386 (Part 4)
4	Los Angeles abrasion test	Aggregate abrasive value	2386 (Part 4)

(Continued)

S. no.	Test name	Property to be determined	Relevant IS code
5	Water absorption test	Quantity of water absorption	2386 (Part 4)
6	Shape test	Angularity, flakiness and elongation indices	383 and 2386 (Part 1)
7	Soundness test	Durability	2386 (Part 5)
8	Specific gravity test	Specific gravities	2386 (Part 3)
9	Bulk density	Bulk density or bulkage of sand	2386 (Part 3)
10	Deleterious substance test	Impurities	2386 (Part 2) and 383

The specific gravity of natural mineral aggregates generally varies from 2.55 to 2.75. Therefore, density of concrete will be in the range of 22.5 to 24.5 kN/m^3.

8.3.1 Gradation or Particle Size Distribution

The strength of a compacted and hardened cement concrete mix depends upon its density or air void content. Suitable proportioning of different sizes of aggregates has to be made to get a dense mix. If the mineral aggregates are not properly proportioned, the resulting mix may contain more air voids or the cement paste content may not be sufficient to bind all the aggregates. This results in low strength of concrete. To avoid this unwanted mix proportion, a suitable standard gradation curve should be adopted for a specified mix and flexural strength. The basic definitions and details of gradation have already been discussed in Section 6.2.

Standard guidelines have been specified by the Bureau of Indian Standards, MORTH and IRC for construction of concrete pavements (IS: 383-1970; MOSRTH 2006; *Pocketbook* 2002). According to IS: 383-1970, the following gradation has been prescribed for different sizes of aggregates.

1. Coarse aggregates
2. Fine aggregates
3. All-in-one aggregates

The coarse aggregates are graded according to the nominal size of aggregates.

In the case of *all-in-aggregates*, suitable blending of different locally available aggregates may be carried out according to the limits of gradation given in IS: 383. All-in aggregates are classified into two types based on the nominal size of aggregates, viz., 40 and 20 mm, for which the sieve sizes range between 80 to 0.15 mm and 40 to 0.15 mm, respectively.

Mineral aggregates of different sizes obtained from different sources may be blended with reference to the above specifications of gradation to achieve the desired density and number of air voids.

8.4 WATER

Generally, potable water is used for concrete works. However, in extreme situations or at inaccessible sites, deserts or sea shores, the quality of water may be checked against tolerable level of impurities, viz. turbidity, algae, salts, fluorides, sulphates, chlorides, etc. The pH value of the water should be preferably in between 6 to 8. The properties of water and permissible limits of contaminants are mentioned in IS: 456.

If soil or water consists of sulphates (SO_4), the sulphate ions penetrate into the concrete. A chemical reaction takes place between the sulphate ions and cement in the presence of water to form gypsum and ettringite. Thereby, the volume of the concrete increases and results in the formation of surface cracks. This effect is considered significant if the sulphate content in water is in excess of 150 ppm. It can be prevented by reducing the permeability of concrete using high content of slag as mineral admixture (Bakker 1983) or lowering the water- cement ratio (<0.4) (Nawy, 1997). It can also be prevented by using sulphate-resistant cement.

According to the Indian practice, if sub-grade soils are contaminated with soluble salts like sulphates (SO_4) by more than 0.5 per cent, it is preferable to use sulphate resistant cement (IS:6909; MOSRTH 2001; IRC: SP:62-2004).

8.5 ADMIXTURES

The ingredients which are admixed in cement concrete mix may be categorised into three types, as:

1. Chemical admixtures: Retarders, Accelerators, plasticizers, Super-plasticisers, Pigments and other chemical compounds.
2. Mineral admixtures: Fly ash – all types, Silica fumes, blast furnace slag, rice husk ash, calcined clay (metakaolin), diatomaceous earth and other
3. Polymer-based admixtures: Resins and synthetic polymers – all types
4. Non-polymer admixtures: Fibers/beads/powder of metal and glass – all types.

The above admixtures, alone or in combination, can be blended with cement concrete mix during the process of batching to modify properties. Setting time, strength, workability and durability can be modified by reducing the water and cement quantities and introducing admixtures. A cement concrete mixed with an admixture should be evaluated according to the guidelines in codes of practices (IS: 9103 and IS: 6925) before it is used for construction works.

In the following sections, the different additives or admixtures generally used in cement concrete works are described. The economics of the use of these have to be ascertained before using them.

8.5.1 Retarders

Retarders increase the setting time of cement concrete by retarding the process of hydration with minimal loss of the final strength. They are useful in construction works during the hot summer season (IS: 9103). Retarders include lignin, borax, sugars, molasses etc and slow down the setting time by hours. For example, about half per cent of sugar by weight of cement retards the setting time of cement by 4 hours (Ghosh et al., 1973).

8.5.2 Accelerators (IS: 9103)

This type of admixture is blend with cement concrete mix to accelerate the process of hydration which is useful during construction in winter seasons or where rapid strength gain is required for early removal of forms or for road repair works. Calcium salts or triethanol-amine may be used as accelerators (Tucker, 1936). Calcium chloride ($CaCl_2$) of not more than 1.5 per cent by weight of cement is mixed in concrete for reinforced and un-reinforced concreting works (IS: 7861 (Part 2)).

8.5.3 Plasticisers and Super-plasticisers

Plasticisers increase the workability of cement concrete even at lower water content (water–cement ratio of less than 0.3). It is also used to reduce the heat of hydration. Plasticisers are also called *workability aids*.

The reduction of mixing water content varies from 5 to 15 per cent depending upon the type and quality of the concrete ingredients (Bagui, 2004). These admixtures are extensively used when cement concrete has to be pumped to great heights for constructing high-storeyed structures.

Super-plasticisers are also used to improve plasticity of fresh concrete. They are often high range water-reducing admixtures which are used in manufacturing high-strength concrete (above 60 MPa). They reduce the requirement of water by 15 to 30 per cent with no loss of viscosity, strength and improve flow characteristics. This admixture is also used to prevent loss of slump when ready-mix concrete is to be transported for long distances. Super-plasticiser is generally limited to 1 to 2% by weight of cement as the compressive strength of concrete may be reduced if higher quantity of super-plasticiser is used.

There are mainly three types of super-plasticisers (Jagadish et al., 1997; Ramachandran et al., 1998; Chaubal and Hans, 2000) that are commonly used.

- Modified ligno-sulphonates
- Melamine formaldehyde sulphonates
- Naphthalene formaldehyde sulphonates

Water-soluble polymers, referred to as polyacrylate esters (PAE) or polycarboxylates are also used as super-plasticisers at lower dosages to yield very high strength concrete (40–100 MPa) at low water–cement ratio (< 0.3). Proper care should be taken while fixing dosages; otherwise it may entrain air and cause problems related to set-retarding (Nelu and Carmel, 2006). Plasticisers are also very costly and only specific commercial firms deal with these chemicals.

8.5.4 Air-entraining Agents

Minute air bubbles (< 0.1 mm dia.) at uniform inter-bubble spacing are introduced in the concrete mix to improve workability and increase resistance to frost action. Frost action causes the water molecules in cement concrete to expand and during the process of thawing, they decrease in volume. This phenomena increases alternative stresses in the mass of concrete due to frost expansion. The stress can be relieved by the presence of very small pores or cavities which bring into play a stress diffusion mechanism. They act as barriers for preventing propagation of micro-cracks (Jackson et al., 1944). Care should be taken not to use air-entraining agents in excess quantity as it may result in highly porous concrete (IS: 9103). Animal or vegetable oils, surfactants (soaps or synthetic analogs of tall oil fatty acids), wood resins, etc may be used as air-entraining agents in concrete mix. Generally, in cold countries like the Antarctic region, the use of air-entraining agents is fairly common.

8.5.5 Water Reducers

Pozzolana (IS: 3812 (Part 2), burnt clay (IS: 1344), polycarboxylic acids and their salts improve workability at low water content and improve flow properties of fresh concrete by dispersing the flocculated cement particles. Pozzolana or blast furnace slag is used as replacement for cement up to a certain extent to prevent alkali–aggregate expansion. Use of super-plasticisers also reduce the required quantity of water by improving workability and reducing porosity. Extracts from pulp and paper industry in the form of derivatives of lignins acts as retarders and dispersants (Mark, 1938). Polynapthalene sulphonate is also used as a water reducing as well as a pigment dispersion agent in concrete (Nelu and Carmel, 2006). The economics of these admixtures are to be worked out before using them.

8.5.6 Pigments

Mineral oxide pigments are also used as admixtures to impart colour to the cement concrete mix for aesthetic and beautification purposes. Generally, pigments are added for crosswalks, pedestrian ways, footpaths and for any specific purpose of paved surfaces. The quantity of mineral oxide is limited to not more than 10% by weight of cement since it may considerably reduce the strength of the concrete (PCA, 1995). Natural or synthetic iron oxides of different colours, chromium oxide and cobalt oxide are some widely used pavement pigments, to name a few. More details and standard specifications of pavement pigments can be found in ASTM C 979 or the manufacturer's recommendations.

8.5.7 Others

Silica fumes, fly ash, blast furnace slag, rice husk ash, calcined clay (metakaolin) and diatomaceous earth are other mineral additives used in cement concrete. They are called supplementing cementitious materials. They modify the mix chemistry and thereby influence its strength, durability and workability properties (Nelu and Carmel, 2006). Use of silica fumes and blast furnace slag makes the cement concrete mix dense.

High-strength and low-permeability concrete is produced by using silica fumes (having a diameter $\approx$ 0.1 μm) since it has a high specific surface area (13,000 to 30,000 m^2/kg) and brings forth high pozzolanic activity (Nawy, 1997; Arora, 2003). Silica fumes are obtained from the blast furnaces of iron and steel manufacturing units. The workability of the concrete mix can be considerably increased as silica fumes act as powerful dispersants and water reducers at acceptable water content (Spiratos et al., 2003). Depending upon the required strength and workability, silica fume in the order of 5 to 30% by weight of cement is used (Nawy, 1997). Silica fumes were used in Srisailam dam to achieve high strength cement concrete.

Fly ash is used as a partial substitute for cement. It improves workability, reduces water content and thermal cracking of concrete. Fly ash concrete pavements are less susceptible to thermal cracking since the fly ash enables the hydration process to occur with the liberation of less heat. This lower hydraulic activity of fly ash may result in concrete bleeding and segregation (Khayat, 1998). A study reveals that the static flexural and flexural fatigue strength of concrete decreases when the cement replaced with fly ash increases from 15 to 50% (Naik and Singh, 1994). In addition to this, excess quantity of carbon (1 to 10%) in fly ash behaves like activated carbon which strongly affects the performance of air-entraining admixtures by erratic generation of air voids (Hill et al., 1997). The severity of this erratic formation of air voids depends on the fly ash constitution, viz., amount of residual carbon, surface area and chemical composition. These problems should be carefully addressed and investigated in the laboratory before selecting fly ash from different sources. Suitable admixtures should be used to balance the ill-effects of residual carbon in fly ash and to improve water retention, stability and rate of strength gain.

Fly ash should be selected according to the following standard specifications (IS: 3812-1981; IRC: SP: 62-2004)

- Minimum surface area: 320 m^2/kg
- Minimum lime reactivity: 4.5 MPa
- Compressive strength at 28 days: Above 80% of the standard cement mortar cubes
- Maximum drying shrinkage: 0.15%
- Maximum expansion by autoclave test: 0.8
- Maximum loss on ignition: 5%

Any one of the above admixtures or admixtures in multiple combinations may be used in a cement concrete mix to serve the desired purpose based upon scientific analysis. Rigorous testing is needed as these admixtures may become problematic in the hardening or hydration stage. They may cause mix disproportions and poor or non-performance due to their adverse reactions with the ingredients of the mix.

Uniformly dispersed alkali-resistant glass fibres in homogeneous concrete make the concrete mass resistant to a high degree of stress distribution. They improve the modulus of rupture and impact strength and act as a crack arrestor. The following commercially available glass fibres are used for concrete works (Grimer and Ali, 1969).

- Strands (200 or 400 single filaments of 10 μm dia. each).
- Rovings (20 or 60 single filaments of 10 μm dia. each).
- Woven or chopped strand mat.

For more details on different types of admixtures and their usability in cement concrete please refer to the *Concrete Admixtures Handbook* (Ramachandran, 1995).

8.6 FACTORS CONSIDERED FOR DURABLE CONCRETE

Besides the quality of materials, compaction methods, etc., the following factors also should be considered for the design of durable concrete mix.

- The effect of freezing and thawing on concrete
- Salt and chemical actions
- Differential thermal stresses
- Structurally weak and alkali reactive aggregates
- Water quality
- Leaching and efflorescence
- The action of abrasive and impact loads of traffic
- Shrinkage of concrete

8.7 DESIRABLE PROPERTIES OF FRESH CONCRETE

For any kind of concreting work, mixing, placing and compaction of concrete are the important operations to be performed until the final setting time is reached. The engineer desires to perform these tasks smoothly and in a cost-effective manner without compromising on the strength. The ease with which the above-mentioned operations can be performed is called workability. The workability of concrete is one of the most important factors to consider while handling fresh concrete (Venable, 1930). It decides the overall cost of the project as well as the performance in the post-construction period.

Details on workability of concrete are given below.

8.7.1 Workability

Workability is an important consideration while deciding on the various design parameters and the relative proportion of the ingredients in the mix economically. A concrete mix may be said to have good workability if we are able to perform the following operations with ease:

- mixing
- transportation and placement without segregation of coarse aggregate particles
- adjustment of mix in the desired shape of form or placing
- compacting
- achieving the desired surface finish, strength and durability.

The following important factors affect the workability of cement concrete mix:

- maximum size of coarse mineral aggregates and their gradation
- flakiness and elongation indices of aggregates

- water–cement ratio (WCR)
- aggregate–cement (a/c) ratio
- fine aggregate–coarse aggregate (f/c) ratio
- type and quantity of admixtures used
- type of concreting work—mixing, placing, compaction and finishing
- temperature control
- quality of work.

Strength of the fully compacted concrete is related to the water–cement ratio (WCR). According to Abrams' law (Abrams, 1918), the relation can be stated as

$$\text{Strength of concrete} = P/Q^{\text{WCR}} \tag{8.1}$$

where P and Q are empirical constants which depend on cement type, aggregates type, curing method, age and test method. WCR is the ratio between water and cement by weight. Abrams' law indicates that reduction of WCR beyond the minimum requirement does not improve concrete strength and results in poor workability.

The workability of cement concrete mix is generally determined by

1. slump test
2. compacting factor test
3. veebee (VB) time test
4. flow test.

Depending on the field conditions of the placement of cement concrete, an appropriate test may be followed. Standard guidelines for selecting a suitable type of test for determining the workability of cement concrete are discussed below.

For construction of highway pavements with a desired target strength of cement concrete, a very low degree of workability, less than or equal to 25 mm of slump, is desirable. However, for paving works, a slump value of 35 ± 15 mm may be suitable, depending on the site requirement (MOSRTH, 2001). To achieve lower degree of workability, great care should be taken to control the proportioning of ingredients during mixing. For low-workability cement, the compacting factor test is preferred and found to be more relevant than the slump. The values of the compaction factor prescribed are of the order 0.75 to 0.80 (IS: 456-2000). If the value of the compacting factor is lower than 0.7, it indicates an extremely stiff mix. In such cases, the veebee time test is more suitable (SP: 23-1985).

Although the slump test is suitable for determining low to high workability when the slump value is between 25 to 125 mm (SP: 23-1982), it is more useful and simple for controlling the manufacturing of concrete in field at all field temperatures. In case of very high workability of concrete, when the slump value is expected to be more than 150 mm, the flow test method is more suitable.

The detailed standard procedures for testing the workability of cement concrete by the above methods are given in IS: 1199-1959.

8.8 TESTS ON HARDENED CONCRETE

A brief discussion on the important tests conducted on hardened concrete is given below.

8.8.1 Compressive Strength Test

The value of the compressive strength of hardened concrete is an important property used while designing cement concrete mixes. Determination of compressive strength is very simple. Due to this reason, many other strength parameters of concrete, viz. tensile strength, shear strength, elastic modulus value, bond strength etc., are correlated with the values of compressive strength to determine its overall structural strength.

The compression test may be performed by any one of the following standards adopted in different countries, viz. the Indian practice (IS: 516-1959; 1999), the British practice (BS: 1881 (Part 108)-1983) and the American practice (ASTM C 39-86). The compression test procedure according to the Indian practice is described below.

The compressive strength of hardened cement concrete is determined by preparing a specimen of standard size (a cube of 150 mm size or a cylinder having a diameter of 150 mm and a height of 300 mm). It is cured by submerging in water for 28 days or steam curing for 7 days. After the given period of curing, the specimen is tested in a compression machine (Fig. 8.2)—a compressive load is applied at a uniform rate of approximately 14 N/mm²/minute on to the specimen (IS: 516-1959; 1999). The specimen is loaded gradually until it fails by cracking or breakdown. The failure stress is calculated from the maximum failure load (or peak value) and is recorded as the compressive strength of the concrete. An average value of three such test results is reported as the final compressive strength of the concrete. For approximate calculation of the equivalent compressive strength of a cylinder, about 80 per cent of the cube strength may be taken.

Figure 8.2 Cube compression test in progress.

Strength tests are also performed on cement concrete cubes cured under water for 3 days and 7 days—this is only to determine the trend of strength gain during short periods. These test values are projected and compared to the probable strength gain at 28 days. Generally, the ratio of strength gain for 7 to 28 days is calculated to be 2/3rd value approximately.

Long-term strength evaluation tests of 90 and 365 days are also conducted for important works like concrete for dams and hydraulic structures.

The characteristic compressive strength of concrete is defined as the value below which not more than 5% of the test results are expected to fail (IS: 456). It is also called as the 95% characteristic value. The target mean strength for a mix design takes in to account the characteristic strength of concrete plus the standard deviation times the desired tolerance limit.

8.8.2 Elastic Modulus Test

The elastic modulus (E) value of concrete is an indicative property for predicting many aspects of its strength and durability. It also helps in evaluating various mathematical models in engineering practices. It describes the stress-and-strain relationship. It is also used as an important input parameter along with Poisson's ratio (μ) for evaluation of the strength of a structural member.

The modulus of elasticity of cement concrete may be defined as a linear relationship (i.e., slope) between applied stresses to the corresponding recoverable or elastic strains. The value is determined by many methods. A conventional static load method for determination of modulus of elasticity is described below.

A cylindrical cement concrete specimen having standard size (150 mm diameter and 300 mm height) is suitably placed in an extensometer–compressometer and tested in a compression testing machine. The extensometer–compressometer consists of two dial gauges—one dial gauge is used to measure change in original length of specimen and the other is used to measure the diametrical expansion of the specimen under loading. A loading machine similar to the compression testing machine is used for loading the specimen in three cycles. About 30 percent of the ultimate compressive strength of the stress is gradually applied on to the specimen in the axial direction (IS: 516-1959). About 25 and 40 percent of the ultimate compressive stress may also be taken (ASTM C 469). Changes in lateral and diametric dimensions are recorded under loading and unloading processes. If the test results are consistent, the average values of Poisson's ratio and different types of modulus values based on the slope of the stress–strain cure (Fig. 8.3) will be calculated. For Indian conditions, the Poisson's ratio value for normal weight of concrete may be taken as 0.15.

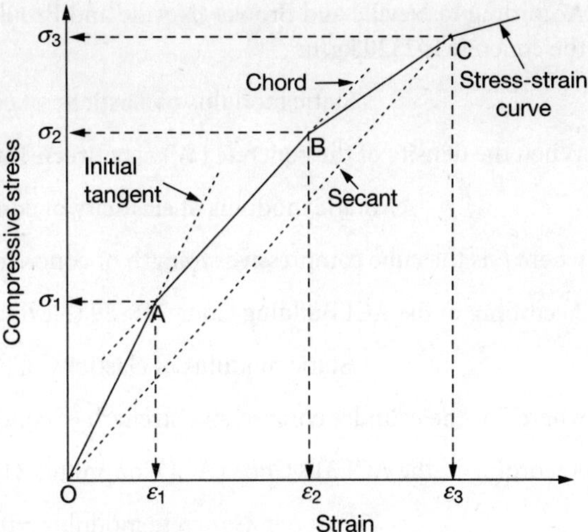

Figure 8.3 Types of elastic moduli.

The details of the calculation of different types of static modulus values are as follows.

With reference to Fig. 8.3:

By considering OA,

$$\text{Initial tangent modulus} = \frac{\sigma_1}{\varepsilon_1} \tag{8.2}$$

By considering AB,

$$\text{Chord modulus} = \frac{\sigma_2 - \sigma_1}{\varepsilon_2 - \varepsilon_1} \tag{8.3}$$

By considering OC,

$$\text{Secant modulus} = \frac{\sigma_3}{\varepsilon_3} \tag{8.4}$$

Dynamic modulus of concrete is also determined by the non-destructive testing (NDT) method called electro-dynamic or ultrasonic pulse velocity method. In general, the dynamic modulus values are more than the static modulus of elasticity values. The ratio between static and dynamic modulus may be taken as 0.8 (IRC: 58 2002). Details of the above test procedure may be obtained from IS: 516-1959 and ASTM C 469.

The following relationships may suitably be used to estimate the elastic modulus value of cement concrete for a typical value of normal weight and compressive strength.

According to Carse and Behan (Carse and Behan, 1980),

$$\text{Elastic modulus of concrete} = C \times W^{1.5}\,(f_c)^{0.5}\ \text{MPa} \tag{8.5}$$

where

C = a constant varying between 0.04 to 0.06 depending on the type of concrete
f_c = characteristic compressive strength of cylinder in MPa
W = density of concrete in kg/m^3

According to IS: 456-2000,

$$\text{Static modulus of elasticity of concrete, } E_s = 5000\,(f_{ck})^{0.5}\ \text{MPa} \tag{8.6}$$

where f_{ck} is the characteristic compressive strength of concrete in MPa.

According to Neville and Brooks (Neville and Brooks, 2002), for normal concrete, when the density of the concrete is 2320 kg/m^3,

$$\text{Static modulus of elasticity of concrete, } E_s = 9100\ (f_c)^{0.33} \text{ MPa} \quad (8.7)$$

When the density of the concrete (W) is between 1400 to 2320 kg/m^3

$$\text{Static modulus of elasticity of concrete, } E_s = 1700\ W^2\ (f_c)^{0.33} \text{ MPa} \quad (8.8)$$

where f_c is the cube compressive strength of concrete in MPa.

According to the ACI Building Code 318-89 (Neville and Brooks, 2002),

$$\text{Static modulus of elasticity of concrete, } E_s = 4700\ (f_c)^{0.5} \text{ MPa} \quad (8.9)$$

where f_c is the cylinder compressive strength of concrete in MPa.

According to the ACI 318 Code (ACI Committee 318-1996; Nawy 1997),

$$\text{Concrete modulus} = 0.043\ W^{1.5}\ (f_c)^{0.5} \text{ MPa} \quad (8.10)$$

where compressive strength (f_c) up to 42 MPa and density of concrete (W) for normal and light weight concrete is 2400 and 1765 kg/m^3 respectively.

According to the ACI 435 Code (ACI Committee 435-1995; Nawy 1997),

$$\text{Concrete modulus} = (6895 + 3.32\ (f_c)^{0.5}) \times (W/2320)^{1.5} \text{ MPa} \quad (8.11)$$

where compressive strength (f_c) ranges from 42 to 83 MPa and density of concrete (W) for normal and light weight concrete is 2400 and 1765 kg/cm^3 respectively.

According to the UK practice of BS: 8110 (Part 2)-1985 (Neville and Brooks, 2002),

$$\text{Modulus of elasticity at 28 days} = 20{,}000 + 0.2 \times f_c \quad (8.12)$$

where f_c is the cube strength at 28 days. Eq. (8.12) is valid for normal weight aggregate concrete.

For the thickness design and prediction of deformations of concrete pavements, the average value of the static modulus of elasticity of concrete is required. It depends upon the cement type, the ingredients of the concrete, the proportion, the prevailing construction practices and the testing procedures adopted. Different organisations prescribe different values (Table 8.3). When actual test results are not available, an appropriate selection of a suitable static elastic value may be made.

Table 8.3 Values of static modulus of elasticity of concrete (IRC: 58-2002; Neville and Brooks, 2002).

S. No.	Static modulus of elasticity of concrete (MPa)	Concrete grade	According to
1	Mean values: 28,000	M 40	BS: 8110 (Part 2)-1985
2	Mean values: 32,000	M 40	The ACI Building Code 318-89
3	Mean values: 28,000		PCA, USA
4	21,000; 28,000; 35,000; 42,000; 49,000		AASHTO
5	Between 35,000 to 40,000		Croney and Croney
6	Mean values: 29,430 (3×10^5 kg/cm^2)	M 40	IRC: 58-2002

8.8.3 Flexural Strength Test

This test is conducted to determine the extreme fibre tensile strength of concrete in pure bending by indirect means. In this test, a concrete beam (150 × 150 × 450 mm), simply supported at its ends, is loaded at two points. When concentrated loads are applied, the bottom fibres of the beam experience maximum tensile stresses up to failure. The maximum tensile stress developed may be termed as modulus of rupture. If an unequal settlement of the base course occurs, it causes loss of support by creating voids underneath the pavement. Therefore, the flexural strength or tensile strength in bending of concrete pavement slabs play an important role in resisting traffic wheel loads. This is true even when there is loss of support for distributing the load over a wider area. Hence, the flexural strength property of concrete is a very relevant property to evaluate critical stresses and strains in thickness design of rigid pavements.

In the flexural strength test, a cement concrete beam having standard cross-sectional area (150 × 150 mm) is moulded and compacted according to a given standard specification. The prepared beam is cured for 28 days and tested by applying equal loads at one-third points of the beam (whose effective length of span ranges from 450 or 600 mm) (Fig. 8.4). The beam is uniformly loaded at a constant rate (4 to 4.5 kN/minute) until it fails due to fracture. Tensile stresses are induced along the bottom fibre of the beam due to loading. The beam fails by fracture when the bending moment in the beam is unable to withstand the induced tensile stresses. The failure load at fracture and the distance from a nearest support of the fracture is recorded to calculate the flexural strength of the concrete beam.

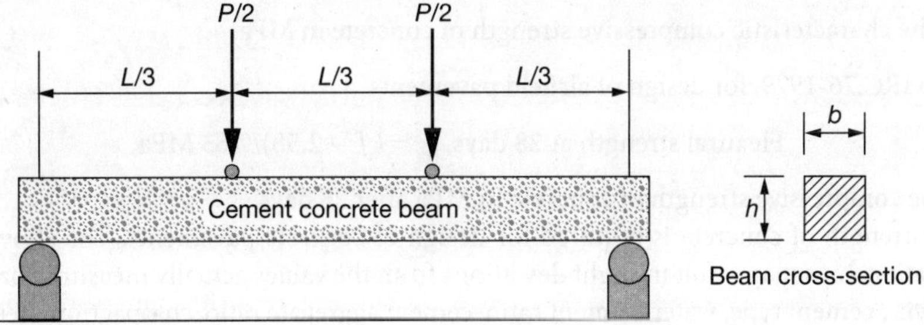

Figure 8.4 Flexural strength test on cement concrete beam (schematic).

The following equations are used to calculate the flexural strength of modulus of rupture (MR) of the concrete.

(i) If the fracture occurs in the middle third of the beam (i.e., the horizontal distance in between the application of loads)

$$MR = PL/(bh^2) \quad (8.13)$$

(ii) If the fracture occurs at one-third distance from any support (i.e. along one end of the beam, one-third distance from a support)

$$MR = P \times 3\,a/(bh^2) \quad (8.14)$$

where

P = total load ($P/2 + P/2$)
L = span of the beam
b = average breadth of the beam
h = average depth of the beam
a = occurrence of fracture at a minimum distance from the nearest support (the value should not be closer than 5 percent of the span) or $a = (L/3) - (0.05 \times L)$

For more details of the flexural strength test procedure, readers may refer to Codes ASTM C 78-84, IS: 516-1959 and BS: 1881 (Part 118)-1983.

The tensile strength of the concrete may also be estimated by using following equations.

According to IS:456: 2000 and IRC:SP:62-2004,

$$\text{Flexural strength or modulus of rupture, } f_{cr} = 0.7\,(f_{ck})^{0.5} \text{ MPa} \quad (8.15)$$

According to PCA (SP: 23-1985, 2001),

$$\text{Flexural strength of crushed quartzite aggregates, } f_{cr} = 0.73\,(f_c)^{0.5} \text{ MPa} \quad (8.16)$$

$$\text{Flexural strength of rounded river pebbles, } f_{cr} = 0.48\,(f_c)^{0.5} \text{ MPa} \quad (8.17)$$

According to ACI 318-77 (SP: 23-1985, 2001),

$$\text{Flexural strength of crushed quartzite aggregates, } f_{cr} = 0.63\,(f_c)^{0.5} \text{ MPa} \quad (8.18)$$

where f_c is cube compressive strength of concrete in MPa.

According to Croney and Croney (IRC: 58-2002),

$$\text{Flexural strength or modulus of rupture of gravel aggregates, } f_{cr} = 0.49\,(f_{ck})^{0.55} \text{ MPa} \quad (8.19)$$

$$\text{Flexural strength or modulus of rupture of crushed aggregates, } f_{cr} = 0.36\,(f_{ck})^{0.7} \text{ MPa} \quad (8.20)$$

where f_{ck} is the characteristic compressive strength of concrete in MPa.

According to IRC:76-1979, for design of airfield pavements,

$$\text{Flexural strength at 28 days, } f_{cr} = (f_c - 2.58)/7.63 \text{ MPa} \quad (8.21)$$

where f_c is the compressive strength of concrete in MPa after 28 days.

Flexural strength of concrete is required for design of concrete pavements. The above-mentioned statistical relationships may result in slight deviations from the values actually measured based on many parameters, viz., cement type, water–cement ratio, cement aggregate ratio, compaction method, density, type of admixtures used in concrete, aggregate size, shape and type. Hence, it is more appropriate to determine the flexural strength by experimental methods. However, if such experimental data is not available, a suitable approximate estimation may be made using the above equations. For Indian conditions, based on the type of aggregate, a flexural strength value of 45 kg/cm^2 should be taken for M 40 grade of concrete (IRC: 58-2002).

EXAMPLE 8.1

A few samples of cement concrete mix beams are prepared for determination of their flexural strength. The beams are tested at one-third loadings. Calculate the modulus of rupture of concrete beams from the following results of the flexural strength test.

Specimen 1

Span of the beam = 450.5 mm

Average breadth of beam = 150.2 mm

Average depth of beam = 150.2 mm

Total failure load = 35,126 N

Location of fracture = 200 mm from a nearest support

Curing period = 7 days

Specimen 2

Span of the beam = 450 mm

Average breadth of beam = 150.0 mm
Average depth of beam = 149.99 mm
Total failure load = 46,709 N
Location of fracture = 135 mm from a nearest support
Curing period = 28 days

Specimen 3

Span of the beam = 450 mm

Average breadth of beam = 150.0 mm
Average depth of beam = 150.0 mm
Total failure load = 36,947 N
Location of fracture = 90 mm from a nearest support
Curing period = 14 days

Answer

Specimen 1

On substituting the given values in equation (8.13), we get

$$MR = 35{,}126 \times 450/(150.1 \times 150.2^2) = 4.67 \text{ MPa}.$$

Specimen 2

On substituting the given values in equation (8.14), we get

$$MR = 46{,}709 \times 3 \times 135/(150.0 \times 149.99^2) = 5.60 \text{ MPa}.$$

Specimen 3

The minimum distance should be calculated as follows.

$$a = (450/3) - \{(5/100) \times 450\} = 150 - 22.5 = 127.5 \text{ mm}$$

The value of a is within 5% of the span from a nearest support or 90 mm < 127.5 mm. Hence, the test value may be discarded.

Generally, the ratio of flexural to compressive strengths is determined for correlation purposes. As an example, concrete of grade M25 has a flexural strength of 8 to 13% of compressive strength.

8.8.4 Split Tensile Strength Test

This test is conducted to determine the tensile strength of concrete by indirect means. A concrete test specimen is made into a cylindrical shape of diameter 150 mm and length 300 mm (or a cubical shape whose size is ≥ 150 mm). It is loaded diametrically (Fig. 8.5) at a uniform rate of 1.2–2.4 N/mm²/minute (IS: 5816-1999). On loading, the specimen fails by splitting across the diameter due to the development of tensile stresses on its diametrical plane. The maximum failure load is recorded and the following equation is used to calculate the splitting tensile strength of concrete (IS: 5816-1999).

$$\text{Splitting tensile strength } (f_{st}) = \frac{2 \times P}{\pi \times D \times L} \tag{8.22}$$

Where,

P = Maximum failure load(N)
D = Diameter or depth of the specimen(mm)
L = Length or depth of the specimen(mm)

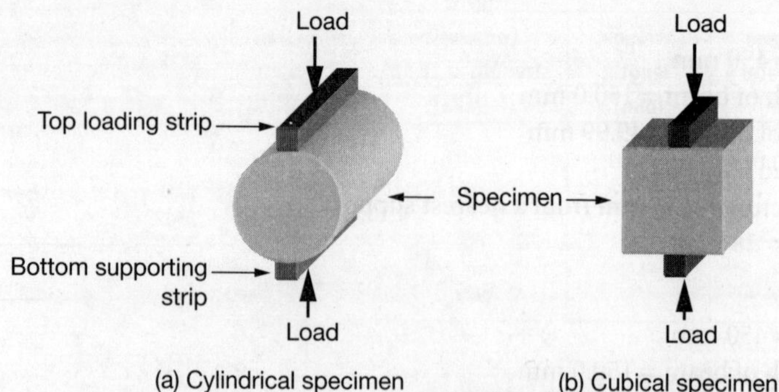

Figure 8.5 Principle of indirect tensile strength test (schematic).

The relevant codes may be referred to for detailed test procedures (IS: 5816-1999; BS: 1881 (Part 117)-1983 and ASTM C 496-90). This test may also be performed on extracted cores from the field. Suitable care should be taken during testing of these cores. The standard test specifications should be strictly followed to obtain consistent and reliable results.

The ratio of the split tensile to compressive strengths of cement concrete is also evaluated to correlate its strength properties. To determine the tensile splitting strength for normal-weight concrete, the following equation is recommended by the ACI Committee 363, 1992 and ACI 1997 (Nawy, 1997).

$$f_{st} = 0.59 \times (f_c)^{0.5} \quad (8.23)$$

where,

f_{st} = Tensile splitting strength of concrete in MPa

f_c = Cylinder compressive strength in MPa.

EXAMPLE 8.2

Calculate the splitting tensile strength of cement concrete from the following test results:

Specimen	Average diameter (mm)	Average length (mm)	Compressive strength at 28 days (MPa)	Failure load (N)
1	150.3	300.2	40	223376
2	150.4	300.3	40	242065
3	150.1	300.4	40	238950

Answer

Specimen	Average diameter (mm)	Average length (mm)	Compressive strength at 28 days (MPa)	Failure load (N)	Splitting tensile strength (MPa)	Percent of compressive strength
1	150.0	300.0	40	223376	$\dfrac{2 \times 223376}{\pi \times 150.0 \times 300.0}$ $= 3.16$	$\dfrac{3.16}{40} \times 100$ $= 7.90$
2	149.99	300.0	40	242065	$\dfrac{2 \times 242065}{\pi \times 149.99 \times 300.0}$ $= 3.42$	$\dfrac{3.42}{40} \times 100$ $= 8.55$

Specimen	Average diameter (mm)	Average length (mm)	Compressive strength at 28 days (MPa)	Failure load (N)	Splitting tensile strength (MPa)	Percent of compressive strength
3	150.0	300.0	40	238950	$\frac{2\times 238950}{\pi\times 150.0\times 300.0} = 3.38$	$\frac{3.38}{40}\times 100 = 8.45$
				Average	3.32 MPa	8.30%

8.9 DESIGN OF CEMENT CONCRETE MIX

Basically, design of cement concrete mix consists of proportioning its ingredients in order to attain the desired workability, strength, durability and economy. A few of the methods are enumerated below.

1. Arbitrary method
2. Minimum void method
3. Fineness modulus method
4. Talbot–Richart method
5. American Concrete Institute (ACI) method
6. US Bureau of Reclamation method
7. Road note no. 4 method
8. British method
9. Indian method according to the
   - Bureau of Indian Standards (IS:10262)
   - Indian Roads Congress (IRC:44)

The first four methods are purely empirical and need many trial tests to attain the desired degree of workability and strength even though there is not much change in the quality of the ingredients. The economy as well as the durability of the hardened concrete are also not given due consideration.

The rest of the mix design methods were developed in different countries and are based on standard formulae, correlations, charts and tables which were developed from experience, extensive research and practical considerations based on the ingredients, proportioning, compaction methods, density, curing, content of air voids, admixtures, temperature, climatic conditions, quality of materials, durability, strength and finally, cost effectiveness. The basic principles of the different mix design philosophies are similar when the proportions, strength, durability, performance and cost effectiveness are considered. However, they differ slightly with respect to variations in mix ingredients, compaction methods, air void content, temperature, corrections due to change in material properties, target strength and purpose of construction by employing standard mathematical and probabilistic methods. The Bureau of Indian Standard method of concrete mix design (IS: 10262) and IRC:44 are the most widely used methods. The details of these mix proportioning methods are out of the scope of this book and they can be obtained from the Chapter 5 of Text Book of Highway Engineering by R Srinivasa Kumar, Universities Press.

8.10 DRY LEAN CEMENT CONCRETE

The concrete mix used for construction of dry lean cement concrete sub-base should be proportioned with a maximum cement aggregate ratio of 1:15 in which the maximum size of coarse aggregate should not be more than 25 mm. The aggregate should be blended according to the standard gradation prescribed by IRC (MOSRTH 2001). The minimum content of cement should not be less than 150 kg per cubic metre of concrete. At least five trial mixes should be made by varying the water content—ranging from 5.0 to 7.0 per cent in increments of 0.5 per cent. They should be checked for compactability and strength. The compressive strength of these cubes of standard dimension should not be less than 10 N/mm^2 at 7 days. A similar procedure may be adopted for proportioning concrete mix for construction of rolled cement concrete base.

8.11 CONCRETE MIX DESIGN FOR RURAL ROADS

For rural roads or low-volume roads, the characteristic compressive and flexural strengths at 28 days should not be less than 30 MPa and 3.8 MPa respectively. Flexural strength of cement concrete based on characteristic compressive strength is determined using Eq. (8.15). The target strength to be achieved should be taken as the minimum required flexural strength in the field. Similar to the conventional concrete mix design, the following equation is used to determine the target average flexural strength of cement concrete (IRC: SP: 62-2004; MORD, 2004).

$$F_{\mathrm{TF}} = f_{\mathrm{cf}} + K \times \sigma \tag{8.24}$$

where

F_{TF} = Average target flexural strength at 28 days in $\mathrm{N/mm^2}$.

f_{cf} = Characteristic flexural strength of concrete in $\mathrm{N/mm^2}$; it may be defined as the average value of flexural strength of standard beams below which not more than five percent of the test results are found to fail.

K = A statistical constant representing the tolerance limits desired. From statistical consideration, for a tolerance limit of 1:20, the value is 1.65.

σ = Standard deviation of field test values which depend upon degree of quality control during concrete production and construction. A suitable value may be taken from Table 8.4.

Mix design for cement concrete rural roads is made based upon the type of equipment used for compaction. There are two methods for compaction, viz., roller compaction and compacting using other tools. Roller compacted concrete is a zero-slump concrete which is transported, placed, spread and compacted using vibratory rollers with a 10 tonne capacity low-amplitude and high frequency vibration (Schrader, 1997; Seehra, 2001 and Ashraf, 2004).

Table 8.4 Aggregate gradation for cement concrete rural roads (Table 1500.3 of MORD 2004).

Sl. No.	IS sieve size (mm)	Per cent passing by weight
1	26.5	100
2	19.00	80–100
3	9.50	55–75
4	4.75	35–60
5	0.60	10–35
6	0.075	0–8

Roller-compacted concrete does not relate to the Abram's rule of water– cement ratio since it is a no-slump concrete (Seehra, 2001 and IRC: SP: 62-2004). To minimise segregation, the maximum size of aggregates in roller compacted concrete is restricted to 25 mm (Seehra, 2001 and IRC: SP: 62-2004). Roller-compacted pavements are used for low-volume village, mandal or sub-arterial roads. The concrete is designed by weight of all ingredients to satisfy Eq. (8.24). Concrete trail mixes should be made with water content ranging from 5 to 7 percent at an interval of 1 or 0.5 per cent. Optimum moisture content (OMC) should be determined from the above trial samples which give maximum dry density (AASHTO T-180).

Veebee consistometer apparatus may be used with some modifications for evaluation of consistency of the concrete (Seehra 2001). Suitable adjustments in the OMC are to be made to achieve the desired dry density. At this optimum moisture content, samples are prepared and flexural strength tests are conducted at 7 and 28 days. Necessary final adjustments in the proportions of water–cement ratio or cement or fly ash (if used) are made to achieve the desired consolidation and to meet the target flexural strength.

The other compaction tools are vibratory screed, needle vibrator, hand temper and plate compactor. Trial mixes should be prepared on a conventional basis according to IRC: 44-1976 or IS: 10262-1982. Suitable adjustments to the proportion of ingredients should be made to meet the target strength requirement. Table 8.4 shows the gradation applicable for the design of concrete mix in the above two cases.

The maximum and minimum quantity of cement or cement with fly ash or cement with slag or cementitious materials per cubic metre of concrete should be 425 and 310 kg respectively. If fly ash is replaced with cement content, the cement content should not be less than 250 kg per cubic metre of concrete (MORD 2004).

Review Questions

1. What is the chemical composition of cement? Explain the significance of each component element.
2. What are the different types of cements available? State how they are different from OPC?
3. State the classification of cement according to ASTM.
4. Discuss the various properties of cement and briefly explain the related test methods.
5. Distinguish between initial and final setting times of cement. Explain the methods of determining these setting times according to the Bureau of Indian Standards.
6. State the significance of initial and final setting times of cement used for pavement construction.
7. State the importance of the soundness test of cement and write its basic principle.
8. Explain the significance and principle of testing cement strength by the cube compressive test method.
9. What is the chemical composition test of cement? Define hydration of cement. State the factors affecting rate of liberation of heat during the hydration process.
10. State the measures to be taken to control excess liberation of heat due to hydration in hot weather concreting.
11. Classify the admixtures used in cement concrete and write an overview on their suitability.
12. What are the different types of admixtures useful in cement concreting? Explain their effect on strength gain and workability of cement concrete mix.
13. What is meant by workability of cement concrete mix? State the factors affecting workability of the concrete mix.
14. Briefly explain the following tests which are conducted on hardened concrete:
    (i) Compressive strength test, (ii) elastic modulus test, (iii) flexural strength test and (iv) split tensile strength test
15. Enumerate the different methods available for the design of cement concrete mix.
16. State the factors affecting durability of hardened cement concrete.
17. State the significance of air voids in hardened cement concrete.
18. Write a short note on cement concrete mix design for rural roads according to IRC: SP: 6.

Multiple Choice Questions

1. The first artificial cement was patented by Joseph Aspdin, in England, in the year ________.
   (a) 1824 (b) 1856
   (c) 1902 (d) 1910
2. For rapid-hardening Portland cement (RHPC) (as per IS: 8041), the proportion of dicalcium silicate compound is kept ________ than that of tricalcium aluminate, in terms of chemical composition.
   (a) lesser (b) greater
   (c) twice (d) one-and-a-half times.
3. In the Blain air permeability test of cement (as per IS: 269), the surface area of a sample of OPC should not be ________ m^2/kg.
   (a) less than 225
   (b) greater than 225
   (c) equal to less than 225
   (d) between 225 and 325
4. In the autoclave test (as per IS: 269 and ASTM C 151-89), under standard test conditions of humidity and room temperature of 27 °C, the

cement cube samples are tested for 60 to 75 minutes at a steam pressure of ________ MPa.
(a) 1.0 (b) 1.5
(c) 2.1 (d) 3.3

5. In case of cement mortar cubes as per IS: 650-1991, the standard sand comprises three equal proportions (33.33%) of sizes of ________
(a) 2 to 1 mm, 1 mm to 500 microns and 500 to 90 microns
(b) 1 to 2 mm, 500 microns to 1.5 mm and 90 to 500 microns
(c) 2 to 1 mm, 1 mm to 500 microns and 900 to 500 microns
(d) 0.075 to 2 mm, 500 microns to 1.5 mm and 100 to 500 microns.

6. Identify the retarder used in cement concrete mixes
(a) Sugar
(b) Calcium salt
(c) Soap
(d) Naphthalene formaldehyde sulphonates.

7. Identify the pigment used in cement concrete mixes
(a) Polycarboxylic acids
(b) Calcium chloride
(c) Melamine formaldehyde sulphonates
(d) Cobalt oxide

8. As per IS: 516-1959, a compressive load at what uniform rate is applied on to a cement concrete cube having size of 15 cm?
(a) 1.4 N/mm^2/minute
(b) 14 N/mm^2/minute
(c) 1/4th N/mm^2/second
(d) 28 N/mm^2/minute.

9. In the flexural strength test, a cement concrete beam having standard cross-sectional area (150 × 150 mm) is moulded, cured for 28 days and tested by applying equal loads at one-third points of the beam at a constant rate of ________
(a) 4 to 4.5 kN/minute
(b) 50 to 55 kN/minute
(c) 14 to 14.5 kN/minute
(d) 1.2 to 2.4 kN/minute.

10. To determine the tensile strength of cement concrete sample by indirect means (as per IS: 5816), a cylindrical shaped concrete test specimen having diameter 150 mm and length 300 mm is tested by applying load at a uniform rate of ________
(a) 4 to 4.5 kN/minute
(b) 50 to 55 kN/minute
(c) 14 to 14.5 kN/minute
(d) 1.2 to 2.4 kN/minute

Answers to Multiple Choice Questions

1. (a)	3. (a)	5. (a)	7. (d)	9. (a)
2. (a)	4. (c)	6. (c)	8. (b)	10. (d)

CHAPTER 9 Factors Affecting Pavement Design and Performance

9.0 INTRODUCTION

Pavement design consists of determining the pavement thickness or the thickness of individual layers of the pavement in order to serve traffic for an anticipated design period based upon various factors, viz., intensity of traffic wheel loads, repetitions, temperature, rainfall, humidity, frost, drainage conditions, sub-grade soil strength, properties of materials used, criteria of mix design and economic considerations. The factors affecting pavement design and performance are discussed in this chapter.

9.1 FUNCTIONS OF PAVEMENT

The following are the main functions of pavements:

- They provide a firm, strong, smooth and adequately skid-resistant riding surface with low noise generation in all seasons.
- They receive and distribute traffic wheel loads safely over a large area of the sub-grade so that stresses and strains are kept within allowable limits.

The basic structure of a pavement comprises of a bound surface or wearing course, which transmits wheel loads onto subsequent layers safely. Under a wheel load, the surface course experiences compressive and tensile stresses at the top and bottom fibres respectively (Fig. 9.1). The wheel load is distributed further onto the subsequent bottom layers in the form of a cone having an angle of about 45 degrees.

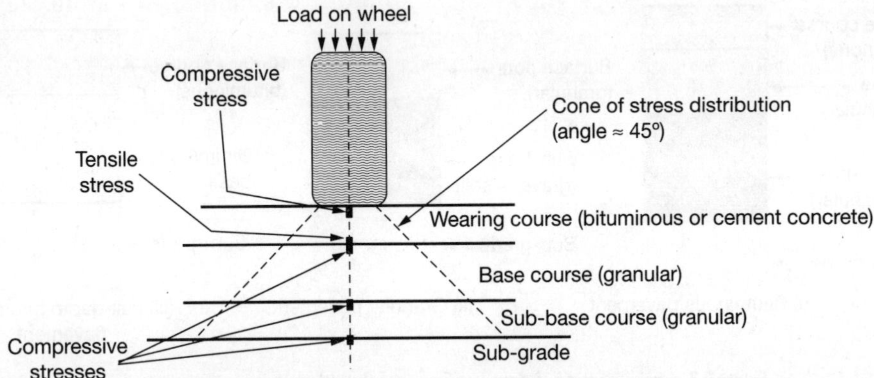

Figure 9.1 Structural function of pavement.

The sub-grade will experience compressive stresses over the conical spread area with lower intensity of stress as compared to the stress on the surface course. The over-all crust thickness of a pavement should be sufficient to keep the stresses or strains occurring due to wheel loads within allowable limits so that the sub-grade can withstand traffic loads without any plastic deformation. Therefore, each layer of pavement structure should be strong enough to resist wheel loads for a given design period under prevailing seasonal climatic conditions.

9.2 TYPE OF PAVEMENTS

Based on the type of materials used for construction, pavements may be classified into three types, viz., flexible pavements, rigid pavements and composite pavements. The salient features of these pavements are discussed below.

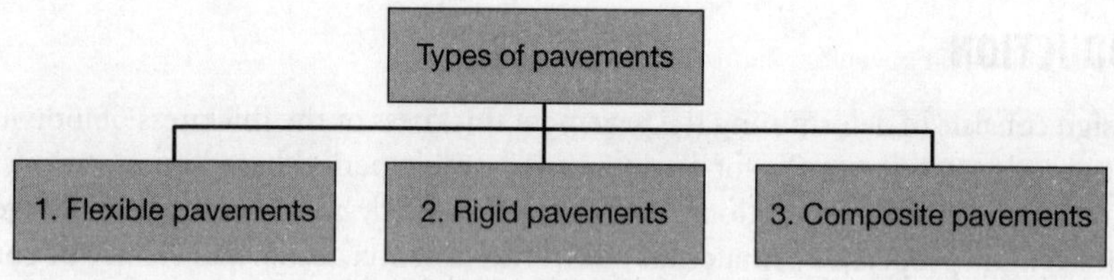

Figure 9.2 Types of pavements.

9.2.1 Flexible Pavements

Flexible pavement consists of multiple layers constructed using granular materials. To serve in all weather conditions, its top layer is constructed to be impervious by using different types of binders, viz., bitumen, tar and modified bituminous materials. Such layers are termed as bound layers and resist loads by developing tensile stresses at its bottom fibres. The particles that make up the granular layers are not bound together by any binder. Such layers are termed as unbound layers. The unbound layers are either stabilised by mechanical action or treated with additives so that they possess very low tensile strength. Unbound layers are constructed as sub-base and base courses since they lie under the impervious surface course. Cross-sectional details of typical flexible pavement structures are shown in Figs. 9.3 (i)–(iii).

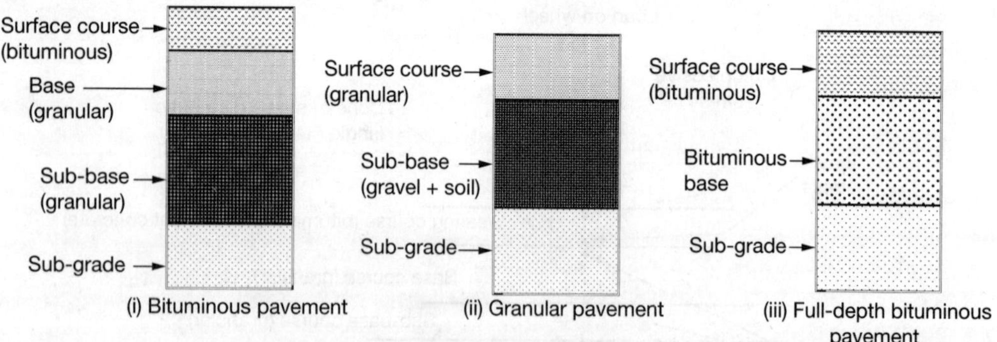

Figure 9.3 Cross-sectional details of typical flexible pavement structures.

The surface course of flexible pavements acts as a flexible layer. On wheel-loading, the surface layer bends with elasticity into a concave shape. This bowl-like surface moves in the longitudinal direction under the moving wheel (on the load path). Every application of wheel load on the pavement surface causes elastic deflection (or elastic strain). Repetitive application of wheel loads in different climatic conditions causes unrecoverable deformations in pavement structure, which are termed as plastic deformations (or uneven surface). Plastic deformations will occur in the vertical and horizontal directions due to vertical compressive stresses and horizontal tensile stresses respectively. Occurrence of such plastic deformations in the vertical direction, viz., rutting, consolidation, shear failure, upheaval or any kind of undulation in any one of the layers, causes a similar trend of deformation in the rest of the layers. These deformations may propagate in any vertical direction ranging from the sub-grade to the surface course and vice versa (Figs. 9.4 and 9.5). The plastic deformations caused by horizontal tensile stresses are generally observed as cracking in bound layers and displacement of materials in case of unbound layers.

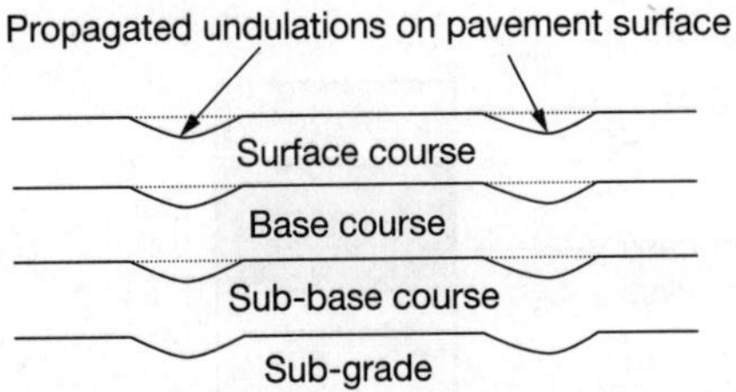

Figure 9.4 Pavement surface undulations caused due to plastic deformation of the sub-grade.

Propagated undulations at different layers
Surface course
Base course
Sub-base course
Sub-grade

Figure 9.5 Undulations caused due to rutting of surface course.

Surface courses have to be impervious, firm, strong, durable and skid-proof to provide resistance to abrasion, compression, tension, repetitive action of wheel loads and weathering action without undergoing fatigue (cracking) failure during its design life time. Because of these requirements, materials used for constructing surface courses are of higher quality than those used for other layers below it.

The base course is the foundation of the pavement structure that makes the pavement capable of bearing, transferring and spreading wheel loads onto wider area of the sub-base course and sub-grade. Therefore, to withstand loads, base and sub-base layers should be designed from the structural strength point of view. The base and sub-base courses should be adequately thick to reduce the intensity of stresses caused by wheel loads on to the sub-grade. Further, the sub-base layer should be designed to improve the sub-grade soil strength, control capillary rise of ground water and act as a drainage layer too.

The sub-grade may be natural ground or a treated soil layer having a thickness of 500 mm or more. Requirement of overall crust thickness will depend on the strength of the sub-grade. If the sub-grade soil is weak (clay soil), the required pavement crust thickness is more and vice versa. A satisfactory sub-grade should posses the following requirements (Mullis, 1939):

- Adequate compressive strength to withstand maximum wheel load applied
- Limited volume of effective pore space so that entry of water, which causes plasticity, is restricted
- Be protected to withstand abrasive and impact forces of traffic and weathering action.

9.2.2 Rigid Pavements

In rigid pavements, the surface course consists of a cement concrete slab with or without reinforcements. Under wheel loads, the top and bottom fibres of the slab experience compressive and tensile stresses respectively. The stresses developed are resisted by beam action or flexural strength of the cement concrete slab, which bridges over soft spots in the supporting layers of the slab.

The cement concrete slab may be directly placed over the sub-grade if the sub-grade has favourable minimum desirable strength, drainage and on-site conditions (Fig. 9.6 (i)). Sub-base layer is provided for the following reasons/conditions (Fig. 9.6 (ii)), (Collins and Sharp, 1958; Childs et al., 1959):

- To decrease deflections of slab with an accompanying reduced tendency for pumping mud
- If the plastic sub-grade soil is susceptible to pumping and has a tendency to weaken severely in the presence of melting frost
- If the sub-grade tends to change its volume significantly with varying moisture content.

Use of sub-base which is thicker than normal to prevent pumping of mud does not greatly enhance the load-carrying capacity of the pavement. Moreover, thicker sub-bases may not be structurally economical.

The structural functions of the remaining layers below the surface course are similar to those of the layers of a flexible pavement.

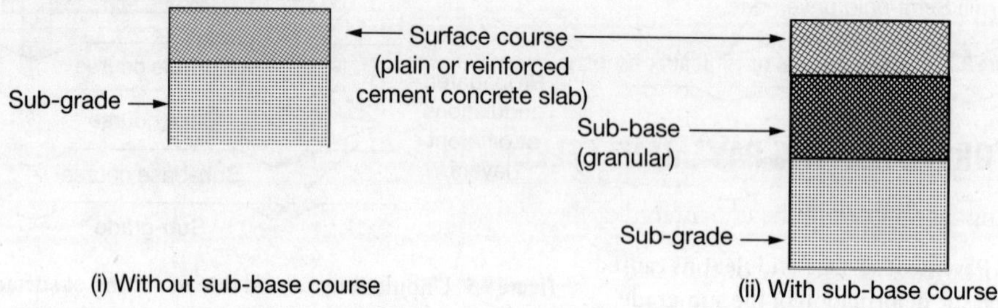

Figure 9.6 Cross-sections of typical rigid pavement structures.

9.2.3 Composite Pavements

In this category, the wearing course of the pavement is constructed by spreading bricks or pre-cast rectangular cement concrete blocks, or interlocking concrete blocks (Fig. 9.7 (i)). The joints between them are filled with fine sand of standard gradation. These pieces of blocks act as small rigid plates. Hence, the wearing course constructed of these blocks is neither truly flexible nor a truly rigid layer. It possesses dual properties and is suitable for low-volume village roads or footpaths.

A composite pavement consists of cement concrete or cement-treated granular and bituminous layers. These kinds of pavements are observed when an old concrete pavement is overlaid with a bituminous layer or any granular layer below a bituminous wearing course is treated with cement or lime (Fig. 9.7 (ii) and (iii)). While designing composite pavement structures, the flexural strength of the base slab or treated granular should be taken into consideration for resisting tensile stresses.

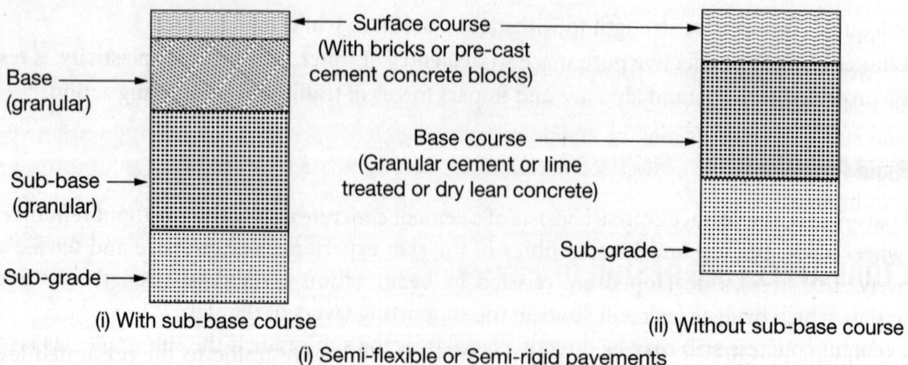

(i) Semi-flexible or Semi-rigid pavements

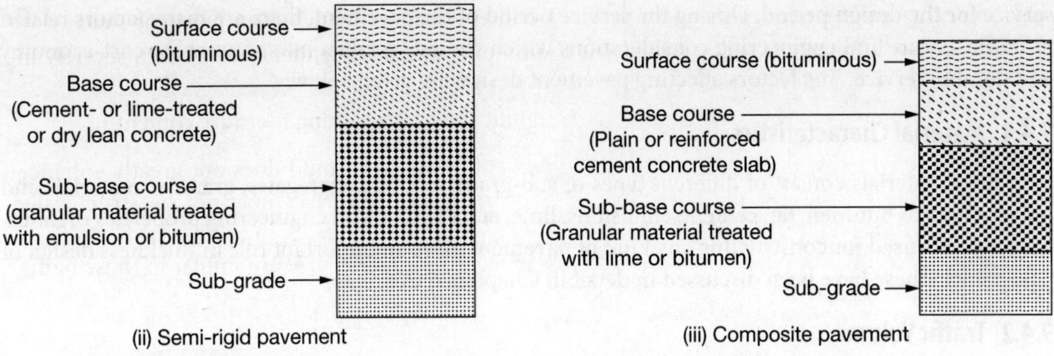

Figure 9.7 Cross-sections of typical composite or semi-flexible or semi-rigid pavement structures.

9.3 FACTORS AFFECTING PAVEMENT SELECTION

The following factors should be considered at the planning stage of constructing a pavement:

- Importance and purpose of the road
- Location, climate and natural ground conditions
- Traffic intensity and growth rate
- Type of vehicles and design speed
- Nature of wheel loads
- Local availability of materials
- Proposed frequency of maintenance
- Stage construction
- Funds availability at the initial stage and for regular post-construction maintenance.

Flexible pavements are preferred when initial funds allotted are less or stage construction is proposed. Regular maintenance is necessary for flexible pavements, which involves more maintenance costs. On the other hand, the initial cost of full-depth bituminous pavements is higher than that for traditional flexible pavements since the base course is constructed with good-quality bituminous layer. Full-depth bituminous pavements are very strong and cost-effective when considering their life-cycle cost. They are also preferred when total pavement crust thickness is restricted to avoid drainage related problems, especially in well-developed urban localities with higher volume of traffic.

For rigid pavements, the initial cost is more compared to the cost of flexible pavements. However, the maintenance cost of rigid pavements is low and the life-cycle cost is economical and justified with reference to its performance during its serviceable period of life.

Brick or concrete block pavements may be used for low-volume roads (like cycle tracks) or village roads or roads of less importance where the funds allotted are less. New construction of composite pavements involves higher cost. To improve the structural capacity of an old concrete pavement, a bituminous layer may be overlaid as a maintenance measure.

Base and sub-base courses may be stabilised with suitable admixtures or binders to improve their structural capacity. A suitable cost-effective treatment may be made to these layers to meet minimum strength requirement.

9.4 FACTORS AFFECTING DESIGN OF PAVEMENT

Pavement design comprises thickness determination of different layers for projected volume of traffic, in situ and climatic conditions. The designed pavement should serve traffic to the presumed level of service for the design period. During the service period of the pavement, there are many factors related to traffic, nature and engineering considerations which should be taken into account for cost-economy and durable service. The factors affecting pavement design are given below.

9.4.1 Material Characteristics

Pavement materials consist of different types of sub-grade soil, fine aggregates, granular materials and binders such as bitumen, tar, cement, emulsions, lime, etc. Physical and engineering properties of different materials used for constructing any kind of pavement plays an important role in thickness design of pavements. These have been discussed in detail in Chapters 6 to 8.

9.4.2 Traffic Volume

Traffic volume is one of the main factors which decide pavement thickness. Traffic count surveys are conducted at road junctions or imaginary sections of roads to determine passage of a number of vehicles per unit time period. According to Indian practice (IRC: 9-1972), the average traffic volume per day is calculated for a 7-day, 24-hour traffic count. If it is not possible to conduct a count survey for 7 days, a 3-day (Saturday, Sunday and Monday) count survey may be conducted. The final traffic volume estimate should also incorporate the additional traffic volume that may be expected in the near future during the design life by considering local economic developmental activities.

Estimation of future traffic volume for a design period is calculated in terms of cumulative standard axles per year (csa/year). The following equations are used to estimate the traffic volume for a given period (say *n* years) and design value of traffic (IRC: 37-2001 and IRC: 58-2002).

$$A = P(1+r)^{nc} \tag{9.1}$$

$$N_{flexible} = 365 \times TGF \times A \times VDF \times LDF \times 10^{-6} \tag{9.2}$$

$$N_{rigid} = 365 \times TGF \times A \times 10^{-6} \tag{9.3}$$

where,

A = Estimated traffic volume during the road construction period(Commercial vehicles/day)

P = present traffic volume from traffic count survey(Commercial vehicles/day)

r = growth rate of traffic(decimal value)

nc = no. of years from the date of traffic count to the date of opening of road to traffic.

$N_{flexible}$ and N_{rigid} = Cumulative number of standard axles in million standard axles(m sa) which is used for design of flexible and rigid pavements respectively.

$\text{TGF} = \left(\dfrac{(1+r)^{nd}-1}{r}\right)$ = Cumulative Traffic Growth Factor based on geometric growth.

nd = design period(no.of.years); 30 to20 years for rigid pavements; above 5 year for flexible pavements.

VDF = Vehicle damage factor

LDF = lane distribution factor

Traffic Growth Rate

This value is used to estimate future traffic (A) after n number of years based on present traffic volume (P). The rate of growth of traffic is generally calculated from the growth rate of vehicle registrations, fuel consumption and socio-economic developments in a locality for the previous few (10 or 20) years. In the absence of such a data bank, an average traffic growth rate of 5.0 % may be assumed.

Design Life

Scientifically, the term 'design life' indicates the cumulative number of standard axle load repetitions for which a pavement is supposed to provide the desired level of service before needing strengthening operations. The design life in terms of cumulative standard axles (csa) is calculated based on the desirable service period (in number of years).

For flexible pavements, the design service period (*nd*), may be taken as 10 to 20 years (IRC: 37). A higher service period is taken in case of expressways and national highways and other economically important roads. A value between 5 to 10 years may also be taken if stage construction of a road is envisaged.

For rigid pavements, the design service period varies—in case of roads with low and high traffic volume, it may be taken as 20 to 30 years respectively (IRC: 58). However, any suitable number of years of design service period may be taken according to the pavement's purpose and economic point of view.

Lane Distribution Factor (LDF)

Lane distribution factor is a decimal value which indicates the concentration of placement of wheel load repetitions along a road stretch. As a general tendency, on a multi-lane undivided one-way pavement, a driver places his vehicle towards the centre or away from the edge. Therefore, lower concentration of wheel load repetitions is observed near the edges. Thus, based on the number of lanes, only a part of the commercial traffic volume is taken as the design traffic volume, as an economic consideration. The lane distribution factor (LDF) should be determined from realistic and appropriate field surveys based on special repeatability of wheel paths of different vehicles in the transverse direction of a pavement surface. In the absence of such data, the following LDF may be used to design flexible pavements (Table 9.1).

Table 9.1 Lane distribution factors.

S. No.	Type of carriageway	LDF*	Total number of commercial vehicles considered in
1	Dual two-lane	0.75	Each direction
2	Dual three-lane	0.60	
3	Dual four-lane	0.45	
4	Single-lane	1.00	Both directions
5	Two-lane two-way	0.50	
6	Four-lane single	0.40	

*(Source: IRC:37)

EXAMPLE 9.1

Determine the traffic volume just after construction of a road project using the following data:

Present traffic volume = 1090 commercial vehicles per day

Growth rate of traffic = 7.1 %

Duration of road construction = four-and-a-half years.

Answer

Using Eq. (9.1),

$A = P(1 + r)^{nc} = 1090\,(1 + 0.071)^{4.5} = 1484$ commercial vehicles per day (CVPD).

9.4.3 Factors Related to Axle Loads and Vehicle Gross Loads

The input values relating to axle load, configuration of axle with wheels, and gross weight of vehicles are useful for the purpose of designing pavement thickness.

Standard axle loads

Under mixed traffic conditions, repetitions of different axles having different loads plying on a road will not indicate any meaningful value relating to how much damage has been caused to the pavement due to their combined action. Different axle loads will cause different degrees of damage. Therefore, it is customary to convert repetitions of axles having different loads to an equivalent standard axle load. Configuration of the Indian standard single axle with dual wheel assembly is presented in Fig. 9.8.

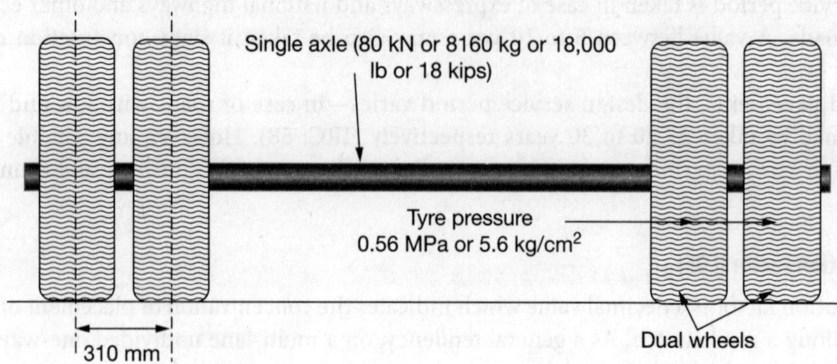

Figure 9.8 Details of the Indian standard axle.

According to the Indian practice (IRC: 37-2018), standard axle load is prescribed for different axle load configurations, as follows:

- Single axle with single wheel on either side: 65 kN
- Single axle with dual wheels on either side: 80 kN
- Tandem axle with dual wheels on either side: 148 kN
- Tridem axles with dual wheels on either side: 224 kN.

IRC Charts (IRC: 37, IRCSP: 72, IRC: 58 and IRC: SP: 62) have been prepared for a specified number of repetitions of such equivalent standard axle loads on the pavement layers.

Configuration of Axle with Wheels The total weight of a vehicle is carried by its axles. The load on the axles is transferred to the wheels and this load is ultimately transferred on to the surface of the pavement in contact with the tyres.

To keep wheel-load-induced stresses on pavements within allowable limits, the total vehicle load is distributed onto wider areas of pavement by using more axles and wheels. This is why more axles and wheels are fitted onto heavy load-carrying trucks. Fig. 9.9 shows the different types of configurations of axles and wheels used in the manufacture of road vehicles.

Legal Axle Load Limit The maximum allowable axle load limit is referred to as the legal axle load limit. This means that no axle of a vehicle should carry more than the legal axle load. The details of the legal axle load limits adopted in India are given in Table 9.2.

Table 9.2 Legal axle load (tonnes) limits adopted in India.

Single axle	Tandem axle	Tridem axle
10.2	19.0	24.0

Measurement of Axle Load and Truck Weight

There are two methods to weigh trucks or axle loads, viz., (i) static weighing and (ii) weight in motion (WIM). In the static weighing method, vehicles are stopped and weighed. But in the weight-in-motion

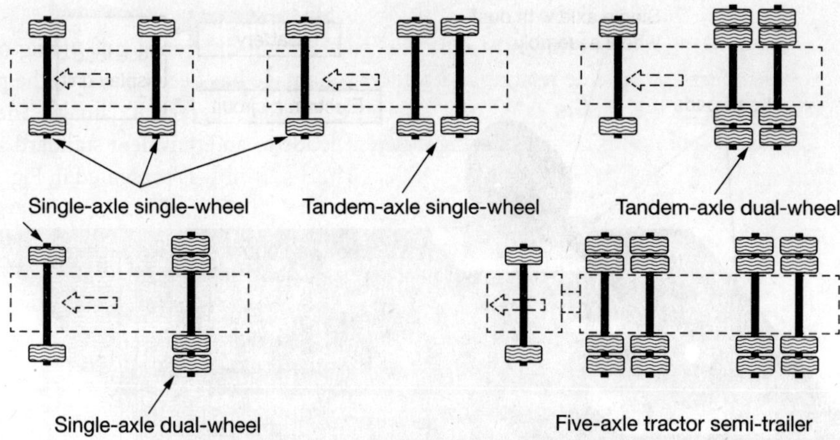

Figure 9.9 Different axle-and-wheel configurations.

method, vehicles are weighed dynamically, while in motion. A comparison of the two methods is given below.

S. No.	Static weighing method	Weight-in-motion (WIM)
1	Vehicles are stopped and weighed	Vehicles are weighed automatically, while in motion, without disturbing the driver
2	Accurate weight measurement	Weight measurements may be influenced by parameters related to vehicle speed, suspension system of the vehicle, tyre pressure, acceleration and deceleration of the vehicle and dynamic forces produced due to pavement roughness, wind velocity, etc.
3	Takes more time and interrupts free flow of traffic; may pose problems related to safety.	There is no such interruption to flow of traffic; can weigh high volumes of traffic.
4	Other information such as body type of vehicles, loading type etc can be physically ascertained	Collection of other information is not possible by automated WIM equipment.
5	Less number of vehicles can be measured; selected vehicles are weighed; need more personnel, time and space to weigh all vehicles	More number of vehicles can be measured; better coverage of all vehicles since it is automatic.
6	Less installation and maintenance cost	High installation and maintenance cost
7	The weigh pads can be installed at any location	The WIM equipment can be installed at fixed locations only

In the static method, the axle load of a vehicle is weighed using portable weights or load-pads (Fig. 9.10) or a weighing platform. Only axle loads having more than or equal to 3 tonnes are taken into account for analysis since the damage caused by axles weighing less than 3 tonnes is negligible. For traffic analysis, vehicles having axle weight ≥ 3 tonnes are referred to as commercial vehicles.

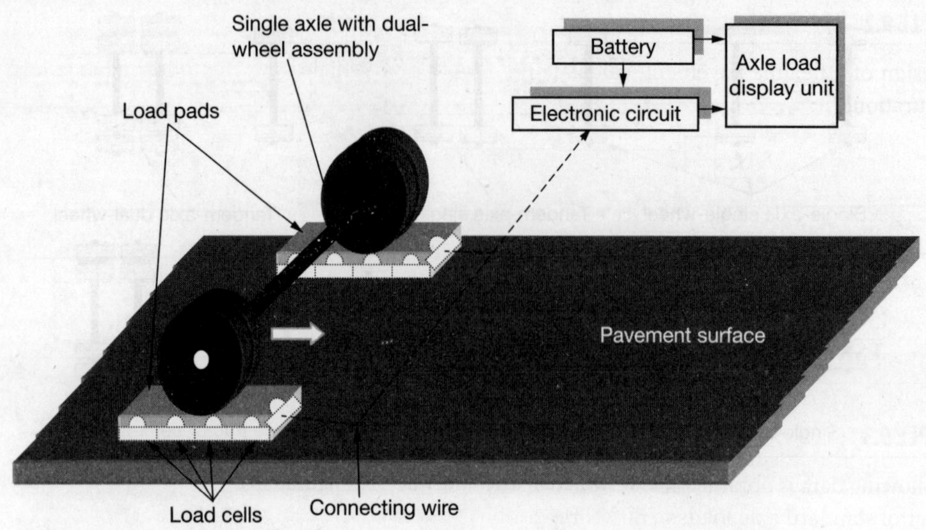

Figure 9.10 Axle load measurement using load-pads (schematic).

9.4.4 Load Equivalency Factor (LEF)

Load equivalency factor (LEF) is a number which relates the amount of equivalent damage caused by a given load of axle to the standard axle load. Based on field test data, AASHTO has recommended specific load equivalency factors for single and tandem axles. These LEFs were developed based on the effect of wheel load on serviceability of pavements.

Equivalent standard axle loads (ESALs) may be calculated from the following relationship:

Repetitions of equivalent standard axle loads (ESALs) = Load equivalency factor (LEF) × No. of axles observed (9.4)

Fourth Power Damage Rule The following relationships (Eqs. 9.5(a) and (b)) developed from the AASHO road test data are widely used to determine equivalent standard axle load (ESAL) repetitions when axle load data of different axles is available. This relationship is popularly known as the fourth power damage rule.

$$\frac{N_1}{N_2} = \left(\frac{W_2}{W_1}\right)^4 \qquad (9.5(a))$$

$$\text{Load equivalency factor (LEF)} = \left(\frac{W_m}{W_{std}}\right)^4 \qquad (9.5(b))$$

Where,

N_1 is a number representing repetitions of axle loads each having a weight equal to W_1
N_2 is a number representing repetitions of axle loads each having a weight equal to W_2
W_m = Measured axle load obtained from axle load(survey data)
W_{std} = Standard axle load

Eq. 9.5(a) may also be explained as stating that the damage caused by N_1 number of repetitions of axle having weight W_1 is equal to the damage caused by N_2 number of repetitions of axle having weight W_2.

EXAMPLE 9.2

For design of a flexible pavement, calculate the ESALs of a vehicle with the following axle-and-wheel configuration.

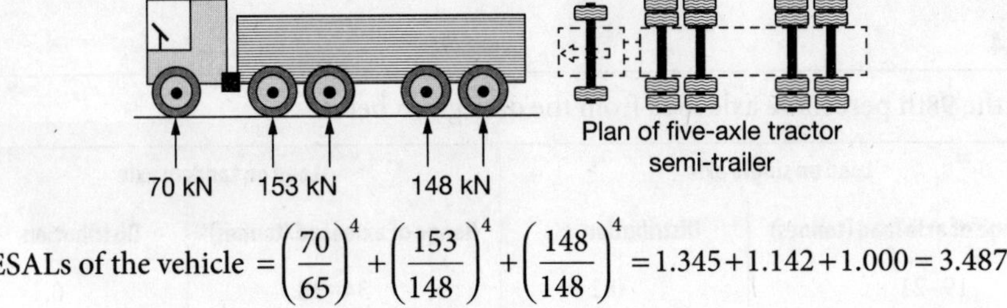

$$\text{ESALs of the vehicle} = \left(\frac{70}{65}\right)^4 + \left(\frac{153}{148}\right)^4 + \left(\frac{148}{148}\right)^4 = 1.345 + 1.142 + 1.000 = 3.487$$

EXAMPLE 9.3

The following data is obtained from the axle load survey conducted for 3 days. Determine the equivalent number of standard axle loads of 80 kN repetitions per year.

Axle load (kN)	30–40	40–50	50–60	60–70	70–80	80–90
No. of axles	54	65	64	78	103	98
Axle load (kN)	90–100	100–110	110–120	120–130	130–140	140–150
No. of axles	110	98	78	89	67	79

Answer

LEFs are calculated using Eq. (9.5(b)). The details are given below.

Average axle class (1)	Average axle load (W_m) in kN (2)	No. of axles (3)	LEF = $(W_m/80)^4$ (4)	$ESAL_{80kN}$ (5) = (3) × (4)
30–40	35	54	0.0366	1.9784
40–50	45	65	0.1001	6.5073
50–60	55	64	0.2234	14.2979
60–70	65	78	0.4358	33.9929
70–80	75	103	0.7725	79.5650
80–90	85	98	1.2744	124.8941
90–100	95	110	1.9885	218.7395
100–110	105	98	2.9675	290.8194
110–120	115	78	4.2700	333.0627
120–130	125	89	5.9605	530.4813
130–140	135	67	8.1091	543.3128
140–150	145	79	10.7923	852.5879
	Total =	983		3030.239

80 kN ESAL repetitions for 3 days = 3030.239
80 kN ESAL repetitions per year ≈ (365/3) × 3030.239 = 368679
= 0.368679 million standard axles or
= 0.368679 msa

EXAMPLE 9.4

Determine the 98th percentile axle load from the data given below.

Load on single axle		Load on tandem axle	
Range of axle load (tonnes)	**Distribution, %**	**Range of axle load (tonnes)**	**Distribution, %**
19–21	0.2	34–38	0.1
17–19	0.6	30–34	0.2
15–17	0.9	26–30	2.1
13–15	30.5	22–26	4.3
11–13	20.1	18–22	1.5
9–11	19.2	14–18	0.5
Less than 9	18.1	Less than 14	1.7
Total	89.6	–	10.4

Answer

Single-axle load		Tandem-axle load					
Average value of axle load (t) (1)	**Distr. % (2)**	**Average value of axle load (t) (3)**	**Equivalent single-axle load (ESAL) (4)**	**Distr. % (5)**	**All single-axle loads (6)**	**Distr. % (7) = (2) + (5)**	**Percentile value (8)**
20	0.2	36	18	0.1	20	0.2 + 0.0 = 0.2	100
18	0.6	32	16	0.2	18	0.6 + 0.1 = 0.7	99.8
16	0.9	28	14	2.1	16	0.9 + 0.2 = 1.1	99.1
14	30.5	24	12	4.3	14	30.5 + 2.1 = 32.6	98.0
12	20.1	20	10	1.5	12	20.1 + 4.3 = 24.2	65.4
10	19.2	16	8	0.5	10	19.2 + 1.5 = 20.7	41.0
8	18.1	14	6	1.7	8	18.1 + 0.5 = 18.6	20.3
Total	89.6	Total	–	10.4	6	0.0 + 1.7 = 1.7	1.7

After plotting the values in Columns (6) and (8), we get the 98th percentile axle load = 14 *t*.

9.4.5 Legal Axle Load Limit

To safeguard pavements against overloading, different agencies (government or any competent authority) in different countries have prescribed maximum allowable limits of axle load depending on their method of compaction, materials used, considerations related to mix design as well as pavement crust thickness. The maximum allowable axle load limit is referred to as the legal axle load limit. This means that no axle of a vehicle should carry more than the legal axle load. The details of the legal axle load limits adopted in different countries are given in Table 9.3. Maximum permissible loads for different type of vehicles in India are given in Table 9.4.

Table 9.3 Legal axle load limits followed in different countries (Kanchan et al., 2006; CSIR Report, 2001).

Sl. No.	Country	Single axle load (tonnes)	Tandem axle load (tonnes)	Tridem axle load (tonnes)	Maximum permissible gross load (tonnes)
1	India	10.2	19.0	24.0	45.0
2	UK	10.5	20.3	22.5	38.0
3	South Africa	9.0	18.0	24.0	56.0
4	USA*	9.0	18.0	20.0	36.0
5	Canada*	9.1	17.1	24.0	53.5
6	Germany	10.0	11.5	21.0	40.0
7	Australia	9.0	16.5	20.0	44.0

Note: * May vary in different states/provinces.

9.4.6 Vehicle Damage Factor

Vehicle passage on a pavement will cause a certain amount of damage or distress to the pavement in some form. The degree of damage caused by a vehicle depends on its gross weight, number of axles as well as configuration of wheels. For example, if two vehicles have equal gross load, one with a single-axle–single-wheel and the other with a tandem-axle–dual-wheel assembly, the damage caused by the former vehicle will be greater. This is because the gross load is transferred onto the pavement surface over a wider area by more number of axles and wheels. The damage caused by different vehicles is calculated using the vehicle damage factor, which is a parameter used for performance modelling, design and maintenance of pavements.

Table 9.4 Maximum permissible loads of different types of vehicles in India (Kanchan et al., 2006).

S. No.	Type of vehicle	Legal limit of load (tonnes)
1	Truck 2 axles	16.20
2	Truck – Semi-trailer	25.0
3	Truck – Trailer	35.0
4	Truck – 3 axles (rear tandem)	25.0
5	Truck – Multi-axle (rear tandem)	35.0
6	Truck – Multi-axle (both rear tandem)	45.0

The vehicle damage factor (VDF) is a numerical value which represents the equivalent number of standard axles per truck. It is a multiplication factor used to convert different commercial vehicles with varying axle load repetitions to standard axle-load repetitions. From axle-load survey data, VDF is calculated using the following equation.

$$VDF = \frac{\sum_{i=1}^{N}[V_i \times LEF_i]}{N} \qquad (9.6)$$

where,

V_i = Traffic volume of the *i*th vehicle load-class
LEF_i = Load equivalency factor of *i*th vehicle load-class
N = Total number of vehicles weighed

Factors affecting VDF

VDF is affected by numerous factors which are related to damage or distress caused to the pavement. The factors may be related to traffic composition at the time of survey, load on axles, possible occasional (or seasonal) overloading, number of axles, wheels configuration, terrain, type of pavement, region, pavement condition, temperature, rainfall, etc.

In pavement design, VDF obtained from axle-load survey data should be used rather than an assumed value because the VDF value obtained from survey data represent a realistic value by considering actual traffic loading and other factors related to the region (or pavement) considered.

EXAMPLE 9.5

The following data were obtained from an axle load survey. Determine VDF based on a standard axle load of 80 kN.

Axle load (kN)	30–40	40–50	50–60	60–70	70–80	80–90
Frequency (%)	5.76	6.45	7.22	6.96	9.63	12.04
Axle load (kN)	90–100	100–110	110–120	120–130	130–140	140–150
Frequency (%)	14.02	10.49	8.43	7.82	5.25	5.93

Answer

LEF is calculated using Eq. 6.1 (b). The calculation details are given below.

Average axle class (1)	Average axle load (W_m) in kN (2)	Frequency (%) (3)	LEF = $(W_m/80)^4$ (4)	$ESAL_{80kN}$ (5) = (3) × (4)
30–40	35	5.76	0.0366	0.21
40–50	45	6.45	0.1001	0.65
50–60	55	7.22	0.2234	1.61
60–70	65	6.96	0.4358	3.04
70–80	75	9.63	0.7725	7.44
80–90	85	12.04	1.2744	15.34
90–100	95	14.02	1.9885	27.87
100–110	105	10.49	2.9675	31.13

(continued)

Average axle class (1)	Average axle load (W_m) in kN (2)	Frequency (%) (3)	LEF = $(W_m/80)^4$ (4)	$ESAL_{80kN}$ (5) = (3) × (4)
110–120	115	8.43	4.2700	35.98
120–130	125	7.82	5.9605	46.64
130–140	135	5.25	8.1091	42.53
140–150	145	5.93	10.7923	64.03
	Total =	100.00		276.46

VDF = 276.46 ÷ 100 = 2.76

EXAMPLE 9.6

The following data was obtained from an axle load survey:

Total number of vehicles observed (N) = 98.

Assume standard axle load (W_{std}) for

(i) single axle–single wheel = 6.6 tonnes
(ii) single axle–dual wheel = 8.16 tonnes
(iii) tandem axle– four wheel = 15.1 tonnes

Axle load class in tonnes (1)	Number of passes of axles with		
	Single-axle–single wheel (2)	Single-axle– dual-wheel (3)	Tandem-axle–four-wheel (4)
5–6	47	1	–
6–7	33	3	–
7–8	18	3	–
8–9	–	14	2
9–10	–	13	5
10–11	–	3	9
11–12	–	3	12
12–13	–	–	14
13–14	–	–	3
14–15	–	–	4
16–17	–	–	6
17–18	–	–	3
Total	98	40	58

Calculate the value of VDF based on the fourth power damage rule.

Answer

The fourth power damage rule is used for calculating load equivalency factors (LEF). The calculation procedure is given below.

Average axle load (W_m)	Repetitions of single axle–single wheel	LEF for single axle single wheel $= (W_m/6.6)^4$	Repetitions of single axle–dual wheel	LEF for single axle–dual wheel $= (W_m/8.16)^4$	Repetitions of tandem axle–four wheel	LEF for tandem axle four wheel $= (W_m/15.1)^4$
5.5	47	0.48	1	0.21	–	–
6.5	33	0.94	3	0.40	–	–
7.5	18	1.67	3	0.71	–	–
8.5	–	–	14	1.18	–	–
9.5	–	–	13	1.83	6	0.02
10.5	–	–	3	2.74	6	0.02
11.5	–	–	3	3.94	9	0.13
12.5	–	–	–	–	10	0.19
13.5	–	–	–	–	10	0.19
14.5	–	–	–	–	7	0.05
15.5	–	–	–	–	5	0.01
16.5	–	–	–	–	5	0.01
Total	98	3.09	40	11.00	58	0.63

$$\text{VDF} = \left[\frac{\begin{gathered}(47\times0.48)+(33\times0.94)+(18\times1.67)\\+(1\times0.21)+(3\times0.4)+(3\times0.71)+(14\times1.18)+(13\times1.83)+(3\times2.74)+(3\times3.94)\\+(6\times0.02)+(6\times0.02)+(9\times0.13)+(10\times0.19)+(10\times0.19)+(7\times0.05)+(5\times0.01)\\+(5\times0.01)\end{gathered}}{98}\right]$$

$$VDF = \left[\frac{(83.64)+(63.89)+(5.66)}{98}\right] = \left[\frac{153.19}{98}\right] = 1.563$$

9.4.7 Tyre Pressure and Contact Pressure

The pressure in an inflated rubber tube of a tyre is referred to as tyre pressure or inflation pressure. In the AASHTO road test, tyre pressures were used in the range of 0.515 to 0.55 MPa. However, as of today, the tyre pressures on commercial highway vehicles vary from 0.7 to 1.0 MPa. For design purposes, an average tyre pressure of 0.8 MPa (8.15 kg/cm^2) may be assumed for Indian conditions (IRC: 58-2002).

Tyre contact pressure may be defined as a pressure or stress developed due to wheel load over the tyre imprint area of the pavement surface (Fig. 9.11). The tyre contact pressure is used to calculate stresses induced in the pavement structure and hence, it can be related to pavement damages.

As tyre pressure increases, its contact area decreases and vice versa. But the load on the tyre and the contact area also increases. As a worst case scenario, as tyre pressure and axle load increases, the contact pressure also increases. This phenomenon is described in Fig. 9.12. This kind of situation is generally observed on a summer day, when trucks are overloaded, since elevated inflation pressures are observed due to higher temperatures of air and pavement. However, for design purposes, tyre contact pressure may be assumed to be equal to tyre pressure.

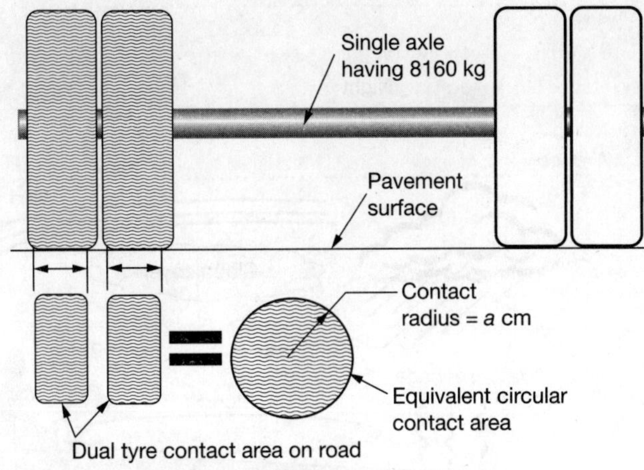

Figure 9.11 Contact area of a dual tyre.

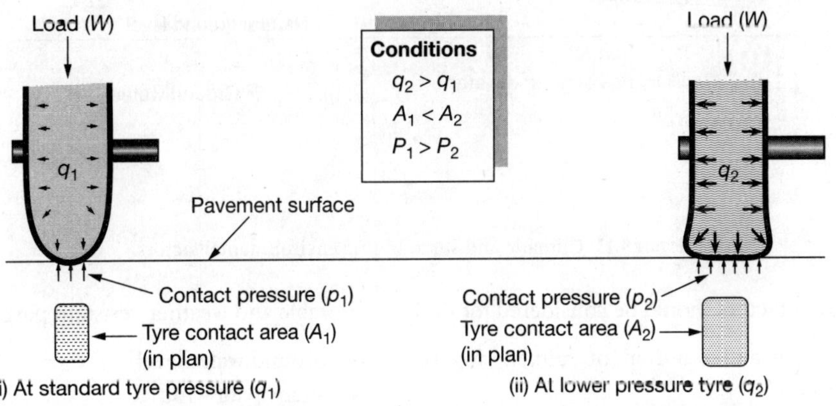

Figure 9.12 Interactions between tyre and pavement surface.

EXAMPLE 9.7

Calculate the radius of tyre contact area if tyre pressure is 0.56 MPa and wheel load is 40 kN?

Answer

Let tyre contact radius = a

$$\text{Tyre pressure} = \frac{\text{Wheel load}}{\text{Contact pressure}} = \frac{\text{Wheel load}}{\pi a^2}$$

$$\text{Tyre contact radius} = a = \sqrt{\frac{\text{Wheel load}}{\pi \times \text{Tyre pressure}}} = \sqrt{\frac{40 \times 10^3 \text{ N}}{\pi \times 0.56 \times 10^6 \text{ N/m}^2}} = 0.1507 \text{ m} = 15.07 \text{ cm}$$

9.4.8 Factors Related to Climatic and Surrounding Environment

In pavement design, climatic and environmental factors (Fig. 9.13) must be given as much importance as factors related to traffic because the designed pavement will undergo numerous physical (volumetric) changes in different climatic and environmental conditions during different seasons at various locations.

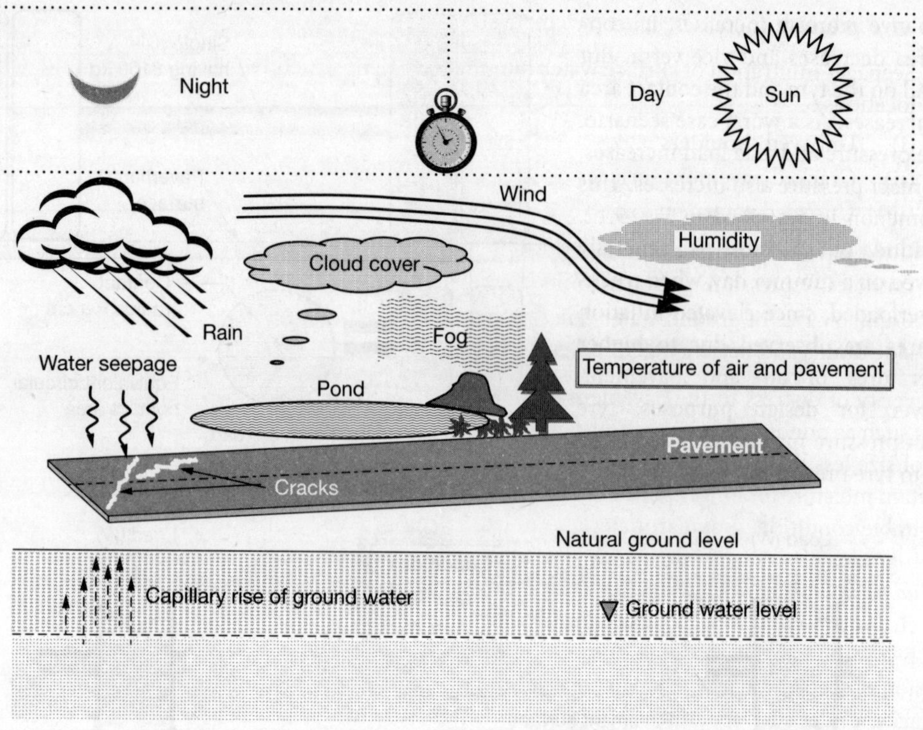

Figure 9.13 Climatic and surrounding environmental factors.

The following factors should be considered for design of durable and weather-resistant pavements

- Intensity and duration of rainfall, fog or snowfall
- Temperatures of air and pavement
- Surface and sub-surface drainage conditions
- Type of landscape and soil type
- Water bodies nearby
- Ground water level
- Intensity of wind
- Cloud coverage
- Relative humidity
- Solar radiation

9.4.9 Moisture

The presence of moisture (other than OMC) in the pavement structure is an undesirable condition which causes volumetric changes in compacted layers at the affected locations and thereby results in failures in different modes. In particular, parameters related to strength of sub-grade and granular layers are highly sensitive to any little change in OMC. The strength of sub-grades or sub-bases, in terms of CBR or modulus or any such strength value, decreases with change in moisture content.

The surface water due to rainfall and snowfall should be suitably drained off to the side or underground drains and further on to a safe disposable location. Intensity and duration of rainfall data is more helpful than total annual rainfall data to analyse pavement performance. High-intensity rainfall of short durations will not cause as much damage as low-intensity rainfall over a longer period of time since in the latter case the pavement will be continuously wet for long duration and absorb more moisture. Data related to precipitation, viz., rainfall intensity, duration, wind, humidity and air temperature should be collected from the closest meteorological department and used as inputs for design of pavements.

Change in moisture content in the pavement structure is possible under the following two conditions.

1. Seepage/ infiltration of surface water into the pavement structure takes place from the following locations.
   (i) Unpaved shoulders
   (ii) Cracked pavement surfaces
   (iii) Unsealed pavement joints

   If the pavement surface is not completely impermeable due to the presence of cracks, hotspots, potholes, depressions and open joints of cement concrete slabs, the rate of infiltration will increase.
2. Capillary rise of ground water due to a higher water table and movement of moisture through side drains and nearby water bodies

The severity of distress to the pavement is more if ground water level is raised by surrounding water bodies such as ponds, lakes, rivers and canals. Certain type of soils especially clay-rich soils having particles of size less than 75 microns are more prone to become water-logged. Such soils generally absorb and retain moisture for longer periods. This condition is termed as *poor drainage condition*. To avoid this undesirable condition, granular materials are used to form a water-filtering layer which is placed over the sub-grade.

Mud pumping is a widely known distress condition in cement concrete pavements. It is a phenomenon in which, water seeps through cracks or enters into joints of concrete slabs and decreases the stiffness of the layers below the slab. Due to weak support, on wheel loading, the slab deflects downward and causes the expulsion of seeped water with small particles of fine-grained sands onto the surface of the slab through the cracks. Under such repetitive application of wheel loads, bigger size voids or cavities will be formed beneath of the slab. Ultimately, the slab breaks into small pieces and settles down forming depressions.

9.4.10 Frost Action

Frost action is a phenomenon of distress which is observed in pavements where the sub-grade and granular layers, consisting of fine-grained soil, freeze over (Bergen and Fredlund, 1973). At freezing temperatures, the moisture present in the base, sub-base and sub-grade layers of the pavement become ice crystals (or ice lenses). These ice crystals increase in volume (or swell) and thereby, occupy more space than soil particles. Further, growth of such ice lenses is maintained in a zone of freezing if there is a continuous supply of moisture either by the capillary rise of moisture or raised water table in the vicinity (Taber, 1918). An uneven surface formation is observed on the top surface of the pavement and is termed as frost heave (Fig. 9.14) (Johnson et al., 1982). Frost action is possible when poor drainage conditions and favourable soil characteristics exist (Sourwine, 1930).

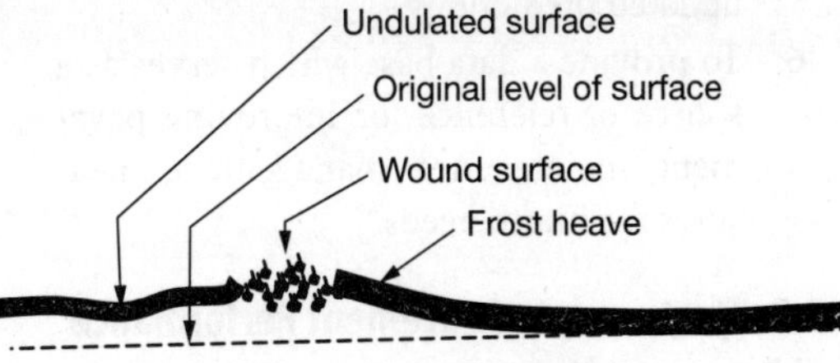

Figure 9.14 Frost heave.

As the degree of saturation increases, soil expansion takes place, thus significantly affecting the resilient modulus of granular soils (Rada and Witczak, 1981). This crystallised water (ice lenses) melts when the pavement temperature rises (during thawing) and migrate either in the upward direction or infiltrate into the layers below. Voids will, thus, develop in the pavement due to movement of water in different directions. Under traffic loading, displacement of material particles takes place to fill these voids. Ultimately, compactness of the pavement structure will be disturbed leading to premature failure. The above phenomenon is termed as *frost action* or thaw weakening in pavements.

9.5 PAVEMENT PERFORMANCE

Performance of pavement and its serviceable life are subjected to the design conditions, geological environment, water table fluctuations, construction variability in terms of material and compaction, and prevailing factors such as intensity of traffic, drainage and climatic conditions. Assessment of deformations/distresses and other influencing parameters is complex because, occurrences of such effects are uncertain and random in nature. As inevitable phenomena, these factors influence to degrade the predetermined performance of pavement and finally reduce its serviceable life.

Because of the above reasons/factors affecting serviceability, pavement evaluation has gained importance and it is imperative to diagnose/investigate/examine the existing conditions of pavement in terms of its utility values and structural adequacies. Thus, study of pavement performance is essential to make strategic decisions for addressing maintenance needs, with available resources management.

9.5.1 Purpose of Pavement Performance Evaluation

Functional ability of pavement in terms of strength and riding quality during its serviceable period is a subjective measurement of its stiffness and roughness in different forms. Evaluation of pavement performance by various methods is essential for the following purposes:

1. Assess existing condition in terms of riding quality and structural adequacy of pavement structure.
2. To establish maintenance prioritisation by comparing the design parameters with allowable values of distresses/loss of stiffness factors.
3. To evolve maintenance/repair strategy that best suits the related distresses found, with reference to economic parameters. Further, it may be related to life-cycle cost or for estimating cost associated with long-term and short-term budget allocations.
4. To estimate the remaining life of pavement and develop performance/distress models that are intended to be used for checking and updating existing design and rehabilitation methods.
5. For rescheduling maintenance/rehabilitation activities based on the above updated predictions.
6. To provide a data base which serves as a source of reference for improving pavement maintenance management measures for future needs.

9.5.2 The Concept of Pavement Performance and Pavement Maintenance

In-service pavement's existing condition/performance data over a period is collected by different methods and converted to an index value or parameter. Rate of loss of performance in terms of an index value is evaluated with reference to minimum allowable index value and appropriate maintenance measure is initiated (Fig. 9.15). With reference to each interval of maintenance

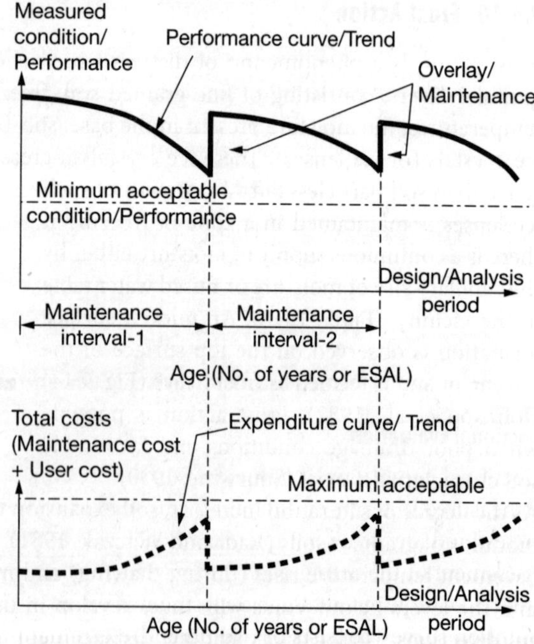

Figure 9.15 Basic concept of pavement evaluation and management.

activity on the age scale, total cost of maintenance (comprises of user, operating, testing and maintenance costs) is estimated and models are developed to estimate future needs of funding based on the anticipated rate of loss of performance. Evaluation and handling of such data collected and its manipulation is a complex issue (at network level) which governs efficiency of the pavement maintenance management system (PMMS). However, with the advent of suitable software, scheduling of evaluation activities and the associated maintenance management proceedings have been made simple and rational.

9.5.3 Classification of Pavement Performance Evaluation Methods

Several methods are available for the evaluation of pavements and these methods use different equipment, depending on the purpose and application. As a broad classification, the methods may be categorised as (*a*) existing condition evaluation and (*b*) economic evaluation. The existing condition evaluation methods are classified as (Fig. 9.16):

(i) Functional evaluation and
(ii) Structural evaluation

The functional evaluation methods may be further classified into distress and riding quality evaluation methods. The riding quality may be evaluated in terms of surface roughness measurement and safety evaluation by measuring skid resistance.

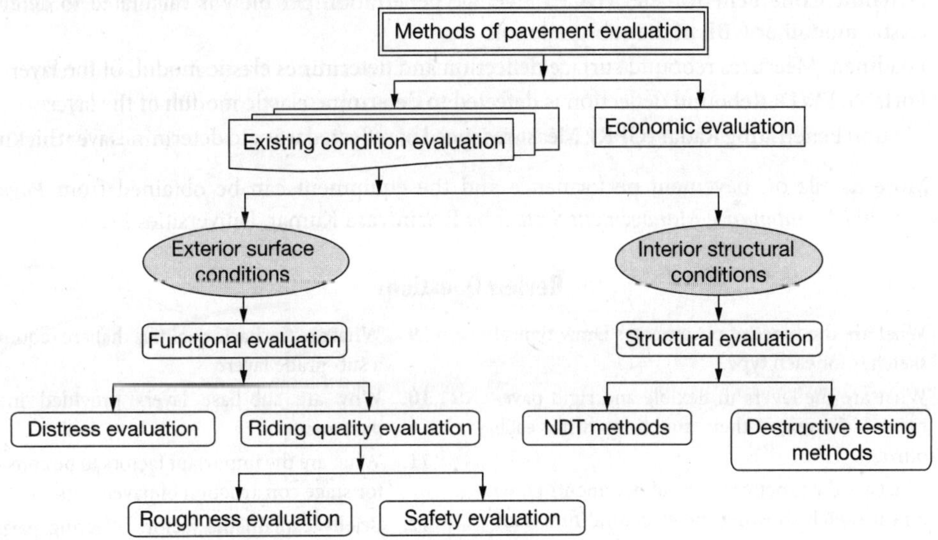

Figure 9.16 Classification of pavement evaluation methods.

Functional Evaluation

Functional conditions of pavement surface are concerned essentially with its riding quality which may be measured in terms of different forms such as roughness/smoothness, skid resistance/surface texture, distress and any other parameters/coefficients that relate to unevenness measured by NDT equipment. Assessment of the riding quality of pavement surface can be carried out by various methods using different NDT equipment. Such assessment is often referred to as pavement surface *condition surveys*. The condition survey consists of two parts. Determination of distress by visual inspection along surface of a road is considered as preliminary part of the survey and later, it is followed by suitable mechanized measurement procedure. Thus, road inventory data is often supplemented with this technique.

The following is a list of equipment used and their test results which are considered for determination of pavement performance in terms of surface profile characteristics:

1. Bump integrator: Unevenness index
2. MERLIN: D-value
3. Profilometers (Laser or non-laser based): International roughness index (IRI)
4. Network vehicle: IRI and distress type and severity in terms of pavement condition index (PCI)
5. Locked wheel tester and Grip Tester: Skid resistance

Structural Evaluation

Structural strength/stiffness of pavement is determined by destructive and NDT methods. The strength parameters determined by different methods are related to structural adequacy of pavement structure as a whole or to any individual component of the pavement in question.

The following is a list of most widely used equipment and their test results, which are considered to determine pavement performance in terms of structural adequacy/stiffness/strength:

1. Benkelman Beam (BB): Rebound surface deflection is analysed for overlay thickness.
2. Falling Weight Deflectometer (FWD): Rebound surface deflection bowls are used to determine elastic moduli of the pavement layers
3. Dynamic Cone Penetrometer (DCP): Average penetration per blow is calculated to determine elastic moduli or CBR of the sub-grade soil
4. Loadman: Measures rebound surface deflection and determines elastic moduli of the layer
5. Portable FWD: Rebound deflection is detected to determine elastic moduli of the layer
6. Ground Penetrating Radar (GPR): Measures speed of reflected wave to determine layer thicknesses

Note: More details on pavement performance and the equipment can be obtained from *Pavement Evaluation and Maintenance Management System* by R Srinivasa Kumar, Universities Press.

Review Questions

1. What are the types of pavements? Draw typical sketches for each type.
2. What are the layers in flexible and rigid pavements? Compare their functions with each other.
3. What are the functions of road pavement? Draw a neat sketch showing the structural functions of a pavement.
4. Explain the behaviour of a pavement under a wheel load with a neat sketch.
5. How are road pavements classified based on the materials used? Draw typical sketches for each case.
6. How do flexible pavement layers undergo plastic deformation due to wheel loading?
7. What are the functions of the individual layers of flexible and rigid pavements?
8. Draw typical cross-sections of flexible and rigid pavements.
9. What are the basic qualities that are required for a sub-grade layer?
10. Why are sub-base layers provided in rigid pavements?
11. What are the important factors to be considered for stage construction of pavements?
12. Briefly explain the factors affecting pavement design and give some examples.
13. Explain how design traffic is calculated from the data obtained from traffic surveys.
14. Define and explain the significance of the following: (i) Traffic growth rate, (ii) Design life and (iii) lane distribution factors.
15. Draw a few typical configurations of axle and wheels of vehicles. Explain why tandem axles are preferred to single axles for truck load transfer on to the pavement surface.
16. Distinguish between static weighing and weigh-in-motion methods.

17. Explain how axle load is weighed using weigh pads with a neat sketch.
18. What do you understand by standard axle load? Draw a sketch showing the salient features of a standard axle according to IRC.
19. What is the standard axle load value of a single-axle–single-wheel, single-axle–dual-wheel and tandem axle–four wheels according to IRC recommendation?
20. What is the value of the standard axle load considered by other agencies for pavement design?
21. Define load equivalency factor. Explain its significance for design of pavements.
22. Define fourth-power rule. Explain its significance for design of pavements.
23. What are the different models available for the determination of equivalent single axle loads (ESAL)?
24. Define legal axle load. What is its significance for design of pavements? Write the maximum permissible loads for different types of vehicles in India.
25. Define vehicle damage factor (VDF). What are the factors affecting VDF? Explain the significance of VDF in the design of pavements.
26. Define tyre pressure and contact pressure.
27. Enumerate the factors that should be considered for the design of durable and weather resistant pavements
28. What is the influence of moisture other than OMC value on the pavement structure? Explain how moisture content changes in the lower layers of the pavement.
29. Define poor drainage conditions and mud pumping in rigid pavements.
30. Explain how pavements are damaged due to frost action.
31. Write the factors relating to axle load and vehicle gross load that affect the performance of vehicles on the road.
32. Define standard axle and legal axle load. Give the values of different types of standard axle and legal axle loads.
33. What is meant by pavement performance? State its significance in pavement design.
34. What are the purposes of pavement performance evaluation?
35. Explain the concept of pavement performance and pavement maintenance with a sketch.
36. Classify the pavement performance evaluation methods with a sketch.
37. Define functional and structural pavement performance evaluation methods. Mention a few equipments used for pavement evaluation and their expected test results, as applicable?

Multiple Choice Questions

1. The axle load (tonnes) limit for tandem axle adopted in India is
   (a) 8.0 (b) 10.2
   (c) 19.0 (d) 24.0
2. According to the Indian practice (IRC), the standard axle load prescribed for a configuration with tandem axle with dual wheels on either side is
   (a) 224 kN (b) 148 kN
   (c) 80 kN (d) 190 kN
3. According to the Indian practice (IRC), the standard centre-to-centre spacing between dual wheels is
   (a) 50 mm (b) 510 mm
   (c) 210 mm (d) 310 mm
4. For traffic analysis of pavement design, the vehicles having axle weight ________ tonnes are referred to as commercial vehicles.
   (a) ≥ 3 (b) ≤ 3
   (c) ≥ 3.5 (d) ≤ 3.5
5. For pavement design purposes, an average tyre pressure of ________ may be assumed for Indian conditions.
   (a) 0.8 MPa (b) 8 MPa
   (c) 80 MPa (d) 8.8 MPa
6. The equipment used for measurement of rebound deflection of pavements is
   (a) Dynamic cone penetrometer
   (b) Ground penetrating radar
   (c) Benkelman beam
   (d) Bump integrator
7. Functional evaluation of pavements is conducted by using ________.
   (a) dynamic cone penetrometer
   (b) ground penetrating radar

(c) Benkelman beam
(d) bump integrator

8. Structural evaluation of pavements is conducted by using ________.
(a) profilometers
(b) grip tester
(c) falling weight deflectometer
(d) MERLIN

9. Which one of the parameter, index or unit is applicable as a part of functional evaluation of pavements?
(a) Any deflection value
(b) Pavement condition index
(c) Any stress value
(d) Pavement modulus index

10. The vehicle damage factor (VDF) can be defined as a numerical value which represents
(a) the equivalent number of standard axles per truck.
(b) the equivalent number of legal axles per truck.
(c) the equivalent number of standard axles measured during a day.
(d) the equivalent number of standard axles measured during a design period.

Answers to Multiple Choice Questions

1. (c)	**3.** (d)	**5.** (a)	**7.** (d)	**9.** (b)
2. (b)	**4.** (a)	**6.** (c)	**8.** (c)	**10.** (a)

CHAPTER 10

Analysis and Design of Flexible Pavements

10.0 INTRODUCTION

Pavement layers are made of different kinds of materials. To withstand traffic loading under prevailing environmental and *in situ* conditions, the pavement should be designed based on allowable deflections, stresses and strains. For these reasons all pavement structures should be analysed for different conditions of loading, material characteristics and boundary conditions.

10.1 STRESS ANALYSIS OF FLEXIBLE PAVEMENT

The following sections present a brief review of the equations used to calculate the stresses and surface deflections of an elastic half-space based on linear elastic principles (Boussinesq, 1885; Fadum, 1948; Foster and Ahlvin, 1954; Ahlvin and Ulery, 1962).

10.1.1 Boussinesq's Theory (Single-layer Theory)

In 1885, Boussinesq presented a theory for calculating the stresses in soil mass based on the following assumptions.

- The soil is homogenous (without any stratification) and isotropically linear elastic, which means that it abides by Hooke's law.
- The soil mass boundary is a semi-infinite elastic half-space.
- The load is applied on to a level surface.
- The soil mass is weightless.

The following equations are based on Boussinesq's theory and are widely used to calculate stresses in soil mass.

Case 1: Point load: Consider a point load on the soil mass as depicted in Fig.10.1.

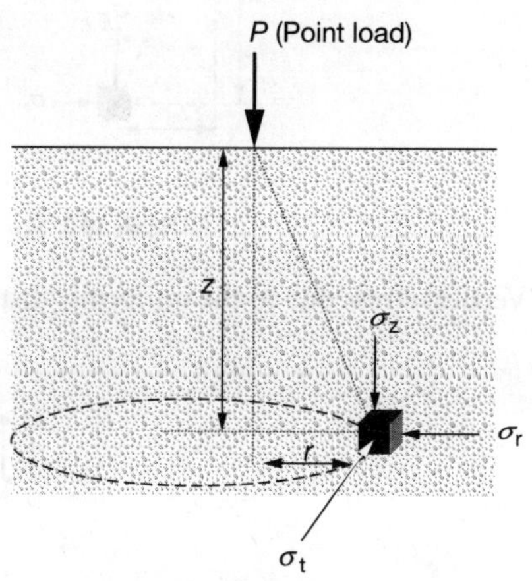

Figure 10.1 Stress due to point load.

$$\text{Vertical stress due to a point load } (\sigma_z) = \frac{3P}{2\pi z^2}\frac{1}{\left[1+\left(\frac{r}{z}\right)^2\right]^{5/2}} \tag{10.1}$$

where,

P = point load
z = depth from surface
r = radial distance from the axis of loading
σ_z = vertical stress

The above equation is also used to calculate stresses in a soil mass where the contact pressure is not uniform. Non-uniform pressures are calculated as equivalent point loads at their centre. The total stress that the soil mass is under can be calculated by super-positioning of the stresses.

EXAMPLE 10.1

Calculate the intensity of vertical pressure due to a point load of 4200 kg at a depth of 4 m from the surface and a distance of 2 m from the axis of loading.

Answer

At 4 m depth and distance 2 m from the axis of loading: $z = 4$ m and $r = 2$ m. By Eq. (10.1),

$$\text{Vertical stress due to a point load } (\sigma_z) = \frac{3\times 4200}{2\pi\times 4^2}\frac{1}{\left[1+\left(\frac{2}{4}\right)^2\right]^{5/2}} = 71.74 \text{ kg/m}^2$$

Case 2: Uniformly distributed load over a circular area:
Generally, for pavement analysis, the equivalent circular contact area of a tyre on pavement surfaces is taken. For this purpose, a uniformly loaded circular area is considered for calculating the stresses in the soil mass. To calculate stresses under a uniformly loaded area using flexible plates (similar to a rubber tyre), Eq. (10.1) may be integrated over the circular area (Fig.10.2):

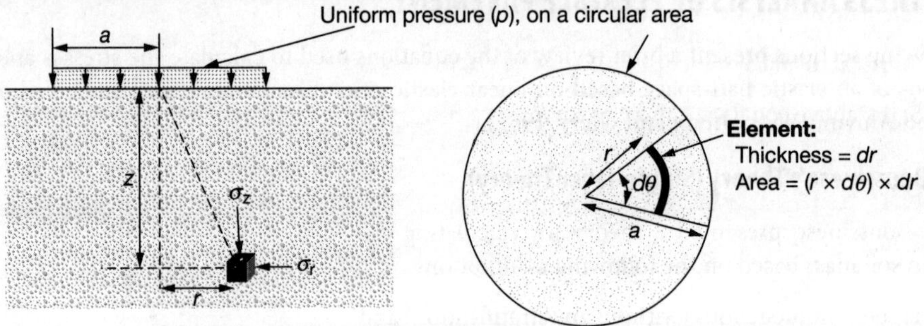

Figure 10.2 Stress under a uniformly loaded circular area.

Vertical stress due to the uniformly distributed circular are (σ_z)

$$= \int_0^{2\pi}\int_0^{a}\left(\frac{3p((rd\theta)dr)}{2\pi z^2}\left[\frac{1}{1+\left(\frac{r}{z}\right)^2}\right]^{5/2}\right) \tag{10.2}$$

The above equation can be simplified as (Burmister 1938)

$$\text{Vertical stress under the center of the circular are } (\sigma_z) = p\left(1-\left[\frac{1}{1+\left(\frac{a}{z}\right)^2}\right]^{3/2}\right) \tag{10.3}$$

Horizontal orradial stress beneath the centre of circular area (σ_r),

$$\sigma_r = 0.5p\left[(1+2\mu)-\left(\frac{2z(1+\mu)}{\sqrt{(a^2+z^2)}}\right)+\left(\frac{z^3}{(a^2+z^2)^{3/2}}\right)\right] \tag{10.4}$$

Elastic vertical displacement at z = 0, under a circular flexible plate $(\Delta s) = \dfrac{2pa(1-\mu^2)}{E}$ (10.5)

If $\mu = 0.5$, the above equation can be written as

$$\Delta s = \frac{1.5pa}{E} \tag{10.6}$$

where,

p = uniformly distributed pressure
a = radius of the circularly loaded area
μ = Poisson's ratio of soil
E = Elastic modulus of soil (assumed as single layer)

The stress distribution under the flexible plate is uniform since the flexible plate bends (or flexes) under loading, similar to a rubber sheet. Therefore, the elastic surface deflection under the flexible plate is non-uniform.

If a circular rigid plate (or steel plate) is used to apply load on to the soil, the elastic surface displacement is uniform but the stress distribution is non-uniform (Ullidtz, 1987). Analysis of plate load test data is one of the best examples of an application of Boussinesq's theory. The following formula may be used to calculate the elastic surface deflection under a rigid plate.

Vertical displacement at $z = 0$, under a circular rigid plate $(\Delta s) = \dfrac{\pi pa(1-\mu^2)}{2E}$ (10.7)

If $\mu = 0.5$, the above equation can be written as

$$\Delta s = \frac{1.18pa}{E} \tag{10.8}$$

EXAMPLE 10.2

A semi-infinite soil mass is subjected to stress under a circular plate having a 15 cm radius. The load intensity over the plate is 4000 kg. Calculate the vertical stress in the soil under the axis of the circular plate at 2 m depth.

Answer

Intensity of pressure over plate = 4000/($\pi \times 15^2$) = 5.66 kg/cm²

$$\text{Vertical stress under the center of the circular are } (\sigma_z) = 5.66\left(1 - \left[\frac{1}{1+\left(\frac{15}{2}\right)^2}\right]^{3/2}\right) = 5.64 \text{ kg/cm}^2$$

EXAMPLE 10.3

Calculate the rebound surface deflection on a single-layer pavement under a wheel load of 40 kN with a tyre pressure of 0.8 MPa. The effective elastic modulus of sub-grade may be taken as 40 MPa and Poisson's ratio of the soil as 0.5.

$$A\text{: Tyre pressure} = \frac{\text{Wheel load}}{\text{Tyre contact area}} = \frac{\text{Wheel load}}{\pi a^2}$$

$$\text{Radius of tyre contact area} = a = \sqrt{\left(\frac{\text{Wheel load}}{\pi \times \text{Tyre pressure}}\right)}$$

$$=\sqrt{\left(\frac{40\times10^{3}}{\pi\times0.8\times10^{6}}\right)}=0.126\text{ m}=12.6\text{ cm}$$

$$\text{Intensity of pressure}=p=\frac{\text{Load}}{\text{Area}}=\frac{40\times10^{3}}{\pi\times0.126^{2}}=801990.14\text{ N/m}^{2}$$

If a flexible plate is considered, re-bound surface deflection = Δs

$$\Delta s=\frac{1.5pa}{E}=\frac{1.5\times801990.14\times0.126}{40\times10^{6}}=0.00378\text{ m}=3.78\text{ mm}$$

The theory presented by Boussinesq has resulted in basic fundamental concepts related to the calculation of stresses and deflections of an elastic half-space. However, the equations based on Boussinesq's theory do not provide realistic values since

- The soil mass is never ideally elastic and isotropic.
- The pressure distribution may be non-uniform depending on the loading conditions.
- The soil mass is never weightless and homogenous.
- The pavement consists of multiple layers having different thicknesses and material characteristics.

Therefore, layered theories which consider multiple layers of pavement with different material characteristics have more significance for the design and analysis of pavement structures.

10.1.2 Burmister's Theory

In 1943, Donald M Burmister developed a method of analysing a two-layered soil system which resembled a flexible pavement having its top layer stiffer than its bottom layer. Later, in 1945, he extended the method to include three-layered systems. As an important outcome of Burmister's numerical solutions, the flexible pavement thickness design method was developed for airfield pavements (*ASCE* Vol.115, 1950). In 1962, Ahlvin and Ulery of the US Army Corps of Engineers (USACE) reported solutions for calculating the stresses, strains and deflections on a homogenous half-space withstanding a uniform circular load (Ahlvin and Ulery, 1962). Since then, Burmister's theory with required modifications is being used for various pavement engineering applications. Many researchers have developed different methods for analysing multi-layered systems based on elastic, non-linear, viscoelastic and finite element methods. The solution of various problems related to a multi-layered system is made easy by the use of computers.

Burmister's two-layered system of analysis is briefly explained below.

Assumptions

- Every layer material is homogenous, isotropic and ideally elastic. They are characterised by a unique elastic modulus and Poisson's ratio.
- A pavement consists of two layers. The elastic modulus value of the top layer is more than the bottom layer (i.e., $E_1 > E_2$).
- The layers are weightless and infinite in the horizontal direction.
- The top layer has finite thickness (h) but the bottom layer is infinitely thick.
- The top layer is in full contact with the bottom layer.
- The interface between these layers is rough and there is no loss of applied stress transferred from the top layer to the bottom layer.
- There will be no shear and normal stresses outside the loaded area.
- The applied stress is uniform over a circular area.

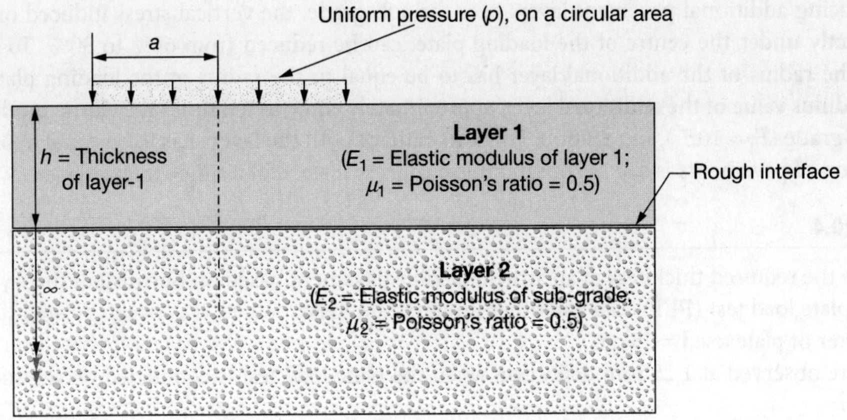

Figure 10.3 Burmister's two-layered system.

The displacement equations given by Burmister for a two-layered system are as follows (Fig.10.3):
Surface deflection under the centre of the flexible plate:

$$\Delta = \frac{(1.5\,pa)}{E_2} \times F_{\rm d} \tag{10.9}$$

Surface deflection under the rigid plate:

$$\Delta = \frac{(1.18\,pa)}{E_2} \times F_{\rm d} \tag{10.10}$$

Where,

D = vertical deflection at surface (when $z = 0$)
p = Uniform pressure on circular plate
a = Radius of the circular plate
E_2 = Elastic modulus of sub-grade
F_d = Displacement factor, depends ratios of $\left(\frac{E_2}{E_1}\right)$ and $\left(\frac{h}{a}\right)$; Refer to Fig.10.4
E_1 = Elastic modulus of the top layer
h = Thickness of the top layer

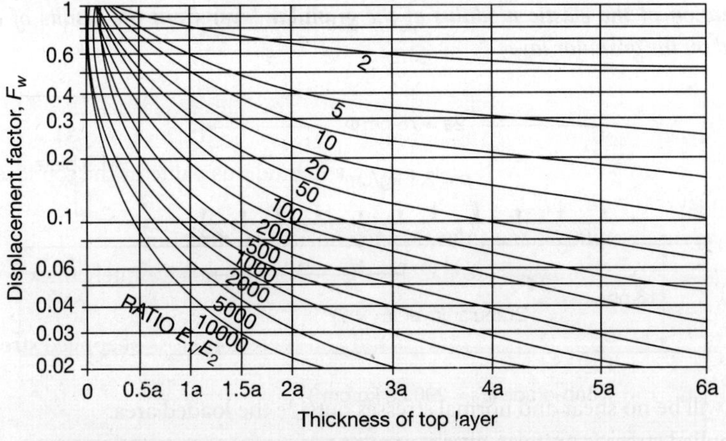

Figure 10.4 Burmister's two-layered system (after Burmister 1943).

By introducing additional pavement layers over the sub-grade, the vertical stress induced on the sub-grade, exactly under the centre of the loading plate, can be reduced from 68% to 30%. To make this possible, the radius of the additional layer has to be equal to the radius of the loading plate and the elastic modulus value of the additional layer, approximately equal to ten times the elastic modulus value of the sub-grade ($E_1 \approx 10E_2$). In addition, Poisson's ratio of both the layers has to be equal to 0.5. Hence, construction of pavements is carried out in multiple layers instead of a single thicker layer.

EXAMPLE 10.4

Determine the required thickness of an airfield flexible pavement based on Burmister's theory using the following plate load test (PLT) data and other input parameters:

Diameter of plate used = 75 cm

Pressure observed at 1.25 mm deflection when the plate load test is conducted on the sub-grade = 0.82 kg/cm²

Pressure observed at 1.25 mm deflection when the plate load test is conducted on a base course of 16 cm thickness = 2.1 kg/cm²

Design wheel load = 23000 kg; Tyre pressure = 15 kg/cm²;

(a) If allowable deflection = 0.125 cm and (b) If allowable deflection = 0.50 cm

Answer

(i) *Determination of the elastic modulus of the sub-grade from a plate load test conducted on the sub-grade*

From the standard relationship,

$$\text{Deflection } (\Delta) = 1.18\frac{p\,a}{E_2}$$

Where,

E_2 is the elastic modulus of the sub-grade

On substitution,

$$0.125 \text{ cm} = 1.18\frac{0.82 \text{ kg/cm}^2 \times 37.5 \text{ cm}}{E_2}$$

$$E_2 = 290.28 \text{ kg/cm}^2$$

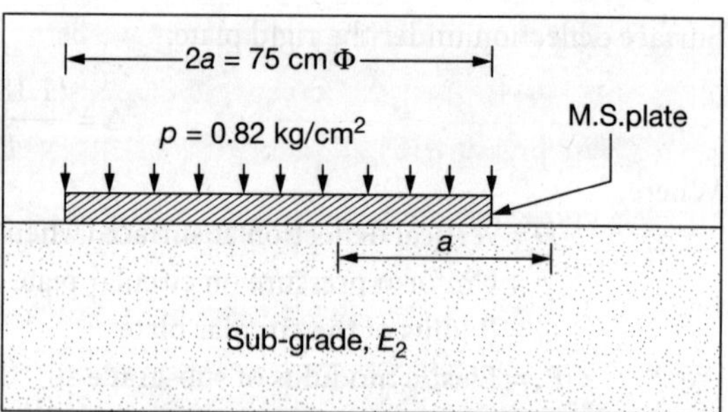

Figure 10.5 Diagram (a) for Example 10.4.

(ii) *Determination of the elastic modulus of the granular layer from the results of a plate load test conducted on the granular layer*

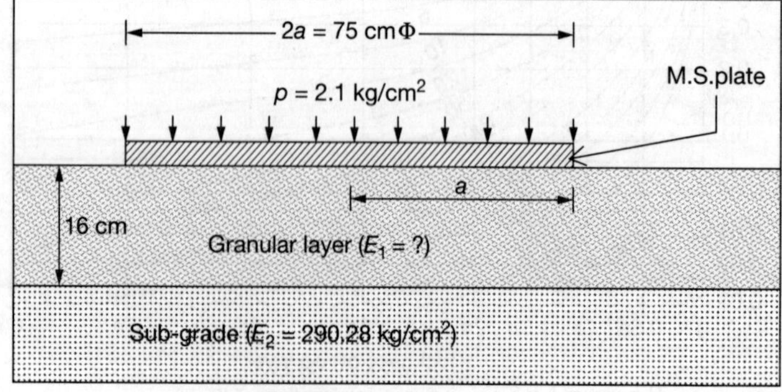

Figure 10.6 Diagram (b) for Example 10.4.

$$\text{Deflection } (\Delta) = 1.18\frac{p.a}{E_2} \times F_{\text{shape}}$$

where,
E_2 = elastic modulus of the sub-grade
F_{shape} = shape factor, based on the ratios of $\frac{h}{a}$ and $\frac{E_1}{E_2}$
On substitution,

$$0.125 \text{ cm} = 1.18 \times \frac{2.1 \text{ kg/cm}^2 \times 37.5 \text{ cm}}{290.28 \text{ kg/cm}^2} \times F_d$$

$$F_{\text{shape}} = 0.39; \text{ ratio of } \frac{h}{a} = \frac{16 \text{ cm}}{37.5 \text{ cm}} = 0.426,$$

Now read the value of $\frac{E_1}{E_2}$ ratio from Fig. 10.4

$$\therefore \frac{E_1}{E_2} = 200, \ E_1 = 200E_2; \ E_1 = 200 \times 290.28 = 58056 \text{ kg/cm}^2$$

(iii) *Design of flexible pavement*

Tire contact are on pavement = $\pi \times a^2$
where a = radius of tire contact on pavement

$$\pi \times a^2 = \frac{\text{Wheel load}}{\text{Tyre pressure}} = \frac{23{,}000 \text{ kg}}{15 \text{ kg/cm}^2}; \ \therefore \ a = 22.09 \text{ cm}$$

We know that the deflection equation for a flexible circular plate is

$$\text{Allowable deflection } (\Delta a) = 1.5\frac{p.a}{E_2} \times F_d$$

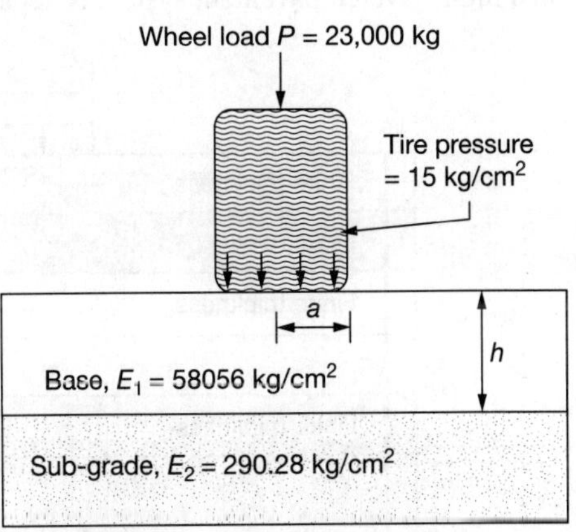

Figure 10.7 Diagram (c) Example 10.4.

(a) If allowable deflection = 0.125 mm
On substitution,

$$0.125 \text{ cm} = 1.5 \times \frac{15 \text{ kg/cm}^2 \times 22.09 \text{ cm}}{290.28 \text{ kg/cm}^2} \times F_d$$

$$F_d = 0.073$$

Now read the ration of $\frac{h}{a}$ from Fig. 10.7

Based on the ratio of $\frac{E_1}{E_2} = 200$

$$\frac{h}{a} = 2.79, h = 2.79 \times a = 2.79 \times 22.09 = 61.63 \text{ cm}$$

∴ For the base course, provide a thickness of 62 cm.

(b) If allowable deflection = 0.50 mm,

$$0.5 \text{ cm} = 1.5 \times \frac{15 \text{ kg/cm}^2 \times 22.09 \text{ cm}}{290.28 \text{ kg/cm}^2} \times F_{\text{shape}}$$

$$F_{shape} = 0.438$$

Now read the ratio of $\frac{h}{a}$ from the ratio of $\frac{E_2}{E_1} = \frac{1}{200}$

$$\frac{h}{a} = 0.36, h = 0.36 \times a = 0.36 \times 22.09 = 7.95 \text{ cm}$$

∴ For the base course, provide a thickness of 8 cm.

In the above design example, the thickness of the base course is calculated based only on the allowable elastic surface deflection. This method does not consider traffic loads and non-linear behaviour of pavement materials. These factors are major influencing parameters for thickness design of pavements. Therefore, the basic philosophy of Burmister's method for the analysis of stresses and strains is used in conjunction with the factors affecting pavement design which are discussed in Chapter 9.

10.1.3 Linear Elastic Multi-layered Pavement System

In reality, a flexible pavement is a multi-layered structure in which each layer is constructed with different thickness (h), each having a respective material characteristic such as the elastic modulus (E) and Poisson's ratio (μ). A typical configuration of a circular plate subjected to a uniform pressure (p) placed on a multi-layered pavement system is depicted in Fig.10.8.

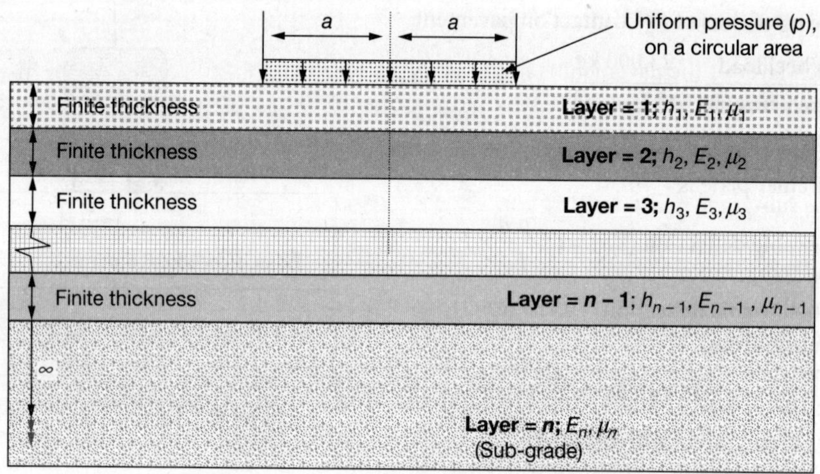

Figure 10.8 Multi-layered pavement system.

The following considerations/ assumptions are generally made for analysis of stresses, strains and deflections in a multi-layered flexible pavement system.

- The bottom (nth) layer is a sub-grade that has infinite thickness.
- Above the sub-grade, each layer has a finite thickness and infinite dimension in the horizontal direction.
- The self-weight of the sub-grade and other layers are considered as weightless.
- There is no loss of stress at each layer interface. The layers are perfectly in contact at the top and bottom surfaces. But for some methods, a smooth layer interface may also be considered.

- Each layer in the pavement system is linearly elastic, homogenous and isotropic.
- There will be no stresses, strains and deflections at infinite depth of the sub-grade.
- Only static loads are considered.
- There is no shear stress or radial displacement of the top layer.
- The surface of the top layer is perfectly horizontal and has no undulations.

The stress function adopted by Burmister for analysis of the three-layered flexible pavement is

$$\Phi(r,z)=\left[\left(A+Bz\right)e^{-mz}+\left(C+Dz\right)e^{mz}\right]J_o\left(mr\right) \tag{10.11}$$

where,

r = radial distance

z = depth

A, B, C and D = (each layer) constants of integration

J_o = Bessel function of the first kind of order zero

With the above stress function and assumptions (boundary conditions) that satisfy the governing differential equation, the Henkel transform is applied to obtain stresses and displacements in terms of numerical integrals. The details of the analysis may be obtained from Jeuffroy and Bachelez, 1962; Verstraeten, 1967, 1972 and elsewhere (Schiffman, 1962; Ashton and Moavenzadeh, 1967).

The analysis methods based on the above-mentioned assumptions are unrealistic in practice due to the following reasons:

- Each layer of the pavement is non-elastic, inhomogeneous and an anisotropic solid. They never truly obey Hooke's law.
- The materials used for the sub-base and base layers are granular materials which show non-linear behaviour on loading.
- The sub-grade may consist of non-linear stiffness with depth.
- The bituminous surface layer shows visco-elastic behaviour, which varies based on loading time and temperature.
- The Poisson's ratio of pavement materials is variable and depends on *in situ* conditions and different loading pressures.
- The wheel loading is generally dynamic and transient in nature.
- The boundary condition of being infinite in the horizontal direction is never a true condition.
- Shear and radial stresses will be developed on the surface layer during rolling wheel loads.
- Cracked bound layers cause discontinued distribution of pressure.

To account for the realistic behaviour of pavements under traffic loading during *in situ* conditions, the multi-layered flexible pavement system has been suitably modified by different researchers. Many software programs (Table 10.1) have also been developed. It can be observed from Table 10.1 that most programs use linear elastic theory since it provides a reasonably good estimate of surface deflections and strains in the bound layers. The linear elastic programs which are simple and easy to perform may be used for an approximate analysis of pavement structures. However, for accurate and realistic modelling of the non-linear behaviour of unbound layers, visco-elastic and plastic behaviour of bituminous pavements by finite element methods are more appropriate. However, these methods involve complex mathematical concepts that would place a huge load on the computer.

Table 10.1 A few programs for analytical solution of multi-layered flexible pavements.

Name of the program	Theory used	Program source	Reference
CHEVRON	LEMLT	Chevron Research Corporation	Warren and Diekmann, 1963; COST 336, Task Group 1.
FEPAVE	FEM (axi-symmetric)	–	Duncan et al., 1968
SLAB 30 and SLAB49	FDM	–	Deacon, 1969
ELSYM	LEMLT	FHWA (UCB)	Ahlborn, 1972
BISAR	LEMLT	Shell International	De Jong et al., 1973
JULEA	LEMLT	USACE WES	Uzan, 1976
VEROAD	MLV	Delft University of Technology	Nilsson et al., 1996
CIRCLE	LEMLT	MINCAD, Australia	Wardle 1977; Wardle and Rodway 1998.
VESYS	LEMLT or MLV	FHWA	Kenis et al. 1982
ABAQUS	FEM	–	Lee et al., 1998
ILLI-PAVE	FEM	University of Illinois	Thompson and Elliott, 1988
FENLAP	FEM	University of Nottingham	Brunton and d'Almedia, 1992
SAPSI-M	MLV with moving loads	Michigan State University	Chatti and Yun, 1996
DAMA	LEMLT	The Asphalt Institute	The Asphalt Institute, 1983
KENLAYER	LEMLT or non-linear elastic or MLV	University of Kentucky	Yang H. Huang, 2004
ELAYER	LEMLT	IIT Kharagpur	Sudhakar, 1993
FPAVE	LEMLT	IIT Kharagpur	Animesh, 1998
NCPA (Non-linear cross-anisotropic pavement analysis)	FEM	–	Jeongho Oh et al., 2007.

LEMLT: Linear elastic multi-layer theory; MLV: Multi-layered visco-elastic; FEM: Finite element method; FDM: Finite difference method

10.2 FLEXIBLE PAVEMENT STRUCTURAL DESIGN METHODS

Designs of flexible pavement have undergone tremendous changes during the past few decades, ranging from empirical procedures to analytical methods. Reaching for perfection, many modifications of pavement structures have been made in different design methods to obtain the desired level of service and structural adequacy using locally available as well as modified materials at a reasonable cost. Different thickness design methods of flexible pavements have been developed by different researchers and organisations around the world. They may be broadly classified as follows.

10.2.1 Empirical Methods

Pavement design factors related to traffic loading, material strength, and climatic and environmental conditions are inevitably variable in nature. Therefore, to analyse pavement performance, the mean effect of these factors in terms of distress is measured and related to pavement crust thickness by using empirical relationships. Many empirical relationships have been developed by relating (or correlating) pavement performance (in terms of different modes of failure) to traffic loading, characteristics of subgrade soil, climatic etc. The following groups explain methods of pavement design based on different empirical relationships.

Group 1 A selection of suitable crust thickness of flexible pavement is made based on soil classification, grain size distribution, moisture content, plastic limit, liquid limit, truck traffic volume and experience record of *in situ* pavements. Group index (GI) (1945) and FAA methods may be considered under this group. These methods have not been found successful because parameters associated with material strength, climatic, environmental and *in situ* conditions are not considered.

Group 2 Thickness design charts were developed by correlating performance of pavements to various design inputs related to material strength parameters, density, truck traffic volume, climate, environment, frost, ground water level and parameters related to *in situ* conditions. The anticipated occurrences of uncertainties of these parameters are accounted for and analysed based on different methods in different countries. The strength parameters of pavement materials are obtained based on laboratory as well as *in situ* tests.

Group 3 Design methods based on the performance serviceability concept may be included in this group. The first of this kind, the AASHO Interim Guide (AASHO 1972) was developed using AASHO test data. In this method, a minimum structural number (SN) (1 to 6) is determined using a nomograph based on sub-grade support value (1 to 10), regional factor (0.2 to 5.0) and equivalent 18-kip (kilo pounds) single-axle load repetitions. The individual layer thicknesses are derived from the SN value using pavement layer equivalent thickness factors. Many modifications were made in later editions (1981, 1986, 1993). However, the design philosophy of all the editions is based on empirical (regression) equations derived from the AASHO road test data related to various characteristics of materials, traffic and different modes of performance. Applicability of the AASHTO design method in different countries should be verified before using it since the empirical relationships incorporated in the design are developed from the AASHO road test data obtained from test tracks constructed in Illinois and other places in USA.

10.2.2 Theoretical or Analytical or Pure Mechanistic Methods

In this group of methods, the stresses, strains and deflections in a pavement system are calculated based on principles of engineering mechanics which are related to elastic, visco-elastic and non-linear or finite element methods. Thickness design charts were developed based on correlations between pavement performances and theoretically calculated performance characteristics of layer materials subjected to loading. The correlations include stress–strain relationships or surface deflection strength parameters. The methods described by Burmister based on elastic theory (in 1943 and 1945) are the foremost examples of this kind of analytical pavement design methods.

So far there is no accurate or direct solution available to analyse a multi-layered pavement system as it exists in the field under repetitive transient load passes, prevailing climatic and *in situ* conditions.

Research is in progress towards development of models to precisely calculate the realistic behaviour of pavements under differential distribution of stresses and strains which are produced due to variability in material characteristics, loading pattern, *in situ* conditions and environment conditions. Due to the complexity associated with incorporating realistic conditions in mathematical analysis or modelling, analytical design methods are used for limited applications. Of late, empirical pavement design methods supported by mechanistic principles are gaining significance.

10.2.3 Mechanistic–Empirical (M–E) or Analytical–Empirical or Rational Design Methods

The M–E design procedure is carried out in two steps. In the first step, principles related to engineering mechanics are used to calculate the response of pavement layers at critical locations in terms of stresses, strains and deflections for a given loading configuration of wheels–axles, material characteristics of each layer (E_i, μ_i and h_i) and environmental conditions. The theoretically calculated pavement response is compared to the allowable maximum values of stresses or strains in the second step. Distress or damage caused by these critical values of pavement response is determined from empirical relationships in terms of rutting, cracking or other forms of distress (or any defined failure). These relationships are often termed as failure criteria. In these approaches, the major modes of pavement failures are identified and the corresponding terminal (or failure) conditions are defined. Some of the failure modes commonly considered are fatigue cracking of bound layers (top-down or bottom-up cracking), rutting (undulations as permanent deformations) along wheel paths and low temperature (or thermal) cracking. To control the load-associated failures of pavements, critical mechanistic parameters (as responses of pavement to applied load), considered to be causative of the failures, are kept within allowable limits during the design process. A similar procedure may be adopted for modelling of pavement performance.

In the design of flexible pavements, the following failure criteria (strain limits) were first used and reported by Dormon (in 1962). This work was based upon the parallel work carried out by Saal and Pell at the University of Nottingham in conjunction with Shell Laboratories in 1960 and by Monismith et al. at the University of California in 1961. These criteria are widely used in many M–E methods of design of flexible pavement including the Indian design method of IRC: 37-2001 (Fig.10.9).

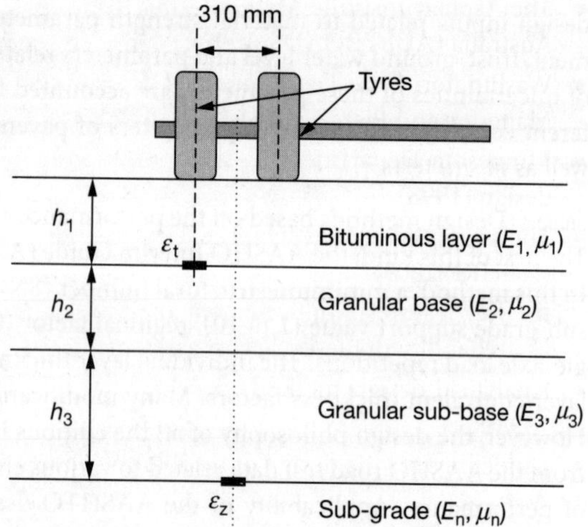

Figure 10.9 Critical strains in a layered flexible pavement structure - A basic model.

1. Vertical compressive strain (ε_v) on top of the sub-grade layer under the centre of dual wheels is correlated with rutting (vertical permanent strain) of pavement.
2. Horizontal tensile strain (ε_t) at the bottom fibre of the bituminous bound surface layer under the centre of a wheel of dual wheels is correlated with the fatigue failure (or cracking) of the bituminous surface layer.
3. In addition to the above design criteria, vertical compressive stresses and strains within the bituminous layer, granular base and sub-base courses are also considered (NCHRP, 2003).

Relationships correlating the limiting values of the mechanistic parameters with the service life of the pavements are known as performance criteria. The performance criteria developed for the design of new pavements are often used for the design of overlays also. While the performance criteria form the core of the pavement design and evaluation methodology, other important issues involved are (i) the selection of an appropriate theory for analysis of pavements and (ii) the selection of inputs, especially material properties, for the analytical model. In the case of overlays, the methodology adopted for evaluation of in-service pavements in terms of structural index or layer properties is important for the design process. A number of empirical and mechanistic methods are available for the design of new pavements and overlays. These methods vary in terms of the failure modes considered, the terminal condition used to define failure, the theory used for the analysis of pavements, the procedure adopted for the selection of material properties (for new or in-service pavements) and performance criteria used. There are also differences in the estimation of traffic loading, reliability concepts used and methods followed for the evaluation of in-service pavements.

A few methods of flexible pavement design based on M–E principles are given below.

- COmparing Mix and Pavement StructureS (COMPASS) by the CROW working group.
- South African mechanistic design method (SAMDM) (Theyse et el., 1996 and 1997).
- CARE (Road and Hydraulic Engineering Institute, Deft) (CROW Report D06-06).
- French Design Manual, LCPC, Paris (CROW Report D06-06).
- Shell Pavement Design Manual (Shell, 1985).
- CROW Design Procedure for Thin Asphalt Pavements: (CROW Report D06-06).
- The Asphalt Institute Method, MS-1, USA (TAI 1991). AUSTROADS Pavement Design Guide, Australia (AUSTROADS, 1992).
- Washington State Department of Transportation (WSDOT) Pavement Guide (WSDOT 1995; Mahoney and Pierce, 1996).
- Minnesota Department of Transportation (MnDOT) mechanistic–empirical flexible pavement design (1998).
- The proposed AASHTO M–E Design Guide for New and Rehabilitated Pavement Structures (NCHRP, 2003).
  - Key features: Finite element analysis, load spectra, roughness evaluation in terms of IRI and reliability in life-cycle cost assessment.
- Analytical design methods developed at IIT Kharagpur, India. (Sudhakar, 1993 and Animesh, 1998).
- Guidelines for the design of flexible pavements according to IRC: 37-2001, 2012 and 2018.

10.3 BENEFITS OF PAVEMENT DESIGN BASED ON M–E METHOD

The following benefits can be obtained using M–E pavement design methods.

- Structural analysis can be made to calculate critical stresses, strains and deflections at any point in a pavement structure at any period of service life.
- Precise structural strength assessment of individual pavement layers is possible using suitable software.
- Better alternative designs can be attained based on variability in locally available material or alternative materials.
- Cost-effective design with strategic rehabilitation philosophy.
- Uncertainties can be accounted for during the design process as an integrated approach related to traffic loading, climatic factors, moisture, seasonal changes etc.
- Different distress models belonging to specific locations can be effectively incorporated in design.

- Increased precision in analysing locality-specific performance failure modes. Such analysis provides insight into the realistic behaviour of pavements during calibration of deterministic or empirical models or transfer functions[1]. Thus, the degree of uncertainty in performance prediction can be considerably reduced to a permissible limit.
- Theoretical modelling can be made to verify and validate new techniques related to design, performance, maintenance and rehabilitation strategies.

From literature, it can be noticed that there is no clear-cut separation between empirical and arbitrary methods as well as semi-mechanistic and empirical methods of flexible pavement design (Waterhouse, 1962; Skok and Finn, 1962).

As our understanding of pavement failure mechanisms improved, many developments took place to modify the design concepts, ranging from empirical to mechanistic–empirical methods. In this process, characterisation of pavement materials played a significant role for design as well as performance evaluation of pavements.

10.4 OVERVIEW ON STRUCTURAL DESIGN OF FLEXIBLE PAVEMENTS AS PER IRC GUIDELINES

Design of new flexible pavements in India is carried out according to the guidelines of the Indian Roads Congress given in IRC:37-2018 (4th Revision), if the traffic volume is 2 msa. The guidelines of IRC:SP:72 may be used in case the traffic volume is less than 2 msa or 450 commercial vehicles per day (CVPD). However, the both the design methods were developed based on mechanistic–empirical (M–E) principles.

10.4.1 A Few Salient Features of the IRC: 37-2018

1. General failure criteria of pavement, as given in Section 10.2.3, were considered for the design of flexible pavement (Fig. 10.9). The details of options provided for pavement layer composition between the compacted sub-grade and bituminous surfacing layers and their failure criteria, with reference to locations of tensile and compressive strains under standard wheel load of 40 kN dual-wheel (having each 20kN) and as recommended by the IRC:37 (2018), are briefly summarized below.

   Case-I: Granular sub-base + Granular base (Fig. 10.10 (a)): This combination may be provided with granular sub-base (conforming to MORTH, 2013) and granular base as WBM or WMM or crusher run macadam or reclaimed concrete or in combination with gravel, moorum kankar, brick-bats (conforming to IRC/MORTH, 2013).

   Case-II: Cement-treated sub-base + Cement-treated base + Granular crack relief layer (Fig. 10.10 (b)): Cracking of cement treated base course is considered as design terminal life of pavement. The aggregate interlayer is provided on the cement treated base course because the cement-treated layer is subjected to cracking due to shrinkage. Hence, provision of an aggregate layer between the cement-treated base and the bituminous layer will relieve the possible propagation of cracks from cement-treated base to the bituminous layer. For the purpose of analysis, elastic modulus value of the cement-treated sub-base and cement-treated base may be taken as 600 MPa and 5000 MPa, respectively.

   The cement treatment sub-base and base courses may be considered for stabilisation of soil + aggregate with lime, fly ash, cement and other chemicals or with some combination thereof. As

1 Transfer functions are relationships used to relate the state of stress or surface deflection or strain in a pavement structure (measured in different forms such as rutting, cracking, roughness and so forth) to its overall performance (Thompson 1996). These relationships are useful for developing distress prediction models subjected to extensive field calibration and verification.

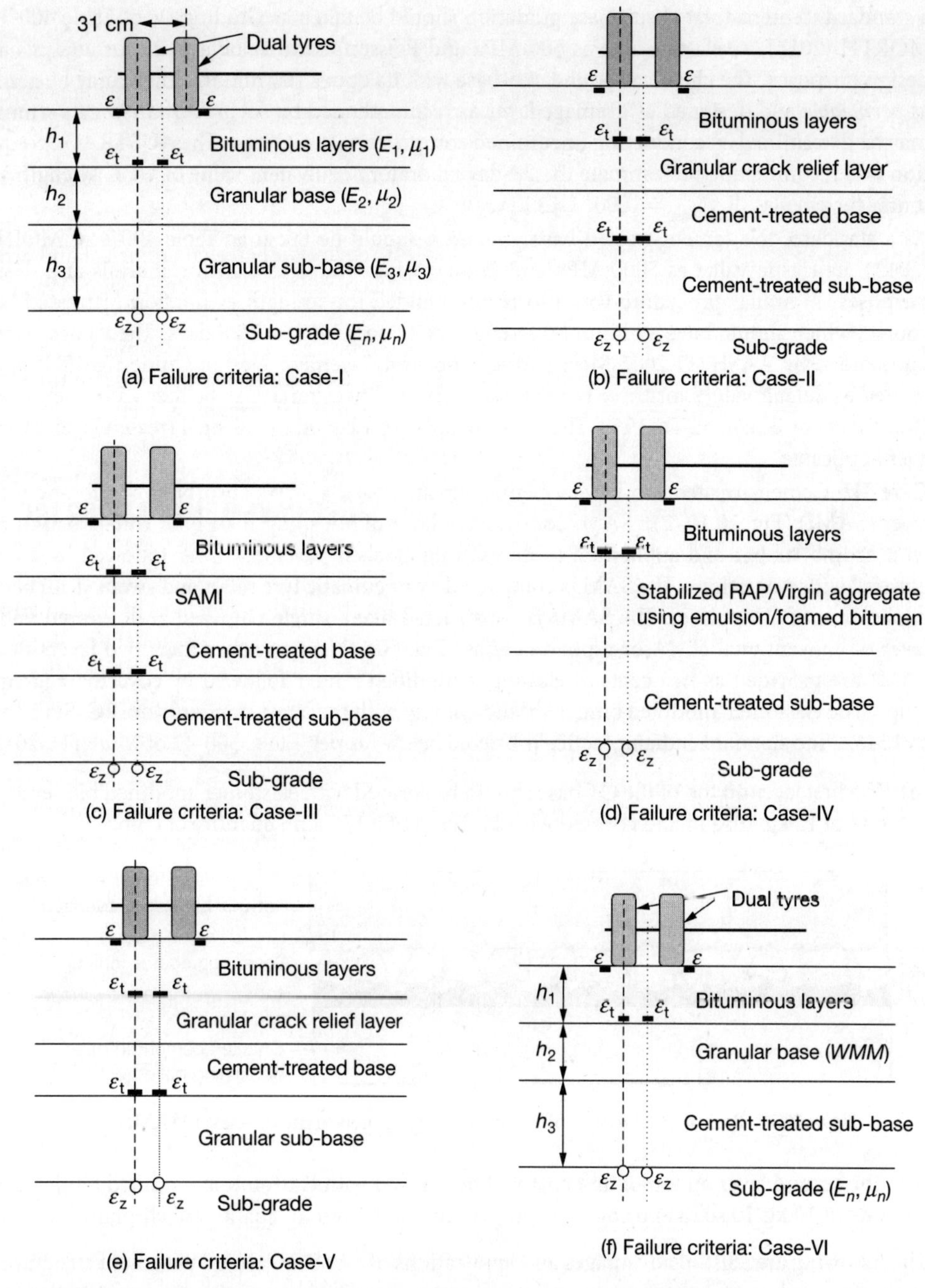

Figure 10.10 Typical combinations with their Critical location of strains considered for structural design of flexible pavements as per IRC: 37-2018 (Redrawn based on IRC: 37, 2018) Courtesy: Indian Roads Congress, India.

a standard, cement-treated sub-base gradation should be taken as Grading IV of Table 400-1 of MORTH (2013), its elastic value as 600 MPa and Poisson's ratio value as 0.25 for analysis and design purposes. The chemical-bound sub-base with its upper 100 mm thickness may be graded as *permeable* and designed as drainage layer as recommended by AASHTO (1993). Its strength may be determined by conducting unconfined compression (UCC) test. The AUSTRAODS equation was recommended to estimate the 28-day laboratory equivalent value of UCC strength and it may be given as, $E_{sub\text{-}base} = 1000^* \; UCC_{28\text{-}Day}$.

As a standard, the cement-treated base gradation should be taken as Table 400-4 of MORTH (2013), its elastic value as 5000 MPa and Poisson's ratio value as 0.25 for analysis and design purposes. A similar procedure was also recommended for strength estimation of treated base course, which should have a minimum strength of 4.5 to 7 MPa in 7/28 days. The values, as recommended by AASHTO 2002 Mechanistic Empirical Pavement Design Guide (MEPDG), are treated as default values for these treated bases. Durability criteria may be met with reference to IRC: SP: 89 or BIS: 4332-IV-1968. The above details are adapted for all such treated layers, wherever applicable.

Case-III: Cement-treated sub-base + Cement-treated base + Stress absorbing membrane interlayer (SAMI) (Fig. 10.10 (c)): SAMI consists of a layer of hot sprayed asphalt, which is modified with crumb rubber and applied on to an existing cracked pavement layer followed by a layer covered with stone chips. The SAM is compacted by pneumatic tyre roller and swept clean before application of HMA layer. The SAMI is constructed like a single-chip seal and covered with a layer of conventional HMA or asphalt overlay (Fig. 10.11). As an alternative, two layers of the SAMI are provided as two coats of elastomer-modified binder followed by covering aggregate chips. The elastomer-modified binder should comply with the latest standards of IRC: SP: 53 and IS:15462. The standard Indian practice is briefed below (as per Table. 500-43 of MORTH, 2013).

(a) The first layer on top of the CT base should be sprayed with elastomer-modified binder at the rate of 12 kg/10sq.m and covered with 11.2 mm of 0.10 cu.m. quantity of chipping.

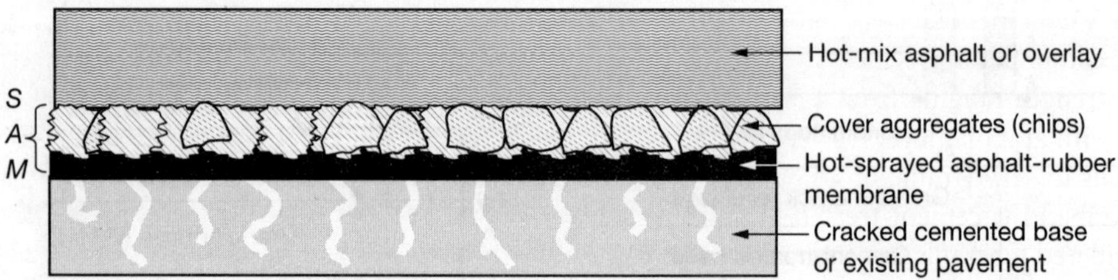

Figure 10.11 Typical crust details of a stress absorbing membrane interlayer (SAMI).

(b) The second layer on top of first coat will be sprayed with elastomeric modified binder at the rate of 10 kg/10 sq.m and covered with 5.6 mm of 0.10 cu.m. quantity of chipping.

The following are a list of advantages and applications of SAMI in layered pavement structure:

- Provides additional tensile strength to the proposed surface course to resist reflective cracking that is originating from its supporting layer
- Acts as a resilient membrane and water-proofing barrier for the ingress of water from the surface to its supporting (i.e., existing) layer such as (cement treated or granular) base or sub-base courses
- Seals the existing layer and acts as a rejuvenator of oxidised hot-mixed asphalt (HMA) surface

- Consolidates movement of cracks and fills distresses of existing asphalt and concrete layers
- Improves resilience when mixed with chopped fibre glass strands
- This membrane can be placed as a flexible, resilient inter-layer prior to micro-surfacing or conventional HMA and used for pavement rehabilitation by sealing cracks and restricting further propagation of cracks.
- Contributes to extra structural capacity of pavement and extends the life of the subsequent overlay
- Provides good bonding between proposed new asphalt layer and the existing concrete pavement
- Suitable for milled and non-milled surfaces of treated granular base, asphalt and concrete surfaces

Case-IV: Cement-treated sub-base + Reclaimed asphalt pavement (RAP) or virgin aggregate stabilised with emulsion/foamed bitumen (Fig. 10.10 (d)): The RAP layer is constructed with pre-processed asphalt mix which uses a small amount of recycled asphalt pavement material. The term reclaimed indicates that asphalt layer material is removed from an existing pavement, screened and pre-processed according to the re-usability of the existing constituents and combined with a certain proportion of virgin materials. The pre-processing is generally carried out according to the guidelines and recommendations of any standard agency. Usage of reclaimed materials has become a common practice and the RAP proportion with virgin material mix range between 10 to 80 %. This practice has been proved to be economic and energy saving apart from being environment-friendly (Al-Qadi et al, 2007).

Case-V: Granular sub-base + Cemented treated base + Granular crack relief layer (Fig. 10.**10** (e)): The granular crack relief layer avoids reflective cracking from cement-treated base. A 100 mm thick layer is constructed to meet the specifications of WMM and its elastic modulus value is taken as 450MPa and the Poisson's ratio as 0.35. About 1 to 2 % bitumen emulsion may be used if there is a possibility of disturbance to the granular layer due to the movement of construction traffic. The sub-base should meet the standards as applicable for Case-I.

Case VI: Cement-treated Sub-base + Wet-mix Macadam (WMM) (Fig. 10.10 (f)): This kind of layer combination is widely used in South African dry areas with cement treated sub-base having low strength ranging from 1.5 to 3 MPa (These, 1995). Such low-strength cement-treated sub-base with granular layer (as WMM) and bituminous concrete wearing course may be used in wet regions to serve traffic up to 30 msa. The standard specifications (i.e., strength, 0.75–1.5 MPa) prescribed for cement-treated granular sub-bases as per Grading IV of IRC: SP-89 should be considered for traffic less than 10 msa only.

2. The bituminous mixes used for surface courses should be of VG 30 with 80% reliability and are considered for traffic less than 20 msa. Modified bitumen is used in the bituminous base/binder courses (DBM) and should be of VG 40 for traffic beyond 20 msa with 90% reliability. However, in case of design of expressways, VG40 and 90 % reliability should be considered.
3. The following empirical relationships were used to convert CBR values to resilient modulus of sub-grade (Shell, 1978):

$$E_{\text{subgrade}}\,(\text{MPa}) = 10 \times \text{CBR} \qquad \text{for lab-CBR} \leq 5\,\% \tag{10.12}$$

$$E_{\text{subgrade}}\,(\text{MPa}) = 17.6 \times (\text{CBR})^{0.64} \qquad \text{for lab-CBR} > 5\,\% \tag{10.13}$$

4. Elastic modulus of granular sub-base and base courses (After, Dorman and Metcalf):

$$E_{\text{granular layer}}\,(\text{MPa}) = E_{\text{Effective-subgrade}} \times 0.2 \times (\text{thickness of granular layer in mm})^{0.45} \tag{10.14}$$

Guidelines are also given for estimation of elastic modulus of cement-treated sub-base courses based on 7 day UCS values.

5. Pavement thickness is determined based on mechanistic principles for given values of design (or estimated) traffic volume in terms of cumulative standard axle load repetitions (by Eq. 6.2) and soaked CBR value.
6. In case of arid regions where the annual rain fall is less than 1000 mm, instead of a 4-day soaked CBR value, the design sub-grade CBR value may be determined at the field moisture content that is prevailing immediately after the monsoon season. However, judicious judgment may be made with reference to the local weather and drainage conditions (IRC: 37, 2012).

   Minimum design sub-grade CBR, i.e., effective sub-grade CBR value should be greater than 5% if estimated traffic is more than 450 CVPD two-way by the time of construction. If the CBR value is less than 5%, the sub-grade should be strengthened with a capping layer of at least 500 mm of select soil/stabilised soil, which is compacted to a minimum of 97 % of the density as obtained in lab (as per IS2720-8).
7. The effective sub-grade modulus of a sub-grade with capping layer or sub-grade with embankment or such any combination of multi-layers can be determined by knowing the maximum surface deflection (Δs) value in Eq. 10.5.

   Effective sub-grade elastic modulus value (MPa) = $E_{\text{Effective-subgrade}} = \dfrac{2(1-\mu^2)pa}{\Delta s}$

   The maximum surface deflection value (δ) can be computed from IITPAVE software by inputting respective layer moduli values (by using Eq. 7.19 and Eq.7.20) and their Poisson's ratio values. By substituting the standard assumed values in the above equation, Poisson's ratio of soil (μ) = 0.35; contact pressure (p) = 0.56 MPa; radius of circular contact area of tyre (a) = 150.8mm, the above equation reduces to:

$$E_{\text{Effective-subgrade}}\ (\text{MPa}) = \frac{2(1-0.35^2)0.56\times 150.8}{\Delta s} = \frac{148.2062}{\Delta s} \tag{10.14}$$

8. For calculating the tensile strains at the bottom fibre of the bituminous layer, dense bituminous macadam (DBM) with bitumen grade of VG 30 or above is used.
9. A fatigue criterion is calibrated based on different research programs and the model is given below

$$N_f = K\times C\times 10^{-4}\times(1/\varepsilon_t)^{3.89}\times(1/E_{bit})^{0.854} \tag{10.15}$$

where,

N_f = Number of cumulative standard axles (80 kN) to cause 20% and more cracked surface area

K = regression coefficient value considered as 1.6064 for 80% reliability; or 0.5161for 90% reliability with VG 40 grade of bitumen

ε_t = Tensile strain at bottom fibre of the bituminous layer ($\times 10^{-6}$) and

E_{bit} = Resilient modulus of the bituminous surfacing (MPa). Standard values of E_{bit} at different temperatures applicable for Indian conditions may be referred to in IRC: 37.

$C = 10^M$

$M = 4.84\times\{[V_{be}/(V_a+V_{be})] - 0.69\}$

V_{be} = Volume of effective bitumen binder in the bottom bituminous layer (%)

V_a = Volume of air voids in the bottom bituminous layer (%)

10. The following rutting criterion was calibrated to Indian conditions by considering maximum allowable rut depth as 20 mm.

$$N_R = K \times 10^{-8} \times (1/\varepsilon_z)^{4.5337} \quad (10.16)$$

where,

N_R = Number of cumulative standard axles to produce rut depth of 20 mm which is applicable to the value of design traffic.

K = A regression coefficient calculated as 4.1656 with 80 % reliability, and 1.41 with 90% reliability

ε_z = Vertical compressive sub-grade strain ($\times 10^{-6}$)

11. Model equations are also recommended for calculation of cumulative fatigue damage to cement treated base by using (i) VDF and (ii) axle load spectrum data.
12. The bituminous layers can be comprised of two layers, viz., Base + Surface course. The Base may be proffered as DBM and the surfacing may be optionally paved as BC/SMA with modified bitumen of VG40 in case of design traffic exceeding 20msa, including for Expressways; in case the design traffic is less than 20 msa, BC/SDBC/Mix seal surfacing (MSS)/Pre-mix carpet (PMC)/ Surface dressing (SD) is used with unmodified binders of VG30/VG40. However, such thin layers should not be accounted in IITPAVE analysis. The DBM may be laid as double or single layer. The target void contents recommended are 3.5% and 3% for single DBM and bottom DBM layer, respectively. The details of field compaction densities of DBM and the VG grades applicable for the pavement temperatures may be obtained from IRC: 37-2018. The resilient modulus values of BC + DBM layer with VG30 (for < 20 msa) and VG 40 (for 20–50msa) may be taken as 2000 MPa and 3000 MPa, respectively.
13. Minimum thickness of bituminous layer (DBM + BC) required playing traffic greater than 30 msa (as major highways) is considered as 100 mm (AASTHO, 1993), with aggregate interlayer having minimum thickness, as 100 mm. The author recommends that this aggregate interlayer should be well-graded and dense, so that any possibility of accommodation of water pockets, due to air voids in the interlayer is minimised.

10.4.2 Sub-surface Drainage Considerations

1. To avoid capillary rise of water, the difference in elevation between the bottom surface level of the sub-grade and the level of the ground water table or high flood level should not be less than 1.0 or 0.6 m (IRC:37, 2001). A bituminous impermeable layer as cut-off layer may be constructed beneath of the sub-grade, in case of possible capillary rise of such water.
2. Due to poor drainage conditions onsite, there may be a chance of accumulation of water inside of the pavement crust, especially in the sub-base course. The water accumulated in the sub-base course may be due to capillary rise of ground water as well as ingress of surface rain water through cracks from top

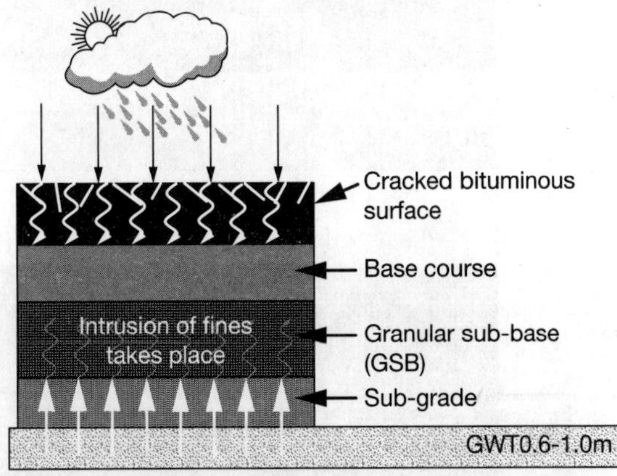

Figure 10.12 Accumulation of sub-surface water and intrusion of fines into the GBS due to capillary rise of ground water and ingress of surface water (schematic).

layers (Fig.10.12). As an ideal sub-surface drainage condition, the sub-base course is designed as open-graded granular sub-base (GSB) so as to facilitate horizontal flow of such accumulated water away from the SGB (Fig.10.13). However, intrusion of fines takes place from subgrade to GSB when there is capillary rise of the ground water. This undesirable condition makes the GSB layer clogged with fines and the horizontal drain-off of the accumulated water from the GSB does not happen.

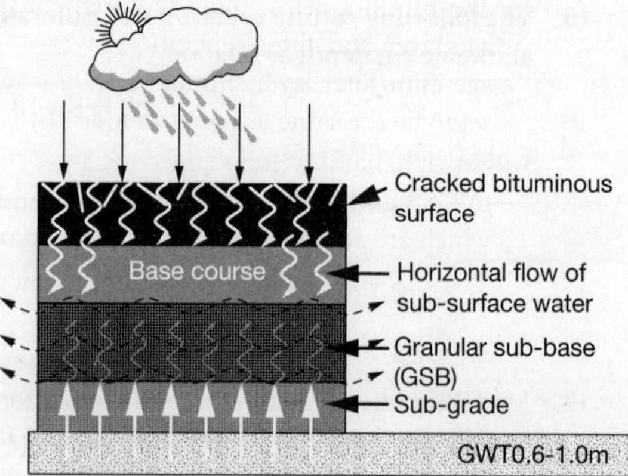

Figure 10.13 The desirable condition of horizontal flow of sub-surface water collected in the granular sub-base (schematic).

Appreciable research has been conducted in the USA, Australia and other countries to improve the drainage of water from the sub-surface layers. As an example, AASHTO has recommended gradation of GSB based on water permeability. This recommendation has also been cited in IRC: 37-2012, which states that for optimal serviceability, the rate of permeability of water (i.e., *k*-value) through the GSB should be in the range of 15 to 35 m/day.

3. In this connection, MORTH (2013) recommends the GSB layer should be comprised of two layers, viz., 1. Upper GSB as drainage layer and 2. Lower GSB as filter/separation layer (Fig.10.14). MORTH recommends six types of GSB gradations. Of these, Grade III or IV should be adopted as filter/separation layer and the remaining grades may be suitably adopted as drainage layer, provided the minimum thickness of the lower and upper layers of each grade is not less than 150 mm. If the total GSB thickness criterion is not satisfied, the GSB is designed as a single layer only, as a drainage-cum-filter layer by adopting Grade V or VI.

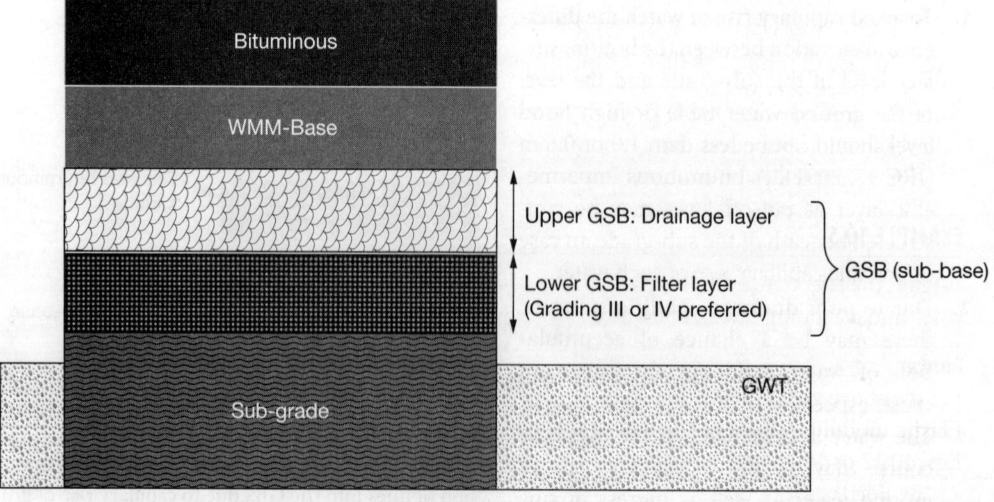

Figure 10.14 Composition of Granular Sub-base (GSB) as Upper GSB and Lower GSB.

4. The minimum thickness requirement (as per IRC:37-2018) for design of GSB layer as drainage layer and filter/separation layer is set as 100mm each; the minimum thickness of drainage-cum-filter layer should be 150 for functional requirement and the remaining conditions are set to be the same as recommended by MORTH, 2013.
5. Terzaghi has developed the concept of checking of possibility of intrusion of fines from sub-grade to GSB and finally clogging of GSB. The granular material should satisfy the following criteria which are adopted in India and several other countries.

   D_{15} of drainage layer ≥ (5×D_{15} of sub-grade) and D_{50} of drainage layer ≤ (5×D_{50} of sub-grade) (10.17)

   where, D_{15} is the size of the sieve that allows 15% weight of the material to pass through. Similar definition holds good for D_{85}, D_{50} and D_2. These values can be obtained by plotting the particle size distribution curve. With reference to the curve of the sub-grade soil, the gradation of granular material used for the drainage layer can be marked on the plot.

   The granular material used for construction of drainage layer is treated as very good when (i) $D_{85} < 4 \times D_{15}$ and (ii) $D_2 > 2.5$ mm (10.18)
6. Additionally, to prevent intrusion of the sub-grade soil fines into the open-graded drainage layer, a filter layer made up of non-oven geo-synthetic fabric may also be placed in between the sub-grade and the drainage layer.
7. In case of large inflows of sub-surface water, an open-graded drainage layer connected with longitudinal porous pipes and connected pipe system should be constructed.
8. Shoulders should be constructed and maintained properly, as per standard specifications of cross fall and level difference between carriageway and shoulder.
9. Shoulders should be made impervious and stabilised with lime and their compacted thickness shall be in the range of 150 to 200 mm.

10.4.3 General Features

(a) Design charts are used to determine the individual layer thicknesses of a flexible pavement for traffic 2–150 msa, with design sub-grade CBR values ranging from 5 to 15%.

(b) The design can be carried out for higher traffic also by using IITPAVE software, which is based on multi-layer elastic analysis program.

(c) In case of traffic less than 2 msa, the design may be carried out based IRC: SP: 72.

Note: Readers may use the IITPAVE software for design of flexible pavement with user- defined input values and choice of pavement layer composition (Refer Fig. 10.10).

A few sample design problems are solved below.

EXAMPLE 10.5

What is the effective modulus of the sub-grade layer which comprises of a natural sub-grade having CBR = 4% and a capping layer of 500 mm thickness, having CBR = 12%.

Answer

Let the above given inputs and the standard loading condition be represented shown in Fig.10.15. Elastic modulus values of the sub-grade and capping layer can be calculated by using CBR values and Eqs. 10.12 or 10.13.

By Eq.10.12,

As the CBR ≤ 5 %, $E_{subgrade}$ (MPa) = 10 × CBR = 10 × 4 = 40 MPa

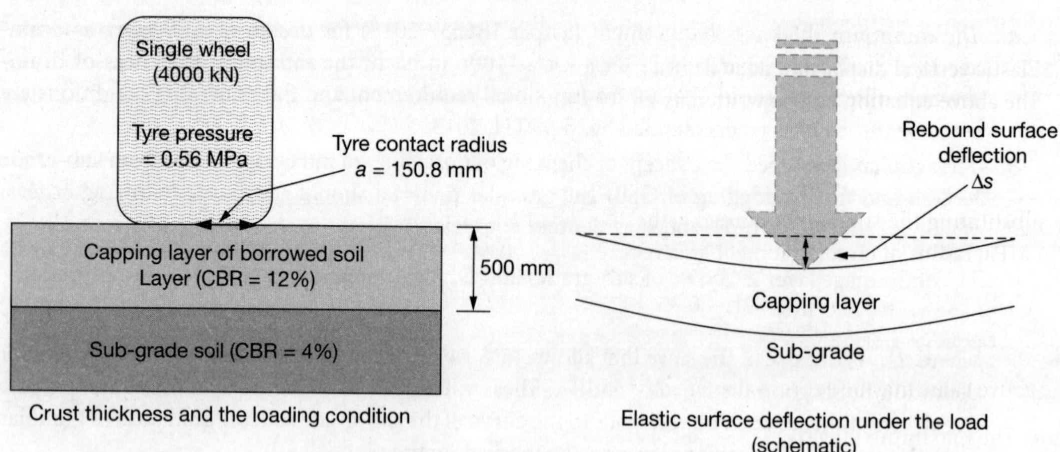

Figure 10.15 Diagram for Example 10.5.

By Eq.10.13,

As the capping layer CBR > 5 %

$$E_{\text{capping layer}} \text{ (MPa)} = 17.6 \times (\text{CBR})^{0.64} = 17.6 \times (12)^{0.64} = 86.33 \text{ MPa}$$

The effective elastic modulus of the combined sub-grade with capping layer can be calculated by using the IITPAVE software and the inputs as given below:

Number of layers = 2

Elastic modulus value of capping layer = 86.33 MPa

Thickness of the capping layer = 500 mm

Poisson's ratio of the capping layer = 0.35

Elastic modulus value of sub-grade soil = 40 MPa

Thickness of the sub-grade soil = semi-infinite

Poisson's ratio of the capping layer = 0.35

Tyre pressure = 0.56 MPa

Wheel load = 40000 N (Single-wheel load, wheel set = 1)

Number of analysis points = 1

At depth = 0 mm, Radius = 0 mm

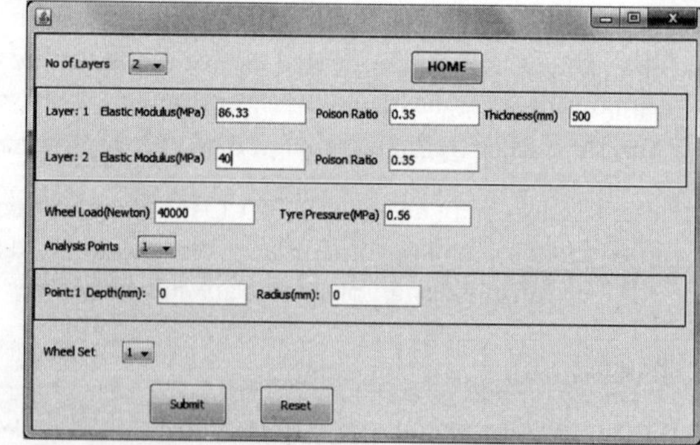

Figure 10.15 (a) IITPAVE screenshot with inputs from Example 10.5.

<u>**Output from the IITPAVE**</u>

No. of layers	2	
E values (MPa)	86.33	40.00
Mu values	0.35	0.35
thicknesses (mm)	500.00	
single wheel load (N)	40000.00	
tyre pressure (MPa)	0.56	

Single Wheel

Z	R	SigmaZ	SigmaT	SigmaR	TaoRZ	**DispZ**	epZ	epT	epR
0.00	0.00	−0.5525E+00	−0.4888E+00	−0.4888E+00	0.0000E+00	**0.2056E+01**	−0.2437E−02	−0.1440E−02	−0.1440E−02

The displacement on top of the capping layer = Δs = 0.2056 E + 01 = 2.056 mm

By using Eq. 10.5, we can calculate the E value, i.e., as effective elastic modulus of the sub-grade with capping layer.

Elastic vertical displacement at(depth, $z = 0$)under a circular flexible plate(Δs) = $\dfrac{2pa(1-\mu^2)}{E}$

The above equation can be written as Eq. 10.14 and written as,

$$E = \frac{2pa(1-\mu^2)}{\Delta s}$$

By substituting the standard assumed values for Poisson's ratio of soil (μ) = 0.35; contact pressure (p) = 0.56 MPa; radius of circular contact area of tyre (a) = 150.8mm, the above equation reduces to:

$$E_{\text{Effective sub-grade}}\,(\text{MPa}) = \frac{2(1-0.35^2)0.56\times150.8}{\Delta s} = \frac{148.2062}{\Delta s} = \frac{148.2062}{2.056} = 72.08473\ \text{MPa}$$

The above value may be expressed as effective CBR of the combined two layers ≈ 7.2% < 15%; hence, OK.

Note: The maximum effective CBR value is restricted to as 100 MPa (≈15.1%) as per IRC: 37.

EXAMPLE 10.6

Determine the thickness adequacy of WMM as a proposed granular sub base (GSB) layer considering a trail thickness of 200mm, against the repetitions of truck-traffic carrying construction material during the road construction work. Take effective elastic modulus of the combined two-layers of capping layer with sub-grade to be 72 MPa. Take the dumping-truck's rear-tandem-axle weight as 25 ton and front-single-axle weight as 7.8 tons. Assume the minimum number standard axle load repetitions on the proposed layer WMM layer during the construction-period as 10000.

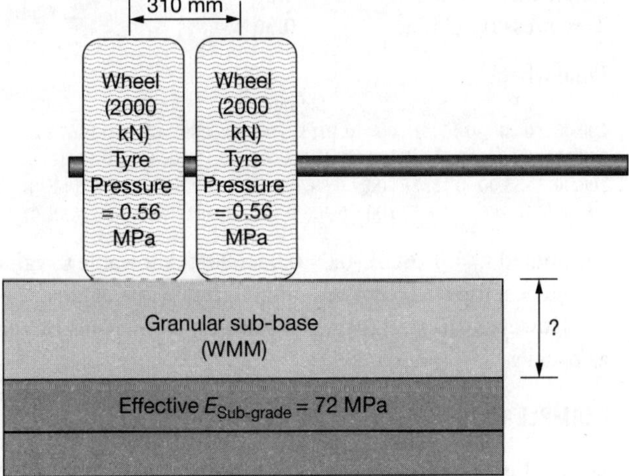

Figure 10.16 Diagram for Example 10.6.

Answer

Let the above given inputs can be represented as shown in Fig.10.16.

Allowable vertical compressive sub-grade strain ($\times10^{-6}$) (ε_z) can be determined by Eq. 10.16 with 90% reliability by considering maximum allowable rut depth as 20 mm:

$$N_R = 1.41\times 10^{-8} \times (1/\varepsilon_z)^{4.5337}$$

The above equation can be written as,

$$\varepsilon_z = \left(\frac{1.41\times10^{-8}}{N_R}\right)^{\frac{1}{4.5337}} = \left(\frac{1.41\times10^{-8}}{10000}\right)^{\frac{1}{4.5337}} = 0.002433$$

We know, N_R = Number of cumulative standard axles = 10000 (given)

Vertical compressive sub-grade strain due to the trail thickness of 200 mm for WMM under the given condition can be computed by using the IITPAVE software and the inputs are given as below:

Number of layers = 2

Trail thickness of the WMM GSB layer = 200 mm

By Eq. 10.14,

Elastic modulus value of WMM GSB Layer (MPa) :

$$= E_{\text{Effective sub-grade}} \times 0.2 \times (\text{thickness of granular layer in mm})^{0.45} = 72 \times 0.2 \times (200)^{0.45} = 156.25 \text{ MPa}$$

Poisson's ratio of the WMM GSB layer = 0.35

Elastic modulus value of the effective sub-grade = 72 MPa

Thickness of the sub-grade soil = semi-infinite

Poisson's ratio of the capping layer = 0.35

Tyre pressure = 0.56 MPa

Wheel load = 20000 N

(Dual wheel load, wheel set = 2)

Number of analysis points = 2

Point1: At depth = 200mm, Radius = 0mm

Point2: At depth = 200mm, Radius = 155mm

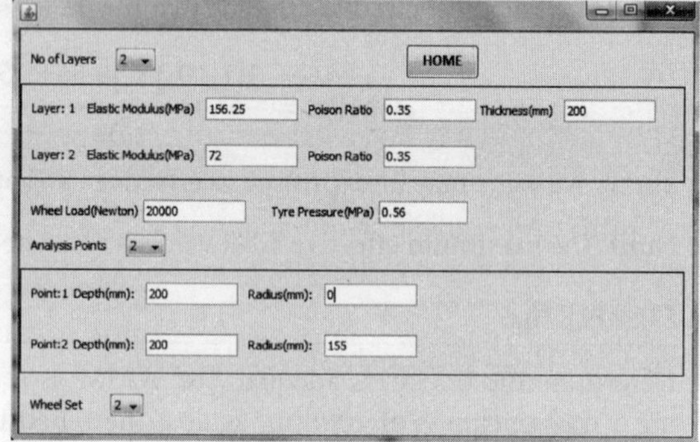

Figure 10.16 (a) IITPAVE screenshot with inputs from Example 10.6.

Output from the IITPAVE

No. of layers	2	
E values (MPa)	156.25	72.00
Mu values	0.35	0.35
Thicknesses (mm)	200.00	
Single wheel load (N)	20000.00	
Tyre pressure (MPa)	0.56	

Dual wheel

Z	R	SigmaZ	SigmaT	SigmaR	TaoRZ	DispZ	epZ	epT	epR
200.00	0.00	-0.1543E+00	0.1013E+00	0.6289E-01	-0.1761E-01	0.7870E+00	-0.1355E-02	0.8531E-03	0.5213E-03
200.00L	0.00	-0.1543E+00	0.1871E-02	-0.1583E-01	-0.1761E-01	0.7870E+00	-0.2076E-02	0.8531E-03	0.5213E-03
200.00	155.00	-0.1337E+00	0.8870E-01	0.1125E-02	-0.7706E-01	0.8133E+00	-0.1057E-02	0.8646E-03	1079E-03
200.00L	155.00	-0.1337E+00	0.2068E-02	-0.3829E-01	-0.7705E-01	0.8133E+00	**-0.1680E-02**	0.8646E-03	0.1079E-03

Computed ε_z = 0.1680E-02 = 0.001680 < Allowable ε_z value: 0.002433

Hence, the trial thickness of 200 mm is safe against the movement of construction traffic.

However, the designers may also do a few trials by reducing the thickness and check its adequacy, as usually.

EXAMPLE 10.7

Design the thickness of a flexible pavement with granular base and granular sub-base courses as per the guidelines of IRC: 37-2018 with 90% reliability for four-lane divided carriageway with the following data.

Traffic expected by the end of construction = 3970 commercial vehicles/day (in both directions)

Design life = 20 years

Traffic growth rate = 5.8 % (as determined from the actual data)

Effective elastic modulus of sub-grade layers = 60 MPa

Vehicle damage factor = 5.1 (from axle load data)

Take lane distribution factor = 0.75 for each direction

Assume 50 % of the traffic flow in each direction

Assume that Marshall stability tests conducted on DBM found the air void content (V_a) to be 3% and the volume of effective bitumen in the DBM mix (V_{be}) to be 11.5%

Answer

(i) Initial traffic in each lane = 50 % of 3970 = 1985 commercial vehicles/day

Using Eq. (9.2),

$N_{flexible} = 365 \times TGF \times A \times VDF \times LDF \times 10^{-6}$

$$= 365 \times \left(\frac{(1.058)^{20} - 1}{0.058}\right) \times 1985 \times 5.1 \times 0.75 \times 10^{-6} = 99.779 \text{ msa,}$$

Say $N_{flexible} = 100$ msa

(ii) Allowable vertical compressive strain in the top fibre of the sub-grade is obtained by using Eq. 10.16:

$$\varepsilon_z = \left(\frac{1.41 \times 10^{-8}}{N_R}\right)^{\frac{1}{4.5337}} = \left(\frac{1.41 \times 10^{-8}}{10 \times 10^6}\right)^{\frac{1}{4.5337}} = 0.000318997$$

$$= 0.318.997 \times 10^{-3}$$

(iii) Allowable horizontal tensile strain ($\varepsilon_t \times 10^{-6}$) in the bottom fibre of the bituminous base is now calculated by the given /calculated inputs, by using Eq. 10.15:

$N_f = 0.5161 \times C \times 10^{-4} \times (1/\varepsilon_t)^{3.89} \times (1/E_{bit})^{0.854}$.

By rewriting the above equation and after substitution, we get,

$$\varepsilon_t = \left(\frac{05161 \times C \times 10^{-4} \times \frac{1^{0.854}}{E_t}}{N_f}\right)^{\frac{1}{3.89}} = \left(\frac{05161 \times 3.155 \times 10^{-4} \times \frac{1^{0.854}}{3000}}{100 \times 10^6}\right)^{\frac{1}{3.89}} = 0.000160785$$

Assume, resilient modulus of the bituminous surfacing at $35^0 = 3000$ MPa

$M = 4.84 \times \{[V_{be} / (V_a + V_{be})] - 0.69\} = 4.84 \times \{[11.5 / (3 + 11.5)] - 0.69\} = 0.499$

$C = 10^M = 10^{0.499} = 3.155$

(iv) Let us consider the layer composition as Criteria-I (Fig. 10.10 (a)). The layer characteristics are calculated as shown below:

As the traffic is greater than 50msa, provide trail thicknesses of 40m BC, 70mm of DBM-I and 70mm of DBM-II with VG-40.

Trail thickness of WMM = 250 mm

Trail thickness of GSB = 200 mm

Thickness of total granular layer = 250 + 200 = 450 mm

Design resilient modulus of the granular layer = $(0.2 \times 450^{0.45}) \times 60.02 = 187.63$ MPa.

Standard values of E_{bit} at different temperatures applicable for Indian conditions may be referred to in IRC: 37.

The above layer thicknesses (Fig.10.17) are considered for computing critical strains by ITPAV

(v) The IITPAVE software and the inputs are given as below:

Number of layers = 3

Resilient modulus of bituminous layers = 3000 MPa (Assumed)

Trail thickness of the 40 mm BC +70 mm DBM+70mm DBM = 180 mm (Assumed)

Poisson's ratio of the WMM GSB layer = 0.35 (Assumed)

Trail thickness of the 250 mm WMM + 200mm GSB layer = 450 mm (Assumed)

By Eq. 10.41, elastic modulus value of WMM GSB Layer (MPa) :

$= E_{Effective\ sub\text{-}grade} \times 0.2 \times$ (thickness of granular layer in mm)$^{0.45} = 72 \times 0.2 \times (200)^{0.45} = 156.25$ MPa

Poisson's ratio of the WMM GSB layer = 0.35 (Assumed)

Elastic modulus value of the effective sub-grade = 60 MPa (Given)

Thickness of the sub-grade soil = semi-infinite (Assumed)

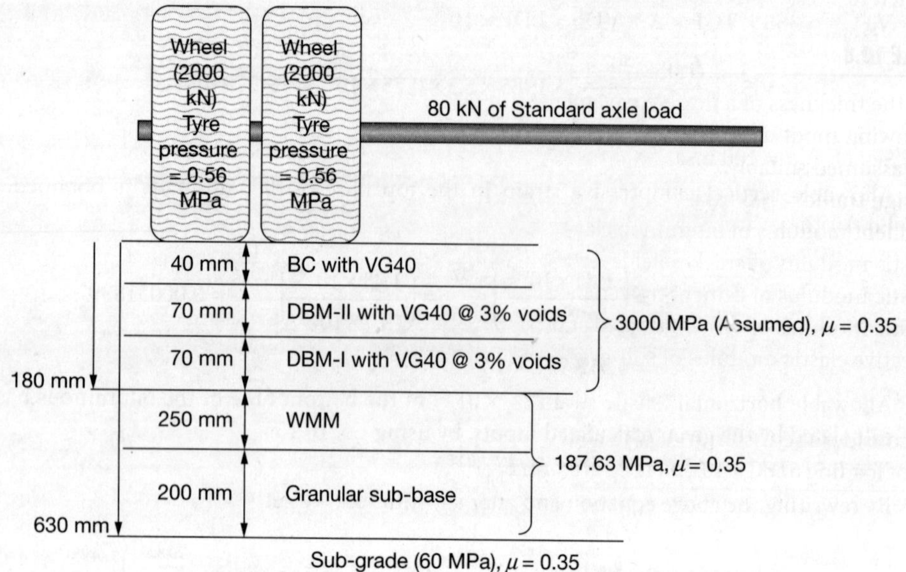

Figure 10.17 Data for Example 10.7: Trail thicknesses and the standard loading.

Poisson's ratio of the capping layer = 0.35 (Assumed)
Tyre pressure = 0.56 MPa
Wheel load = 20000 N (Dual wheel load, wheel set = 2)
Number of analysis points = 4
Point 1: At depth = 180 mm, Radius = 0 mm
Point 2: At depth = 180 mm, Radius = 155 mm
Point 3: At depth = 630 mm, Radius = 0 mm
Point 4: At depth = 630 mm, Radius = 155 mm

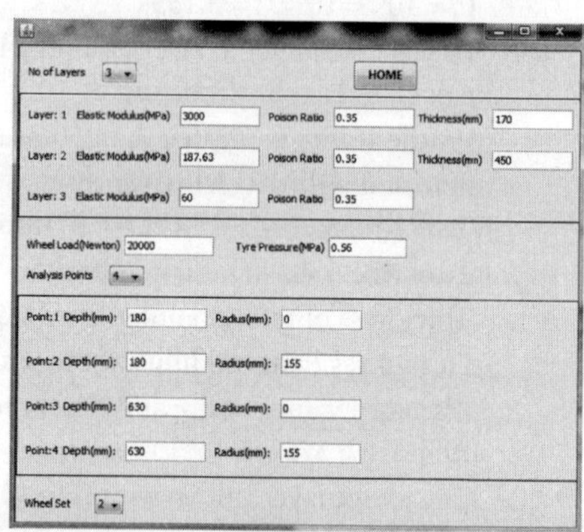

Figure 10.17 (a) IITPAVE screenshot with inputs from Example 10.7.

(vi) Output from the IITPAVE:

No. of layers	3		
E values (MPa)	3000.00	187.63	60.00
Mu values	0.350.350.35		
thicknesses (mm)	180.00	450.00	
single wheel load (N)	20000.00		
tyre pressure (MPa)	0.56		

Dual Wheel

Z	R	SigmaZ	SigmaT	SigmaR	TaoRZ	DispZ	epZ	epT	epR
180.00	0.00	-0.7490E-01	0.6118E+00	0.4951E+00	-0.1185E-01	0.3845E+00	-0.1541E-03	0.1549E-03	0.1024E-03
180.00L	0.00	-0.7490E-01	0.4578E-03	-0.6841E-02	-0.1185E-01	0.3845E+00	-0.3873E-03	0.1549E-03	0.1024E-03
180.00	155.00	-0.7154E-01	0.5801E+00	0.3610E+00	-0.3188E-01	0.3949E+00	-0.1336E-03	0.1596E-03	0.6098E-04
180.00L	155.00	-0.7154E-01	0.1694E-03	-0.1354E-01	-0.3188E-01	0.3949E+00	-0.3564E-03	**0.1596E-03**	0.6098E-04
630.00	0.00	-0.1524E-01	0.2137E-01	0.1913E-01	-0.2166E-02	0.2883E+00	-0.1568E-03	0.1067E-03	0.9050E-04
630.00L	0.00	-0.1524E-01	0.1249E-02	0.5378E-03	-0.2166E-02	0.2883E+00	-0.2644E-03	0.1066E-03	0.9057E-04
630.00	155.00	-0.1608E-01	0.2245E-01	0.2117E-01	-0.2657E-02	0.2938E+00	-0.1670E-03	0.1102E-03	0.1009E-03
630.00L	155.00	-0.1608E-01	0.1292E-02	0.8801E-03	-0.2657E-02	0.2938E+00	**-0.2806E-03**	0.1102E-03	0.1009E-03

Computed ($\varepsilon_t \times 10^{-6}$) = 0.1596E-03 = 0.0001596 < Allowable (0.00016078), Hence, OK.
Computed ($\varepsilon_z \times 10^{-6}$) = 0.2806E-03 = 0.0002806 < Allowable (0.000318997), Hence, OK.

EXAMPLE 10.8

Design the thickness of a flexible pavement with 90% reliability for a dual-lane single carriageway with the following input data. The IRC: 37-2018 design guidelines may be adopted and any missing values may be assumed suitably.

Design traffic = 100 msa
Resilient modulus of bituminous layer at 35^0 C having VG40 grade = 3000 MPa
Elastic modulus of crack relief layer = 450 MPa
Elastic modulus of cement treated base = 5000 MPa
Elastic Modulus of cement treated sub-base = 600 MPa
Effective elastic modulus of sub-grades = 77 MPa

Answer

Let us consider the Criterion V (Fig. 10.10 (e)) and assume the following thicknesses and the standard values of loading and Poission's ratio are taken as:

Layer	Trial thickness	Poisson's ratio	Elastic modulus
Bituminous layers (50mm BC + 60mm DMB)	110 mm	0.35	3000 MPa
Crack relief layer	100 mm	0.35	450 MPa
CT Base	100 mm	0.25	5000 MPa
CT Sub-base	200 mm	0.25	600 MPa
Sub-grade layers	-	0.35	77 MPa

(i) Allowable horizontal tensile strain ($\varepsilon_t \times 10^{-6}$) in the bottom fibre of the bituminous base is now calculated used given /assumed inputs, by using Eq. 10.15:
$N_f = 0.5161 \times C \times 10^{-4} \times (1/\varepsilon_t)^{3.89} \times (1/E_{bit})^{0.854}$.
By rewriting the above equation and after substitution, we get,

$$\varepsilon_t = \left(\frac{05161 \times C \times 10^{-4} \times \frac{1^{0.854}}{E_t}}{N_f}\right)^{\frac{1}{3.89}} = \left(\frac{05161 \times 3.155 \times 10^{-4} \times \frac{1^{0.854}}{3000}}{100 \times 10^6}\right)^{\frac{1}{3.89}} = 0.000160785$$

Assume mix voids of the bituminous base at 35^0 having resilient modulus 3000 MPa
$M = 4.84 \times \{[V_{be} / (V_a + V_{be})] - 0.69\} = 4.84 \times \{[11.5 / (3 + 11.5)] - 0.69\} = 0.499$
$C = 10^M = 10^{0.499} = 3.155$

(ii) Allowable vertical compressive sub-grade strain ($\times 10^{-6}$) (ε_z) can be determined by Eq. 10.16 with 90% reliability by considering maximum allowable rut depth as 20 mm as:
$N_R = 1.41 \times 10^{-8} \times (1/\varepsilon_z)^{4.5337}$
The above equation can be written as,

$$\varepsilon_z = \left(\frac{1.41 \times 10^{-8}}{N_R}\right)^{\frac{1}{4.5337}} = \left(\frac{1.41 \times 10^{-8}}{100 \times 10^6}\right)^{\frac{1}{4.5337}} = 0.000318997 = 0.318.997 \times 10^{-3}$$

(iii) Allowable horizontal tensile in cement treated base (CTB): The following equation has been in use in Australia (AP-T102-08) and also recommended in IRC: 37-2018.

N_f= RF {[(113000/ $E_{CTB^{\wedge}0.804}$)+191]/ ε_{tCTB} }^{12}

RF= Reliability factor; 1 for Expressways, NH, SH, MDR and any other roads having design traffic > 10msa; and 2 for other cases

By rewriting the above equation and after substitution, we get,

$$\varepsilon_{tCTB} = RF\left(\frac{\dfrac{113000}{E_{CTB}^{0.804}}+191}{N_f^{\left(\frac{1}{12}\right)}}\right) = \varepsilon_t = 1\times\left(\frac{\dfrac{113000}{450^{0.804}}+191}{(100\times10^6)^{\left(\frac{1}{12}\right)}}\right) = 220.3\times10^{-6}$$

(iv) The output obtained from the IITPAVE is presented below:

No. of layers 5

E values (MPa) 3000.00 450.00 5000.00 600.00 77.00

Mu values 0.35 0.35 0.25 0.25 0.35

Thicknesses (mm) 110.00 100.00 100.00 200.00

Single-wheel load (N) 20000.00

Tyre pressure (MPa) 0.80

Dual-wheel

Z	R	SigmaZ	SigmaT	SigmaR	TaoRZ	DispZ	epZ	epT	epR
110.00	0.00	-0.2632E+00	0.5142E+00	0.4255E+00	-0.1932E-01	0.2853E+00	-0.1974E-03	0.1524E-03	0.1126E-03
110.00L	0.00	-0.2632E+00	-0.4334E-01	-0.5663E-01	-0.1932E-01	0.2853E+00	-0.5071E-03	**0.1524E-03**	0.1126E-03
110.00	155.00	-0.1638E+00	0.2204E+00	-0.2065E+00	-0.8547E-01	0.2795E+00	-0.5623E-04	0.1167E-03	-0.7543E-04
110.00L	155.00	-0.1638E+00	-0.4190E-01	-0.1059E+00	-0.8547E-01	0.2795E+00	-0.2490E-03	0.1167E-03	-0.7543E-04
310.00	0.00	-0.5499E-01	0.3053E+00	0.2347E+00	-0.1773E-01	0.2478E+00	-0.3800E-04	0.5208E-04	0.3442E-04
310.00L	0.00	-0.5499E-01	0.2050E-01	0.1203E-01	-0.1773E-01	0.2478E+00	-0.1052E-03	0.5207E-04	0.3443E-04
310.00	155.00	-0.5969E-01	0.3283E+00	0.2510E+00	-0.3609E-01	0.2547E+00	-0.4090E-04	**0.5609E-04**	0.3676E-04
310.00L	155.00	-0.5969E-01	0.2188E-01	0.1260E -01	-0.3609E-01	0.2547E+00	-0.1139E-03	0.5609E-04	0.3676E-04
510.00	0.00	-0.1705E-01	0.6391E-01	0.5388E-01	-0.2714E-02	0.2316E+00	-0.7750E-04	0.9117E-04	0.7027E-04
510.00L	0.00	-0.1702E-01	0.9923E-03	-0.2049E-03	-0.2714E-02	0.2316E+00	**-0.2246E-03**	0.9117E-04	0.7018E-04
510.00	155.00	-0.1825E-01	0.6846E-01	0.6108E-01	-0.3966E-02	0.2370E+00	-0.8439E-04	0.9626E-04	0.8088E-04
510.00L	155.00	-0.1825E-01	0.1106E-02	0.2285E-03	-0.3965E-02	0.2370E+00	-0.2430E-03	0.9626E-04	0.8088E-04

Concluding remarks:

Computed ($\varepsilon_t \times10^{-6}$) = 0.1524E-03 = 0.0001524 < Allowable (0.00016078), Hence, OK.

Computed ($\varepsilon_z \times10^{-6}$) = 0.2246E-03 = 0.0002246 < Allowable (0.000318997), Hence, OK.

Computed ($\varepsilon_{zCTB} \times10^{-6}$) = 0.5609E-04 = 0.00005609 < Allowable (0.000220), Hence, OK.

Hence, the trail thickness is adequate against the allowable strain values.

EXAMPLE 10.9

The designed pavement crust details after satisfying the strain criteria of respective layers are presented below (Fig. 10.18):

Now compute and verify the adequacy of cumulative damage (CFD) in cement-treated base (CTB) due to (i) Single-axle loads (ii) Tandem-axle loads and (iii) Tridem Axle loads. The following data is summarised from the axle load survey conducted at a highway section.

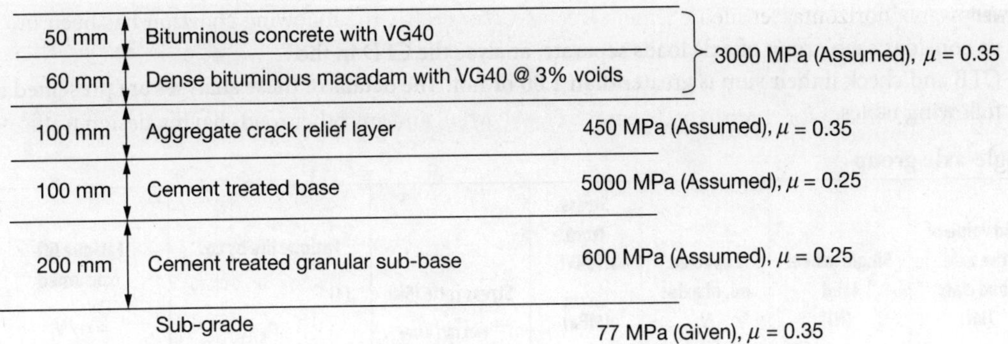

Figure 10.18 Data for Example 10.9: Trail thicknesses, with assumed and given input values

Single axles		**Tandem axles**		**Tridem axles**	
Axle load class	**Expected no. of axles**	**Axle load class**	**Expected no. of axles**	**Axle load class**	**Expected no. of axles**
[1]	[2]	[3]	[4]	[5]	[6]
185–195	0.0	390–410	18925.1	585–615	0.0
175–185	9462.5	370–390	37850.1	555–585	22710.1
165–175	18925.1	350–370	41635.1	525–555	32172.6
155–165	22710.1	330–350	56775.2	495–525	34065.1
145–155	287660.8	310–330	87055.2	465–495	32172.6
135–145	446631.2	290–310	227100.6	435–465	60560.2
125–135	613171.6	270–290	230885.6	405–435	105980.3
115–125	1116578.0	250–270	454201.2	375–405	170325.5
105–115	1286903.4	230–250	478803.8	345–375	227100.6
95 –105	1305828.5	210–230	662376.8	315–345	227100.6
85 – 95	1741104.6	190–210	1078727.9	285–315	246025.7
< 85	2441331.5	170–190	1468583.9	255–285	681301.8
–	–	< 170	1911430.1	< 255	1040877.8
Sum = 9290307.0 (49.09%)		Sum = 6754350.3 (35.69%)		Sum = 2880392.6 (15.22%)	
				Grand sum = 18925050.0 (100%)	

Assume flexural strength of the cement treated base = 1.4 MPa and other missing values suitably.

Answer

Let us consider each group of axle loads separate, analyse the CFD in the

CTB and check if their sum is greater than 1.00 or not. The details of these analyses are presented in the following tables.

Single-axle group

Mid value of the axle load class (kN) [1]	Single-wheel load (N) [2] = [1]/[4]	Expected no. of axles N_f [3]	Stress from IITPAVE epT CTB (MPa) [4]	Stress ratio (SR) = epTCTB/1.4 MPa [5] = [4]/1.4MPa	Fatigue life by Eq. $10^{((0.972-SR)/0.0825)}$ n_i [6]	Fatigue life consumed $= n_i/N_f$ [7] = [6]/[3]
190	47500	0.0	0.7342	0.7342/1.4 = 0.524	$10^{\left(\frac{(0.972-0.524)}{/0.0825}\right)}$ = 266138.8	266138.8 /0.00 = 0.000
180	45000	9462.5	0.6993	0.500	533669.9	0.018
170	42500	18925.1	0.664	0.474	1078699.4	0.018
160	40000	22710.1	0.6285	0.449	2189070.5	0.010
150	37500	287660.8	0.5924	0.423	4495871.1	0.064
140	35000	446631.2	0.5559	0.397	9307459.8	0.048
130	32500	613171.6	0.5197	0.371	19153629.5	0.032
120	30000	1116578	0.4817	0.344	40855959.2	0.027
110	27500	1286903.4	0.4439	0.317	86801681.5	0.015
100	25000	1305828.5	0.4058	0.290	185523217.8	0.007
90	22500	1741104.6	0.3672	0.262	400495280.4	0.004
85	21250	2441331.5	0.3478	0.248	589607189.5	0.004
					CFD =	**0.247**

Output 1 for the axle load of 190 kN (Wheel load = 190/4 = 47.5 kN)

No. of layers	5				
E values (MPa)	3000.00	450.00	5000.00	600.00	77.00
Mu values	0.35	0.35	0.25	0.25	0.35
Thicknesses (mm)	110.00	100.00	100.00	200.00	
single wheel load (N)	**47500.00**				
Tyre pressure (MPa)	0.80				

Dual wheel

Z	R	SigmaZ	**SigmaT**	SigmaR	TaoRZ	DispZ	epZ	epT	epR
310.00	0.00	-0.1220E+00	0.6764E+00	0.5257E+00	-0.4046E-01	0.5835E+00	-0.8450E-04	0.1151E-03	0.7742E-04
310.00L	0.00	-0.1220E+00	0.4539E-01	0.2730E-01	-0.4046E-01	0.5835E+00	-0.2336E-03	0.1151E-03	0.7742E-04

310.00	155.00	-0.1351E+00	**0.7342E+00**	0.5828E+00	-0.7716E-01	0.5998E+00	-0.9286E-04	0.1244E-03	0.8661E-04
310.00L	155.00	-0.1351E+00	0.4847E-01	0.3031E-01	-0.7716E-01	0.5998E+00	-0.2580E-03	0.1244E-03	0.8661E-04
510.00	0.00	-0.3958E-01	0.1465E+00	0.1251E+00	-0.6226E-02	0.5466E+00	-0.1791E-03	0.2086E-03	0.1639E-03
510.00L	0.00	-0.3958E-01	0.2023E-02	-0.5245E-03	-0.6226E-02	0.5466E+00	-0.5208E-03	0.2086E-03	0.1639E-03
510.00	155.00	-0.4242E-01	0.1573E+00	0.1413E+00	-0.8892E-02	0.5593E+00	-0.1951E-03	0.2209E-03	0.1877E-03
510.00L	155.00	-0.4242E-01	0.2307E-02	0.4146E-03	-0.8888E-02	0.5593E+00	-0.5633E-03	0.2209E-03	0.1877E-03

Critical SigmaT: 0.7342E + 00MPa. This value is placed in 1st row in the Column [4].

Output 2 for the Axle load of 180 kN (Wheel load = 180/4 = 45kN)

No. of layers	5				
E values (MPa)	3000.00	450.00	5000.00	600.00	77.00
Mu values	0.35	0.35	0.25	0.25	0.35
Thicknesses (mm)	110.00	100.00	100.00	200.00	
Single-wheel load (N)	**45000.00**				
Tyre pressure (MPa)	0.80				

Dual wheel

Z	R	SigmaZ	**SigmaT**	SigmaR	TaoRZ	DispZ	epZ	epT	epR
310.00	0.00	-0.1162E+00	0.6445E+00	0.5001E+00	-0.3847E-01	0.5532E+00	-0.8047E-04	0.1097E-03	0.7360E-04
310.00L	0.00	-0.1162E+00	0.4326E-01	0.2592E-01	-0.3847E-01	0.5532E+00	-0.2225E-03	0.1097E-03	0.7360E-04
310.00	155.00	-0.1285E+00	**0.6993E+00**	0.5533E+00	-0.7379E-01	0.5687E+00	-0.8834E-04	0.1186E-03	0.8213E-04
310.00L	155.00	-0.1285E+00	0.4621E-01	0.2870E-01	-0.7379E-01	0.5687E+00	-0.2454E-03	0.1186E-03	0.8213E-04
510.00	0.00	-0.3757E-01	0.1393E+00	0.1187E+00	-0.5917E-02	0.5181E+00	-0.1701E-03	0.1983E-03	0.1555E-03
510.00L	0.00	-0.3757E-01	0.1947E-02	-0.4935E-03	-0.5917E-02	0.5181E+00	-0.4945E-03	0.1983E-03	0.1555E-03
510.00	155.00	-0.4026E-01	0.1494E+00	0.1342E+00	-0.8462E-02	0.5302E+00	-0.1853E-03	0.2099E-03	0.1782E-03
510.00L	155.00	-0.4026E-01	0.2213E-02	0.4037E-03	-0.8464E-02	0.5302E+00	-0.5348E-03	0.2099E-03	0.1782E-03

Critical SigmaT: 0.6993E+00MPa. This value is placed in the 2nd row of Column [4].

With reference to column [6], the fatigue damage (n_i) value is calculated by using the following equation (AASHTO):

$$\text{Log}_{10} n_i = \frac{0.972 - SR}{0.0825} \tag{10.19}$$

Where, n_i = Maximum allowable repetitions of axle load Class I on the CTB layer

SR = Stress ratio = Tangential stress at bottom fibre of CTB (SigmaT)/ Flexural strength of CTB

The above equation can be re-written as,

$$n_i = 10^{\left(\frac{0.972-SR}{0.0825}\right)} \tag{10.20}$$

By substituting the computed values from IITPAVE and given flexural strength of CTB, we get, SR = (0.7342MPa/1.4MPa = 0.524

By substituting the SR value in Eq. 10.20, we get,

Maximum allowable repitions of 190 kN of single axle = $n_{190\text{kN}}$

$$= 10^{\left(\frac{0.972-0.524}{0.0825}\right)} = 266138.8$$

Similarly, the above procedure of computation is carried out for the Tandem and Tridel axles and the details are presented below:

Tandem axle group

Mid-value of the axle load class (kN) [1]	Single* wheel load (N) [2] = [1]/4	Expected no. of axles N_f [3]	Stress from IITPAVE SigmaT CTB (MPa) [4]	Stress ratio (SR) = SigmaT CTB/1.4 MPa [5]	Fatigue life by Eq. $10^{((0.972-SR)/0.0825)}$ n_i [6]	Fatigue life consumed $= n_i/N_f$ [7] = [6]/[3]
400	50000	7570.0	0.7687	0.549	133784.8	0.057
380	47500	5677.5	0.7342	0.524	266138.8	0.021
360	45000	37850.1	0.6993	0.500	533669.9	0.071
340	42500	56775.2	0.664	0.474	1078699.4	0.053
320	40000	75700.2	0.6285	0.449	2189070.5	0.035
300	37500	227100.6	0.5924	0.423	4495871.1	0.051
280	35000	227100.6	0.5559	0.397	9307459.8	0.024
260	32500	378501.0	0.5197	0.371	19153629.5	0.020
240	30000	454201.2	0.4817	0.344	40855959.2	0.011
220	27500	662376.8	0.4439	0.317	86801681.5	0.008
200	25000	1078727.9	0.4058	0.290	185523217.8	0.006
180	22500	1631339.3	0.3672	0.262	400495280.4	0.004
170	21250	1911430.1	0.3478	0.248	589607189.5	0.003
					CFD in CTB	0.359

* Assume, one tandem axle load may be equal to two equal single axles of load.

Tridem axle group

Mid value of the axle load class (kN) [1]	Single* wheel Load (N) [2] = [1]/[4]	Expected no. of axles N_f [3]	Stress from IITPAVE epT CTB (MPa) [4]	Stress Ratio (SR) = epTCTB/1.4 MPa [5]	Fatigue life by Eq. $10^{((0.972-SR)/0.0825)}$ n_i [6]	Fatigue life consumed $= n_i/N_f$ [7] = [6]/[3]
600	50000	0.0	0.7687	0.549	133784.8	0.000
570	47500	22710.1	0.7342	0.524	266138.8	0.085
540	45000	32172.6	0.6993	0.500	533669.9	0.060
510	42500	34065.1	0.664	0.474	1078699.4	0.032
480	40000	32172.6	0.6285	0.449	2189070.5	0.015
450	37500	60560.2	0.5924	0.423	4495871.1	0.013
420	35000	105980.3	0.5559	0.397	9307459.8	0.011
390	32500	151400.4	0.5197	0.371	19153629.5	0.008
360	30000	151400.4	0.4817	0.344	40855959.2	0.004
330	27500	170325.5	0.4439	0.317	86801681.5	0.002

Mid value of the axle load class (kN)	Single* wheel Load (N)	Expected no. of axles N_f	Stress from IITPAVE epT CTB (MPa)	Stress Ratio (SR) = epTCTB/1.4 MPa	Fatigue life by Eq. $10^{((0.972-SR)/0.0825)}$ n_i	Fatigue life consumed $=n_i/N_f$
[1]	[2]=[1]/[4]	[3]	[4]	[5]	[6]	[7]=[6]/[3]
300	25000	189250.5	0.4058	0.290	185523217.8	0.001
270	22500	510976.4	0.3672	0.262	400495280.4	0.001
255	21250	1419378.8	0.3478	0.248	589607189.5	0.002
					CFD in CTB	0.233

* Assume one tridem axle load may be equal to three equal single axles of load.

Total CFD in CTB = CFD due to (Single + Tandem + Tridem) Axles = 0.247 + 0.359 + 0.233 = 0.539 < 1.00, Hence safe.

Conclusion: As the sum of CFD is less than 1.00, the provided thickness of cement treated base (CTB) is *safe* against fatigue damage due to tensile stress at its bottom fibre, based on the given axle-load data.

Note: If the above criterion is not satisfied (i.e., Total CFD in CTB is >1.00), we need to increase the thickness of the CTB and re-do the design by verifying the strain criteria and then check the adequacy of the stress criteria by CFD due the axle loads.

EXAMPLE 10.10

Design a perpetual pavement (i.e., long life pavement) for a design traffic, say 300 msa or greater. Consider the endurance limits of horizontal tensile strain at the bottom fibre of the bituminous pavement as 80 micro-strains at 35^0 C and that of vertical compressive strain at the top fibre of the sub-grade as 200 micro-strains. This criteria have been related and recommended in IRC: 37-2018, based on the Asphalt Institute-USA, MS-4, 7th Edition of 2007.

Let us consider the following a flexible pavement with trail thickness of individual layers as give in Fig. 10.19, with the assumed standard input values:

40 mm Bituminous concrete with VG40
160 mm Dense bituminous macadam with VG40 @ 3% voids
} 3000 MPa (Assumed), μ = 0.35

100 mm WMM with crushed aggregates — 350 MPa (Assumed), μ = 0.35

180 mm Cement treated base — 5000 MPa (Assumed), μ = 0.25

260 mm Cement treated granular sub-base — 600 MPa (Assumed), μ = 0.25

Sub-grade — 77 MPa (Given), μ = 0.35

Answer

The following OUTPUT is obtained based on the above inputs:

No. of layers	5				
E values (MPa)	3000.00	350.00	5000.00	600.00	77.00
Mu values	0.35	0.35	0.25	0.25	0.35
thicknesses (mm)	200.00	100.00	180.00	260.00	
single wheel load (N)	20000.00				
Tyre pressure (MPa)	0.56				

Dual wheel

Z	R	SigmaZ	SigmaT	SigmaR	TaoRZ	DispZ	epZ	epT	epR
200.00	0.00	-0.1168E+00	0.2569E+00	0.1866E+00	-0.1233E-01	0.1924E+00	-0.9068E-04	0.7748E-04	0.4587E-04
200.00L	0.00	-0.1168E+00	-0.2559E-01	-0.3378E-01	-0.1233E-01	0.1924E+00	-0.2744E-03	0.7748E-04	0.4587E-04
200.00	155.00	-0.1162E+00	0.2272E+00	0.8213E-01	-0.3405E-01	0.1956E+00	-0.7481E-04	0.7971E-04	0.1442E-04
200.00L	155.00	-0.1162E+00	-0.2874E-01	-0.4567E-01	-0.3405E-01	0.1956E+00	-0.2575E-03	**0.7971E-04**	0.1442E-04
480.00	0.00	-0.2502E-01	0.1587E+00	0.1362E+00	-0.7834E-02	0.1690E+00	-0.1975E-04	0.2617E-04	0.2056E-04
480.00L	0.00	-0.2502E-01	0.1170E-01	0.9005E-02	-0.7834E-02	0.1690E+00	-0.5033E-04	0.2617E-04	0.2056E-04
480.00	155.00	-0.2763E-01	0.1696E+00	0.1523E+00	-0.1122E-01	0.1719E+00	-0.2162E-04	0.2768E-04	0.2337E-04
480.00L	155.00	-0.2763E-01	0.1224E-01	0.1018E-01	-0.1122E-01	0.1719E+00	-0.5539E-04	**0.2768E-04**	0.2337E-04
740.00	0.00	-0.8222E-02	0.3045E-01	0.2780E-01	-0.1040E-02	0.1587E+00	-0.3797E-04	0.4259E-04	0.3707E-04
740.00L	0.00	-0.8222E-02	0.4480E-03	0.1331E-03	-0.1040E-02	0.1587E+00	-0.1094E-03	0.4259E-04	0.3707E-04
740.00	155.00	-0.8583E-02	0.3181E-01	0.3010E-01	-0.1254E-02	0.1609E+00	-0.4010E-04	0.4405E-04	0.4049E-04
740.00L	155.00	-0.8583E-02	0.4871E-03	0.2840E-03	-0.1254E-02	0.1609E+00	**-0.1150E-03**	0.4405E-04	0.4049E-04

Conclusion

Computed ($\varepsilon_t \times 10^{-6}$) = 0.7971E-04 = 79 micro-strains < Allowable (80micro-strains). OK.
Computed ($\varepsilon_z \times 10^{-6}$) = 0.1150E-03 = 115 micro-strains < Allowable (200 micro-strains), OK.

It is observed that, the above trial thickness of the flexible pavement is satisfying the strains criteria as related from Asphalt Institute Manual of MS-4, 7th edition of 2007. However, as the axle-load data is missing, it is also required to check the adequacy of the thickness of CTB by the CFD criterion.

The designers can opt any combination of layers and trail thickness with a wide variety of materials having different resilient modulus values. For example, 5500 MPa of bituminous layer having less thickness at 35^0 C may be a challenging job to a pavement engineer considering the in-service conditions prevailing in India. Also, production of such high-strength resilient modulus mixes with 5500 MPa may not be economical in India, until the superpave technology is implemented. Due to this reason, selection of such high modulus of bituminous layers may not produce an economical model even though the designer may conduct trial with less thickness.

Review Questions

1. What is the significance of deflection and stress analysis of pavements?
2. State and write the assumptions used in Boussinesq's theory of analysis of single layer pavement.
3. What are the limitations of Boussinesq's theory of stress analysis?
4. Derive an expression for determination of vertical stress under a uniformly loaded circular area with a neat sketch.
5. Explain why the pressure distribution is different under flexible and rigid plates.
6. Write the salient features and assumptions used in Burmister's theory of analysis of flexible pavements. With a neat sketch, show the details of a two-layer pavement system.
7. What is the significance of multi-layered flexible pavement analysis? Explain the considerations/assumptions made for analysis of stresses, strains and deflections in a multi-layered flexible pavement system.
8. Define how the assumptions made in a multi-layered flexible pavement system are unrealistic in practice or in-service conditions of pavements.
9. List a few programs with salient details of the analytical solution/programs of multi-layered flexible pavements.
10. In detail, explain the different methods for design of flexible pavements.
11. Write the broad classification of design methods for flexible pavements. Explain each method with suitable examples.
12. What are the limitations of empirical methods of pavement design? Explain the M–E methods widely used for design of pavements.
13. Distinguish between empirical and M–E methods of pavement designs.
14. Write the design criteria used in many M-E methods of design of flexible pavement? Draw a neat sketch showing details of critical strains in a layered flexible pavement structure?

15. What are the critical strains considered in design of M-E methods of flexible pavement design? Write how performance evaluation criteria are decided based on the relationships correlating the limiting values of the mechanistic parameters?
16. List a few methods of flexible pavement design based on M–E principles?
17. What are the benefits of pavement design based on M–E method?
18. Write how pavement failure mechanism is related with critical strains in different layers of flexible pavement.
19. State why sub-base layer is constructed as drainage layer. Also write why open-graded granular material is used for construction of drainage layer.
20. Write the significance of providing a filter layer in between the drainage layer and sub-grade.
21. Write the salient features of flexible pavement design as per IRC: 37-2018, describing with applicable sketches.
22. State the failure criteria adopted for thickness design of flexible pavement as per IRC: 37,-2018. Explain the composition o a typical pavement layer with reference to the considered failure criteria.
23. Write the typical pavement layer composition proposed in IRC: 37-2018 and explain the details with functionality of the each layer with reference to top down cracking/fatigue cracking and other type of failures.
24. Explain the salient factors to be considered for selecting cement-treated base/sub/base courses. State the precautions recommended by the IRC: 37-2018 with reference to material mix characteristics and gradation requirements.
25. What is SAMI layer? Write its significance of use in pavement. Draw a typical sketch and show its location in the pavement crust thickness.
26. List the advantages and disadvantages of providing SAMI in pavements.
27. What is RAP and how it is useful for construction/design of new highway pavements?
28. Differentiate between the functionalities of SAMI and aggregate layer placed as interlayer in the pavement.
29. In detail, write the standard characteristics of pavement materials (soil, aggregate blend, bituminous mix) considered for design of flexible pavement as per IRC: 37-2018.
30. State the significance of improving the viscosity grade of concrete and considering volumes of air and binder content for estimation of maximum allowable value of cumulative standard axle-load repetitions.
31. State the standard requirement of air with binder content in a bituminous mix with reference to traffic intensity as per IRC: 37-2018.
32. Write a short note on drainage considerations that should be adopted as per the guidelines of IRC: 37-2018.
33. Write the general procedure adopted for thick ness design of flexible pavement as per the guidelines of IRC: 37-2018.

Multiple Choice Questions

1. Select a correct statement related to assumptions made for analysis of stresses, strains and deflections in a linear elastic multi-layered flexible pavement system.
   (a) The bottom (nth) layer is a sub-grade that has finite thickness.
   (b) There will be no stresses, strains and deflections at finite depth of the sub-grade
   (c) Only static loads are considered
   (d) The pavement layers are perfectly independent at the top and bottom surfaces.
2. The linear elastic multi-layered flexible pavement system analysis is based on an assumption which is unrealistic in practice due to the reason:
   (a) Each layer of the pavement is non-elastic, inhomogeneous and an anisotropic semi-solid and they never truly obey Hooke's law
   (b) The sub-grade may consist of non-linear stiffness with depth and breadth
   (c) The bituminous surface layer shows nonvisco-elastic behaviour, which varies based on loading time and temperature
   (d) The boundary condition of being finite in the horizontal direction is never a true condition.
3. The Group Index method of pavement design can be categorised under
   (a) Empirical methods
   (b) Pure mechanistic method

(c) Semi-empirical method
(d) Mechanistic-empirical method.

4. What is the benefit that can be obtained using mechanistic empirical pavement design methods?
   (a) Structural analysis can be made to calculate critical stresses, strains and deflections at any point in a pavement structure at any period of service life.
   (b) Certainties can be accounted for during the design process as an integrated approach related to traffic loading, climatic factors, moisture, seasonal changes etc.
   (c) Different distress models belonging to specific locations cannot be effectively incorporated in design.
   (d) Theoretical modelling cannot be made to verify and validate new techniques related to design, performance, maintenance and rehabilitation strategies.
5. For the purpose of analysis of flexible pavement (as per IRC: 37-2018), elastic modulus value of the cement-treated sub-base and cement-treated base may be taken as, ________ respectively.
   (a) 0.600 MPa and 50 MPa
   (b) 500 MPa and 5000 MPa
   (c) 600 MPa and 6000 MPa
   (d) 600 MPa and 5000 MPa
6. For the purpose of analysis of flexible pavement (as per IRC:37-2018), the granular crack relief layer having 100mm thickness should meet the specifications of WMM and have an elastic modulus value of ________ MPa.
   (a) 100 (b) 300
   (c) 450 (d) 600
7. For the purpose of analysis of flexible pavement (as per IRC:37-2018), the bituminous mixes used for surface courses should be of VG 30 with 80% reliability and are considered for traffic less than
   (a) 2 msa
   (b) 10 msa
   (c) 20 msa
   (d) 450 commercial vehicles per day.
8. The minimum thickness requirement (as per IRC: 37-2018) for design of GSB layer as drainage layer and filter/separation layer is set as ________ for each layer.
   (a) 100 mm (b) 150 mm
   (c) 200 mm (d) 250 mm
9. The minimum thickness requirement (as per IRC: 37-2018) for the design of GSB layer as drainage-cum-filter layer is ________
   (a) 100 mm (b) 150 mm
   (c) 200 mm (d) 250 mm.
10. The granular material used for construction of drainage layer is considered as very good when
   (a) D_{15} of drainage layer ≤ 5×D_{50} of sub-grade and D_{50} of drainage layer ≥ 25×D_{50} of sub-grade.
   (b) D_{15} of drainage layer ≤ 5×D_{15} of sub-grade and D_{50} of drainage layer ≥ 5×D_{50} of sub-grade.
   (c) D_{15} of drainage layer ≥ 5×D_{15} of sub-grade) and D_{50} of drainage layer ≤ (5×D_{50} of sub-grade).
   (d) D_{85} of drainage layer < 4× D_{50} of subgrade and D_2 of drainage layer ≥ 2.5 mm.

Answers to Multiple Choice Questions

1. (c)	3. (a)	5. (d)	7. (c)	9. (b)
2. (b)	4. (a)	6. (b)	8. (a)	10. (c)

CHAPTER 11

Analysis and Design of Rigid Pavements

11.0 INTRODUCTION

The design phase of rigid pavement should incorporate numerous factors such as sub-grade support, material properties, traffic loads, drainage and prevailing climatic/environmental conditions and ensure that the pavement withstands these aspects during its design life. Due to this reason, for any given boundary condition, the rigid pavement structure should be analysed to determine allowable deflections or stresses or strains.

11.1 ANALYSIS OF RIGID PAVEMENT

Stresses in concrete pavement slabs are induced by wheel loads and by variation in temperature. The intensity and nature of stresses developed at different locations of the slab are dependent upon various factors such as magnitude of load, lateral placement of wheel loads, flexural strength of the slab, temperature changes, moisture, humidity and location of joints. Excessive stresses at the bottom of the concrete slab pavement produce cracks. The major types of stresses in a concrete pavement may be classified as (Moore, 1956)

- Direct tensile stresses which are produced due to slab contractions
- Flexural tensile stresses which are developed by wheel loads
- Flexural tensile stresses caused by restraint due to warping of slab

In this analysis, preparation of sub-grade and base course layers are assumed to be carried out according to the specifications of relevant standard codes. The minimum and maximum requirements of *k*-values (or equivalent CBR values) are assumed to be taken care of. It is also assumed that the necessary stabilisation methods are followed.

Further, it is worth mentioning that, however satisfactory a rigid pavement design is developed, it becomes an ineffective exercise if proper drainage facilities are not provided for the sub-surface and surface layers.

Hence, before designing and constructing the rigid pavement, the above two aspects must be assured by the client and the contractor.

The basic concepts of stress analysis in concrete slabs are given below.

11.1.1 Stresses Produced by Wheel Load

Unlike flexible pavement design, for rigid pavements, mechanistic principles have to be used in the primitive stages of design development itself. The stress analysis of cement concrete slabs developed by Westergaard has evolved into a scientifically based procedure to address problems related to rigid pavement design. His work published in several research papers (1926, 1927, 1933 and 1948) show that certain considerations were taken into account while performing the stress analysis. They are presented below.

- Three critical locations of wheel loadings were considered, viz., interior, along the longitudinal edge and the corner of the slab (Fig. 11.1).
- Elliptical, semi-elliptical and circular areas of loadings were considered. In this book, we will consider stresses due to circular loading in the interior and at the corner; and stresses due to circular and semi-circular loading at the edge.

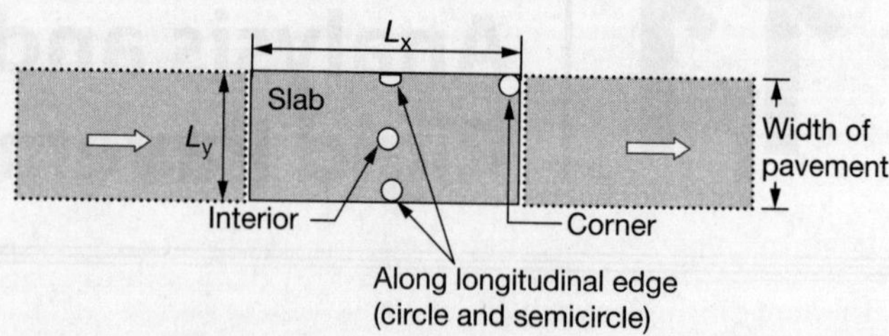

Figure 11.1 Plan view of critical locations of wheel loads on a rigid pavement.

- The cement concrete slab is assumed to be homogenous, isotropic, ideally elastic, uniform in thickness and without cracks. It is also taken that the pavement slab is in full contact with the sub-grade, which is assumed to be a dense liquid.
- The sub-grade support is in the vertical direction and is proportional to the deflection of the slab. The sub-grade does not possess any shear strength.

$$\text{Sub-grade support } (P) = k \times \text{Deflection of the slab } (\Delta)$$

where k is called the modulus of the sub-grade reaction.
- The ratio of pressure to deflection is constant everywhere under the concrete slab.
- Calculated load stresses are not very sensitive even to fairly great variations of the modulus of sub-grade reaction (k).

Westergaard Equations

In 1925 and 1926, Westergaard developed the earliest equations to calculate the wheel-load stresses at salient locations of a slab.

$$\sigma_{\text{corner}} = \frac{3P}{h^2}\left[1 - \left(\frac{a\sqrt{2}}{l}\right)^{0.6}\right] \tag{11.1}$$

$$\sigma_{\text{interior}} = \frac{0.3162P}{h^2}\left[4\log_{10}\left(\frac{l}{b}\right) + 1.069\right] \tag{11.2}$$

$$\sigma_{\text{edge}} = 0.57185\frac{P}{h^2}\left[4\log_{10}\left(\frac{l}{b}\right) + 0.3593\right] \tag{11.3}$$

where,

σ_{corner} = stress produced at the corner by the wheel load spread over a circular area

σ_{interior} = stress produced on the interior by the wheel load spread over a circular area

σ_{edge} = stress produced at the edge by the wheel load spread over a semi-circular area

P = design wheel load (i.e., half of the single-axle load)

h = thickness of the concrete slab

a = radius of the circular area of the wheel loading

$$l = \text{radius of relative stiffness} = \left(\frac{Eh^3}{12\,k\left(1-\mu^2\right)}\right)^{0.25} \tag{11.4}$$

E = Young's modulus of the cement concrete slab

k = modulus of the sub-grade reaction or sub-grade support value (may be obtained from plate load test data)

μ = Poisson's ratio of the cement concrete slab (taken as 0.15 for all the equations in the book)

b = radius of equivalent pressure distribution at the bottom surface of the slab

If $a \geq 1.724h$, $b = a$, else, $b = \sqrt{1.6a^2 + h^2} - 0.675h$ (11.5)

The above formulae were modified by Westergaard in 1933 and again later based on the field experiments conducted by Teller and Sutherland (1935–1943). According to IRC: 58-2002, the following equation of stress due to load at the critical edge region is adopted. It is based on Westergaard's formulae modified by Teller and Sutherland.

$$\sigma_{edge} = 0.529\frac{P}{h^2}(1+0.54\mu)\left[4\log_{10}\left(\frac{l}{b}\right)+\log_{10}b-0.4048\right] \quad (11.6)$$

To calculate the stress at the corner due to loading, Westergaard's Eq. (11.1) was modified and the following formulae were recommended by other researchers (Fwa et al., 1996).

$$\text{By Bradbury (1938): } \sigma_{corner} = \frac{3P}{h^2}\left[1-\left(\frac{a}{l}\right)^{0.6}\right] \quad (11.7)$$

$$\text{By Kelly (1939): } \sigma_{corner} = \frac{3P}{h^2}\left[1-\left(\frac{a\sqrt{2}}{l}\right)^{1.2}\right] \quad (11.8)$$

$$\text{By Spangler (1942): } \sigma_{corner} = \frac{3.2P}{h^2}\left[1-\left(\frac{a\sqrt{2}}{l}\right)\right] \quad (11.9)$$

Pickett made a comparative study of the equations recommended by Kelly, Bradbury and Spangler and suggested a formula which has the form of Westergaard's theoretical equation and was found to be more closely matching the larger stress values measured in the Arlington road tests (Moore, 1956).

$$\text{By Pickett (1946): } \sigma_{corner} = \frac{4.2P}{h^2}\left[1-\sqrt{\left(\frac{a}{l}\right)} \times \left(\frac{1}{0.925+0.22\left(\frac{a}{l}\right)}\right)\right] \quad (11.10)$$

EXAMPLE 11.1

Using original Westergaard equations determine the maximum wheel-load stress at interior, edge and corner locations of a slab, if a 44.5 kN of load is applied on a set of dual tyres spaced at 310 mm. Take the tyre pressure of 0.8 Mpa; modulus of sub-grade reaction of 27.2 MN/m^3; slab thickness = 310 mm; elastic modulus of cement concrete = 27.6 GPa; Poission's ratio of cement concrete = 0.15.

Answer

We know the conversion values as,

1 kN = 1000 N

1 GPa = 1000 N/mm^2;

1 MPa = 1N/mm^2;

1MN/m^3 = 0.001 N/mm^3

The given values are written below:
Wheel load, $P = 44.5 \times 1000$ N
Tyre pressure = 0.8 MPa = 0.8 N/mm²
Modulus of subgrade reaction, $k = 27.2$ MN/m³ = 0.0276 N/mm³
Elastic modulus of concrete, $E = 27.6 \times 1000 = 27600$ N/mm²
Poission's ratio of cement concrete, $\mu = 0.15$

$$\text{Radius of relative stiffness} = l = \left(\frac{Eh^3}{12\,k\left(1-\mu^2\right)}\right)^{0.25} = \left(\frac{27600\times 310^3}{12\times 0.0276\left(1-0.15^2\right)}\right)^{0.25} = 1262.4 \text{ mm}$$

$$\text{Radius of tyre contact area} = a = \sqrt{\left(\frac{\text{wheel load}}{\pi \times \text{Tyre Pressure}}\right)}\sqrt{\left(\frac{44.5\times 10^3}{\pi \times 0.8}\right)} = 133 \text{ mm}$$

As a-value of 133 < (1.724×310),

$$b = \sqrt{1.6a^2 + h^2} - 0.675h\;143.45$$

$$= \sqrt{1.6\times 133^2 + 310^2} - 0.675\times 310 = 143.45 \text{ mm}$$

The above-given values are substituted in the Westergaard's equations as:

$$\sigma_{\text{corner}} = \frac{3P}{h^2}\left[1-\left(\frac{a\sqrt{2}}{l}\right)^{0.6}\right] = \frac{3\times 44.5\times 1000}{310^2}\left[1-\left(\frac{133\times\sqrt{2}}{1262.4}\right)^{0.6}\right] = 0.9459 \text{ N/mm}^2$$

$$\sigma_{\text{interior}} = \frac{0.3162P}{h^2}\left[4\log_{10}\left(\frac{l}{b}\right)+1.069\right]$$

$$= \frac{0.3162\times 44.5\times 1000}{310^2}\left[4\log_{10}\left(\frac{1262.4}{143.45}\right)+1.069\right] = 0.7096 \text{ N/mm}^2$$

$$\sigma_{\text{edge}} = 0.57185\frac{P}{h^2}\left[4\log_{10}\left(\frac{l}{b}\right)+0.3593\right]$$

$$= 0.57185\frac{44.5\times 1000}{310^2}\left[4\log_{10}\left(\frac{1262.4}{143.45}\right)+0.3593\right] = 1.096 \text{ N/mm}^2$$

Based on Westergaard's revised formulae (1948), Ioannides et al. (1985), re-analysed the rigid pavement using the finite element method and recommended the following stress equations.

$$\sigma_{\text{corner}} = \frac{3P}{h^2}\left[1\left(\frac{a\sqrt{\pi}}{l}\right)^{0.72}\right] \tag{11.11}$$

For circular area of loading:

$$\sigma_{\text{edge}} = \frac{0.803P}{h^2}\left[4\log_{10}\left(\frac{l}{a}\right)+0.666\left(\frac{a}{l}\right)-0.034\right] \tag{11.12}$$

For semi-circular area of loading:

$$\sigma_{\text{edge}} = \frac{0.803P}{h^2}\left[4\log_{10}\left(\frac{l}{a}\right)+0.282\left(\frac{a}{l}\right)+0.65\right] \tag{11.13}$$

To obtain stresses induced in a slab by dual-wheel loads, the equivalent circular area (Fig. 11.2) of the dual-tyre contact area is calculated to be equal to the contact area of the dual-wheel plus the area between them (Huang, 2004). The radius of contact area is calculated based on this concept. The equation given in IRC: 58-2002 is as follows.

$$a=\left[\left(\frac{0.8521P_1}{q\pi}\right)+\frac{S}{\pi}\left(\frac{P_1}{0.5227q}\right)^{0.5}\right]^{0.5} \quad (11.14)$$

where,

a = equivalent radius of contact area, assumed circular, for single-axle dual-wheels

P_1 = load on one tyre

S = centre-to-centre distance of dual tyres (310 mm may taken as the standard value)

q = tyre pressure

With reference to the original equations of Westergaard (1925, 1926 and 1933), the above modified formulae are empirical since the field test conditions are dependent on the test location and time (Santosh et al., 2006). However, accurate and closed-form equations developed based on FEM analysis can be obtained from literature (Ioannides et al., 1985).

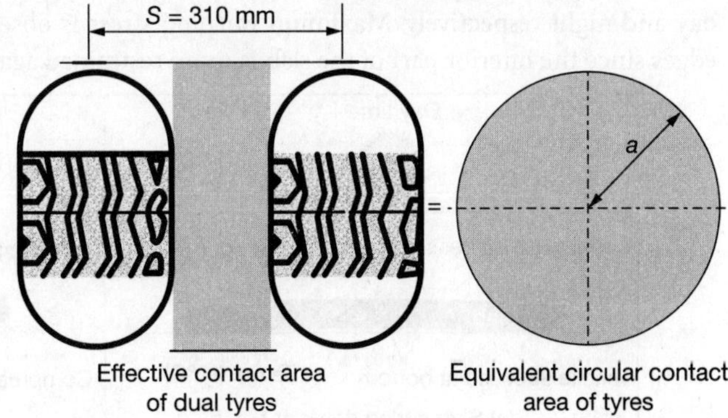

Figure 11.2 Equivalent circular contact area considered for rigid pavement design.

EXAMPLE 11.2

A 40 kN wheel loading is applied on a slab of 240 mm thickness. The radius of the wheel contact area is 150 mm and the radius of the relative stiffness is 0.96 m. (i) Determine maximum stress when loaded at the corner based on the formula developed by Westergaard and the formula modified by Ioannides et al. (ii) Determine the percentage increase of maximum stress when using the circular loaded area method and the semi-circular loaded area method, using the re-analysed equations of Ioannides et al.

Answer

(i) Corner wheel load stress: According to Westergaard (Eq. 11.7),

$$\sigma_{\text{corner}}=\frac{3\times 40}{(0.24)^2}\left[1-\left(\frac{0.15\times\sqrt{2}}{0.96}\right)^{0.6}\right]=1241.2\text{ kPa}$$

According to Ioannides et al. (Eq. 11.10), $\sigma_{corner}=\frac{3\times 40}{(0.24)^2}\left[1-\left(\frac{0.15\times\sqrt{\pi}}{0.96}\right)^{0.72}\right]=1256.9\text{ kPa}$

(ii) From Eqs. (11.11) and (11.12):

For circular area of loading:

$$\sigma_{\text{edge}}=\frac{0.803\times 40}{0.24^2}\left[4\log_{10}\left(\frac{0.96}{0.15}\right)+0.666\left(\frac{0.15}{0.96}\right)-0.034\right]=1837.3\text{ kPa}$$

For semi-circular area of loading:

$$\sigma_{\text{edge}}=\frac{0.803\times 40}{0.24^2}\left[4\log_{10}\left(\frac{0.96}{0.15}\right)+0.282\left(\frac{0.15}{0.96}\right)+0.65\right]=2185.2\text{ kPa}$$

The semi-circular area of loading method gives 18.9% more value of stress than the circular area of loading.

11.1.2 Warping Stresses

Temperature differential between the top and bottom surfaces of a cement concrete slab is a common phenomenon, whether it is day or night. Expansion and contraction of the slab as a result of temperature difference, causes geometric deformation—either curling up or down. Warping or temperature stresses will be produced in the slab when these geometric deformations are completely restrained by its self-weight. Two critical conditions of warping stresses in a cement concrete slab are presented in Fig. 11.3. Due to curling of the slab, tensile and compressive stresses are produced in its bottom fibres during the day and night respectively. Maximum warping stress is observed at the interior of the slab than at its edges since the interior part of the slab is more restrained against curling than the edges.

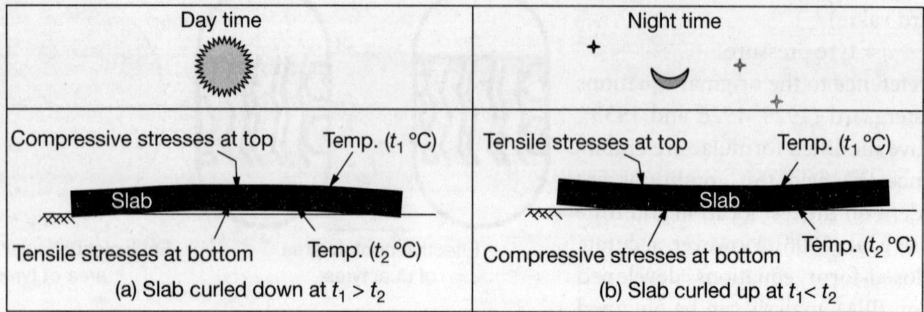

Figure 11.3 Warping stresses in concrete slab when curling is restrained at different times.

Based on plate theory, Westergaard (1926) developed formulae for calculating the warping stresses in a concrete slab. In 1938, Bradbury modified Westergaard's formulae and developed the following equations for calculating the maximum warping stress at the interior and the edge of the slab having finite dimensions.

$$\sigma_{\text{tx\_interior}} = \frac{E\alpha\Delta_t}{2(1-\mu^2)}(C_x + \mu C_y) \tag{11.15}$$

$$\sigma_{\text{ty\_interior}} = \frac{E\alpha\Delta_t}{2(1-\mu^2)}(\mu C_x + C_y) \tag{11.16}$$

$$\sigma_{\text{tx\_edge}} = \frac{E\alpha\Delta_t}{2}C_x \tag{11.17}$$

$$\sigma_{\text{ty\_edge}} = \frac{E\alpha\Delta_t}{2}C_y \tag{11.18}$$

where,

$\sigma_{\text{tx\_interior}}$ = Interior warping stress at the slab's mid-span in the *x*-direction
$\sigma_{\text{ty\_interior}}$ = Interior warping stress at the slab's mid-span in the *y*-direction
$\sigma_{\text{tx\_edge}}$ = Edge warping stress at the slab's mid-span in the *x*-direction
$\sigma_{\text{ty\_edge}}$ = Edge warping stress at the slab's mid-span in the *y*-direction
E = Young's modulus of cement concrete
α = Coefficient of thermal expansion of concrete/ °C
$\Delta t = (t_1 - t_2)$ °C = Temperature differential between the top and bottom of the slab. The temperature gradient across the depth of the slab is assumed to be linear but in real life, it is non-linear.
μ = Poisson's ratio of concrete slab (≈ 0.15)
L_x and L_y = Slab lengths in the *x* and *y* directions respectively (refer to Fig. 11.1)
$C_x = L_x/l$ Bradbury's coefficients of warping stress in the *x* and *y* directions (refer to Table

$C_y = L_y/l$ 11.1)
l = Radius of relative stiffness (refer to Eq. 11.4)

Table 11.1 Bradbury's coefficients (From Bradbury, 1938).

L_x/l or L_y/l	1	2	3	4	5	6	7	8	8.5	9	10	11	≥12
C_x or C_y	0.000	0.040	0.175	0.440	0.720	0.920	1.030	1.077	1.084	1.080	1.075	1.050	1.000

Influence of Temperature and Moisture

Warping stresses in a cement concrete slab are mainly developed due to temperature gradient across the thickness of the slab and differences in moisture within the concrete slab. When the bottom surface of the slab is drier than the top, there is downward warping and vice versa (Hatt, 1925). Warping of the slab (or slab curling) was also observed by Hveem during an investigation of the performance of concrete pavements in California. He concluded that the combined action of temperature and moisture gradients across the thickness of the slab were responsible for the effect (Hveem, 1951). The temperature gradient is not proportional to the thickness of slab (*AASHO Road Test*, Report 5, 1962; Thompson et al., 1987). It was observed from a few field studies that the temperature gradients are non-linear and that the daily fluctuations in temperature are more in the vicinity of the slab surface than at its bottom. Environmental and climatic factors such as air temperature, solar radiation, cloud cover, humidity and precipitation affect the temperature gradient in the concrete slab (Armaghani et al., 1986 and Yu et al., 1998). In addition to this, warping stresses will be more in longer slabs than in shorter slabs as warping results in greater loss of contact along longer slab edges.

The warping stresses determined from the Westergaard–Bradbury equations based on the Winkler foundation model were found to be higher than the full-size experimental results obtained in China. Therefore, the values of stress coefficients were modified by Shi ct al. (1993) using the Pasternak foundation model.

EXAMPLE 11.3

Calculate maximum warping stresses at the edge and interior of a slab of 4.5 m length and 3.5 m width. Take the elastic modulus of concrete as 3×10^4 MPa, radius of relative stiffness as 0.92 m, temperature difference between the top and bottom surface of the slab as 17 °C, coefficient of thermal expansion of concrete as 10×10^{-6} per °C and Poisson's ratio of concrete as 0.15.

Answer

$L_x/l = 4.5/.92 = 4.89$; from Table 8.1, $C_x = 0.6892$
$L_y/l = 3.5/.92 = 3.80$; $C_y = 0.3870$

The value of C_x is more than C_y. Therefore, maximum stress will be observed in the *x*-direction. From Eq. 11.14, maximum stress at the interior is

$$\sigma_{\text{tx\_interior}} = \frac{(3\times10^4\ \text{MPa})\times(10\times10^{-6})\times17}{2\left(1-0.15^2\right)}(0.6892+0.15\times0.3870) = 1.949\ \text{MPa}$$

From Eq. 11.16, maximum stress at the edge is

$$\sigma_{\text{tx\_edge}} = \frac{(3\times10^4\ \text{MPa})\times(10\times10^{-6})\times17}{2}\times0.6892 = 1.757\ \text{MPa}$$

Thus, the warping stress at the interior is 10.9% higher than at the edge.

11.1.3 Stresses due to Frictional Forces

Contraction and expansion (movements in the horizontal direction) of the slab takes place due to uniform temperature changes in the slab. The uniform change may be caused due to variation of temperature or moisture content or both. Slab movements are restrained by its self-weight caused by the inter-surface frictional forces between the slab and the supporting layer (sub-base or base). For example, when the slab contracts, its movements are restrained by frictional forces and tensile stresses are developed (Fig. 11.4) (Sparkes, 1939 and 1954). This is the reason cracks are generally formed at the centre of longer slabs.

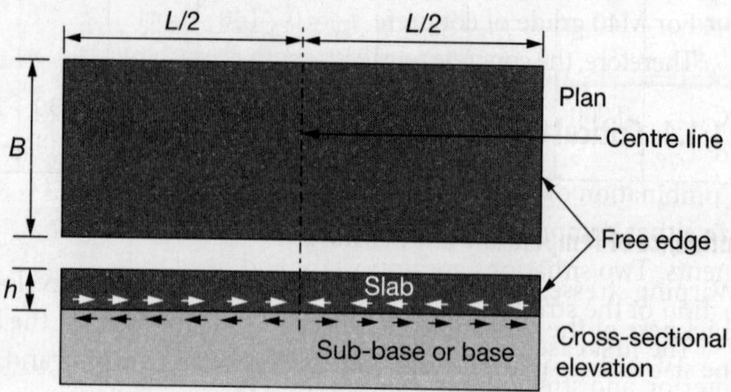

Figure 11.4 Frictional stresses in the slab.

The maximum movement, with respect to the centre line of the slab, is observed at the free edge region and from there it gradually decreases to zero at the centre line (Kelly, 1939). Hence, maximum frictional stress will be developed at the centre of the slab and can be determined by equating the force developed by the slab due to the movement with the frictional resistance offered by the supporting layer on the half-length of the slab across its unit width.

$$\sigma_f \times B \times h = \frac{L}{2} \times B \times h \times W \times f$$

On canceling the common factors, we get,

$$\sigma_f = \frac{W L f}{2} \tag{11.19}$$

where,

σ_f = Tensile stress in concrete slab due to friction
W = Unit weight of concrete
B = Unit width of concrete
h = Thickness of slab
L = Length of slab
f = Average coefficient of friction (often assumed to be ≈1.5)

The stress due to friction depends on the slab length, unit weight of concrete and coefficient of friction.

EXAMPLE 11.4

Determine the stress in concrete due to friction between the slab and the sub-base layer and verify its tensile strength adequacy. Use the following data.

Transverse joint spacing or slab length = 5.0 m
Unit weight of plain concrete = 24.0 kN/m^3
Coefficient of friction = 1.5
Grade of concrete used for slab = M 40

Answer

By Eq. (11.19), $\sigma_f = \frac{WLf}{2} = \frac{24.0 \times 5.0 \times 1.5}{2} = 90 \text{ kN/m}^2 = 90 \text{ kPa}$

According to IS: 456: 2000 and IRC: SP: 62-2004:
Flexural strength or modulus of rupture, $f_{cr} = 0.7\ (f_{ck})^{0.5}$ MPa.
For M40 grade of concrete, $f_{cr} = 0.7\ (40)^{0.5} = 4.42$ MPa = 4420 kPa.
Therefore, the concrete tensile strength is much more than the stress due to friction. Hence, the slab is safe.

11.1.4 Critical Combined Stress due to Warping and Load

Combination of the stresses due to warping and load is found to be critical only when both the stresses are either compressive or tensile. The maximum combined stress is considered for design of rigid pavements. Two situations are important and need to be studied while taking into account a critical combination of the stresses.

The first case (Fig 11.5(a)) is when stresses at the bottom of the slab due to warping at the edge and interior, and stress due to load are both tensile during daytime (i.e., both the values are positive) and the accumulated stress intensity increases.

In the second case (Fig. 11.5(b)), the stress at the bottom of the slab due to warping at the corner is compressive but the stress due to load is tensile during the night. As a result, the accumulated stress intensity will be reduced.

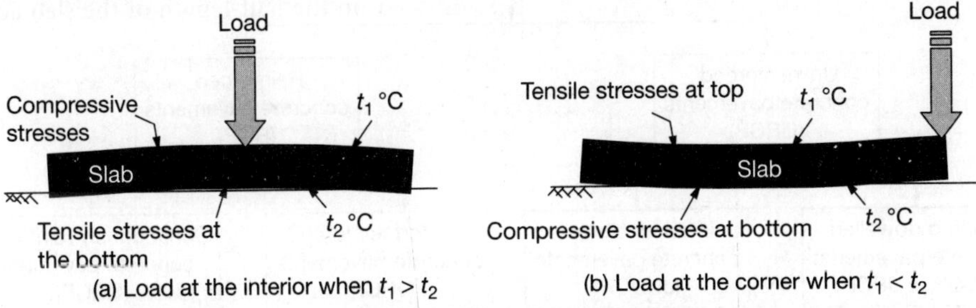

Figure 11.5 Critical combinations of stresses due to load and warping.

For most places in India, the first case holds good for design of rigid pavements. This is because higher temperature differentials (i.e., temperature difference between the top and bottom fibres of the pavement slab) are observed across the country (in six zones). The temperature differential ranges between 12.5 and 21 °C for slab thickness ranging from 15 to 30 cm (IRC: 58). The higher temperature differential causes a greater magnitude of stresses to develop in the slab.

11.2 SOFTWARE/CHARTS FOR ANALYSIS OF RIGID PAVEMENTS

The software/charts which are commonly used to analyse rigid pavements are listed in Table 11.2.

Table 11.2 Programs for analysis of rigid pavement.

Name of the Program/chart	Reference / source	Based on
Influence charts	Pickett and Ray, 1951	Theoretical equations derived by Westergaard and Hogg
PCCHWY	Warnes 1972 (used by PCA)	Westergaard's theory
H-51	Packard 1973	

(continued)

Name of the Program/chart	Reference / source	Based on
WESLIQUID and WESLAYER	Huang and Wang, 1973; Chou, 1980	Finite element method with linear deformation response; Winkler foundation method
ILLI-SLAB	Tabatabai and Barenberg, 1980; Foxworthy and Darter, 1989	
JSLAB	Tayabji and Colley, 1984	
N-S	Tang, 1993	
KENSLABS	Huang, 2004	Finite element method
MSLAB_LOAD and WARP-NONLINEAR	Fwa and Wei, 2006	Thick-plate model using Winkler foundation

11.3 TYPES OF RIGID PAVEMENTS

Concrete pavements may be classified into different groups based on the use of joints, extent of reinforcement and the load transfer mechanism adopted. The different types of rigid pavements are shown in Fig. 11.6.

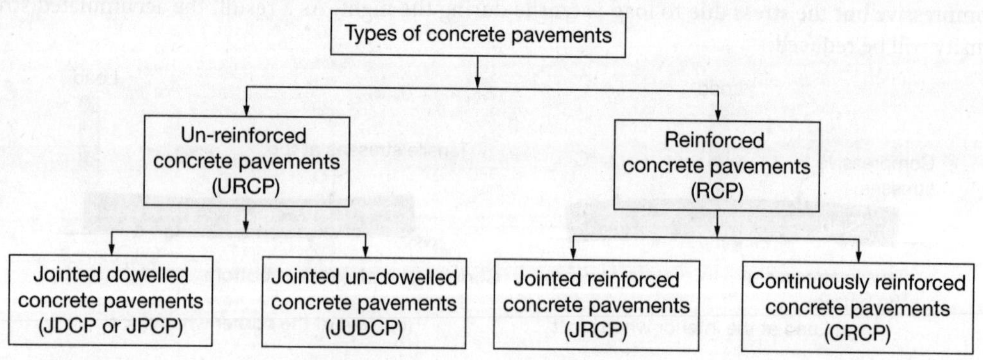

Figure 11.6 Types of concrete pavements.

Un-reinforced concrete pavements (URCP) are also called plain-cement concrete pavements (PCCP). In almost all jointed pavements, load transfer mechanism is implemented using dowel bars placed in transverse joints. Such pavements are called jointed dowelled concrete pavements (JDCP). JDCPs are also termed as jointed plain concrete pavements (JPCP) by a few agencies. They are the most popularly used rigid pavements in India.

Dowel bars may not be used for roads where traffic is very less such as remote rural roads. In such roads, aggregate interlocking of joints should be provided as a low-cost means of load transfer. Such pavements are termed as jointed un-dowelled concrete pavements (JUDCP). Load transfer by aggregate interlocking is not effective when the slabs contract and widen the joint. However, tie bars can be placed across longitudinal joints of dual-lane roads to tie up the adjacent slabs at the same level.

Occurrence of cracks in concrete slabs is inevitable due to repeated application of axle loads and weathering action in different seasons. Steel reinforcement in slabs is provided to inhibit widening of cracks and keep the cracked faces in intimate contact. This extends the useful life of the pavement. In jointed reinforced concrete pavements (JRCP), a steel mesh or mat is placed at the middle of each slab. It is not meant for structural strength improvement of the slab but is provided to control crack width. As usual practice, for load transfer and tie-up of slabs, the dowel and tie bars are placed in transverse and longitudinal joints respectively.

In case of continuously reinforced concrete pavements (CRCP), steel bars are distributed continuously in the longitudinal direction (in the range of 0.5 to 0.7% cross-sectional area of slab) so that construction of transverse joints can be eliminated. Construction of CRCP are preferred in (i) main heavy-traffic corridors such as expressways, (ii) adverse climatic conditions and (iii) weak sub-grades. CRCPs can be maintained at low cost because they have less/ no joints.

In addition to the above basic types, pre-stressed and concrete-blocks are also used for the construction of pavements.

11.4 PAVEMENT JOINTS

Joints are provided in concrete pavement at optimal intervals to allow shrinkage, contraction, expansion, and curl in the up- and downward directions. The slab shrinks due to sudden loss of moisture from fresh concrete before it hardens completely. The rest of the geometric changes take place due to production of warping stresses in the slab caused by variations in temperature and moisture content during different seasons.

The degree to which installing joints brings relief of critical (tensile) stress in concrete slabs by load transfer is called efficiency of the joint action (Teller, 1938). Joints should be spaced optimally to maintain structural efficiency, to relieve (tensile) warping stresses and prevent occurrence of cracks during temperature displacements of slab.

Based on the function and purpose of joints, they may be classified into different types (Fig. 11.7). The joints provided in the transverse direction are called transverse joints (Fig. 11.8(i)). Transverse joints permit the slab to displace in the horizontal direction. Contraction joints allow only contraction whereas expansion joints allow both expansion and contraction. Contraction joints are provided either as (i) dummy joints with saw cuts to a depth of 1/3rd or 1/4th thickness of the slab from the top or as (ii) dowel bars placed at its mid-depth for load transfer. As a guideline, spacing of contraction joints in pavements (URCP) may be taken as 4.5 m and 5.0 m when the thickness of the pavement is up to 25 cm and from 25 to 35 cm respectively (IRC:15).

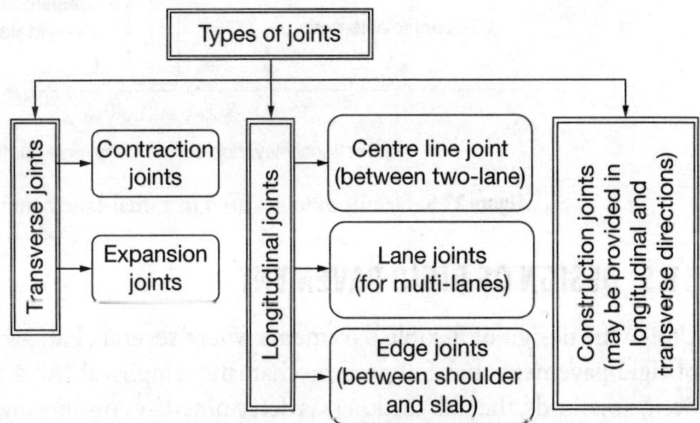

Figure 11.7 Types of joints in concrete pavements.

For expansion joints, dowel bars with metal caps that are partially filled with cotton are used (Fig. 11.8(ii)). Half the length of the dowel bar near the metal cap is lubricated to ease movement of the slab due to temperature variation (Finn, 1961). The gap between the joint faces should be filled with a filler and at the top, an elastomeric material should be used as a joint-sealant to prevent ingress of rain water and grit. The sealant material must be compressible, elastic and strong enough to resist repeated weathering action. Sealing of joints can be achieved by different methods, viz., (i) by hot-pouring bitumen, PVC or modified bitumen, (ii) by cold-pouring polysulphide, polyurethane or silicone and (iii) by compression using polychloroprene (Akoto and Nilles, 2007).

The joints provided in the longitudinal direction are termed as longitudinal joints (Fig. 11.8 (i)). These joints are mainly responsible for relieving warping stresses by hinge action and preventing irregular longitudinal cracks (Older, 1925). Due to this reason, these joints are also known as warping joints. Based on location, these joints may be classified into three types: (i) centre-line joints, (ii) lane joints and (iii) edge joints. For longitudinal joints, tie-bars are placed at the mid-depth of the slab (Fig 11.8 (iii). A plain butt joint or a tie-bar with tongue and groove type can also be used. The main purpose of tie-bars in longitudinal joints is to hold the adjacent slabs together and keep the slabs at the same level. Longitudinal joints are required when the width of the pavement (URCP) exceeds 4.5 m (IRC: 58).

While constructing pavements, joints are required to connect new slabs and the end of a slab which was constructed about a day earlier. These joints may be in the longitudinal or transverse directions or both.

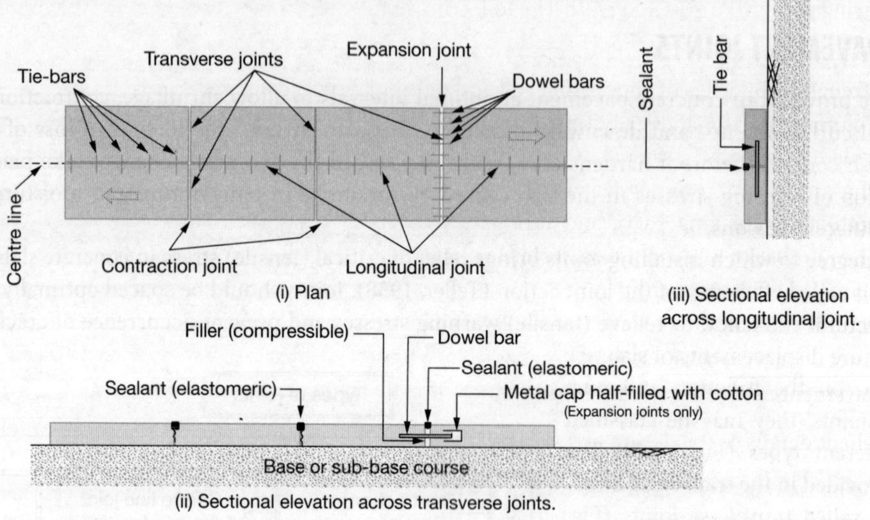

Figure 11.8 Details of joints used in a dual-lane concrete pavement (not to scale).

11.5 DESIGN OF RIGID PAVEMENT

Unlike the design of flexible pavements where several changes took place through the years, the design of rigid pavements has followed mechanistic–empirical (M–E) principles from the beginning. In M–E design methods, the slab thickness is determined by considering the fatigue failure of the slab due to damage caused by cumulative axle load passes. The input parameters relate to sub-grade support, material characteristics, traffic loads and climate. Several researchers (Older, 1925; Goldbeck, 1925; Westergaard, 1925; Teller and Sutherland, 1935; Spangler, 1937; Bradbury, 1938; Pickett and Ray, 1951; Friberg, 1955; Ioannides et al., 1985; and Shi et al., 1993) significantly contributed to thickness design of rigid pavements. However, Westergaard's theory (1925, 1933 and 1948) is the basis on which many modifications took place in the analysis and design of rigid pavements. These modifications have provided reasonably good agreement with the performance of in-service pavements.

The AASHTO method of design of rigid pavement (1993), however, uses empirical principles and serviceability performance concepts. Pavement performance is measured in terms of distresses or damages which are caused due to repetitive traffic loads in different seasons.

11.6 OVERVIEW ON DESIGN METHOD OF PLAIN JOINTED RIGID PAVEMENTS FOR HIGHWAYS AS PER IRC GUIDELINES

This design method is based on mechanistic–empirical principles and the guidelines are applicable to roads plying with an average of 450 commercial vehicles per day (CVPD) and above, where the design traffic is more than 2 msa. A few of the salient features of this design methodology are discussed below:

Rigid pavements can be designed using any of the following options, viz.,

(i) concrete pavement with tied concrete shoulder and doweled transverse joints
(ii) concrete pavement without concrete shoulder and with tied concrete shoulders
(iii) concrete pavement with widened outer lanes and
(iv) concrete pavement slab with unbound base and bonded to dry lean concrete (DLC) layer.

The proposed DLC should have minimum thickness of 100 mm and minimum compressive strength @ 7-day curing as 10 MPa (IRC: SP: 49). The following main factors are considered for the design of rigid pavements:

1. Strength properties of supporting soil and other materials characteristics
2. Volume, directional distribution and composition of commercial traffic with single, tandem and tridem axle loads
3. The number of commercial vehicle repetitions in traffic during the day hours (responsible for bottom–up cracking, BUC) and during the night hours (responsible for top–down cracking, TDC)
4. The number of axles in a vehicle with their spacing less than the spacing of transverse joint (say, 4.5m) and tyre pressure
5. Annual growth rate of traffic (Min. 5%) and design period (≥ 30 years)
6. Lateral placement of wheel paths across the width of carriageway
7. Pavement temperature, rainfall and other climatic factors.

A few salient details of the design are discussed below (IRC: 58-2015).

1. *Composition of Pavement*: The typical pavement layered structure is considered as a 500 mm thick capping sub-grade (min. 8% CBR) laid with:
   - Filter/separation layer + Granular Sub-Base (GSB) as drainage layer + DLC/cement treated sub-base + 150 micron PVC sheet + Pavement quality concrete (PQC)

     or
   - GSB as filter/separation layer + GSB as drainage layer + DLC/cement treated sub-base + Bituminous concrete (BC) + Pavement quality concrete (PQC),

     or
   - GSB filter/separation layer + GSB as drainage layer + Dense bituminous macadam-1 + Pavement quality concrete (PQC).

   The BC and DBM may be considered as de-bonding layers with the PQC. If the sub-grade CBR > 8%, a synthetic geo-composite may be used in place of both the layers of granular separation layer and drainage layers.
2. Modulus of sub-grade reaction (k-value) of supporting layers is used for thickness design of pavement crust. In this process, equivalent k-value of sub-grade may be estimated from soaked sub-grade CBR (by using Table 11.3), where minimum CBR value of sub-grade is restricted to 8%. The sub-grade *in-situ* CBR may also be determined by using DCP value by using an equation as recommended by ASTM D 6951 (Eq.11.20) or in case of FWD test results, the obtained dynamic in-situ k-values of sub-grade may be reduced to 50% and considered as design sub-grade k-value.

Table 11.3 Relationship between soaked CBR and modulus of sub-grade reaction (k-value) recommended for homogeneous soil sub-grade (Partial table from IRC: 58, 2015).

Soaked CBR (%)	7	10	15	20	50	100
k-value (MPa/m)	48	55	62	69	140	220

$$\log_{10}\text{CBR} = 2.465 - 1.12 \times \log_{10} \times (\text{Average value of DCP}_{60}{}^{0}\text{ penetration in mm/blow}) \quad (11.20)$$

Effective k-values of different combinations of sub-grade with GSB or cement-treated sub-base may be determined for the given values of sub-grade k-value and the proposed thickness value of sub-grade layer (Fig. 11.9). Effective k-values of different combinations of sub-grade with DLC (100 and 150 mm thick, laid over GSB as filter layer and drainage layers) may be determined (Fig. 11.10).

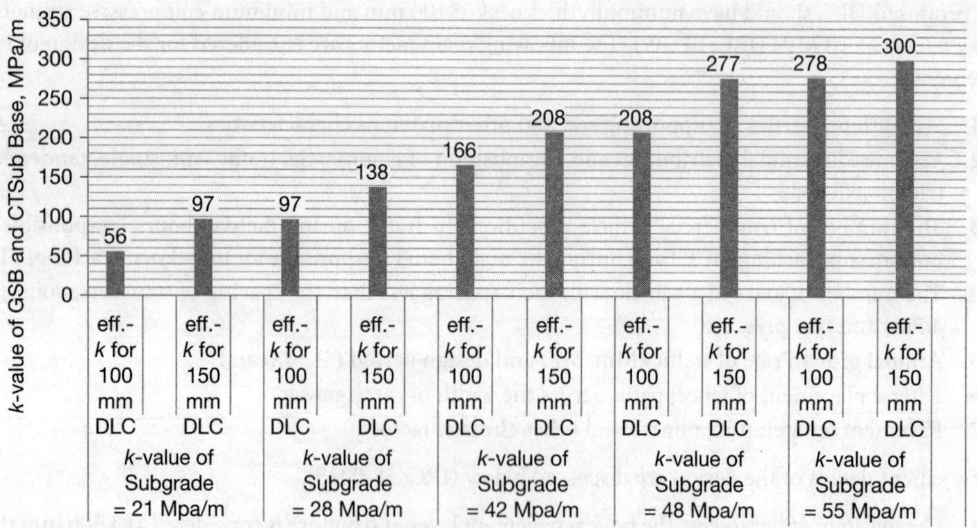

Figure 11.9 The IRC: 58 recommended k-values for granular and cement-treated sub-bases.

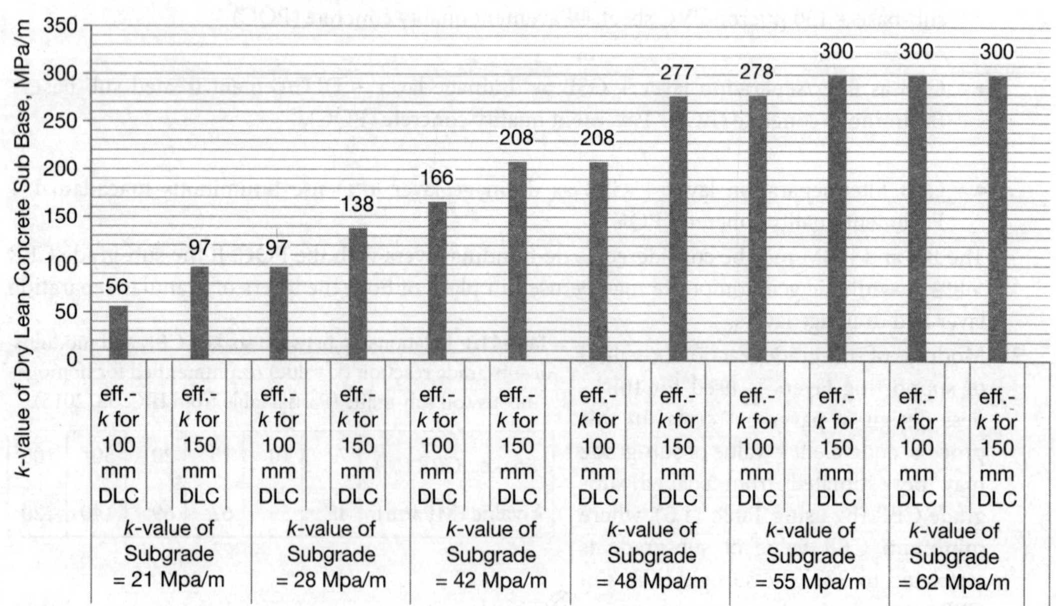

Figure 11.10 The IRC: 58 Recommended k-values (in MPa/m) for DLC sub-bases.

Note in Figs. 11.9 and 11.10 k-values are extracted from AASHTO-1993. The maximum recommended value may be taken as 300 MPa/m for the these cases.

3. Characteristic compressive strength of cement concrete may be estimated as per IS: 456-2000 (Eq.5.16) with a minimum grade as M40. The design flexural strength is considered as 1.1 × flexural strength obtained at the end of 28-day curing.
4. The concrete slab thickness is determined based on combined flexural stresses due to axle-load repetitions and temperature gradients, applicable for typical climatic conditions prevailing in India, with a broad categorisation of seven zones (Fig. 11.11).

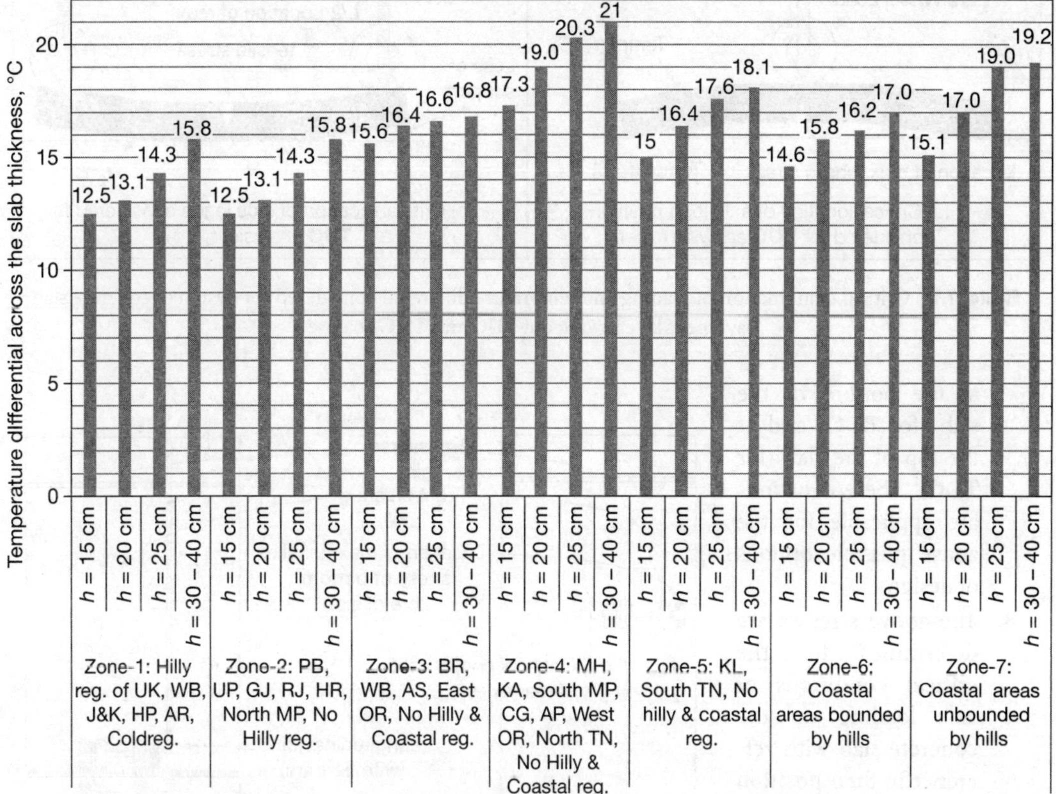

Figure 11.11 Maximum temperature differential for concrete slabs considered across seven zones in India (as per IRC: 58).

5. Single and tandem axle (without front steering axle) loads design repetitions are considered for calculation of cumulative fatigue damage analysis for bottom–up cracking (BUC) analysis (Fig. 11.12 (a)), for day (6-hour, i.e., 10 am – 6 pm) traffic and positive temperature differential observed in a given locality.
6. As the concrete typically shrinks when curing, *shrinkage cracks* may be induced on the top surface of the slab. In course of time, these cracks widen and propagate down under the effect of loading (due to self-weight and axle loading), and due to curling-up at night. Single, tandem axle and tridem axle (without front steering axle) load design repetitions are considered for calculation of cumulative fatigue damage analysis for top–down cracking (TDC) analysis (Fig. 11.12 b), for night (6-hour, i.e., 0 AM–6 AM) traffic, and negative temperature differential is observed.
The above 6-hour time may slightly vary depending on the geographic position of the site.
7. Finite element based analysis was carried out for development of regression equations (see Appendix-V, pp.73–75 of IRC: 58-2015) prescribed for calculation of maximum tensile stresses

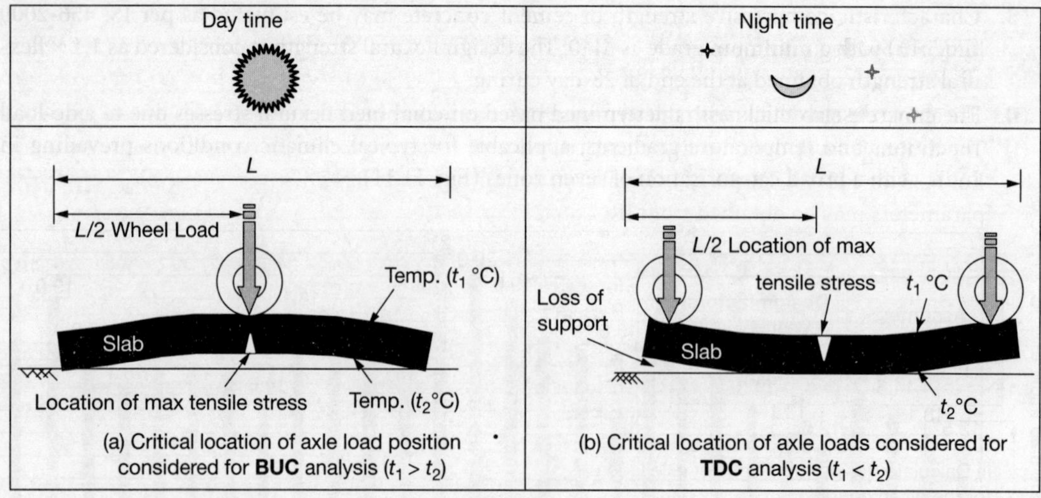

Figure 11.12 Critical combination of loading and temprature differetial considered for design of concrete slab pavement by carrying out BUC and TDC analyses.

at the bottom of the slab (for BUC) and at the top of the slab (for TDC). These equations are applicable for the above possible options of design.

8. The above stresses are determined for the critical conditions of axle loading on the concrete slab with reference to their position adjacent to edge and joints (Figs. 11.13 (a) and (b)).

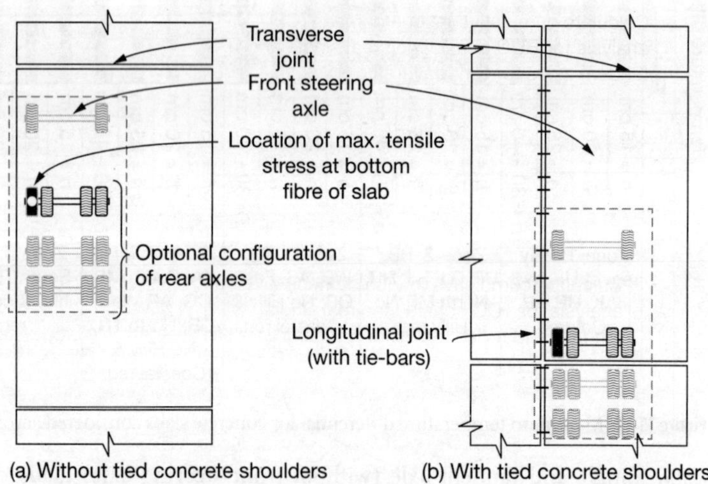

Figure 11.13 Critical location of placement of first rear axle considered for determination of (a) Max. edge flexural stress for BUC case and (b) Max. tensile stress for TDC case (A sketch elaborated based on IRC: 58, 2015).

The expected number of standard axle load repetitions (N_i) for each stress ratio (SR) should be obtained by standard relationships (Eqs. 11.21 (a)–(c)) for each group of axle load. The resulting consumed cumulative fatigue damage induced by different groups of axle load for the each case of BUC and TDC should not exceed 1.0

If $SR < 0.45$, N_f = Infinite (11.21a)

$$\text{If } 0.45 \le SR \le 0.55,\ N_f = \left[\frac{4.2577}{SR - 0.4325}\right]^{3.268} \quad (11.21b)$$

$$\text{If } SR > 0.55,\ \log_{10} N_f = \left(\frac{0.9718 - SR}{0.0828}\right) \tag{11.21c}$$

(i) The above design details with the step-wise procedure may be summarised as shown in Fig.11.14

(j) The drainage layer design may be carried out with reference to its rate of permeability by using Eq. 6.33 of FHWA-TS-80-224. Details of design of bonded rigid pavement and other design parameters may be obtained from IRC: 58, 2015.

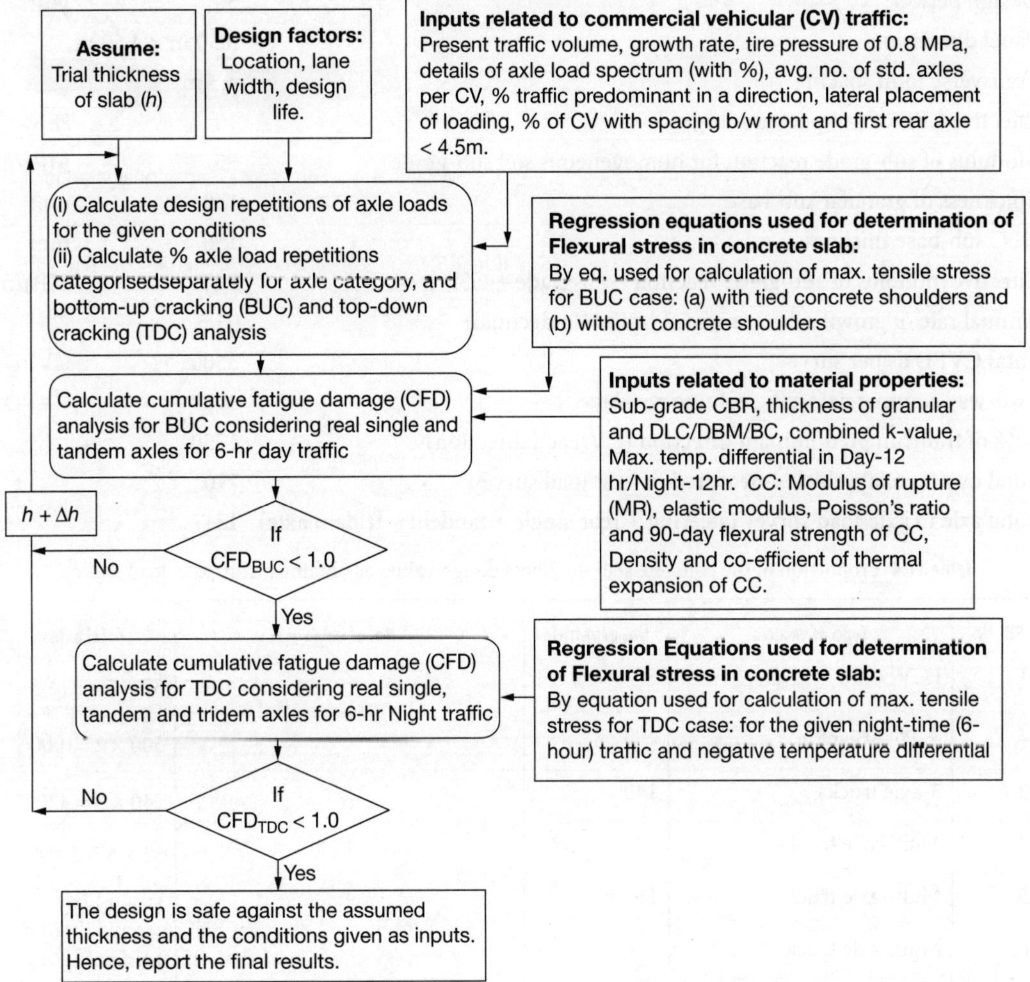

Figure 11.14 Typical step-wise procedure recommended for thickness design of cement concrete slab pavement (a) with tied concrete shoulder with doweled transverse joints and (b) without tied concrete shoulder and without doweled transverse joint, according to IRC: 58.

Thickness design of cement concrete slab pavement according to the recommendation of IRC: 58 is presented in the example given below.

EXAMPLE 11.5

The following is a sample thickness design of a cement concrete slab according to the guidelines of IRC: 58-2015:

The proposed location of the road construction	Karnataka	
No. of carriageways	2	
No. of lanes in each carriageway	2	
Design period	30	years
Panel dimension	3.50 m × 4.50 m	
Transverse joint spacing	4.5 m	
Effective CBR of compacted sub-grade	10	%
Modulus of sub-grade reaction for homogeneous soil sub-grade	55	MPa/m
Thickness of granular sub-base	150	mm
DLC sub-base thickness	150	mm
Effective modulus of sub-grade reaction (sub-grade + GSB + DLC)	300	MPa/m
Annual rate of growth of commercial traffic (in decimal)	0.05	
Total CVPD as per survey	3500	CVPD
Two-way commercial traffic volume per day	3500	CVPD
50% of traffic in predominant direction (CV/each direction)	1750	
Total commercial vehicles measured in axle-load survey	770	
Total axle in axle-load survey (steering + rear single + tandem + tridem axle)	1817	

Table 11.4 Proportion of the concrete slab thickness design values as obtained from axle-load survey.

SR. No.	Type of vehicle	No. of vehicle	No. of axles	Total axles
1	LCV	57	2	57 × 2 = 114
2	2-axle trucks	500	2	500 × 2 =1000
3	3-axle trucks	140	3	140 × 3 = 420
4	Multi-axle trucks	52	4	52 × 4 = 208
5	Multi-axle trucks	15	4	15 × 4 = 60
6	Multi-axle trucks	1	5	1 × 5 = 5
7	Bus	5	2	5 × 5 = 10
	Total commercial vehicles in axle-load survey	= 770	Total axle in axle-load survey (Steering + Rear single + Tandem + Tridum)	= 1817

Average number of axles (steering/single/tandem/tridem) per commercial vehicle = 1817/770 ≈ 2.36

Front-steering axles	770	nos. (42.38%)
Rear single axles	562	nos. (30.93%)
Rear tandem axles	470	nos. (25.87%)
Rear tridem axles	15	nos. (0.82%)
Total		(100.00%)

Percent of vehicle with front-to-rear axle spacing < 4.5m, for BUC 100.00%
Percent of vehicles traveling during night time (12 Hr: 6PM to 6AM) 35.00%

Total two-way commercial vehicles during 30-yr. design period @5.23%

$$= 365 \times 3500 \times \left(\frac{(1+0.0523)^{30} - 1}{0.0523} \right) = 88305385 \text{ nos.}$$

Total two-way axle repetitions during the design period = 88305385 × 2.36 = 208400709 nos.
No. of axles in predominant direction = ½ × 208400709 = 104200355 nos.
Design traffic after adjusting for lateral placement of axles = 0.25 × 208400709 = 52100177 nos.
Night-time design axle repetitions (12 hours) = 35.00 % of 52100177 = 18235062 nos.
Day-time design axle repetitions (12 hours) = 52100177 – 18,235,062 – 33,865,115 nos.
Day-time traffic (6 Hours)(10 AM to 4 PM) = 55.00%
Day-time design axle repetitions (6 hours) = 55.00 % of 33,865,115 = 18,625,813 nos.
Design number of axle-load repetitions for bottom-up cracking analysis = 18,625,813 nos.
Night-time traffic (6 Hr: 12 midnight to 6 AM) = 45.00%
Percent of commercial vehicles having the spacing between the front axle and the first axle unit less than 4.5 m = 100.00 %
Night-time 6 Hr. axle-load repetitions for top-down cracking analysis = (45/100) × (100/100) × 18,235,062 = 8,205,778 nos.

Table 11.5 Axle-load repetitions.

S. No.	Axle category	Proportion of axle category	Category-wise axle repetitions for bottom-up cracking (BUC)	Category-wise axle repetitions for top-down cracking (TDC)
1	Front steering single	42.38	42.38% of 18,625,813 = 7,893,20	42.38% of 8,205,778 = 3,477,609
2	Rear single	30.93	30.93% of 18,625,813 = 5,760,964	30.93% of 8,205,778 = 2,538,047
3	Tandem	25.87	25.87% of 18,625,813 = 4,818,498	25.87% of 8,205,778 = 2,122,835
4	Tridem	0.82	0.82% of 18,625,813 =152,732	0.82% of 8,205,778 = 67,287
	Total	= 100.00 %	≈18,625,813	8,205,778

Cumulative fatigue damage analysis for bottom-up cracking (BUC) and top-down cracking (TDC) and selection of slab thickness (h):

Assume: Concrete pavement with tied concrete shoulder with dowel bars across transverse joints

Trial thickness of slab $= h =$ 0.29 m

Effective modulus of sub-grade reaction (Sub-grade + GSB + DLC) $= k = 300$ MPa/m

Elastic modulus of concrete = E = 30000 MPa

Poisson's ratio of concrete $= m =$ 0.15

Unit weight of concrete $= g =$ 24 kN/m^3

28 days compressive strength $= f_{ck} =$ 40 MPa

90 days compressive strength $= f_{ck} =$ 48 MPa

28 days flexural strength $= f_{cr} = 0.7 \times \sqrt{40} \approx 4.5$ MPa

90 days flexural strength $= f_{cr} = 1.1 \times 4.5 = 4.95$ MPa

Max. day-time differential in slab (in Karnataka) for BUC = DT = 21 °C

Night-time temperature differential in slab for TDC = DT = (21/2) + 5 = 15.5 °C

$$\text{Radius of relative stiffness(l)} = \left[\frac{30000 \times 0.29^3}{12 \times 300 \times \left(1 - 0.15^2\right)}\right]^{0.25} = 0.675 \text{ m}$$

Table 11.6 Axle-load spectrum.

Single-axle loads			**Tandem-axle loads**			**Tridem-axle loads**		
Axle-load class, kN		**Percentage of axle-loads**	**Axle-load class, kN**		**Percentage of axle-loads**	**Axle-load class, kN**		**Percentage of axle-loads**
205	215	0.00	420	440	0.00	590	620	0.00
195	205	0.50	400	420	0.00	560	590	0.00
185	195	1.00	380	400	0.00	530	560	0.00
175	185	1.00	360	380	0.00	500	530	0.00
165	175	0.50	340	360	0.01	470	500	0.00
155	165	2.00	320	340	0.02	440	470	0.00
145	155	3.00	300	320	0.59	410	440	0.00
135	145	4.00	280	300	0.60	380	410	5.00
125	135	5.00	260	280	3.18	350	380	6.00
115	125	6.00	240	260	6.60	320	350	7.00
105	115	7.00	220	240	10.00	290	320	34.00
95	105	8.00	200	220	7.50	260	290	27.00
85	95	7.00	180	200	10.00	230	260	0.00
0	85	55.00	0	180	61.50	0	230	21.00
	Total	= 100.00 %		Total	= 100.00 %		Total	= 100.00 %

Use the above inputs, the calculated values and axle-load spectrum values are now required to be substituted in the Excel file, as supplied by the IRC: 58-2015. The obtained output and the results are presented below:

Cumulative fatigue damage analysis for bottom-up cracking (BUC)

$gh^2 = 2.02$
$kl^2 = 136.79$
$h/(kl^4) = 0.0046$

k>150 MPa/m						k> 150 MPa/m					
Rear single axles						Rear tandem axles					
Expected repetitions (n_i)	Axle load KN (P)	Flex. stress in MPa (S)	Stress ratio (SR)	Allowable Repetition (N_i)	Fatigue damage (n_i/N_i)	Expected repetitions (n_i)	Axle load KN (P)	Flex. stress in MPa (S)	Stress ratio (SR)	Allowable repetition (N_i)	Fatigue damage (n_i/N_i)
1	2	3	4	5	6	7	8	9	10	11	12
0	210	0.000	0	Unlimited	0.0000	0	430	2.369	0.479	2575774	0.0000
28,805	200	2.693	0.544	147792	0.1949	0	410	2.301	0.465	8304419	0.0000
57,610	190	2.617	0.529	236978	0.2431	0	390	2.233	0.451	52363284	0.0000
57,610	180	2.542	0.514	411605	0.1400	0	370	2.166	0.438	Unlimited	0.0000
28,805	170	2.467	0.498	840751	0.0343	482	350	2.098	0.424	Unlimited	0.0000
115,219	160	2.391	0.483	1966925	0.0586	964	330	2.03	0.41	Unlimited	0.0000
172,829	150	2.316	0.468	6222972	0.0278	28,429	310	1.962	0.396	Unlimited	0.0000
230,439	140	2.241	0.453	37439709	0.0062	28,911	290	1.894	0.383	Unlimited	0.0000
288,048	130	2.165	0.437	Unlimited	0.0000	153,228	270	1.826	0.369	Unlimited	0.0000
345,658	120	2.090	0.422	Unlimited	0.0000	318,021	250	1.758	0.355	Unlimited	0.0000
403,267	110	2.015	0.407	Unlimited	0.0000	481,850	230	1.69	0.341	Unlimited	0.0000
460,877	100	1.939	0.392	Unlimited	0.0000	361,387	210	1.623	0.328	Unlimited	0.0000
403,267	90	1.864	0.377	Unlimited	0.0000	481,850	190	1.555	0.314	Unlimited	0.0000
3,168,530	85	1.826	0.369	Unlimited	0.0000	2,963,376	180	1.521	0.307	Unlimited	0.0000
5,760,964				Fatigue damage from single axles	0.7047	4,818,498				Fatigue damage from tandem axles	0.0000
Cumulative fatigue damage for BUC = 0.7047											

Cumulative fatigue damage analysis for top-down cracking (TDC)

"Beta" factor for transverse joint efficiency in top-down cracking $B = 0.66$
$Bh/(kl^4) = 0.0031$
$h^2/kl^2 = 0.0006$

Rear single axles						Rear tandem axles						Rear tridem axles					
Expected repetitions (n_i)	Axle load KN (*P*)	Flex. stress in MPa (*S*)	Stress ratio (SR)	Allowable repetition (N_i)	Fatigue damage (n_i/N_i)	Expected repetitions (n_i)	Axle load KN (*P*)	Flex. stress in MPa (*S*)	Stress ratio (SR)	Allowable repetition (N_i)	Fatigue damage (n_i/N_i)	Expected repetitions (ni)	Axle load KN (*P*)	Flex. stress in MPa (*S*)	Stress ratio (SR)	Allowable repetition (N_i)	Fatigue damage (n_i/N_i)
1	2	3	4	5	6	7	8	9	10	11	12	1	2	3	4	5	6
0	210	2.659	0.537	182672	0.0000	0	430	2.685	0.542	156797	0.0000	0	605	2.605	0.526	262743	0.0000
12,690	200	2.607	0.527	253766	0.0500	0	410	2.633	0.532	214417	0.0000	0	575	2.554	0.516	380253	0.0000
25,380	190	2.555	0.516	380253	0.0667	0	390	2.581	0.521	314437	0.0000	0	545	2.503	0.506	576922	0.0000
25,380	180	2.504	0.506	576922	0.0440	0	370	2.53	0.511	465275	0.0000	0	515	2.452	0.495	979960	0.0000
12,690	170	2.452	0.495	979960	0.0129	212	350	2.478	0.501	726285	0.0003	0	485	2.401	0.485	1732461	0.0000
50,761	160	2.4	0.485	1732461	0.0293	425	330	2.426	0.49	1286914	0.0003	0	455	2.349	0.475	3455963	0.0000
76,141	150	2.349	0.475	3455963	0.0220	12,525	310	2.374	0.48	2402754	0.0052	0	425	2.298	0.464	9197403	0.0000
101,522	140	2.297	0.464	9197403	0.0110	12,737	290	2.323	0.469	5682913	0.0022	3,364	395	2.247	0.454	32043150	0.0001
126,902	130	2.245	0.454	32043150	0.0040	67,506	270	2.271	0.459	16179928	0.0042	4,037	365	2.196	0.444	Unlimited	0.0000
152,283	120	2.193	0.443	Unlimited	0.0000	140,107	250	2.219	0.448	Unlimited	0.0000	4,710	335	2.144	0.433	Unlimited	0.0000
177,663	110	2.142	0.433	Unlimited	0.0000	212,284	230	2.167	0.438	Unlimited	0.0000	22,878	305	2.093	0.423	Unlimited	0.0000
203,044	100	2.09	0.422	Unlimited	0.0000	159,213	210	2.116	0.427	Unlimited	0.0000	18,167	275	2.042	0.413	Unlimited	0.0000
177,663	90	2.038	0.412	Unlimited	0.0000	212,284	190	2.064	0.417	Unlimited	0.0000	0	245	1.991	0.402	Unlimited	0.0000
1,395,926	80	1.986	0.401	Unlimited	0.0000	1,305,544	180	2.038	0.412	Unlimited	0.0000	14,130	180	1.88	0.380	Unlimited	0.0000
2,538,045	**Fatigue damage from single axles**				0.2400	2,122,837	**Fatigue damage from tandem axles**				0.0122	67,287	**Fatigue damage from tridem axles**				**0.0001**

Cumulative fatigue damage for TDC = 0.2400 + 0.0122 + 0.0001 = 0.2523Cumulative fatigue damage for BUC + TDC = 0.7047 + 0.2523 = 0.957 < 1.000 OK. Hence, the assumed thickness of 0.29 m is OK.

Add 0.01m for tinning purpose and finalise the design slab thickness = 30 cm

EXAMPLE 11.6

Design a cement concrete pavement with widened outer lanes based on the above-given input values and assumed site conditions.

Answer

The outer lanes may be considered as construction of additional width of slab on the outer sides of a lane with 0.5 to 0.6 m width and need to be tied similar to tied concrete shoulders. The design procedure is adopted as in the case of cement concrete pavement with tied concrete shoulder and dowel bars across transverse joints (IRC: 58, 2011). Hence, the same thickness of the slab should be provided for the outer lanes also.

EXAMPLE 11.7

Design the cement concrete pavement bonded to dry lean concrete (DLC) layer based on the above given input values and assumed site conditions.

Answer

The following procedure may be adopted for design of cement concrete pavement for the above-given condition (IRC: 58, 2011).

Step 1: Select to provide suitable thickness of a granular sub-base over sub-grade, with reference to traffic intensity.

Step 2: Determine effective modulus of *k*-value due to combined effect of sub-grade CBR and the selected thickness of granular sub-base

Step 3: Design the thickness of cement concrete slab with the assumption of providing dowel and tie bars

Step 4: Assume suitable minimum thickness of DLC which is intended to be provided over the granular sub-base with appropriate values of elastic modulus and Poisson's ratio.

Step 5: Assume suitable trial thickness with appropriate values of elastic modulus and Poisson's ratio of the proposed cement concrete slab.

Step 6: Determine depth of neutral axis, calculate individual stiffness of DLC and combined stiffness of slab with DLC.

Step 7: Calculate the stiffness of the slab with design thickness and compare it with combined stiffness of slab with DLC. If the combined stiffness value is more than the design stiffness, the assumed condition is considered as safe.

For more details and the equations used for calculation of the stiffness values, the reader may refer to IRC: 58-2011.

11.7 DESIGN OF DOWEL BARS

A few salient features of dowel design, according to IRC: 58, are presented below.

Permissible bearing stress of concrete should be greater than the maximum bearing stress in the dowel bar. The bearing stress in concrete and the maximum bearing stress in the dowel bar are calculated using the following equations.

$$\text{Permissible bearing stress in concrete(PBS)} = \frac{(101.6 - d_0)}{95.25} \times f_{ck} \quad (11.22)$$

where,

d_0 = diameter of dowel bar(mm)

f_{ck} = characteristic compressive strength of cement concrete(MPa) defined in Section 8.8.1

For M40 grade, take f_{ck} value at 28-day as 40 MPa, and 90-day as 48 MPa.

Equation (11.21) has been adapted from the Committee 325 of the American Concrete Institute (ACI 1956). The bearing stress of concrete around the dowel bar is calculated based on Friberg's analysis using Eq. (11.22)

The number of dowel bars effectively participating as a group in load transfer is to be determined in a distance equal to 1.0 × radius of relative stiffness (l), which begins from the point of load application close to the edge of the slab (Fig. 11.15). Therefore, number of dowel bars in the effective distance (N) = 1.0 + (l /spacing between dowel bars).

When applied wheel-load is close to the outer edge of the slab (Fig. 11.15), the maximum load will be carried by the dowel bar that is exactly under the load and no load is carried by the dowel bar which is away from the point of loading at a distance of 1.0 × l. Within this effective distance (1.0 × l), the load carried by dowel bars vary linearly. The load transferred across the joint is assumed to be 40%. Installation of dowel bars in slabs is restricted at places where the slab thickness is less than 15 cm.

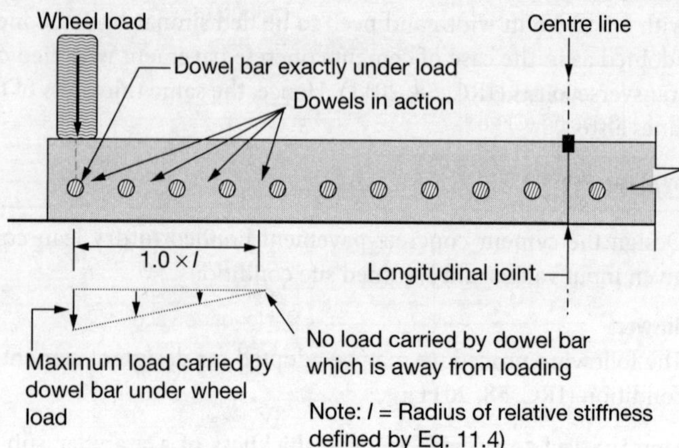

Figure 11.15 Typical sketch showing details of dowel bars participating in load transfer at transverse joint.

The IRC has recommended guidelines for suitable dimensions of dowel bars based on slab thickness. However, for the following conditions, smaller spacing between dowel bars may be assumed.

- To reduce the magnitude of load carried by dowel bars in the effective length.
- For thinner slabs
- When load on axles is heavy or abnormal and
- When the sub-grade strength is poor.

The design procedure for dowel bars is presented in Fig. 11.16 and explained in the following example.

EXAMPLE 11.8

Design dowel bars for the following design inputs: Thickness of slab = 30 mm; Dual-wheel load = 188 kN; Efficiency of load transfer across the joint = 40%; Width of expansion joint = 20 mm; Radius of relative stiffness = 0.994 m; Characteristic compressive strength of cement concrete cube (150 mm) after a 28-day curing = 40 MPa; Elastic modulus of dowel bar = 2.02×10^5 MPa and Modulus of dowel bar support = 415200 MPa/m. Assume no load transfer on to the tied concrete shoulder **Note**: Assume 30% of load transfer at terminal stage if tied shoulder is considered (IRC: 58).

Answer

Wheel-load considered for design of dowels = 70 % of (dual – wheel load/2)

= 0.7 (188/2) = 65.8 kN

However, for safety, as per IRC: 58, let us consider the design wheel load = 80 kN

Assume that diameter of dowel bar (d_o) = 38 mm, spacing between them (C) = 300 mm, length of dowel bar = 500 mm and that the first dowel bar is placed at a distance of 15 cm from the free edge of the pavement. Assumption of the above values may be taken from IRC: 58 or by experience.

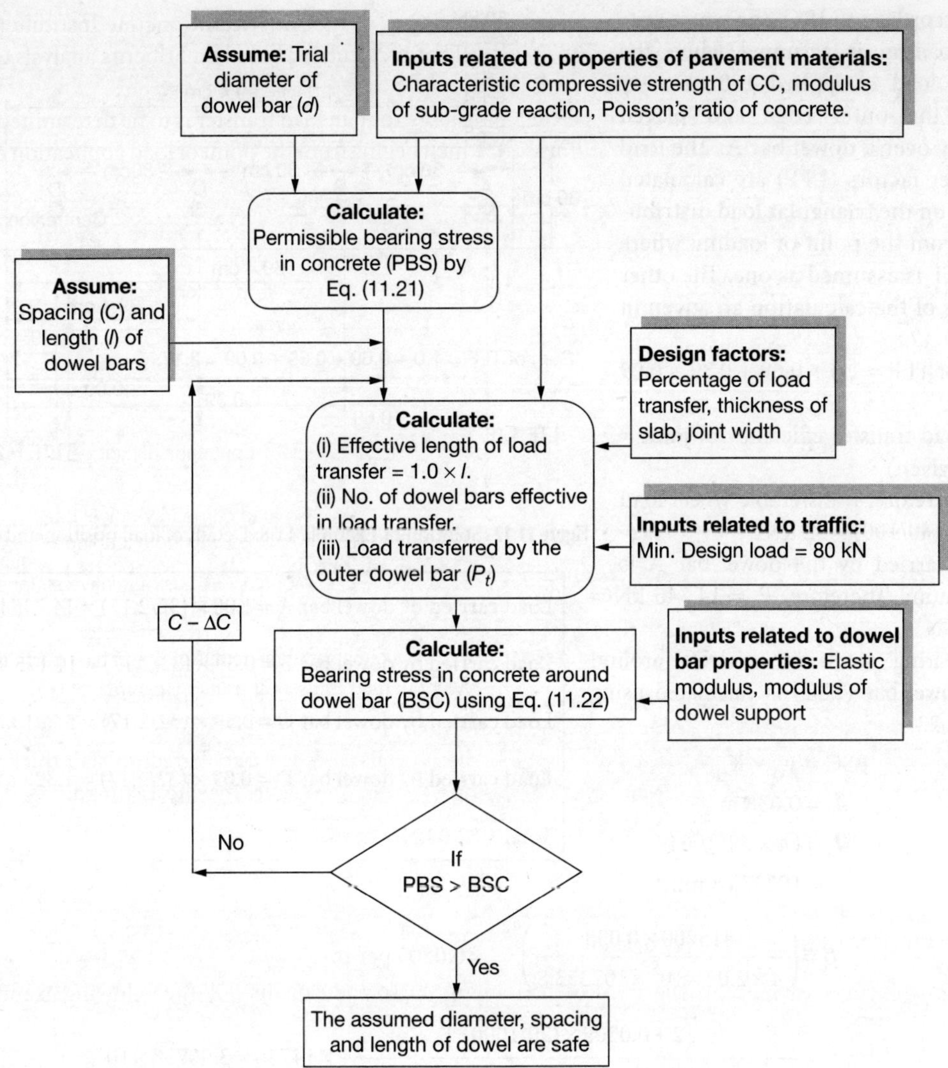

Figure 11.16 Design of dowel bars according to IRC: 58

Remarks: Dowel bars are not provided for thin slabs whose thickness is less than 200 mm. Dowel bars should be provided at the contraction joints for heavy traffic where the CVPD is greater than 450. The doweled joint width is recommended as 20 mm and 5 mm in case of expansion and contraction joints, respectively. As the maintenance of the expansion joints is difficult, with a need to be frequently cleaned and filled with compressible material, it was decided to restrict their usage only near permanent structures like bridges and culverts (IRC: 58).

Permissible bearing stress (PBS) in concrete is calculated by using Eq. (11.21),

$$\text{PBS} = \frac{(101.6 - 38)}{95.25} \times 40 = 26.7 \text{ MPa}$$

Effective distance in which the dowel bars act as a group = $1.0 \times l = 99.4$ cm.

According to IRC, the worst critical loading is assumed when the wheel load is applied 150 m away from the outer edge and placed exactly over a dowel bar A. The load transfer factors (LTF) are calculated based on the triangular load distribution from the point of loading where the LTF is assumed as one. The other details of the calculation are given in Fig. 11.17.

Sum of TLF = 1.0 + 0.69 + 0.39 + 0.09 = 2.17.

Load transfer efficiency of joint = 40% (given)

Therefore, transferable wheel load = 80 × (40/100) = 32 kN

Load carried by the dowel bar A is maximum. Therefore, P_t = 14.746 kN = 14746 N.

Bearing stress in concrete around the dowel bar (BSC) is calculated using Eq. 11.22

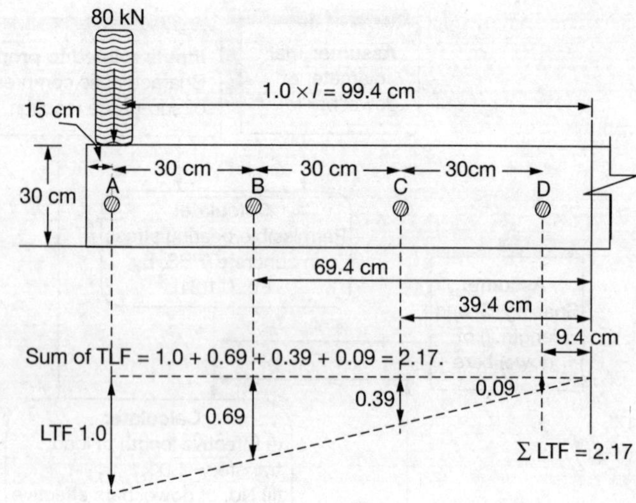

Figure 11.17 Solution of Example 11.8: Details of load position and LTF.

Load carried by dowel bar A = 1.00 × (32/2.17) = 14.746 kN
Load carried by dowel bar B = 0.69 × (32/2.17) = 10.175 kN
Load carried by dowel bar C = 0.39 × (32/2.17) = 5.751 kN
Load carried by dowel bar D = 0.09 × (32/2.17) = 1.328 kN
Total = 32.0 kN

$$BSC = y_O \times K_O$$

$$d_o = 0.038 \text{ m}$$

$$I_o = (\pi \times 38^4)/64 = 102353.8 \text{ mm}^4$$

$$\beta = \left(\frac{415200 \times 0.038}{4 \times 2.02 \times 10^5 \times 102353.8}\right)^{0.25} = 0.0209 \text{ per m}$$

$$y_O = \frac{(2 + 0.0209 \times (20/1000))}{(4 \times 0.0209^3 \times (2.02 \times 10^5) \times 102353.8 \times 1000)} \times 14746 = 3.90738 \times 10^{-5} \text{ m}$$

$$BSC = y_O \times K_O = 3.90738 \times 10^{-5} \text{ m} \times 415200 \text{ MPa/m} = 16.22 \text{ MPa}$$

Check: BSC (16.22 MPa) < PBS (26.7 MPa). OK.

Hence, provide dowel bar diameter = 38 mm, centre-to-centre spacing = 300 mm and length = 500 mm. If the above criterion is not satisfied, reduce the spacing or increase the diameter of the dowel bar as found suitable to the specific site.

11.8 DESIGN OF TIE BARS

Plain or deformed bars having a circular cross-section are used as tie bars and placed across longitudinal joints. Longitudinal joints should be provided where the lane width exceeds 4.5 m. Suitable diameter, length and type of bars are assumed for design and the assumed parameters are checked for stresses due to tension and bonding. Salient details of tie bar design, according to IRC: 58, are placed below.

The area of steel per metre length of the joint may be calculated using the following equation.

$$A_{st} = \frac{hBW}{S} \times f \quad (11.22)$$

where,

A_{st} = area of steel (in mm^2) required per/ m length of slab
h = thickness of slab (mm)
B = width of lane (m)
W = unit weight of slab (kN/m^3)
S = allowable working stress in steel (for plain bars it is taken as125 MPa; for deformed bars, 200 MPa)

To determine the length of the tie bar, its length must be equal to twice the length needed to develop the required bond strength to resist tensile stress in steel. The following equation is recommended by IRC.

$$L_{Tie} = \left(\frac{S \times d}{2 \times B_s}\right) \quad (11.23)$$

where,

L_{Tie} = length of the tie bar (mm)
S = allowable working stress in steel (for plain bars it is taken as125 MPa; for deformed bars, 200 MPa)
d = diameter of the tie bar (mm)
B_s = allowable bond stress of concrete; according to IRC, the B_s value for plain and deformed bars is equal to1.75 MPa and 2.46 MPa respectively.

The following IRC guidelines are useful for the design of tie bars.

- To relieve warping stresses at the joint, the diameter of tie bars should be restricted to not greater than 16 mm
- To prevent concentration of tensile stresses in tie bars, the centre-to-centre spacing between the tie bars should not be wider than 750 mm.
- To make up for incorrect or erroneous placement of tie bars during construction and to compensate loss of bond due to painting, the length of the tie bar calculated using Eq. (11.23) should be increased by 5 to 10 cm.

Based on the thickness of the slab, standard design values of different tie bars for two-lane pavements with each lane of width 3.5 m

EXAMPLE 11.9

Determine the suitable diameter, spacing and length of tie bars according to the guidelines of IRC for the following design inputs: thickness of slab = 300 mm; width of lane = 3.50 m; coefficient of friction between the bottom surface of the slab and the top surface of the supporting layer = 1.5; unit weight of cement concrete slab = 24 kN/m^3. Take diameter of dowel bar as 12 mm; allowable working stress in steel = 125MPa and permissible bond stress of tie bars = 1.75 MPa.

Answer

Area of steel required is calculated using Eq. 11.22,

$$A_{st} = \frac{(0.3\text{ m} \times 3.5\text{ m} \times 24000\text{ N})}{125\text{ MPa}} \times 1.5 = 302.4\text{ mm}^2/\text{m}$$

Cross-sectional area of single tie bar $(A_o) = \pi \times (12/4)^2 = 113.09$ mm^2

Spacing of ties bars = $1000 \times A_o/A_{st} = 1000 \times 113.09/302.4 = 373$ mm > Max. recommended by IRC: 58-2011 as 370 mm.

Therefore, provide centre-to-centre distance between the tie bars as 370 mm.

Length of tie bars is determined using Eq. 11.23.

$$L_{Tie} = \left(\frac{125 \times 12}{2 \times 1.74}\right) = 431 \text{ mm}$$

Increase the calculated length of the tie bars by 50 mm and 100 mm to make up for incorrect placement and loss of bond due to paint respectively.

Therefore, $L_{Tie} = 431 + 50 + 100 = 581$ mm

11.9 FACTORS AFFECTING PERFORMANCE OF RIGID PAVEMENT

Rigid pavements include plain (non-reinforced) and reinforced Portland cement concrete (PCC) pavements. Reinforcement is usually provided by placing a steel wire mesh at mid-slab depth. The commonly known causes for distresses and failures in rigid pavement are temperature variations across the depth of the slab, moisture changes in supporting granular layers, traffic-induced stress, poor quality of material and construction defects. The following general factors related to the performance of rigid pavement may be kept in mind.

1. The reinforcement provided is intended to limit/control (i) crack opening and (ii) the pavement of slabs of cement concrete pavement.
2. In case of PCC pavement, the formation of cracks due to temperature variations and the slab movements are controlled by constructing contraction joints just after initial curing. These joints are usually cut to one-third the depth of the slab and the groove formed is filled with a compressible material. Expansion joints, on the other hand, allow the expansion of slabs and are constructed as full-depth joints. The width of the expansion joint is more than contraction joints (20-mm wide) and is filled with a material that allows the slabs to expand. Expansion joints are provided only occasionally, e.g., adjacent to structures that cannot move with the pavement such as abutments of bridges and culverts.
3. The load-carrying capacity of rigid pavements is different from that of flexible pavements in that the traffic load is resisted by the bending strength of the cement concrete slab, technically known as flexural strength. Flexural strength plays a major role in the downward transfer of wheel load to the layers beneath over a *wide* area. Adequate transfer of loads across discontinuities such as joints and cracks in the slab is vital to a rigid pavement's performance; poor load transfer creates high stresses in the slab, causing distresses such as faulting, edge drop-off, pumping and corner breaks. In the case of hairline and very narrow cracks, aggregate-interlocking can take care of load transfer across the joint, but for wider cracks, dowel bars are used.
4. Doweled joints are provided to overcome the difficulty of sustaining higher stresses at joints and enable efficient load transfer at joints. Dowel bars placed perpendicular to the face of the joint at mid-depth of the slab can effectively transfer load across the joints and avoid higher stresses due to edge loads. To avoid rusting of dowel bars which fail to perform as desired, epoxy-coated dowel bars are used.
5. *Pumping* is a common phenomenon of loss of support due to the upward migration of subsurface water or squirting up of soil-slurry through cracks or joints. The loss of support under the slab leads to bending of the slab and widening of cracks and joints, and eventually, the portion of the slab disintegrates into small pieces. Due to the pumping problem, draining of the surface and subsurface water is very important to keep the compacted granular layers in position and stable.

Improving/providing sub-surface drainage facility is the solution to this problem, with added provision of slab-jacking or mud-jacking to take care of the loss of support under such slabs.

6. Faulting is also a common type of distress across joints in which the upstream edge of the slab is higher than the downstream edge of the adjacent slab. This step-like deformation occurs due to heavy truck traffic loading, particularly at un-doweled joints. This causes spalling along the joint face and cause discomfort to rides.
7. Timely sealing of joints and cracks with appropriate drainage measures can retard further deterioration and considerably prolong the life of rigid pavements.

Review Questions

1. Determine the crack-free length of a slab if its tensile strength is 2.1 MPa, unit weight is 23.2 kN/m^3 and average coefficient of friction is 1.5.
2. Determine the percentage difference between corner load stress calculated based on the radius of wheel contact area according to (i) Westergaard's analysis and (ii) the equation recommended by IRC: 58-2002 (Eq. (11.14). The design parameters are given as: Single-axle load = 10000 kg; With dual-wheel assembly; tyre pressure = 8.5 kg/cm^2; Spacing between centres of wheels = 310 mm, slab thickness = 30 cm
3. State and define the different types of rigid pavements.
4. State the objectives, functions and different types of joints in concrete pavements.
5. Explain why dowel and tie bars are placed in different road joints.
6. State the basic requirements of joint sealers and justify.
7. What are the different methods of sealing joints? List the materials used.
8. Differentiate between AASHTO and IRC methods of rigid pavement design.
9. Explain why edge load, edge warping stress and corner load stresses are considered and warping stress at the corner region is not considered in the IRC method of rigid pavement design.
10. Draw a flow-chart representing the process of analysis for design of rigid pavements according to (i) IRC: 58-2002 and (ii) AASHTO 1993.
11. What are the salient features of thickness design of concrete slab pavement according to IRC: 58?
12. State the critical placement of wheel loading and critical combination of stresses considered in IRC: 58.
13. Draw a flow-chart representing the process of analysis for design of dowel bars according to IRC: 58.
14. State the worst condition of loading considered for design of dowel bars according to IRC-58-2002. Draw figure wherever necessary.
15. Write an overview on thickness design of cement concrete pavements as per IRC: 58-2015.
16. State the three basic compositions of cement concrete pavement that are considered in IRC: 58-2015.
17. Briefly explain how the effective and combined modulus of sub-grade reaction is estimated from the set of layers considered for design of pavement based on IRC: 58-2015 guidelines.
18. Describe the phenomenon of top-down and bottom-up Cracking in concrete slab pavement. What are the salient design features TDC and BUC cases of analyses, with reference to axle-load spectra and temperature differential with time?
19. Draw the typical sketches showing critical combination of loading and temprature differetial considered for design of concrete slab pavement by carryingout BUC and TDC analyses, as per IRC:58-2015.
20. Draw the critical location of placement of first rear axle considered for determination of (a) Max. edge flexural stress for BUC case and (b) Max. tensile stress for TDC case, as used by the IRC: 58-2015.
21. Define fatigue damage and explain how it is accounted for in the thickness design of pavements.
22. Draw a typical step-wise procedure recommended for thickness design of cement concrete slab pavement (a) with tied concrete shoulder and with doweled transverse joints (b) without tied concrete shoulder and without doweled transverse joint, according to IRC: 58-2015 guidelines.
23. Write a short note on design of reinforcement to control cracking in concrete pavement.

24. Write about the suitability of providing continuously reinforced concrete pavement (CRCP). What are the factors, limiting criteria and design features considered for design of CRCP?
25. Define and state the benefits of precast/pre-stressed concrete pavements (PPCP) as compared with conventional concrete pavements.
26. State the basic concept of utilization of precast/pre-stressed concrete technique and explain how this concept is used for construction of PPC pavements as compared with conventional concrete pavements.
27. Draw the typical layout of precast pre-stressed concrete panels used for construction of pavements. Explain, in detail, the functions of pre-stressed concrete panels with reference to their position in the layout.
28. Define the M-E based pavement design approach. Why it is being widely adopted by most of the countries in the world?
29. Briefly write the details of calibrated mechanistic–empirical (M–E) design process.
30. State the need and significance of calibrated performance models that the used for M-E based analysis and design of pavements. State any recent example of this case.

Multiple Choice Questions

1. What is the equivalent value of flexural strength of a cement concrete if its 28-day cube compressive strength given as 25 MPa?
   (a) 3.5 MPa (b) 17.5 MPa
   (c) 5.5 MPa (d) 35.71 MPa
2. In India, the most commonly observed temperature differential across the thickness of cement concrete slab layer ranges between ________ for slab thickness ranging from 15 cm to 30 cm.
   (a) 12.5° C and 21.0° C
   (b) 15° C and 30.0° C
   (c) 2.5° C and 35.0° C
   (d) 2.5° C and 25.0° C
3. For rigid pavement analysis, what is the representative elastic modulus value of a new cement concrete slab of M40 grade?
   (a) 1000 MPa (b) 2000 MPa
   (c) 30000 Mpa (d) 4000 MPa
4. As per IRC, the allowable bond stress value of concrete for plain and deformed bars is recommended as equal to ________ respectively.
   (a) 1.75 MPa and 2.46 MPa
   (b) 125 MPa and 200 MPa
   (c) 40 MPa and 415200 MPa
   (d) 0.5 MPa and 2.5 MPa
5. What is the value of modulus of dowel bar support recommended for determination of dowel bar deflection at the face of the joint?
   (a) 125 PMa/m (b) 200 MPa/m
   (c) 30000 MPa/m (d) 415200 MPa/m
6. Installation of dowel bars in cement concrete pavement slabs is restricted at places where the slab thickness is less than
   (a) 10 cm (b) 15 cm
   (c) 18 cm (d) 20 cm
7. To prevent concentration of tensile stresses in tie bars, the centre-to-centre spacing between the tie bars should not be wider than
   (a) 500 mm (b) 750 mm
   (c) 1000 mm (d) 1250 mm
8. To relieve warping stresses at the tie bar joints, the diameter of tie bars should be restricted to not greater than
   (a) 8 mm (b) 10 mm
   (c) 14 mm (d) 16 mm
9. Longitudinal joints in India should be provided where the lane width exceeds
   (a) 3.5 (b) 3.75 m
   (c) 4.0 m (d) 4.5
10. Where is the maximum frictional stress developed in a cement concrete slab pavement due to its movement caused by temperature variations?
   (a) At the center of the slab
   (b) At the corner of the slab
   (c) At the longitudinal edge of the slab
   (d) At the transverse edge of the slab

Answers to Multiple Choice Questions

1. (a) 3. (c) 5. (d) 7. (b) 9. (d)
2. (a) 4. (a) 6. (b) 8. (d) 10. (a)

References

AASHTO. A policy on geometric design of highways and streets, *American Association of State Highway and Transportation Officials*, Washington DC, 2004.

AASHTO. AASHTO guide for design of pavement structures, *American Association of State Highway and Transportation Officials*, Washington D C, 1993.

AASHTO. *Policy on Geometric Design of Highways and Streets*, Washington, DC: AASHTO. 1984, 1990, 1994, 2001 and 2004.

AASHTO. *Standard Specifications for Transportation Materials and Methods of Sampling and Testing, Part II: Tests, 7th edition*, Washington, DC, 1995.

AASHTO. *Standard Specifications for Transportation Materials and Methods of Sampling and Testing: Part 1; Specifications and Tests: Part 2*; Washington DC, 2001.

Abrams D A, Design of concrete mixtures, Structure of Materials Research Lab, Lewis Institute, *Bulletin 1*, Chicago, 1918. Also published in *A Selection of Historic American Papers on Concrete, 1876–1926* (Ed. H. Newlon) by ACI SP-52. 309–330, 1976.

ACI 318 Code. Building code requirements for structural concrete, ACI 318-95. Commentary, ACI 318R-95, American Concrete Institute, Farmington Hills, MI, 1996.

ACI 435 Code. Control of deflection in concrete structures, Committee 435 Report, American Concrete Institute, Farmington Hills, MI, 1995.

Ahlvin R G and Ulery H H, Tabulated values for determining the complete pattern of stresses, strains and deflections beneath a uniform circular loan on a homogeneous half space, *Bulletin 342, Highway Research Board*, Washington DC, 1962.

Akoto B K and Nilles E E, Repair and restoration of concrete pavements, *Jl. of the Indian Roads Congress*, 68-1: 91–106, 2007.

Allen J J and Thompson M R, Significance of variably confined tri-axial testing. *ASCE Journal of Transportation Engineering* 100(TE4), 827–843,1974.

AP-T102-08. Austroads technical report: Fatigue performance of cemented materials under accelerated loading – Influence of vertical loading on the performance of unbound and cemented materials, AUSTROADS, 2008.

ARRB. Dickinson E J, *Bituminous Roads in Australia*, Victoria, Australia, 1984.

ARRB. Dickinson E J, Performance of thin bituminous pavement surfacings in Australia, *Proc. 11th ARRB Conf.* 11 (3), 35–51, 1982.

Ashton J E and Moavenzadeh F, Analysis of stresses and displacements in a three-layered visco-elastic System, *Proc. of 2nd Inter. Conf. of Struct. Des. of Asphalt Pavements*, Vol. I, 209–219, 1967

Awadallah F, Theoretical analysis for horizontal curves based on actual discomfort speed, *ASCE J. of Transportation Engineering* 131 (11): 843–850, 2005.

Bagui S K, Development of program for mix design of concrete according to IS: 10262-1982, *Journal of Indian Highways* 32(8), 29–52, 2004.

Baker R F (Ed.), Byrd L G and Mickle D G (Assoc. Ed.), *Handbook of Highway Engineering*, 496–600, Van Nostrand Reinhold, New York, 1975.

Bakker R F M, Permeability of blended cement concrete, in *Use of Fly Ash, Silica Fume, Slag and Other Mineral Byproducts in Concrete Vol I.* Edited by Malhotra, SP-79, American Concrete Institute, 1983.

Barksdale R D, Laboratory evaluation of rutting in base course materials, *Proc. of 3rd International Conference on Structural Design of Asphalt Pavements*,161–174, 1972.

Bennett C R and Greenwood I D, Modelling road user and environmental effects, HDM-4, Volume 7, *The Highway Development and Management Series*, 2001.

Bergen A T and Fredlund D G, Characterization of freeze-thaw effects on subgrade soils, *Symposium on Frost Action in Road, Vol. II*, Organization for Economic Cooperation and Development. Oslo, Norway, 1973.

Bishop A W and Hankel D J, *The Measurement of Soil Properties in the Tri-axial Test 2nd Edition*, Edward Arnold, London, 1962.

Bishop H K, *Discussion on Highway Location: Practical Considerations by Losh H K*, Transactions of the ASCE, Vol. 96, 477–480, 1932.

Blanks R F, Modern concepts applied to concrete aggregate, *Transactions of the ASCE* 115 (2403): 403–431, 1950.

Bonnaure F, Gest G, Gravais A and Uge P, A new method of predicting the stiffness of asphalt paving mixes, *Proc. of the Association of Asphalt Paving Technologists*, 46, 1977.

Bonneson J A, Bridge size and clearance time of single-point urban interchange, *Journal of Transportation Engineering, Vol. 119 (1)*, American Society of Civil Engineers, Washington DC, 1993.

Bonneson J A, Side friction and speed as controls for horizontal curve design, *ASCE J. of Transportation Engineering* 125 (6): 473–480, 1999.

Bose S and Jain P K, Laboratory studies on the use of organic polymers in improvement of bituminous road surfacings, *Highway Research Bulletin 38*,.63–79, IRC, 1984.

Boussinesq V J, Application des potentials a letude de lequilibre at du mouvement des solids elastiques avec les notes etendues sur divers points de physique, *Mathematique et Danalyse*, Gauthier Villais, Paris, 1885.

Boyce I R, Brown S F and Pell P S, The resilient behaviour of a granular material under repeated loading, *ARRB proceedings*, 8: 1–12, 1976.

BPS. Chandrashekar B P, a PPT titled “Overview and Programme Guidelines for Pradhan Mantri Gram Sadak Yojana (PMGSY I and II)”, last accessed on 12th January 2019.

Bradbury R D, *Reinforced Concrete Pavements*, Wire Reinforcement Inst., Washington DC, 1938.

Brown S F and Brunton J M, *An Introduction to the Analytical Design of Bituminous Pavements*, 3e, Univ. of Nottingham, Dept., of Civil Engg., UK, 1992.

Brown S F and Pell P S, A fundamental structural design procedure for flexible pavements, *Proc. of 3rd International Conference on Structural Design of Asphalt Pavements*, I: 369–381,1972.

Brown S F, Cooper K E and Pooley G R, Mechanical properties of bituminous materials for pavement design, *Proc. of the Eurobitume Symposium*, 143–147, 1981.

Bruce G D, Reaction time in automobile driving, *Journal of Applied Psychology* 20 (3): 355, 1936.

Brule B, Brion Y and Tanguy T, Paving polymer blends: Relationships between composition, structure and properties, *Proc. of the Assoc. of Asphalt Paving Techn.* 57, 41–63, 1998.

BS: 1377-1990. Methods of test for soils for civil engineering purposes, British Standard Institute, London.

Burmister D M, Discussion on graphical representation of the mechanical analyses of soils by Campbell F B, *Transactions of the ASCE* 104: 156–159, 1939.

Burmister D M, Graphical distribution of vertical pressure beneath foundations, *Transactions of the ASCE*, 103, 301–33, 1938.

Campbell F B, Graphical representation of the mechanical analyses of soils, *Transactions of the ASCE*, 104: 150–155, 1939.

Cardoso S H and Witczak M W, Permanent deformation for flexible airfield pavement design, *Transportation Research Record No. 1307*,.111–121, Transportation Research Board, 1991.

Carey W N and Irick P E, Performance of flexible pavements in AASHO road test, *Proc. Int. Conf. Struct. Des. of Asphalt Pavements*, University of Michigan, Ann Arbor, Michigan, 1962.

Carpenter J C, *Discussion on Highway Location: Practical Considerations by Losh H K*, Transactions of the ASCE, Vol. 96,.480–482, 1932.

Carse A and Beehan J, Static chord modulus of elasticity of high strength concrete in uniform compression and flexure, *Proc. of the 10th ARRB Conference* 10 (3), 46–56, 1980.

Casagrande A, Classification and identification of soils, *Transactions of the ASCE* 113(2351), 901–930, 1948.

Chadda H S and Cross S E, Speed (road) bumps: issues and opinions. *ASCE J. of Transportation Engineering* 111(4): 410–418, 1985.

Chary S S, *Rheology*, Academic Books, New Delhi.

Chery S, *A Comparison of non-traditional intersection designs using microscopic simulation*, M. Sc. Thesis submitted to the Faculty of College of Engineering and Computer Science, Florida Atlantic University, Boca Raton, FL, May 2010.

Childs L D, Colley B E and Kapernick J W, Tests to evaluate concrete pavement sub bases, *Transactions of ASCE*, Paper no. 2993, Vol.124, 556–587, 1959.

Chou Y T, Structural analysis computer program for rigid multi component pavement structures with discontinuities: WESLIQUID and WESLAYER, *Technical Report GL-80*, Vicksburg, Miss.: US Army of Waterways Experiment Station, 1980.

CITGO. Power point presentation on performance graded (PG) asphalts. Superpave update for NJDOT/NEAUPG Mechanistic Pavement Design Seminar, Feb. 23, Princeton, NJ, 2003.

Clement J P, *Letter to Editor. Journal of ITE*, 6–7, March 1983.

Collins A R and Sharp D R, The design and construction of concrete roads overseas, *Proc., of the Institute of Civil Engineers*, Paper no. 6267, 23–64, 1958

Cooling L F, Skempton A W and Glossop R, Discussion made on a paper titled 'Classification and identification of soils' by Casagrande A, *Transactions of the ASCE* 113(2351), 966–969, 1948.

Craus J and Polus A, Aspects of spiral transition curve design *Transportation Research Record No. 631*, 1–4, Washington, DC: Transportation Research Board, 1977.

D 3282-93. Standard classification of soils and soil-aggregate mixtures for highway construction purposes, American Standard Testing Methods, 1999.

Dabbour E S, Easa M and Abd El Halim A O, Radius requirements for reverse horizontal curves on three-dimensional alignments, *ASCE J. of Transportation Engineering* 130(5): 610–620, 2004.

Dauzats M and Linder R, A method for the evaluation of the structural condition of pavements with thick bituminous road bases, *Proc. of the 5th International Conference on Structural Design of Asphalt Pavements*, 1, 387–409, 1982.

David K M, Frank M B, Burns N H and Robert O R, Construction of the California precast concrete pavement demonstration project, FHWA-IF-06-010, US Department of Transportation – Federal Highway Administration, 2004.

DDA. *Guidelines for Road Markings*, Unified Traffic and Transportation Infrastructure (PLG & ENG) Centre (UTTIPEC), Delhi Development Authority, November 2009.

Delphi M R C, Bulluz L, Morrall J and Smith G, Rural intersection safety handbook, Final Report, *Transport Canada*, March, 2006.

Deme I, Processing of sand–asphalt–sulphur mixes, *Proc. of the Assoc. of Asphalt Paving Tech.* 43, 465–482, 1974.

Despotopoulos P T, A new easement curve in highway work, *Civil Engineering and Public Works Review* XLII (495): 378–381, September 1947.

Di Benedetto H, Partl M N, Franken L and De La Roche Saint Andre C, RILEM TC 182-PEB Performance testing and evaluation of bituminous materials: Stiffness testing for bituminous mixtures, *Materials and Structures* 34, 66–70, 2001.

Duncan J M and Seed H B, Strength variation along failure surfaces in clay, *Journal of ASCE* 92(SM6), 81–104, 1966.

Easa S M and Hassan Y, Design requirements of equal-arc unsymmetrical vertical curves, *ASCE J. of Transportation Engineering* 124 (5): 404–410, 1998

Easa S M, Dabbour E and Ali M Z A, Three-dimensional model for stop control intersection sight distance, *ASCE J. of Transportation Engineering* 130(2): 261–270, 2004.

Easa S M, Distributing super elevation to maximize highway design consistency, *ASCE J. of Transportation Engineering* 129(2): 127–133, 2003.

Easa S M, Exact minimum sight distance on sag curve with centered overpass, *ASCE J. of Transportation Engineering* 118(4): 588–592, 1992.

Easa S M, Lateral clearance needs on compound horizontal curves, *ASCE J. of Transportation Engineering* 119(1): 111–123, 1993.

Easa S M, Lateral clearance to vision obstacles on horizontal curves, *Transportation Research Record No. 1303*, 22–32. Washington, DC: Transportation Research Board. 1991.

EC. Green public procurement: Street lighting and traffic lights, *Technical Background Report*, European Commission, DG Environment-C1, BU 9, 1160, Brussels, 2011. http://ec.europa.eu/environment/gpp/pdf/tbr/street_lighting_tbr.pdf

Eco-3. *Guidelines: Energy Efficient Street Lighting*, USAID ECO-II Project, Version 2.0, New Delhi, June, 2010.

Emery S J, The prediction of moisture content in untreated pavement layers and an application to design in Southern Africa, Division of Roads and Transport Technology, *DRRT Bulletin 20*, CSIR Research Report 644, Pretoria, 1988.

Fadum (R E), Discussion made on a paper titled 'Classification and identification of soils' by Casagrande A, *Transactions of the ASCE* 113(2351), 931–933, 1948.

Fadum F F, Influence values for estimating stresses in elastic foundations, *Proceedings, 2nd Int. Conf. on Soil Mechanics and Foundation Eng.*, 3, Rotterdam, 1948.

Feld J (R E), Discussion made on a paper titled 'Classification and identification of soils' by A. Casagrande. *Transactions of the ASCE* 113(2351), 946. 1948.

Fenwick J M, The durability of reinforced concrete, *Queensland Roads* 21 (42):16–18. 1982.

FHWA IP-82-3. Flora J W and Keitt K M, Access management retrofit techniques, *Access Management for Streets and Highways, Federal Highway Administration*, June 1982.

FHWA. Continuously reinforced concrete pavement, *Technical Advisory*, T 5080.14, 1990, Available Online at www.fhwa.dot.gov/pavement/t508014.cfm

FHWA. Evaluation of design consistency methods for two-lane rural highways. Publication No. FHWA-RD-99-173, August 2000. Executive Summary. http://www.tfhrc.gov/safety/ihsdm/pdfs/99-173.pdf, as visited on 17/10/2007.

FHWA. *Manual of Uniform Traffic Control Devices (MUTCD)*. US Department of Transportation, Federal Highway Administration, Washington, DC, 2003.

FHWA. Predictive design procedures, *VESYS User Manual, Report No. FHWA-RD-77-154*, Federal Highway Administration, 1978.

FHWA. Speed prediction for two-lane rural highways. Publication No. 99-171, August 2000. http://www.tfhrc.gov/safety/ihsdm/pdfs/99-171.pdf, as visited on 17/10/2007.

Fitzpatrick K, Wooldridge M D and Blaschke J D, "Urban Intersection Design Guide: Volume 1—Guidelines and Volume 2—Applications", *Report No. FHWA/TX-05/0-4365-P2, Texas Transportation Institute, The Texas A&M University System*, College Station, Texas, 2005.

Fitzpatrick K., Brewer M., Dorothy P and Park E S, Design guidance for intersection auxiliary lanes, *Project 03-102, NCHRP Report 780, National Cooperative Highway Research Program, Transportation Research Board*, Washington, D. C. 2014.

Flugge W, *Viscoelasticity*, Blaisdell Publishing Company, A Division of Ginn and Company, London, 1975

Foster C R and Ahlvin R G, Stresses and deflections induced by a uniform circular load, *Proc., Highway Research Board*, Vol.33, 467–470, 1954.

Foxworthy P T and Darter M I, ILLI-SLAB and FWD deflection basins for characterization of rigid pavements, Nondestructive testing of pavements and back-calculation of moduli, ASTM STP 1026, Bush III A J and Baladi G Y, Eds., *American Society of Testing and Materials*, Philadelphia, 368–386, 1989.

Freeme C R, Maree J H and Viljoen A W, Mechanistic design of asphalt pavements and verification using the heavy vehicle simulator, *Proc. of the 5th International Conference on Structural Design of Asphalt Pavements*1,156–173, 1982.

Friberg B F, Stresses in concrete pavements by sector analysis, *Proceedings of the Highway Research Board*, 1955.

Fromm H J and Kennepohl G J A, Sulphur asphaltic concrete on three Ontario roads, *Proc. of the Assoc. of Asphalt Paving Tech*, 48, 135–162, 1979.

Fwa T F and Tan L S, Geometric characterization of road humps for speed-control design, *ASCE J. of Transportation Engineering* 118(4): 593–598, 1992.

Fwa T F and Wei L, *Design of Rigid Pavements: The Handbook of Highway Engineering*, edited by Fwa T F, CRC Press, Taylor & Francis Group, 2006.

García G A and Conesa L M (discussion). Unit curve for design of highway transitions by S. K. Rao. *ASCE J. of Transportation Engineering*, 244, May/June, 1997.

García G A and Conesa L M, Improving steep bends as hairpin curves on mountainous roads (Discussion on this section), *ASCE J. of Transportation Engineering* 118 (6): 882–884, 1992.

Gattis J L and Low S T, *Intersection Angles and Driver's Field of View*, Mack–Blackwell Transportation Center, University of Arkansas, Fayetteville, AR, 1997.

Gattis J L, Vinsin B F III and Duncan L K, Low-speed horizontal curve friction factors, *ASCE J. of Transportation Engineering* 131(2): 112–119, 2005.

Gazzar El M G, *A Logical Approach to Roadway Lighting Design*, www. dewberry.com.

www.dewberry.com/docs/default-source/documents/a_logical_approach_to_roadway_lighting_design.pdf

Gibreel G M, Easa S M, Hassan Y and El-Dimeery I A, State of the art of highway geometric design consistency, *ASCE J. of Transportation Engineering* 125(4): 305–313, 1999.

Giffin, *Discussion on Highway Location: Practical Considerations by Losh H K*, Transactions of the ASCE, Vol. 96, 485–487, 1932.

Glennon J C, New and improved model of passing sight distance on two-lane rural highways, *Transportation Research Record No.1195*,132–137. Washington, DC: Transportation Research Board, 1988.

Goldbeck A T, Researches on the structural design of highways by the United States Bureau of Public Roads, *Transactions of ASCE*, 88, 264–300, 1925.

Goldbeck A T, The interrelation of longitudinal steel and transverse cracks in concrete roads, *Public Roads*, 1925.

Gopal Raju S S S and Rengaraju V R, Use of poly-ethylene waste in stabilization of subgrade soil, *Seminar on Integrated Development of Rural and Arterial Road Network for Socio-economic Growth.* II, 199–208, New Delhi, 2003.

Greenstein J, Transition curves for speed changes, *ASCE J. of Transportation Engineering* 102 (No.TE1): 699–713, 1976.

Grimer F J and Ali M A, The strengths of cements reinforced with glass fibers, *Magazine of Concrete Resaerch* 21(66), 1969.

Guidebook on Energy efficient Electric Lighting for Buildings, International Energy Agency (IEA) – Energy Conservation in Buildings and Community Systems (ECBCS) Annex 45, Edited by Liisa Halonen, Eino Tetri and Pramod Bhusal, Aalto University School of Science and Technology, Lighting Unit, http://lightinglab.fi/IEAAnnex45, 2010.

Hall W M, Manufacturing concrete of uniform quality, *Transactions of the ASCE* 96, 1336–1358, 1932.

Harman T, D'Angelo J and Bukowski J (the Asphalt Team), *Superpave Asphalt Mixture Design*, Workshop Workbook Version 8.0, U S Dept. of Transportation, Federal Highway Administration, Updated 2002.

Harwood D, Mason J and Brydia R, Sight distance for stop-controlled intersections based on gap acceptance, *Transportation Research Record No. 1701*,.32–41. Washington, DC: Transportation Research Board,2000.

Hatt W K, The effect of moisture on concrete, *Transaction of ASCE* 157, 270–315, 1925.

Hawkins H G Jr., How modern headlamp performance impacts sag vertical curve design, *ASCE J. of Transportation Engineering* 133(4), 223–231, 2007.

Head K H, *Manual of Soil Laboratory Testing 1–3,* Prentec Press, London, 1982.

Heukelom W and Klomp A J G, Road design and dynamic loading, *Proc. of the Association of Asphalt Paving Technologists*, 33, 92–125, 1964.

Heukelom W and Klomp A J G, Consideration of calculated strains at various depths in connection with the stability of asphalt pavements, *Proc. of the 2nd International Conference on Structural Design of Asphalt Pavements*,.277–290, 1967.

Heukelom W, An improved method of characterising asphaltic bitumens with the aid of their mechanical properties, *Proc. of the Association of Asphalt Paving Technologies*, 42, 62–98, 1973.

Hickerson T F, *Route Location and Design*, McGraw-Hill Book Co. Inc., New York, 1964.

Hicks R G and Monismith C L, Factors influencing the resilient response of granular materials, *HRB Record No. 345,* 1971.

Highway Lighting Guide, Alberta Transportation, August 2003. www.transportation.alberta.ca/Content/docType233/Production/lighting.pdf

Hill R L, Sarkar S L, Rathbone R F and Hower J C, An examination of fly ash carbon and its interaction with air-entraining agents, *Cement Concrete Res*, 27(2),193–204, 1997.

Hinkle A H, The engineer's part in making the highway safe, *Transactions of ASCE* 94: 522–529, 1930.

Hirsh M, Prashker J N and Ben-Akiva M, New approach to geometric design of highways, *Transportation Research Record No. 1100,* 50–57, Washington, DC: Transportation Research Board, 1986.

HRB. *AASHO Road Test*, Pavement Research, Report 5, 1962.

HRB. Westergaard H M, Analysis of stresses in concrete pavement due to variations of temperature, *Proc. of HRB* 6: 201–215, National Research Council, Washington, DC, 1926, 1927.

HRB. Westergaard H M, Computation of stresses in concrete roads *Proc. of HRB.* 5(Part I), 90–112, National Research Council, Washington DC, 1925.

HRB. Westergaard H M, Theory of concrete pavement design, *Proc. of HRB*, Part I 175–181, National Research Council, Washington DC, 1927.

http://www.nra.co.za/content/GDGGeometricDesignGuide.pdf?Session_ID=d02276d67361c1996ca68627b403a105

Huang Y H and Wang T, Finite element analysis of concrete slabs and its implication for rigid pavement design, *Highway Research Record 466*, 55–69, Washington DC, National Research Council, 1973.

Huang Y H, *Pavement Analysis and Design*, 2e, Pearson Prentice Hall, 2004.

Hunting M T, Geology in highway engineering, *Transactions of ASCE* 110(2238), 271–297, 1945.

Hveem F N, Slab warping affects on pavement joint performance, *Proc. of American Concrete Institute* 47, 797–808, 1951.

India Vision 2020, The Report and Background Papers, Planning Commission, Government of India, 2004.

IRC. Arora J, Development of high performance concrete pavements through polymer modification and roller compaction with special reference to cold climatic regions, *Seminar on Integrated Development of Rural and Arterial Roads Network for Socio-economic growth, II:* III, 155–160, New Delhi, 2003.

IRC. Ashraf M, Roller compacted concrete pavements: Technical and economical benefits, *Seminar on Design Construction and Maintenance of Cement Concrete Pavements, III*, 17–25, New Delhi, 2004.

IRC. Chaubal A M and Hans H, New generation superplasticizers for high-performance concrete, *Seminar on Financing, Implementation and Operation of Highways in 21st Century. II*, 199–202, New Delhi, 2000.

IRC. Ghosh R K, Chatterjee M R, Bhatia M L and Bhatnagar R C, Investigation on sugar mixed concrete for pavements, *Road Research Bulletin 17*, 1973.

IRC. *Pocket Book for Highway Engineers*, Indian Roads Congress, New Delhi, 2006.

IRC. Shukla R S and Jain P K, Improvement of waxy bitumen by the addition of synthetic rubbers, polymers and resins, *Highway Research Bulletin 38*, 17–28, 1984.

IRC. Vasudevan R, Saravanavel S, Rajasekaran S and Thirunakkarasu D, Utilisation of waste plastics in construction of flexible pavement, *Jl. of Indian Highways*, 34(7), 5–20, Indian Roads Congress, 2006.

IRC: 104. *Guidelines for Environmental Impact. Assessment of Highway Projects*, Indian Roads Congress, New Delhi, 1988.

IRC: 15-2015. Standard specifications and code of practice for construction of concrete roads, *Indian Roads Congress*, New Delhi, 2015.

IRC: 3. *Dimensions and Weights of Road Design Vehicles*, Indian Roads Congress, New Delhi, 1983

IRC: 3-1983. *Dimensions and Weights of Road Design Vehicles*, Indian Roads Congress, New Delhi, 1983.

IRC: 35. *Code of Practice for Road Markings*, Indian Roads Congress, New Delhi, 2015.

IRC: 35-1970. *Code of Practice for Road Markings (with Paints)*, Indian Roads Congress, New Delhi, 1970.

IRC: 37. *Guidelines for Design of Flexible Pavements*, Indian Roads Congress, New Delhi, 2012.

IRC: 37. *Guidelines for the 2012 design of flexible pavements*, 3rd revision, Indian Roads Congress, New Delhi, 2001, 2012 and 2018.

IRC: 37. *Tentative Guidelines for the Design of Flexible Pavements* (with CD), 2018.

IRC: 39-1986. *Standards for Road–Rail Level Crossings*, Indian Roads Congress, New Delhi, 1986.

IRC: 44-2008. Tentative guidelines for cement concrete mix design for pavements, for non-entraining and continuously graded concrete, New Delhi. 2008.

IRC: 52-2001. *Recommendations about the Alignment Survey and Geometric Design of Hill Roads*, Indian Roads Congress, New Delhi, 2001.

IRC: 58. *Guidelines for the Design of Plain Jointed Rigid Pavements for Highways* (Fourth Revision) (with CD), 2015.

IRC: 58-2002. *Guidelines for the Design of Plain Jointed Rigid Pavements for Highways*, Indian Roads Congress, 2002.

IRC: 58-2011 and 2015. *Guidelines for the Design of Plain Jointed Rigid Pavements for Highways*, Indian Roads Congress, New Delhi, 2011 and 2015.

IRC: 66-1976. *Recommended Practice for Sight Distance on Rural Highways*, Indian Roads Congress, New Delhi. 1976.

IRC: 67. *Code of Practice for Road Signs*, Indian Roads Congress, New Delhi, 2012.

IRC: 69-1977, *Space Standards for Roads in Urban Areas*, Indian Roads Congress, New Delhi, 1977.

IRC: 69-1977. *Space Standards for Roads in Urban Areas*, Indian Roads Congress, 1996.

IRC: 70-1977. *Guidelines on Regulation and Control of Mixed Traffic in Urban Areas*, Indian Roads Congress, New Delhi, 1977.

IRC: 73-1980. *Geometric Design Standards for Rural (Non-Urban) Highways*, Indian Roads Congress, New Delhi, 1980.

IRC: 73-1980. *Geometric Design Standards for Rural Highways*, Indian Roads Congress, New Delhi, 1980.

IRC: 75. *Guidelines for the Design of High Embankments*, 2015.

IRC: 76-1979. Tentative guidelines for the structural evaluation of rigid airfield pavements, New Delhi, 1979.

IRC: 86-1983. *Geometric Design Standards for Urban Roads in Plains*, Indian Roads Congress, New Delhi, 1983.

IRC: 98-1997, *Guidelines on Accommodation of Utility Services on Roads in Urban Areas, First Revision*, Indian Roads Congress, New Delhi, 1997.

IRC: SP: 13. *Guidelines for the Design of Small Bridges and Culverts*, 2004.

IRC: SP: 15. *Ribbon Development along Highways and Its Prevention*, Indian Roads Congress, New Delhi, 1996.

IRC: SP: 19. *Manual for Survey, Investigation and Preparation of Road Projects*, Indian Roads Congress, New Delhi, 2001.

IRC: SP: 23-1993. *Vertical Curves for Highways*, Indian Roads Congress, New Delhi, 1993.

IRC: SP: 41. *Guidelines for the Design of At-Grade Intersections in Rural & Urban Areas*, Indian Roads Congress, New Delhi, 1994.

IRC: SP: 48-1998. *Hill Road Manual*, Indian Roads Congress, New Delhi, 1998.

IRC: SP: 49, *Guidelines for the Use of Dry Lean Concrete as Sub-base for Rigid Pavement*, Indian Roads Congress, New Delhi, 1998.

IRC: SP: 53-2002. *Guidelines on Use of Polymer and Rubber Modified Bitumen in Road Construction*, New Delhi, 2002.

IRC: SP: 55, *Guidelines on Traffic Management in Work Zones*, Indian Roads Congress, New Delhi, 2014.

IRC: SP: 62. *Guidelines for Design and Construction of Cement Concrete Pavements for Low Volume Rural Roads*, Indian Roads Congress, New Delhi, 2014.

IRC: SP: 62-2004, *Guidelines for the Design and Construction of Cement Concrete Pavements for Rural Roads*, Indian Roads Congress, New Delhi, 2004.

IRC: SP: 76-2008, *Tentative Guidelines for Conventional Thin and Ultra-thin Whitetopping*, Indian Roads Congress, New Delhi, 2008.

IRC: SP: 77. *Recommendations about Overlays on Cement Concrete Pavements*, 1977.

IRC: SP: 92. *Guidelines for the Design of Interchanges in Urban Areas*, Indian Roads Congress, New Delhi, 1985.

IRC: SP: 12. *Tentative Recommendations on the Provision of Parking Spaces for Urban Areas*, Indian Roads Congress, New Delhi

IRC: 10. *Recommended Practice for Borrow Pits for Road Embankments Constructed by Manual Operation*, 1961.

IRCSP: 72. *Guidelines for the Design of Flexible Pavements for Low Volume Rural Roads*, Indian Roads Congress, New Delhi, 2014.

IS: 10262. *Indian Standard Guidelines for Concrete Mix Design*, Bureau of Indian Standards, New Delhi, 1998

IS: 11196-1985. *Specification for Equipment for Determination of Liquid Limit of Soils by Cope Penetration Method*, Bureau of Indian Standards, New Delhi, 1985.

IS: 1199-1959. *Methods of Sampling and Analysis of Concrete*, Bureau of Indian Standards, New Delhi, 1999.

IS: 1201 to 1220-1978. *Methods for Testing Tar and Bituminous Materials*, Bureau of Indian Standards, New Delhi, 2002.

IS: 1208-1978. *Methods for Testing Tar and Bituminous Materials*, Bureau of Indian Standards, New Delhi, 1978.

IS: 1498-1970. *Classification and Identification of Soils for General Engineering Purposes*, Bureau of Indian Standards, New Delhi, 1970.

IS: 215-1995. *Road Tar: Specification*, Bureau of Indian Standards, New Delhi, 2000.

IS: 217-1988. *Specification for Cutback Bitumen*, Bureau of Indian Standards, New Delhi, 1993.

IS: 2386 (Part 1)-1963, *Methods of Test for Aggregates for Concrete, Particle Size and Shape*, Bureau of Indian Standards, New Delhi, 2004.

IS: 2386 (Part 3)-1963. *Methods of Test for Aggregates for Concrete, Specific Gravity, Density, Voids, Absorption and Bulking*, Bureau of Indian Standards, New Delhi, 2002.

IS: 2386 (Part 4)-1963. *Indian Standard Method of Test for Aggregate for Concrete*, Bureau of Indian Standards, 2005.

IS: 2720 (Part 10)-1991. *Methods of Test for Soils, Determination of Unconfined Compressive Strength*, Bureau of Indian Standards, New Delhi, 1992.

IS: 2720 (Part 12)-1981. *Methods of Test for Soils, Determination of Shear Strength Parameters of Soil from Consolidated Undrained Tri-axial Compression Test with Measurement of Pore Water Pressure*, Bureau of Indian Standards, New Delhi, 2004.

IS: 2720 (Part 13)-1986. *Methods of Test for Soils, Direct Shear Test*, Bureau of Indian Standards, New Delhi, 1987.

IS: 2720 (Part 16)-1987, *Methods of Tests for Soils, Laboratory Determination of CBR*, Bureau of Indian Standards, New Delhi, 2004.

IS: 2720 (Part 2)-1973. *Methods of Tests for Soils, Determination of Water Content*, Bureau of Indian Standards, New Delhi, 2004.

IS: 2720 (Part 28)-1974. Methods of tests for soils, determination of dry density of soils in-place by the sand replacement method, *Bureau of Indian Standards*, New Delhi, 1974.

IS: 2720 (Part 29)-1975. *Methods of Test for Soils, Determination of Dry Density of Soils In-place by the Core-cutter Method*, Bureau of Indian Standards, New Delhi, 2006.

IS: 2720 (Part 30)-1980. *Methods of Test for Soils, Laboratory Vane Shear Test*, Bureau of Indian Standards, New Delhi, 2004.

IS: 2720 (Part 33)-1991. *Methods of Test for Soils, Determination of the Density In-place by the Ring and Water Replacement Method*, Bureau of Indian Standards, New Delhi 1991.

IS: 2720 (Part 34)-1972. Methods of Tests for Soils, Determination of the Density of Soil In-place by Rubber-balloon Method, Bureau of Indian Standards, New Delhi, 2001.

IS: 2720 (Part 4)-1985. *Methods of Tests for Soils, Grain Size Analysis*, Bureau of Indian Standards, New Delhi, 2004.

IS: 2720 (Part 5)-1985. Methods of Tests for Soils, Determination of Liquid Limit and Plastic Limit, Bureau of Indian Standards, New Delhi, 2004.

IS: 2720 (Part 7)-1980. *Determination of Water Content: Dry Density Relation Using Light Compaction*, Bureau of Indian Standards, New Delhi, 2004.

IS: 2720 (Part 8)-1983. *Determination of Water Content: Dry Density Relation Using Heavy Compaction*. BIS Code, Bureau of Indian Standards, New Delhi, 2004.

IS: 334-2002. *Glossary of Terms Relating to Bitumen and Tar*, Bureau of Indian Standards, New Delhi, 2002.

IS: 3812-1981. *Specification for Fly-ash for Use as Pozzolana and Admixture*, Bureau of Indian Standards, New Delhi, Reaffirmed in 1999.

IS: 456-2000, *Indian Standard Plain and Reinforced Concrete Code of Practice*, Bureau of Indian Standards, New Delhi, 2000.

IS: 516-1959. *Methods of Tests for Strength of Concrete*, Bureau of Indian Standards, New Delhi, 1999.

IS: 5816-1999. *Splitting Tensile Strength of Concrete: Methods of Test*, Bureau of Indian Standards, New Delhi, 1999.

IS: 6241-1971. *Method of Test for Determination of Stripping Value of Road Aggregates*, Bureau of Indian Standards, New Delhi, 1971

IS: 73-2006. *Indian Standard Paving Bitumen: Specifications*, Bureau of Indian Standards, New Delhi, 2006.

IS: 9214-1979. *Indian Standard Method of Determination of Modulus of Sub-grade Reaction (k value) of Soils in Field*, Indian Standards Institution, New Delhi, 1980s.

IUTR. *ITS Developed by Japanese Police*, Japan Traffic Management Technology Association (JTMTA), Published by *Institute of Urban Traffic Research (IUTR)*, 302025 Kitahama Chuo-ku Osaka, Japan, 1998. http://tmt.or.jp/research/pdf/its-e.pdf

Jackson F H et al., Concretes containing air-entraining agents: Proceedings, *American Concrete Institute Journal* 40, 509–572, 1944.

Jagadish R, Sharada Bai H and Basavaraj S H, Modern cement super-plasticizers for durability of concrete structures, *Int. Conf. on Maintenance & Durability of Concrete Structures*, JNT University, Hyderabad, 1997.

Jarvis J R, The off-road testing of road humps for use under Australian conditions. *Proc. of the 10th ARRB Conference Vol. 4.* Traffic Australian Road Research Board. August 1980.

Jervis W H and Eustis J B, Accepted procedure for the CBR test: A symposium on development of CBR flexible pavement design method for airfields, *Transactions of the ASCE* 115(2406), 472–484, 1950.

Jeuffroy G and Bachelez J, Note on a method of analysis for pavements, *Proc. of the 1st International Conference on Structural Design of Asphalt Pavements*,.300–309, 1962.

Johnson T C, Cole D M and Irwin L H, Characterization of frees/thaw granular soils for pavement evaluation, *Proc. of the 5th Int. Conf. on the Struct. Des. of Asphalt Pavements*, 805–817, Univ. of Michigan, Ann Arbor, Michigan, 1982.

Kallas B F and Puzinauskas V P, Flexural fatigue tests on asphalt paving mixtures, Report STP 508, *Proceedings of the Association of Asphalt Paving Technologists* 39, 47–65, American Society for Testing and Materials, 1972.

Kallas B F, Dynamic modulus of asphalt concrete in tension and tension–compression, *Proceedings of the Association of Asphalt Paving Technologists* 39, 1–20, 1970.

Kaluva M, *The effect of skew angle on average queue delay at tee-intersections: A simulation study using the Texas model*, M.Sc. Theses, Civil Engineering, DigitalCommons@University of Nebraska - Lincoln, Nebraska, December 2013.

Kandhal P S, Quality control requirements for using crumb rubber modified bitumen (CRMB) in bituminous mixtures, *Jl. of Indian Roads Congress* 67-1(522), 99–104, 2006.

Kanellaidis G, Aspects of highway super elevation, *ASCE J. of Transportation Engineering* 117 (6), 624–632, 1991.

Kanellaidis G, Human factors in highway geometric design, *ASCE J. of Transportation Engineering* 122(1), 59–66, 1996.

Kelly E F, Application of the results of research to the structural design of concrete pavements, *Public Roads*, 20, 1939.

Kerr H S, *Discussion on Highway Location: Practical Considerations by Losh H K*, Transactions of the ASCE, Vol. 96, 483–485, 1932.

Khayat K H, Viscosity enhancing admixtures for cement based materials: An Overview, *Cement and Concrete Composites* 20, 171–188, 1998.

Khedr S A, Deformation characteristics of granular base course in flexible pavement, *Transportation Research Record No. 1043*,.131–138, Transportation Research Board, 1985.

Kirwan R W and Snaith M S, A simple chart for the prediction of resilient modulus, *Geotechnique* 26(1), 212–215, 1976.

Koonce P, Rodegerdts L, Kevin L, Quayle S, Beaird S, Braud C, Bonneson J, Tarnoff P and Tom Urbanik T, *Traffic Signal Timing Manual*, Report No. FHWA-HOP-08-024, U S Department of Transportation, Federal Highway Administration (FHWA), June 2008.http://ops.fhwa.dot.gov/publications/fhwahop08024/fhwa_hop_08_024.pdf

Kulkarni R B, Burns R L, Wright J, Apper B, Baily T O and Noack S T, Decision analysis of alternative highway alignment, *Journal of Transportation Engineering*, ASCE, Vol. 119 (3), 317–332, 1993.

Kumar P, Chandra S and Bose S, Strength characteristics of polymer modified mixes, *Int. Jl. of Pavement Engineering*, 7(1), 63–71, 2006.

Kumar P, Chandra S and Maurya A K, Computer applications in Marshall method of mix design, *Jl. of Indian Highways*, 31(10), 25–43, Indian Roads Congress, 2003.

Lambe T W, Compacted clay: Structure, *Transactions of the ASCE* 125 Part I (3041), 682–706, 1960.

Lamm R and Choueiri E, Recommendations for evaluating horizontal design consistency based on investigations in the state of New York, *Transportation Research Record No. 1122*, 68–78, Washington, DC: Transportation Research Board. 1987.

Lamm R J, Hayward C and Cargin J G, Comparison of different procedures for evaluating speed consistencies, *Transportation Research Record No. 1100*, 10–20, Washington, DC: Transportation Research Board, 1986.

Lamm R, Choueiri E and Mailaender T, Side friction demanded versus side friction assumed for curve design on two-lane rural highways, 11–21, *Transportation Research Record No. 1303*, Washington, DC: Transportation Research Board. 1991.

Lamm R, Choueiri E, Hayward J and Paluri A, Possible design procedures to promote design consistency in highway geometric design on two lane rural roads, *Transportation Research Record No. 119*,,111–122, Washington, DC: Transportation Research Board, 1988.

Leisch J E and Leisch J P, New concepts in design speed application, *Transportation Research Record No. 631*,.4–14, Washington, DC: Transportation Research Board, 1977.

Lieberman E B, Model for calculating safe passing distance on two-lane rural roads, *Transportation Research Record No. 869*, Washington, DC: Transportation Research Board, 1982.

Lipičnik M, New form of road/railway transition curve, *ASCE J. of Transportation Engineering*, 124(6): 546–556, 1998.

Lister N W and Powell W D, Design practice for bituminous pavements in the United Kingdom *Proc. of the 6th International Conference on Structural Design of Asphalt Pavements* 1, 220–231, 1987.

Losh A R, *Highway Location: Practical Considerations*, Transactions of the ASCE, Vol. 96, Paper no. 1799, 468–476, 1932.

Losh, *Transactions of ASCE*, 1932.

Love G D, Sulphur: Potential pavement binder of the future, *ASCE Jl. of Transportation Engg.* 105, 525–533 1979.

Maccarrone S, Holleran G and Gnanaseelan G P, Properties of polymer modified binders and relationship to mix and pavement performance, *Proc. of the Assoc. of Asphalt Paving Tech.*, 209–240,1995.

Mallela J, Glover L T Darter M I, Quintus H V, Gotlif A, Stanley M and Sadasivam S, Guidelines for implementing NCHRP, 1-37A, *M-E Design Procedures in Ohio*, Volume 1 – Summary of findings, implementation plan, and next steps, Applied Research Associates Inc., Champaign, IL, 2009.

Mark J G, Concrete and hydraulic cement, US Patent 2,052,586, 1938.

May R and Witczak M W, Effective granular modulus to model pavement responses, *Transportation Research Record No. 1022*, 1–9, Transportation Research Board, 1985.

Mcfadden G and Pringle T B, Development and scope of investigation: A symposium on the development of CBR flexible pavement design method for airfields, *Transactions of the ASCE* 115(2406), 456–460, 1950.

McGovern E W and Alderton K L, Development of durable road tar binders from high temperature coke-oven tar, *Proc. of the 6th ARRB Conf,* 6(4), 39–72, 1972.

Medina A M F and Tarko A P T, Speed changes in the vicinity of horizontal curves on two-lane rural roads, *ASCE J. of Transportation Engineering,* 133(4): 215–222, 2007.

Mercer G and Bullet A W, The economics of concrete road pavements in the United Kingdom, Department of Transportation, England, Kent County Council, 1982.

Mitchell J K and Monismith C L, A thickness design procedure for pavements with thin cement stabilized bases and thin asphalt surfacings, *Proc. on the 4th International Conference on Structural Design of Asphalt Pavements*, 1, 409–416, 1977.

Mn/DOT. *Mn/DOT Roadway Lighting Design Manual*, Minnesota Department of Transportation, Office of Traffic, Security and Technology, May 2010.

MnDOT *MnDOT Traffic Signal Timing and Coordination Manual*, Minnesota Department of Transportation, May 2013. www.dot.state.mn.us/trafficeng/publ/signaloperations/2013_Signal_Opt_and_Timing_Manual.pdf

Molenaar A A A, Structural design of pavements, Part III, *Design of Flexible Pavements, Lecture Notes*, 2009.

Monismith C L and Deacon J A, Fatigue of asphalt paving mixtures, *ASCE Jl. of Transportation Engg,* 317– 345, 1969.

Monismith C L, Secor K E and Blackmer E W, Asphalt mixture behaviour in repeated flexure, *Proc. of the Assoc. of Asphalt Paving Techn,* 30, 188–222, 1961.

Monismith C L, Seed, H B, Mitry F G and Chan C K, Prediction of pavement deflection from laboratory tests, *Proc. of 2nd International Conference on Structural Design of Asphalt Pavements*, 109–140, 1967.

Moore J H, Thicknesses of concrete pavements for highways, *Trans. of ASCE, Vol. 121, Paper no. 2834*, 125–1142, 1956.

Morales P Y, Rene S, Discussion made on a paper titled 'Classification and identification of soils' by Casagrande A, *Transactions of the ASCE* 113(2351), 960–963, 1948.

MoRD. *Rural Roads Development Plan: Vision 2025 (Draft)*. Ministry of Rural Development, Government of India. New Delhi, 2006.

MORD. *Specifications for Rural Roads,* Ministry of Rural Development IRC, New Delhi, 2004.

MORTH. *Manual for the Construction and Supervision of Bituminous Works*, IRC, New Delhi, 2001.

MORTH. Specifications for road and bridge works, *Ministry of Road Transport and Highways*, Published by IRC, New Delhi, 2013.

MoSRTH. *Specifications for Road and Bridge Works,* 4th Edition, Ministry of Shipping, Road Transport & Highways, Indian Roads Congress, New Delhi, 2001.

MOTSAM. *Manual of Traffic Signs and Markings, Part-1: Traffic Signs,* NZ Transport Agency, 2010. www.nzta.govt.nz/resources/motsam/part-1/motsam-1.html

Mullis I B, Principles applying to highway road-beds, T*ransactions of the ASCE, Vol. 104, paper no.2040*, 1381–1401, 1939.

Mullis I B, Principles applying to highway road-beds, *Transactions of the ASCE, Vol. 104, Paper no.2040*, 1381–1401, 1939.

NAC. *Prospects for Concrete Pavement for Indian Roads,* National Academy of Construction, Hyderabad, 2001.

Naik T R and Singh S S, Fatigue property of concrete with and without mineral admixture, *Concrete Technology: Past, Present and Future,* Edited by P. K. Mehta, *Proc. of V. Malhotra Symposium*, SP-144: 269–288, University of California, Detroit, Berkley, ACI, 1994.

Nair K P, The chemistry of bitumen, *Jl. of Indian Highways* 5(4), 26–27, 1977.

National Co-operative Highway Research Program, Project 1-37A, Draft Final Report, Washington DC, 2003.

Nawy E G (Ed-in-chief), *Concrete Construction Engineering Hand Book*, CRC Press, 1997.

NCHRCP Report 279. Neuman T R, Intersection channelization design guide, *National Cooperative Highway Research Program, Transportation Research Board,* Washington DC, 1985.

NCHRCP. Gattis J L, Gluck J S, Barlow J M, Eck R W, Hecker W F and Levinson H S, Guide for the geometric design of driveways, *Report 659, National Cooperative Highway Research Program*, Transportation Research Board, Washington DC, 2010.

NCHRCP. Koepke F J and Levinson H S, Access management guidelines for activity centers, Report 348, *National Cooperative Highway Research Program*, Transportation Research Board, Washington DC, 1992.

NCHRP Web-only Document 193. Fitzpatrick K, Brewer M A, Gluck J S, Eisele W L, Zhang Y, Levinson H S, von Zharen W, Lorenz M R, Iragavarapu V and Park E S, Development of left-turn lane warrants for unsignalized intersections, , *Transportation Research Board*, Washington DC, 2013.

NDOR, *Roadway Design Manual*, Nebraska Department of Roads, 2015. http://www.transportation.nebraska.gov/roadway-design/rw-design-man-chapters.htm

NDOR. Chapter Four: Intersections, Driveways and Channelization, *Roadway Design Manual, Nebraska Department of Roads*, April, 2012. www.transportation.nebraska.gov/traffeng/mutcd.htm

Nelu S and Carmel J, Concrete chemical admixtures: Perspective and challenges, *Journal of Indian Concrete Institute* 6(4), 23–32, 2006.

Neuman T R and Glennon J C, Cost-effectiveness of improvements to stopping sight distance, *Transportation Research Record No. 923*, 26–34, Washington, DC: Transportation Research Board. 1984

Neville A M and Brooks J J, *Concrete Technology*, 2nd Indian Reprint, Pearson Education Ltd., 2002.

Neville A M, *Properties of Concrete*, 4th Indian Reprint, Pearson Education Pvt. Ltd., Singapore, 2005.

Nexco. *Annual Report 2008*, Central Nippon Expressway Company Limited, Nagoya, Aichi Prefecture 460-0003, Japan.

Nexco. *Annual Report 2014*, Central Nippon Expressway Company Limited, Nagoya, Aichi Prefecture 460-0003, Japan, 2014

Nicholson A, Super elevation, side friction and roadway consistency, *ASCE J. of Transportation Engineering*, 124 (5): 411–418, 1998.

NJDOT. *Roadway Design Manual-Highway Lighting System*, Section 11, RDM, NJDOT, 2015. www.state.nj.us/transportation/eng/documents/RDM/sec11.shtm

NRP. Nagpur Road Plan, Post-war Road Development in India. In *Proceedings of the Chief Engineers' Conference on Road Development, December 15–18,* 1943.

Older C, Highway research in Illinois, *Transactions of ASCE*, 88, 1925.

Oliver J W H, Rubber bituminous digestions: manufacture and properties, *Program and Papers of the 18th ARRB Reg. Symposium*, 36–53, Traralgon, 1981.

Orchard D F, Factors influencing the wear of coarse aggregate in the Los Angeles test, *Proc. of 2nd ARRB Conf.* 2(2), 963–80, 1964.

Ottesen J and Krammes R, Speed profile model for a design consistency evaluation procedure in the United States, *Transportation Research Record No. 1701*, 76–85, Washington, DC: Transportation Research Board, 2000.

Packard R G, Design of concrete airport pavement, *Engineering Bulletin*, EB050.03P (PCA: Skokie, IL), 1973.

Pandey B B, Recent trends on design of bituminous mix design, *Seminar on Design of Bituminous Mixes*, 45–50, IIT Kanpur, India, 2003.

PCA. *Color and Texture in Architectural Concrete*, SP012.02A, Portland Cement Association, 1995.

Pfeiffer J Ph and Van Doormaal P M, The rheological properties of asphaltic bitumens, *Journal of the Institute of Petroleum Technology*, 22(152), 414–440, 1936.

Pickett G and Ray G K, Influence charts for concrete pavements, *Transactions of ASCE* 116, Paper 2425, 1951.

Pocketbook for Highway Engineers, Ministry of Shipping, Road Transport & Highways, Published by IRC, New Delhi, 2002.

Poole A B, Alkali carbonate reactions in concrete, *Proceedings of the 5th International Conf. on Alkali Aggregate Reaction in Concrete*, Cape Town, South Africa, 1981.

Porter O J, Development of the original method for highway design: A symposium on development of CBR flexible pavement design method for airfields, *Transactions of the ASCE* 115(2406), 461–467, 1950.

Porter O J, The preparation of sub-grades, *Proc. of the Highway Research Board* 18 (2), 324–331, Highway Research Board, Washington DC, 1938.

Prisk C W, Passing practices on rural highways, *Proc. of Highway Research Board*, 21, 366–378, Washington DC, 1941.

Procter E R, Design and construction of rolled earth dams, *Engineering News Record* vol. VIII, 1933.

Prof. Rao D V, A Telephonic Discussion, 2008.

Rada G and Witczak M W, A comprehensive evaluation of laboratory resilient moduli results for granular material, Paper presented at *TRB Annual Meeting*, Washington DC, 1981.

Ramachandran V S (Ed.), *Concrete Admixtures Handbook*, Noyes Publishers, New Jersey, 1995.

Ramachandran V S, Malhotra V M, Jolicoeur C and Spiratos N, *Superplasticizers: Properties and Applications in Concrete*, Materials Technology Laboratory, Canada, 1998.

Rao S K, Unit curve for design of highway transitions, *ASCE J. of Transportation Engineering*, 121(2), 169–175, 1995.

RDP, IRC. Draft approach paper. *Road Development Plan 2000–21*, 1999.

RDPI, IRC. *Road Development Plan for India (1981–2001)*, 1984.

RDPI. *Road Development Plan for India, 1981 to 2001*, published by the Indian Roads Congress, New Delhi, 1984.

Rietveld P, "Some Notes on Interconnectivity in Transport Networks", in Cocossis H. and Nijkamp P., Editors, *Overcoming Isolation: Information and Transportation Networks in Development Strategies for Peripheral Regions*, Berlin, Springer, 18–28, 1995.

Roebuck R, Isola R, Goodrum W, Cebon D and Collop A, Vehicle–pavement interaction (VPI) modelling, HVTT12, *International Symposium on Heavy Vehicle Transport Technology, Setting Future Standards*, Stockholm, Sweden, 2012, September 16–19.

Roy B K, Geometric design of compound horizontal curves (Discussion on this section), *ASCE J. of Transportation Engineering*, 120(4), 674, 1994.

Roy B K, Incorporating transition lengths, (Discussion on this section), *ASCE J. of Transportation Engineering*, 118 (6), 887, 1992.

RRDC, Government of India. *Report of the Road Development Committee, 1927 to 1928.* Calcutta: Central Publication Branch, 1928.

SANRAL. Geometric design guidelines, *South African National Road Agency Limited.*

Schiffman R L, General analysis of stresses and displacements in layered elastic systems, *Proc. of 1st Inter. Conf. of Struct. Des. of Asphalt Pavements*, 365–375, 1962.

Schrader E K, in Edward G. (Ed-in-chief)'s *Concrete Construction Engineering Hand Book,* CRC Press, 1997.

Seed H B, Chan C K and Lee C E, Resilience characteristics of subgrade soil and their relation to fatigue failures in asphalt pavement, *Proc. of the Int. Conf. on the Struct. Des. of Asphalt Pavements*,.611–636, Univ. of Michigan, Ann Arbor, 1963.

Seehra S S, Roller compacted concrete for cost-effective pavements in the 21st century, National Seminar on Challenges for Road Transportation in the 21st Century Leading to Fuel Economy, 107–121, CIRT 2001. TIET, Patiala, 2001.

Setra, *The design of interurban intersections on major roads-at-grade intersections*, Produced and disseminated by the Service d'Etudes Techniques des Routes et Autoroutes Centre de la Sécurité et des Techniques Routières (Setra), France, 1998. http://www.infra-transports-materiaux.cerema.fr/IMG/pdf/US_ACI-P_GB.pdf

Sharma B and Bora P K, A study on correlation between compaction characteristics and plastic limit of fine grained soils, *IRC No. 68*,.87–104, Highway Research Board. 2003.

Shell pavement design manual – Asphalt pavement and overlays for road traffic, Shell International Petroleum Company Ltd., London, 1978.

Shen J S, Amirkhanian S N and Tang B, Influence of accelerated aging test temperature on the properties of binders, *Int, Jl. of Pavement Engineering,* 7(3), 191–198, 2006.

Shergold F A, The percentage voids in compacted as a measure of its angularity, *Magazine of Concrete Research* 5 (13), 3–10, 1953.

Shi X P, Fwa T F and Tan S A, Warping stresses in concrete pavements on Pasternak Foundation, *ASCE Jl. of Transportation Engineering*, 119(6), 905–913, 1993.

Shifley L H and Monismith C L, Test road to determine the influence of subgrade characteristics on the transient deflections of asphalt concrete pavements, *Report No. TE-68-5*, Department of Civil Engineering, Institute of Transportation and Traffic Engineering, University of California, Berkeley, 1968.

Shipway C H, A study of aggregate crushing test, *Proc. of 2nd ARRB Conf.* 2(2), 981–96, 1964.

Shook J F, Construction materials control for the ASSHO road test, *Transactions of the ASCE* 126, Part I, 975–990,,1961.

Skempton A W and Northley A, The sensitivity of clays, *Geotechnique* 2(2), 1952.

Skempton A W, The colloidal activity of clays, *Proc. 3rd Int. Conf. ISSMFE*, ISSMFE, Zurich, 1953.

Skok F L and Finn F N, Theoretical concepts applied to asphalt concrete pavement design, *Proc. of the 1st International Conference on Structural Design of Asphalt Pavements*, 412–440, 1962.

SP: 23-1985. Handbook on concrete mixes, *Bureau of Indian Standards*, New Delhi, 2001.

Spangler M G, Discussion made on a paper titled 'Classification and identification of soils' by Casagrande. A, *Transactions of the ASCE* 113(2351), 978–984, 1948.

Spangler M G, Stresses in the corner region of concrete pavements, *Iowa Eng. Exp. Station Bulletin No. 157*, Iowa State University, 1937.

Sparkes F N, Stresses in concrete road slabs, *Structural Engineer*, 17: 98, 1939.

Sparkes F N, The structural design of concrete roads, *Magazine of Concrete Research* 6(17): 61, 1954.

Sparks G A, Neudorf R D, Robinson J B and Good D, Effect of vehicle length on passing operations, *ASCE J. of Transportation Engineering,* 119(2), 272–283, 1993.

Spectrum Year Book, 2008.

Spence I M, Ramsay D M and Dir R K, A conspectus of aggregate strength and the relevance of their factors as the basis for a physical classification of crushed rock aggregate, *Advances in Rock Mechanics*, 79084, National Academy of Science, Washington, 1974.

Spiratos N, Page M, Mailvaganam N P, Malhotra V M, Jolicoeur C, *Super-plasticizers for Concrete: Fundamentals, Technology and Practice,* 322, Supplementary Cementing Materials for Sustainable Development Inc., Canada, 2003.

Srinivasa Kumar R, Reddy K S, Mazumdar M and Pandey B B, Models for estimation of subgrade moduli from DCP tests, *The International Journal of Pavement Engineering & Asphalt Technology* 4(2), United Kingdom, 2003.

Stanton T E, Expansion of concrete through reaction between cement and aggregate, *Proceedings of the ASCE* 66:1781, 1940.

Steyn W J vd M, Monismith C L, Nokes W A, Harvey J T, Holland T J and Burmas N, Challenges confronting road freight transport and the use of vehicle–pavement interaction analysis in addressing these challenges, *Journal of the South African Institution of Civil Engineering,* 54(1): 814, 14–21.

Szary P, Maher A, Strizki M and Moini N, *Use of LED or Other New Technology to Replace Standard Overhead and Sign Lighting (Mercury and/or Sodium),* FHWA-NJ-2005-029, New Jersey DOT, FHWA, USDOT, 2005. www.state.nj.us/transportation/refdata/research/reports/FHWA-NJ-2005-029.pdf

T 222-81. Standard method of test for non-repetitive static plate load test of soils and flexible pavement components, for use in evaluation and design of airport and highway pavements, *AASHTO Materials Part-II Tests, 17th Edition,* AASHTO, Washington DC, 1995.

T 292. Resilient modulus of sub-grade soils and untreated bases/sub-base materials, standard specifications for transportation materials and methods of sampling and testing, AASHTO, Washington DC, 2001.

Tabatabai A M and Barenburg E J, Structural analysis of concrete pavement systems, *ASCE Jl. of Transportation Engineering,* 106(5), 493–506, 1980.

TAI. *Asphalt Technology and Construction Practices, Instructors Guide,* The Asphalt Institute, 1983.

TAI. *Superpave Mix Design,* 3e, Superpave Series No. 2, The Asphalt Institute, 2001.

TAI. *Superpave Performance Grade Asphalt Binder Specification and Testing,* 2e, The Asphalt Institute, 1997.

Taiganidis J and Kanellaidis G, Method for grading design of highways, *ASCE J. of Transportation Engineering,* 117(6), 706–712, 1991.

Tang B, Structural evaluation of airfield rigid pavements using falling weight deflectometer, *ASCE Jl. of Transportation Engineering,* 119(3), 467–476, 1993.

Tayabji S D and Colley B E, Analysis of jointed concrete pavements, *Research Report.* Washington DC, FHWA, Nat. Tech. Information Service, 1984.

Taylor G D, *Construction Materials,* Longman Scientific & Technical, UK, 1991.

Teller L W and Sutherland E C, The structural design of concrete pavements, Part 2, *Public Roads* 16(9), 1935.

Teller L W, Discussion on paper titled 'Effect of dowel bar misalignment across concrete pavement joints' by Smith A R and Benham S W, *Transactions of ASCE* 103, 1145–1146, 1938.

Todd K B and Kulakowski B T, Simple computer models for predicting ride quality and pavement loading for heavy trucks: In Pavement Management and Rehabilitation, Transportation Research Record 1215, *Transportation Research Board,* Washington, DC, 137–150, 1989.

TRB. Armaghani J M and Tu D, Rehabilitation of Ellaville Weigh Station with ultra-thin white topping, *Transportation Research Record 1654,* 3–11, Washington DC, National Research Council, 1999.

TRB. Armaghani J M, Larsen T J and Smith L L, Temperature response of concrete pavements, *Transportation Research Record No. 1121,* 23–33, Washington DC, 1986.

TRB. Burchett G D and Maze T H, Rural expressway intersection characteristics as factors in reducing safety performance, *Transportation Research Record 1953,* National Research Council, Washington DC, 71–80, 2006.

TRB. Button J W, Jawle M, Jagadam V and Little D, Evaluation and development of a pressure aging vessel for asphalt cement, *Transportation Research Record No. 1391,*11–19, 1993.

TRB. Chiu C T, Tai M, Ruth B E and Page G C, Investigation of laboratory aging process of asphalt binders used in Florida, Transportation research record no. 1436, 60–69, 1994.

TRB. Domke C H, Liu M, Davison R R, Bullin J A and Glover C J, Study of strategic highway research program: Pressure aging vessel procedure using long-term low-temperature aging experiments and asphalt kinetics, *Transportation Research Record No. 1586*, 10–15, 1997.

TRB. Harwood D, Mason J and Brydia R, Sight distance for stop-controlled intersections based on gap acceptance, *Transportation Research Record, No. 1701*, Washington DC, 32–41, 2000.

TRB. *Highway Capacity Manual*, Special Report No. 209, National Academy of Sciences, Washington, 2000

TRB. Ioannides A M, Thompson M R and Barenberg E J, Westergaard solutions reconsidered, *Transportation Research Record No. 1043*, pp.13–23, Washington, DC, 1985.

TRB. Oh J, Washington S and Choi K, Development of accident prediction models for rural highway intersections, *Transportation Research Record 1897,* National Research Council, Washington DC, 18–27, 2004.

TRB. Peterson J C and Harnsberger P M, Asphalt aging dual oxidation mechanism and its inter relationships with asphalt composition and oxidative age hardening, *Transportation Research Record No. 1638*, 47–58, 1998.

TRB. Son T, Kim S G and Lee J K, Methodology to Calculate Sight Distance Available to Drivers at Skewed Intersections, *Transportation Research Record 1796,* National Research Council, Washington DC, 41–47, 2002.

TRB. Thompson M R, Dempsey B J, Hill H and Vogel J, Characterizing temperature effects for pavement analysis and design, *Transportation Research Record No. 1121*, 14–22, 1987.

TRB. Walker R J, Coordination of basic design elements: Overview, *Transportation Research Record 1385*, National Research Council, Washington DC, 51–59, 1981.

TRB. Yu H T, Khazanovich L, Darter M I and Ardani A, Analysis of concrete pavement responses to temperature and wheel loads measured from instrumented slabs, *Transportation Research Record No. 1639*, 94–101, Washington DC, Washington, 1998.

TRL. The use of traffic signals in developing cities, Overseas Road Note 13, Overseas Centre, Transport Research Laboratory, Crowthorne, Berkshire, United Kingdom, 1996.

TRRL. *A Guide to Geometric Design*, Overseas Road Note No. 6, Crowthorne, 1988.

Tucker G R, Amine salts and aromatic sulfonic acids, US Patent 2,141,570, 1936.

Turnbull W J, Appraisal of the CBR method: A symposium on development of CBR flexible pavement design method for airfields, *Transactions of the ASCE* 115(2406), 547–554, 1950.

Ullidtz P and Peattie K R, Pavement analysis by programmable calculators, *ASCE Journal of Transportation Engineering*, 106 (5), 581–597, 1980.

Ullidtz P, *Pavement Analysis*, Elsevier Science, Amsterdam, 1987.

Ullidtz P, *Pavement Analysis*, Elsevier Science, New York, 1987.

Uzan J, Characterization of granular material, *Transportation Research Record No. 1022*, 52–29. Transportation Research Board, 1985.

Uzan J, Witczak M W, Scullion T and Lytton R L, Development and validation of realistic pavement response models, *Proc. of the 3rd International Conference on Structural Design of Asphalt Pavements* I, 334–350, 1972.

Van der Poel C, A general system describing the visco-elastic properties of bitumens and its relation to routine test data, *Jl. of Applied Chemistry*, 4(5), 221–236, 1954.

Vasudevan R, Saravanavel S and Rajasekaran S, Utilization of waste plastics in construction of flexible pavement, *Seminar on Integrated Development of Rural and Arterial Road Network for Socio-economic Growth*, II, 243–257, New Delhi. 2003.

Venable W M, Measuring materials for concrete, *Transactions of the ASCE* 94(1751), 1082–1103, 1930.

Venkat Reddy D, *Engineering Geology for Civil Engineering*, Oxford & IBH Publishing Co. Pvt. Ltd., 1995.

Verstraeten J, Moduli and critical strains in repeated blending of bituminous mixes –Application to pavement design, *Proc. of 3rd Inter. Conf. of Struct. Des. of Asphalt Pavements*, Vol. 1, 729–738, 1972.

Verstraeten J, Stresses and displacements in elastic layered systems, General theory – Numerical stress calculation in four layered systems with continuous interface, *Proc. of 2nd Inter. Conf. of Struct. Des. of Asphalt Pavements*, 277–290, 1967.

Vickers B, *Soil Mechanics: Laboratory Work in Civil Engineering*, Granada Publishing, London, 1983.

Waltham T, *Foundations of Engineering Geology 2nd edition* Spon Press, London, 2002.

Wang Y and Cartmell M P, New model for passing sight distance on two-lane highways, *ASCE J. of Transportation Engineering*, 124(6): 536–545, 1998.

Warnes C E, Structural design of concrete pavement by computer, *ASCE Jl. of Transportation Engineering*, 98, 1972.

Waterhouse A, Stresses in layered systems under static and dynamic loading, *Proc. of 1st Inter. Conf. of Struct. Des. of Asphalt Pavements*, Session II, 291–308, 1962.

Westergaard H M, Analytical tools for judging results of structural tests of concrete pavements, *Public Roads* 14(10), 185–188, 1933.

Westergaard H M, New formulas for stresses in concrete pavements of airfields, *ASCE Transactions* 113, paper no. 2340, 425–444, 1948.

Westergaard H M, Stresses in concrete pavements computed by theoretical analysis, *Public Roads* 7(2), 25–35, 1926.

Wiseman G, Uzan J, Hoffman M S, Ishai I and Livneh M, Simple elastic models for pavement evaluation, *Proc. of the 4th International Conference on Structural Design of Asphalt Pavements*, Vol. 2, 1977.

Wistuba M, Lackner R, Blab R and Spiegl M, Low-temperature performance prediction of asphalt mixtures used for LLP: New approach based on fundamental test methods and numerical modeling, *Int, Jl. of Pavement Engineering*, 7(2), 121–132, 2006.

Witczak M W and Bari J, Development of a mater curve (E*) database for lime modified asphaltic mixtures, *Arizona State University Research Project*, 2004.

Woodson W E and Barry T, *Human Factors Design Handbook: Information and Guidelines for Design of Systems, Facilities, Equipment and Products for Human Use*, McGraw Hill, New York, 93, 1992.

WSDOT. *Design Manual, Volume 2 – Design Criteria, M 22-01.11, Engineering and Regional Operations, Development Division, Design Office, Washington State Department of Transportation*, July, 2014. www.wsdot.wa.gov/Publications/Manuals/M22-01.htm

WSDOT. http://training.ce.washington.edu/WSDOT/Modules/03_materials/03-2_body.htm. 28-09-2006.

Yoder E J and Faiz A, Design and performance of continuously reinforced concrete pavement: Technical paper, Publication FHWA/IN/JHRP-73/17, *Joint Highway Research Project*, Indiana DOT and Purdue University, Indiana, 1973.

Yoder E J and Witczak M W, *Principles of Pavement Design 2nd edition*, John Wiley & Sons, Inc., New York, 1975

Index

NOTE: Page numbers suffixed with *f* and *t* indicate figures and tables, respectively.

A

B

C

D

E

F

G

H

W